Dynamic Power Supply Transmitters

Envelope Tracking, Direct Polar, and Hybrid Combinations

EARL McCUNE

CAMBRIDGE
UNIVERSITY PRESS

CAMBRIDGE
UNIVERSITY PRESS

University Printing House, Cambridge CB2 8BS, United Kingdom

Cambridge University Press is part of the University of Cambridge.

It furthers the University's mission by disseminating knowledge in the pursuit of education, learning and research at the highest international levels of excellence.

www.cambridge.org
Information on this title: www.cambridge.org/9781107059177

First published 2015

Printing in the United Kingdom by TJ International Ltd. Padstow Cornwall

A catalog record for this publication is available from the British Library

Library of Congress Cataloging in Publication data
McCune, Earl.
Dynamic power supply transmitters : envelope tracking, direct polar,
and hybrid combinations / Earl McCune.
  pages   cm. – (The Cambridge RF and microwave engineering series)
ISBN 978-1-107-05917-7 (hardback)
1. Power amplifiers.   2. Amplifiers, Radio frequency – Power supply.
3. Radio – Transmitters and transmission.   4. Electric power
supplies to apparatus.   I. Title.
TK7871.58.P6M38   2015
621.3841'31–dc23

2014048900

ISBN 978-1-107-05917-7 Hardback

**Dedicated to the outstanding teams I have worked
with while learning these details**

Contents

Preface

Energy efficiency in all aspects of modern society is now a widespread desire and an active goal. Whether it manifests as driving a high mileage automobile or using a highly rated energy-efficient refrigerator, there is no aspect of modern life that does not benefit from improvement in the energy efficiency used.

Wireless communications are now one of the ubiquitous technological underpinnings of modern society. Mobile communications devices and smart-phones are always at hand and reliably work for the user. As such, it is particularly important for everything involved in wireless communications to be energy efficient. As communication speeds have increased over recent decades, the unfortunate by-product is that the energy efficiency of wireless communications has actually decreased significantly. It is now overdue and important that this gets fixed.

Standards committees have placed the data rate performance of the wireless signals they adopt ahead of any other concern. This means that the energy efficiency of the wireless communication system has not been part of deliberations during the standardization process. The signals adopted years ago are still with us, and there are no plans to change them anytime soon. It is now therefore important to change how these communications devices are made, and to adopt new architectures that will provide the needed energy efficiency while still generating these old signals that are present within the deployed communications standards.

To get full value from this book, the reader should already have a basic familiarity with electrical engineering and wireless communication concepts, including the Fourier transform relationship between time-domain and frequency-domain operations. It is not necessary to have familiarity with the present communications standards.

The contents of this book are drawn from the nearly 20 years of experience I have with dynamic power supply transmitter technology. Being more of a physics-based person than a mathematician, over the decades considerable effort has been given to developing a thorough understanding of the mathematics that underlie the physical relationships being described. Through these pages, I share the results of my efforts with you.

This book is carefully planned so that readers will have a clear understanding of what is being discussed as they work their way through the chapters. This means that the foundations of any topic discussed will as much as possible have already been laid. There are three major parts:

- principles of the dynamic power supply transmitter techniques,
- circuit implementation and special topics for these designs, and
- new issues for testing and calibration of these designs.

The first two major parts start with the 3-port extensions to linear amplifier operation, and then extend the results first to envelope tracking and then to polar modulation. Hybrid designs that use all of the possible techniques in one product have their own chapter. The extensions needed to explain some unusual results experienced when these techniques are applied at multiple stages in the same transmitter also has its own chapter.

Unique contributions in this book include:

- unification of all dynamic power supply operating modes with the inherent characteristics of transistors of any type;
- a specific definition of knee voltage and how this is measured and used;
- direct calculation of what the optimum envelope tracking profile must be for any RF power transistor and selected load line;
- outline of how the concept of matching network design changes significantly when amplifiers are operated in deep compression for polar modulation;
- investigation of the energy efficiency of the various architectures available to implement the dynamic power supply;
- detailed examination of the new interface: connecting the dynamic power supply to the RF PA;
- description of the inherent instabilities in this new interface and what can be done about them;
- clear, unambiguous, and testable definitions of envelope tracking and polar operation, and how these relate to conventional linear operation;
- description of the new transistor specifications needed for polar operation;
- proof that polar operation has higher PA energy efficiency than envelope tracking, and why this must be;
- details of how the concept of amplifier gain must expand into four separate measures that each provide important and different insights;
- identification of the new circuit design rules needed for successful design of polar operation;
- identification of the P-mode amplifier operating region, why this must be avoided by envelope tracking transmitters, and how it can be successfully used by polar operation;
- description of the dynamic power supply feature extensions, including independent automatic compensation at the PA for low battery voltage and/or output impedance mismatch.

Any technology that is involved in multibillion dollar industries, such as wireless communications, is often first published not at a conference or in a technical journal, but rather through the appropriate government patent office. This is certainly true for dynamic power supply transmitter technology. Knowing this fact is particularly important to graduate students who plan on getting doctorate degrees in this technology area, because an idea not seen in the technical journals is not a guarantee that any particular idea is really a new contribution to the technical arts. References in the following chapters do include representative patents that are already published, to aid in accessing that library for further searching.

I gratefully acknowledge help with my ability to access transistors and amplifiers using the many semiconductor technologies through the support of Skyworks, Freescale, TriQuint Semiconductor, RF Microdevices, ST Microelectronics, RF Micropower, NXP, Avago Technologies, and Cree. All of these companies have been a huge help in making this story complete. For his particular help, I salute Gray Wong of the RF distributor Richardson RFPD (now Arrow) who tirelessly made good things happen for this project when they needed to.

I want to particularly acknowledge the tremendous help provided by National Instruments, mainly through Haydn Nelson and Takao Inoue, in providing the automated measurement system and software support that allowed me to collect all of the data used for the technology survey in Chapter 10 and in validating the testing requirements presented in Chapter 13. Without this support, the completeness of the technology survey would not have been possible.

The patience of my wife Barbara to the seemingly endless hours spent writing, drawing, rewriting, and editing needed for the preparation of this book is beyond measure. My gratitude to you again is boundless!

I fervently hope that all who read this book, and who may use it as an additional reference, will enjoy the information and approach as much as I have enjoyed writing it.

Earl McCune

Abbreviations

3G	Third generation cellular, standardized by the Third Generation Partnership Project (3GPP)
AC	alternating current
ACLR	adjacent channel leakage ratio
ACP	adjacent channel power
ALBC	automatic low battery compensation
AM	amplitude modulation
AMO	AM offset
AMPR	average to minimum power ratio
APT	average power tracking
β	bipolar transistor current ratio
BJT	bipolar junction transistor
Bluetooth EDR	extended data rate mode for Bluetooth™
BPSK	binary phase-shift keying
BW	bandwidth
BW3	3 dB bandwidth
BWn	n dB bandwidth
CCDF	complementary cumulative distribution function
CCS	controlled current source
CDF	cumulative distribution function
CDMA	code division multiple access
CE	constant envelope
CFR	crest factor reduction
CJTF	constant joint transfer function
CMOS	complementary metal oxide semiconductor
CoE	conservation of energy
dBm	decibels relative to 1 mW
DC	direct current
DC-DC	direct current in to direct current out
D-FET	depletion mode field effect transistor
DP	direct polar
DPS	dynamic power supply
DPST	dynamic power supply transmitter

DQPSK	difference quadrature phase-shift keying
DSB-SC	double sideband with suppressed carrier
DWC	digital wireless communications
EDGE	enhanced data rate for GSM evolution
EDR	envelope dynamic range
EER	envelope elimination and restoration
EF	emitter follower
E-FET	enhancement mode field effect transistor
EFF	envelope flooring and filling
ENB	equivalent noise bandwith
EpHEMT	enhancement mode pseudomorphic high electron mobility transistor
ET	envelope tracking
EVM	error vector magnitude
FET	field effect transistor
FM	frequency modulation
FSK	frequency-shift keying
GaAs	gallium arsenide
GaN	gallium nitride
GMSK	Gaussian minimum-shift keying
GPRS	generalized packet radio service
GSM	global system for mobile communications
HBT	heterojunction bipolar transistor
HD	harmonic distortion
HEMT	high electron mobility transistor
HSDPA	high-speed downlink packet access
HSPA	high-speed packet access
HTOL	high temperature operating life test
IMN	input matching network
InMN	interstage matching network
IQ	in-phase/quadrature phase
IV	current-voltage
KCL	Kirchhoff's Current Law
LDMOS	lateral diffused metal oxide semiconductor
LDO	low dropout voltage regulator
LDVR	linear dynamic voltage regulator
LINC	linear amplification with nonlinear components
LOS	line of sight
LPF	lowpass filter
LTE	long-term evolution
MESFET	metal semiconductor field effect transistor
MOSFET	metal oxide semiconductor field effect transistor
NADC	North American digital cellular

NF	noise figure
NLOS	non line of sight
OBO	output back-off
OFDM	orthogonal frequency division modulation
OMN	output matching network
O-QPSK	offset quaternary phase-shift keying
P1dB	-1 dB compression point
PA	power amplifier
papr	peak-to-average-power ratio (linear)
PAPR	peak-to-average power ratio (dB)
PCDR	power control dynamic range
PDF	probability density function
PEP	peak envelope power
pHEMT	pseudomorphic high electron mobility transistor
PM	phase modulation
PMPR	peak-to-minimum power ratio
PSD	power spectral density
PSR	power supply rejection
PSRR	power supply rejection ratio
pss	power supply sensitivity (linear)
PSS	power supply sensitivity (dB)
QAM	quadrature amplitude modulation
QM	quadrature modulation
QPSK	quaternary phase-shift keying
R	resistance
RBW	resolution bandwidth
RC	resistor-capacitor
RF	radio frequency
rms	root-mean-square
RX	receiver
SF	source follower
SiGe	silicon germanium
SMPS	switch-mode power supply
SR	slew rate
SRC	spectral raised-cosine
SSB	single sideband
SSB-SC	single sideband with suppressed carrier
SSR	stage series resistance
TDM	time division multiplex
TETRA	terrestrial trunked radio
TRC	time raised-cosine
TX	transmitter
UMTS	universal mobile telephone service

UTB	uniform transfer boundary
VSWR	voltage standing wave ratio
WCDMA	wideband code-division multiple access
X	reactance
Z	impedance

Part I

Motivations, definitions, and principles

1 Motivations

Energy efficiency in radio frequency (RF) transmitters – this is the entire reason there is any interest at all in transmitter designs that incorporate dynamic power supplies. So what does this mean?

Whenever an amplifier provides a signal into a load resistance R_L, Ohm's Law says that the corresponding root-mean-square (rms) current that the signal has is related to the rms output power (P_{OUT}) according to

$$I_{Signal} = \sqrt{\frac{2P_{OUT}}{R_L}}. \qquad (1.1)$$

The current from (1.1) must flow through the load no matter what the amplifier structure is. So whatever the supply voltage is, this current must flow from it as shown in Figure 1-1.

Power is dissipated (lost) when this current flows through any resistance, which includes the amplifier's transistor. This dissipated power is the product of the current in the load times the voltage difference between the supply voltage to the amplifier and the output signal voltage. When the voltage supplied to the amplifier is a constant value, which is by far the most common design practice, the situation in Figure 1-2(a) results. Power dissipation in the amplifier is at its maximum when the output signal voltage is half of the supply voltage. When the output signal voltage is higher, even though the current value is larger, the voltage drop is less and the power dissipation is lower. Similarly, when the output signal voltage is small, even though the voltage drop is now large, the current in the load is smaller and again the power dissipation is lower.

To keep this internal power loss in the amplifier small, the voltage drop must remain small at all times, because the signal current cannot change for the same output power. Various techniques to achieve this goal are shown in the progression of the charts in Figure 1-2. When the signal peak voltage is known not to go all the way to the available supply voltage, the actual supply voltage to the amplifier can be reduced as shown in Figure 1-2(b). This is called the average power tracking (APT) technique. The voltage drop inside the amplifier is reduced further when the supply voltage to the PA (the dashed line) now varies along with the output signal envelope. This situation was shown in Figure 1-2(c). How closely the dynamic power supply (DPS) follows the signal envelope, the *voltage offset*, has a *huge* impact on the operating properties of the amplifier and the transmitter it is in. Details set by the specific application this transmitter supports dictate design limits on this voltage offset based on the required properties

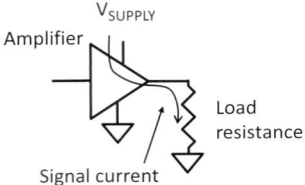

Figure 1-1   Any transmitter must generate a signal current in the load resistance. This current must flow from the amplifier power supply, no matter how it gets modified through the amplifier by the amplifier DC to RF conversion process.

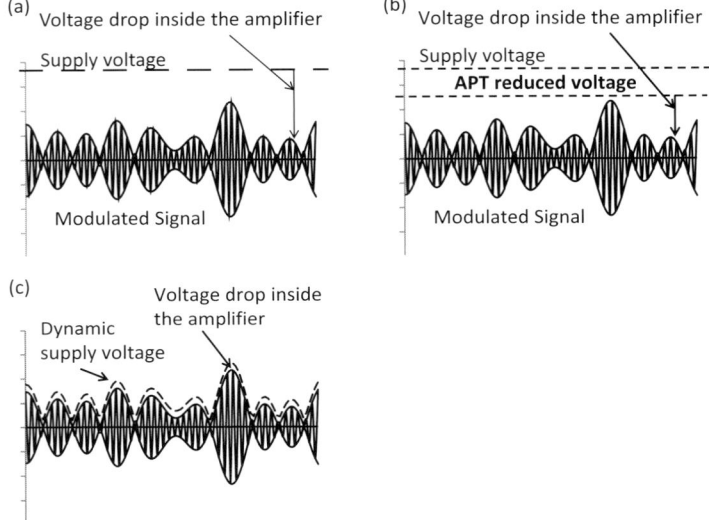

Figure 1-2   Power loss in an amplifier depends on the voltage drop inside the amplifier: (a) voltage drop across the PA transistor with a fixed supply; (b) voltage drop inside the amplifier with a reduced supply voltage from APT operation; (c) voltage drop across the PA transistor with a dynamic supply that follows the signal envelope.

the transmitter must have. In general, this voltage offset must be very carefully controlled.

By adopting a dynamic supply voltage, there are two new problems that need to be solved. One is how to get the waveform needed for the tracking supply, which in modern times is most often directly calculated by the digital signal processor that is calculating the signal itself. Once this waveform exists, it must be properly aligned in time with the signal when they both come together at the final power amplifier. This is discussed in Section 5.5.

This amplifier power dissipation problem is nothing new. Just a few years after vacuum tubes were invented (about one century ago), it was very obvious that as output power was increased, the amplifiers got extremely hot. It was realized before 1920 that if the amplifier could be made to work with a varying supply voltage, then much of this heat would not occur. This problem was solved in a few years to a "good enough" level [1-1], and the resulting transmitter design stayed in wide use for more than 60 years.

It is very important to separate the concepts of instantaneous voltage drop in the amplifier from the peak value of the power supply available to the amplifier. Ohm's Law says that while keeping the instantaneous voltage drop small to gain high efficiency, it is equally important to operate any real amplifier from the highest practical voltage available to further improve its efficiency. Radio transmitters are power-based signal processors, and as for the electric utility, transmission efficiency is greatest when starting from a high voltage. This characteristic is opposite to the energy-based signal processing performed by digital circuitry (particularly complementary metal oxide semiconductor – CMOS) where energy efficiency is improved at lower voltages. These two circuit types optimize at opposite ends of the voltage scale.

1.1   Linearity and linearization

In more modern times, wireless signals are used to communicate a lot of information. These signals are increasingly intolerant of distortion in the transmitter circuitry (and anywhere else, for that matter). This increases the demands on linear performance of the amplifier circuitry. Nothing comes for free – improve the circuit linearity and the power dissipation also increases. And any attempt to reduce the single transistor power dissipation for the same output signal power necessarily results in reduced linearity from the circuit. This has to happen, according to the laws of physics. This sets up a well-known trade-off, between having good energy efficiency or having good circuit linearity. Choose one.

Customers usually do not care about the laws of physics. They want good linearity and high energy efficiency. To meet this demand, we in the product engineering community need to more carefully understand the actual need, which is output signal accuracy. Circuit linearity is not necessary to obtain output signal accuracy, but it is harder to achieve if circuit linearity is not available. The term used for obtaining output signal accuracy, without depending solely on circuit linearity, is called *linearization*.

Fortunately, it is possible to achieve very accurate output signal properties in the complete absence of circuit linearity. This does appear to solve the linearity/efficiency trade-off, but as always there is a cost that may or may not be acceptable. Mainly the costs here are (1) much higher complexity in the necessary implementation along with (2) an inherent incompatibility with several of the high data rate signals used in present communication systems.

Linearity in circuit performance is not necessarily easy to achieve either. Particularly in the CMOS geometries below 100 nanometers, the individual transistors become progressively faster and progressively less linear. Architectures that use this dynamic supply voltage for improved energy efficiency and that also can use the varying supply to improve output signal accuracy become attractive. The amplifier operating mode must change to have the DPS shift from being a source of output signal distortion to being a linearizer.

1.2   Reliability improvement

When power dissipation goes down, temperature goes down. When operating temperature goes down, it is well known that circuit reliability goes up exponentially. Therefore, the most important parameter that predicts long-term reliability of a component is its operating temperature, and the parameter driving the operating temperature is the component power dissipation. Anything that can be done to reduce component power dissipation will improve its reliability. Incorporating a DPS into the transmitter power amplifier (PA) directly reduces power dissipation in the PA.

1.3   High peak-to-average power signal types

Standardization committees have adopted signals in recent decades that have increasingly high peak-to-average power ratio (PAPR) properties, as shown in Figure 1-3 for uplink (mobile to infrastructure) signals [1-2]. The bandwidth efficiency of these signals does not track well with the PAPR; indeed, there are several signals where the bandwidth efficiency decreases while the PAPR increases. This is particularly true for the 3G signal used in the universal mobile telephone service (UMTS), where the spread spectrum chip code needed for code division multiple access (CDMA) operation expands the signal bandwidth with no change in the information data rate. It is widely assumed that in order to achieve high values of bandwidth efficiency, the signal necessarily must have a high PAPR value. This correlation is actually very weak, as the data in Figure 1-3 show.

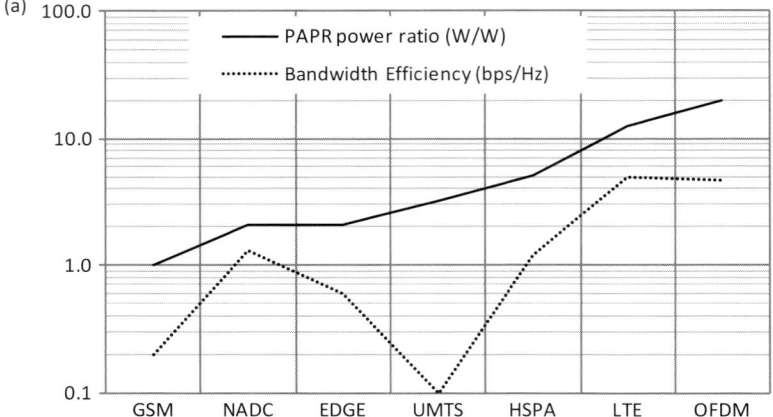

Figure 1-3   PAPR (solid line) and bandwidth efficiency (dotted line) for a progression of uplink signal modulation types in wide use: (a) logarithmic scale; (b) linear scale.

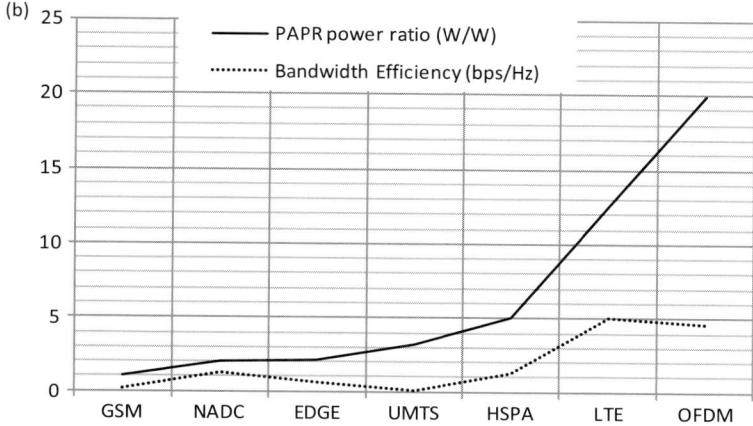

(b)

Figure 1-3   (cont.)

1.4   Energy efficiency

Adding a DPS into any transmitter costs money, so compelling economic reasons and performance gains received from doing so must exist to justify spending this money. The compelling motivation follows from the adoption of communication signals that have high peak-to-average power ratios. Any power amplifier is a peak power limited device, meaning that the PA must be capable of generating the signal peak power. For amplifier linearity, the upper limit of voltage clipping must exceed the peak envelope voltage. Therefore, as the PAPR of the signal modulation being used increases, the average power available from the amplifier decreases proportionally.

This situation is illustrated in Figure 1-4. A perfectly linear amplifier would have its transfer function follow the straight dashed line. But any real amplifier has a finite limit to its maximum output voltage, and this amplifier response curve is normalized to this peak voltage value. The voltage clipping boundary is set where the output voltage no longer increases. The amplifier output efficiency available from a good transistor is also shown by the dotted curve. This output efficiency peaks slightly above the amplifier input signal voltage that reaches the voltage clipping boundary, which corresponds to the square wave shape needed for maximum energy efficiency.

Envelope variations from signal modulation define what the signal PAPR is. And these envelope variations also define what the amplifier linearity requirements must be. Taking four of the signal examples from Figure 1-3 and converting the power values used in the PAPR evaluation to envelope voltage ratios allows the energy efficiency impact of the signal envelope variation to be evaluated in Figure 1-4. The GMSK (Gaussian minimum-shift keying) modulation used for GSM has no envelope variation, and so it can be operated at the amplifier peak efficiency point − here at 61%. The basic UMTS modulation is a special form of quadrature amplitude modulation (QAM), with a PAPR of 3.5 dB. Setting the envelope peak at the amplifier voltage clipping boundary, the signal average moves down to where the amplifier output

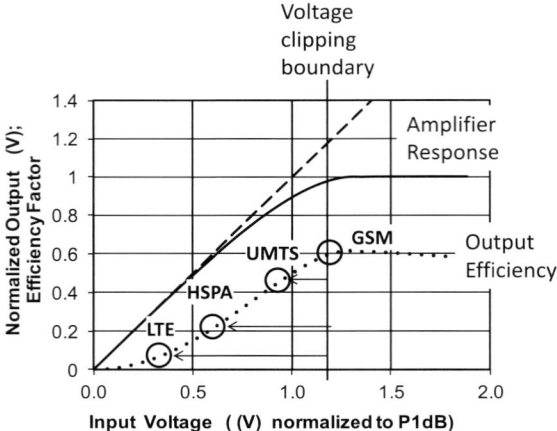

Figure 1-4   Energy efficiency impact on a linear amplifier from signal PAPR for four common signals: higher PAPR necessarily results in lower energy efficiency.

efficiency is still above 40%. The more complicated QAM modulation used for high-speed packet access (HSPA) has a much larger PAPR, and this pushes the amplifier operation down to a maximum operating efficiency just above 20%. Finally, the ortho-gonal frequency division modulation (OFDM) signal that long-term evolution (LTE) is based on has a PAPR large enough to force the operating efficiency of this PA below 10%. It is important to make clear that these efficiencies are the maximum available efficiencies for these signals. At reduced output powers, the PA efficiency drops further along the efficiency curve from these maximum values.

At 10% efficiency, for every ten electrons drawn from the battery only one is useful for making the desired output signal, and the remaining nine stay behind and generate heat through amplifier power dissipation. This is dreadful performance, which forces transmitter designers to find different architectures from this single linear power amplifier that can still provide an accurate output signal, but also provide efficiency closer to 40%.

History holds that products are acceptable when PA efficiency is at or above 40%. For the amplifier shown in Figure 1-4, this efficiency corresponds to a signal PAPR of 6 dB, which is a power ratio of 4:1 and equivalently a voltage ratio of 2:1. The efficiency drop from a conventional linear amplifier when the signal PAPR is less than 6 dB is historically tolerable, and considered to not justify any change to the simple architecture used for simpler signal modulations. As the signal PAPR increases above 6 dB, or 4 W/W, the efficiency drop is less tolerable and does begin to justify the effort to change the transmitter architecture and to accept the resulting production cost and complexity increases. Dynamic power supply transmitters (DPSTs) are one viable option to meet this new requirement.

When the primary energy source is a battery, or particularly some type of energy harvesting mechanism, the electrons from the electron source are best considered as finite. The wireless communication feature exists to communicate, and it is viable to

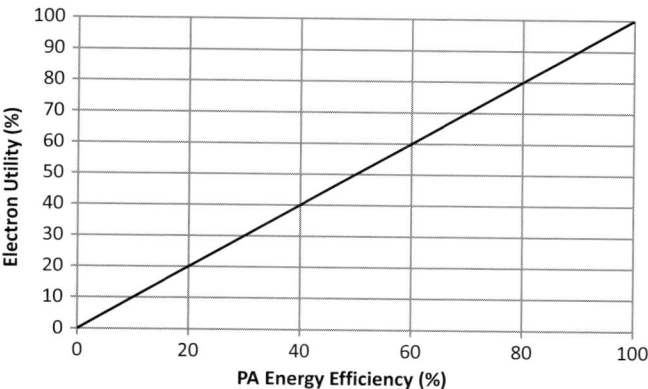

Figure 1-5   Electron utility ratio of a transmitter depends on the PA operating energy efficiency: when the energy source is finite (like a battery), signals and circuits that enable higher electron utility are more viable and valuable.

consider the radio with regard to its effectiveness at using these electrons for the needed communication. Implementation architectures and circuits must consider this electron utility factor – the ratio of electrons drawn from the primary energy source to those that actually result in the needed communication signal – to select the most effective option.

Looking only at the transmitter PA, the electron utility factor is exactly equal to the amplifier efficiency factor, as shown in Figure 1-5. This is obvious, but still useful to describe in this electron utility format because it is electron utility that governs the real design target: battery life.

1.5   Efficiency improvement vs. signal PAPR

From Figure 1-4 it is apparent that output efficiency of an amplifier can improve when used with an envelope-varying signal only when the input drive level is increased, forcing the amplifier into clipping on the signal peaks and therefore no longer supporting the entire range of the envelope variation required of the modulation. By definition, this is a nonlinear operation of the PA. This is another way to view the well-known trade-off between linearity and efficiency in any linear amplifier. Again, we only get to choose one. Fortunately the relationships in Figure 1-4 also illustrate that there is another degree of freedom in the communication system design to improve its efficiency, and that is the selection of signal modulations with lower PAPR values which provide the needed communication properties.

The effectiveness of DPS architectures in providing the needed output efficiencies differ among the available DPS architectures. The simplest DPS architecture is called average power tracking (APT), which actually is a misnomer because what really is happening is peak power tracking. In this DPS architecture, the voltage applied to the PA is set to be slightly above the peak signal envelope voltage as the output power is varied

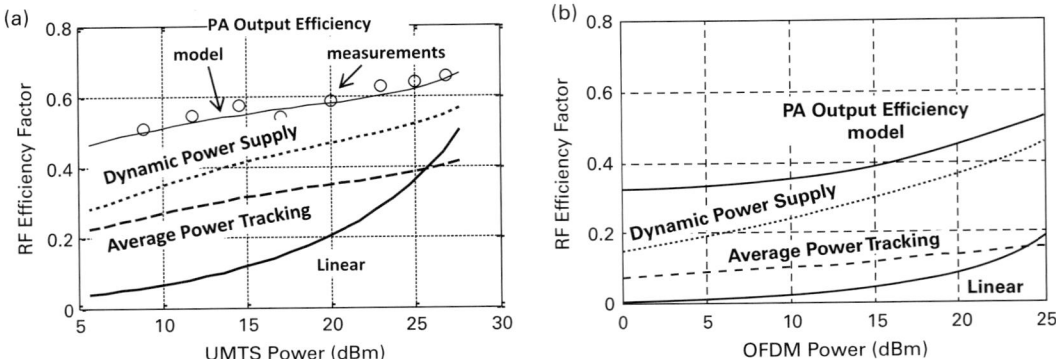

Figure 1-6   PA energy efficiency gains from DPS operation options: (a) UMTS; (b) OFDM.

in an attempt to always operate the PA at its maximum available output efficiency for the signal being used, as shown in Figure 1-4.

At the other extreme of DPS architectures is one that operates the power amplifier in accordance with Figure 1-2(c), but with the voltage drop inside the amplifier set to zero. This provides the highest possible overall energy efficiency the transmitter can have, though at some costs that the application may not want to accept. Setting those costs aside for the moment, the top-level relationships between overall transmitter energy efficiency, PA energy efficiency, and signal PAPR are illustrated in Figure 1-6.

The first step in efficient transmitter design is to make the PA itself maximally efficient. This is done by eliminating all amplifier circuit linearity, with the techniques presented in Chapters 6 and 8. In Figure 1-6, these maximum efficiency results are shown as the top line labeled "PA output efficiency model." Two signal modulation cases are shown in Figure 1-6, a UMTS signal with 3.5 dB PAPR in Figure 1-6(a), and an orthogonal frequency division modulation (OFDM) signal with 10 dB PAPR in Figure 1-6(b). The UMTS design chart also includes direct measurements of the maximum available PA output efficiency at various UMTS output power levels, which validate the model.

It now becomes a task of the adopted architecture to make this efficiency "visible" to the local energy source. The bottom solid line is a measure of the corresponding efficiency of this amplifier when operated as a conventional fixed-supply class A linear amplifier such as that in Figure 1-4. The overall efficiency is low, a well-known problem for linear class A amplifiers. The peak efficiency in this curve corresponds to the open circles in Figure 1-4, as well as the efficiency curve in that figure. The dashed line shows the overall efficiency seen with the average power tracking (APT) technique, which only reduces the power supply voltage to the amplifier to the limit of the highest signal peak as shown in Figure 1-2(b). At the highest output power, APT is actually less efficient than just the linear amplifier itself. This is because the DPS implementing the APT reduced supply voltage has some power dissipation of its own, which is not present when the DPS is not there. The major APT benefit is seen at the lower output power levels where the overall efficiency drops much more slowly, which now corresponds more closely with the efficiency curve available from the PA.

The dotted curve presents the overall output efficiency when the DPS is designed to "go all the way" toward the maximum available efficiency from DPS transmitter operation. The difference between this curve and the available PA output efficiency is due to the efficiency of the DPS itself. At lower output powers, the bias currents required by the DPS become more significant and the operating efficiency falls faster than that of the PA itself.

What is most important to notice from Figure 1-6 is that the efficiency benefit of DPS operation is dramatically different for these two example signals. When UMTS modulation is used, the operating efficiency difference between the simple APT and the more complicated DPS operation is fairly small. This difference is not sufficient to justify adopting the complexity and costs of DPS operation for that signal. When the much higher PAPR of OFDM is used, this conclusion is changed. The operating efficiency from DPS operation is much higher than that from APT operation, which itself is far higher than the very low efficiency of linear operation. For the goal of 40% operating energy efficiency, only DPS operation gets close. This performance improvement now does justify this architectural change to DPS operation, both from marketing and economic perspectives.

The remainder of this book provides detailed investigations into the physical processes involved in all of these DPS circuit and architecture options. Important new design equations are provided, along with new testing requirements needed for this very different way of designing transmitters.

1.6   References

[1-1] F. Terman, *Radio Engineers Handbook*, McGraw-Hill, New York, 1943.
[1-2] E. McCune, *Practical Digital Wireless Signals*, Cambridge University Press, 2010, Appendix G.

2 Definitions

This chapter contains the important definitions needed for the remainder of this book. Beginning at the beginning is the best way to build the foundation upon which this book is built. Topics covered include the basic physics upon which all this work stands. Definitions of important terms are presented so that understanding of discussions within the remainder of this book is clear and unambiguous. The key definitions include meanings of supply and bias, linear vs. polar signal processing, how gain must be interpreted when operating in compression, along with the concepts of power supply rejection (PSR), dynamic range, and bandwidth expansion.

All circuit performance metrics used in this book are derived from device characteristic curves, in order to build physical intuition into what the circuitry is actually doing. Device and block models are secondary in this discussion, as they inherently follow from the device physics. Mathematics always follows the physical discussions. In this book, mathematics is a tool, and not a primary window on to the material.

2.1 Physical foundations

The sinusoidal waveform used in radio communications is not an arbitrary choice, but is a consequence from Maxwell's equations of electromagnetism. Looking at this solution, we see that polar coordinates are the physically natural form of the signal equation. Ohm's Law, itself also a consequence from Maxwell's equations, shows how power dissipation happens in transmitters. Knowing how power dissipation reduces overall energy efficiency provides guidance on how to change designs to improve overall transmitter efficiency.

It is important to use models, both physical and mathematical, that not only describe well what the performance of these transmitters is, but are also descriptive of the physical operations. This joint requirement of the models used here is used consistently.

2.1.1 Maxwell's equations

All electronics, radio included, follow from electromagnetism described by Maxwell's equations (usually as reformulated by Oliver Heaviside) [2-1]. Maxwell's equations are particularly important here because the solution to this linearly polarized propagating electromagnetic wave equation

$$\frac{\partial^2 B}{\partial z^2} - \mu\varepsilon\frac{\partial^2 E}{\partial t^2} = 0 \tag{2.1}$$

is a complex exponential of the form

$$E_x(t,z) = E_0 e^{j\Psi(t,z)} \quad ; \quad \Psi(t,z) = \omega\left(t - z/v\right) + \phi(t) \quad ; \quad v = 1/\sqrt{\mu\varepsilon} \tag{2.2}$$

showing that the magnitude of the electric field is perpendicular to the direction of wave propagation (z) [2-1, p. 247]. Considering only the time-varying aspects of (2.2) by dropping the spatial propagation term and keeping only the real part of the complex exponential, we get the equation

$$s(t) = A\cos(\Psi(t)), \tag{2.3}$$

which is expanded into the familiar signal equation by incorporating modulation (time variations) in the amplitude (A), frequency (ω), and phase-shift (ϕ) parameters

$$s(t) = A(t)\cos(\omega(t)t + \phi(t)) \quad ; \quad \Psi(t) = \omega(t)t + \phi(t). \tag{2.4}$$

This brief review confirms that the sinusoidal signal we use is no accident, but is required by electromagnetic theory. The signal (2.4) is *naturally of polar form*, not quadrature form. This means that our amplifier hardware provides output signals of particular magnitude, and generally does not care what the signal phase is. Physics shows itself to be polar, not quadrature, in operation.

The quadrature signal equation in wide use

$$s(t) = I(t)\cos(\omega(t)) + Q(t)\sin(\omega(t)) \tag{2.5}$$

is therefore *not* intuitively descriptive of how amplifier circuits actually work. Model (2.5) is simply a projection of (2.4) on to Cartesian axes, and is very convenient for mathematical analysis. Historically, while both the solar system models by Ptolemy and Copernicus (the latter beautifully generalized by Newton) succeeded in predicting the observed position of the planets, only one of these models is useful for interplanetary travel. We must be careful to read physical significance into our models only to the extent such physical significance is justified. For radio communications, the quadrature model (2.5) is Ptolemaic in that it beautifully describes all our observations while not describing at all what physically is happening. In this regard, polar models are Copernican, in that they also describe what is happening as well as provide physical intuition on what the electrons are doing.

2.1.2   Ohm's Law

Ohm's Law also follows from Maxwell's equations [2-1]. Here though we are particularly interested in power output and power dissipation. Power output is good, and power dissipation within the PA is bad.

Ohm's Law states that power is developed when current flows through a resistance. When that resistance is the PA load, we desire this output power, so it is important to

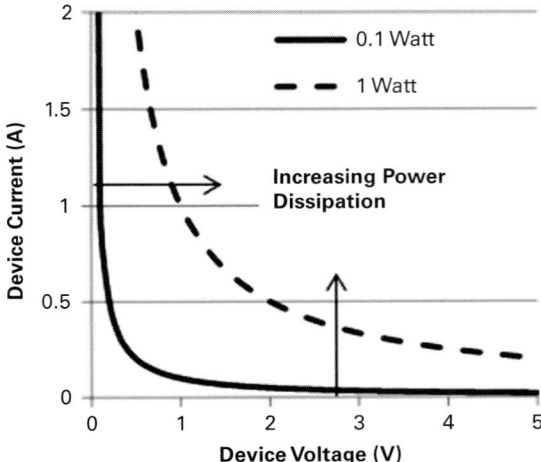

Figure 2-1   Constant power in a circuit branch follows the hyperbolic curve $P = IV$.

have this current (and its corresponding in-phase voltage). But when that resistance is the PA transistor, it is important not to have much in-phase current. This very quest has motivated PA research for nearly a century now.

Constant power contours have hyperbolic shapes on a voltage–current (V–I) plane, as shown in Figure 2-1. These curves show that power dissipation is low whenever the branch conditions (voltage across and current through) are located close to either axis. The greater the distance to both axes becomes, the greater the dissipated power.

We also want all the power to be useful to (dissipated in) the load, so the load must be resistive only. If the source impedance Z_s has a nonzero reactive component X_s ($Z_s = R_s + jX_s$), then we apply a "matching" network to provide a conjugate match $Z_L = R_s - jX_s$ to "tune out" the source reactance. Notice that in this process we have identical resistances, but opposite sign reactances of equal magnitude. This is the very definition of resonance, where reactances "cancel" leaving only the resistances. Conjugate matching, at a particular frequency, is nothing more than building an appropriate resonator.

Power dissipation in the load is good, but power dissipation in the amplifier is not. Ohm's Law is similarly clear here: to eliminate power dissipation, there must not be any in-phase presence of voltage and current in any circuit branch. In particular, this means in the RF transistor, with typical characteristic curves seen in Figure 2-2. All linear amplifier theory depends on simultaneous currents and voltages, so energy efficiency is *fundamentally* a problem with linear amplifiers. This particular point continues to drive power amplifier development as this technology enters its second century – how can in-phase voltage and current be developed in the load while avoiding this same situation within the amplifier? The answer is: by operating the device within the lowest power dissipation regions of its characteristic curves – as near as possible to the current or voltage axes. Much more is explored about this in Chapters 4, 5, and 6.

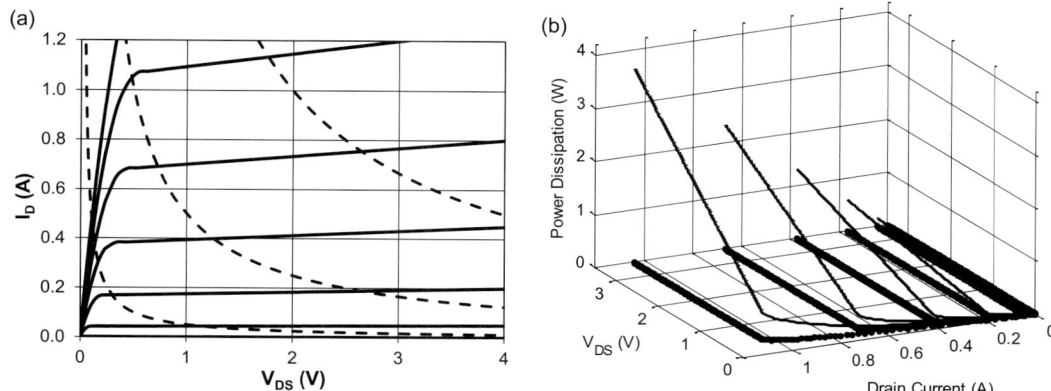

Figure 2-2   Power dissipation in a transistor increases as the operating point moves away from either axis of the characteristic curves: (a) characteristic curves for a power FET, with curves of constant power dissipation (at 0.05, 0.5, and 2 W) shown in dashed lines; (b) characteristic curves showing the corresponding power dissipation of each point in the vertical dimension.

2.2   Supply vs. bias definitions

I clearly remember the beginning of my study in electronics, and my difficulty in understanding what was meant by the word *bias*. Sometimes it was used to mean the entire operating situation of a circuit, e.g. ... with the bias established the amplifier gain is ... Other readings commented ... after biasing the grid to −5 V then the supply can be applied ... For purposes of this book, it is vital to have clear definitions for the terms *supply* and *bias*, because we will encounter many situations where one, the other, or both will be varied or dynamic. After much study, the ambiguity between the terms "supply" and "bias" is resolved by following these definitions:

- **(power) supply**: a power source applied to the *controlled* device port (collector for bipolar transistors, drain for field effect transistors (FET), plate for vacuum tubes (or valves));
- **bias**: a power source applied to the *controlling* device port (base for bipolar transistors, gate for FETs, grid for vacuum tubes) to set quiescent (no input alternating current (AC) signal) currents at the controlled port.

There are publications that use the word bias to refer to the varying power supply, or indeed referring to any power supply. In the spirit of "we should not change the use of a term unless there is an unavoidably good reason," the appropriate term is really supply with an appropriate adjective, such as "agile supply" or "dynamic supply." As noted in the title of this book, I favor using the term "dynamic power supply" to refer to voltages and currents at the controlled port of the RF power transistors.

While the DPS operates, we may or may not also vary the bias at the controlling port of the RF power transistor. It is easiest and lowest cost to set the RF transistor bias once and then leave it alone. In Chapters 4 and 5, situations are presented where additional performance is achieved by allowing the bias to also have some dynamic properties. Fortunately, with these definitions adopted in this book, we are free to treat supply and bias separately and unambiguously.

2.3   Linear vs. polar circuitry

What is a *linear* amplifier? According to one professor I had many years ago, an amplifier is either linear or it is not. According to this definition, no amplifier is linear because all circuits have nonlinear characteristics. This position is not practically useful, so a more useful definition of a linear amplifier from my experience is this:

A **linear amplifier** is a 2-port circuit whose transfer function (voltage, current, or otherwise) can be modeled by a polynomial, where the first-order coefficient dominates the value of all other coefficients.

Thus, when we model the amplifier voltage transfer function with a polynomial of order L

$$v_{OUT}(v_{IN}) \approx \sum_{k=0}^{L} a_k (v_{IN})^k, \tag{2.6}$$

the test for how much the first-order coefficient dominates is performed by defining the amplifier linearity factor L_A as

$$L_A \equiv \frac{a_1}{\displaystyle\sum_{k=1}^{L} a_k}. \tag{2.7}$$

From (2.7), the normalization constrains the range of L_A within $0 \leq L_A \leq 1$. If L_A is close to zero, the amplifier is not linear at all. We have no further information as to the nature of the nonlinearity, but that is not the interest here. As L_A approaches unity, the amplifier is increasingly linear because the first-order coefficient a_1 increasingly dominates the summation in the denominator of (2.7). Defining how close the value of L_A needs to be to unity depends on the amplifier application, which may well have different answers depending on the order of the major nonlinear term(s). For example, a third-order dominant term is usually more of a problem than a fourth-order dominant nonlinearity.

   A **polar signal process** is a circuit block that controls a polar coordinate of the signal passing through it. By definition then, a polar signal process has three ports: two inputs (RF signal, and polar control parameter) and one output. The polar parameter input signal controls a polar parameter (magnitude or phase) of the signal passing from the second input to the circuit output. Since magnitude is one of the polar coordinates, a circuit where the supply voltage sets the output magnitude is *by definition* a polar signal processor. This is explicitly shown in Figure 2-3.

   The fully modulated signal first exists only at the transmitter output at full power. It is therefore not accurate to describe this stage as a power "amplifier." More accurately,

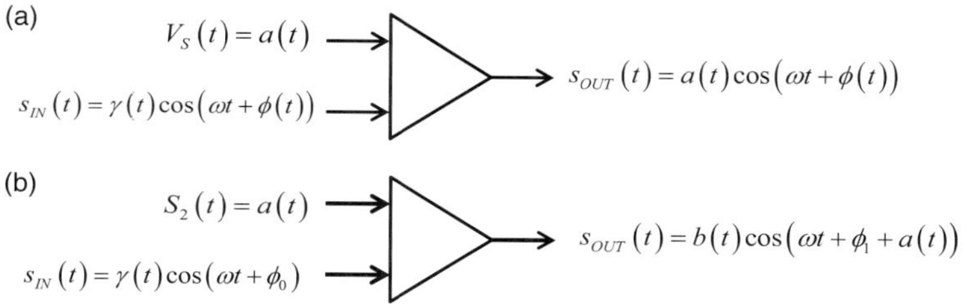

(a)

$V_S(t) = a(t)$

$s_{IN}(t) = \gamma(t)\cos(\omega t + \phi(t))$

$s_{OUT}(t) = a(t)\cos(\omega t + \phi(t))$

(b)

$S_2(t) = a(t)$

$s_{IN}(t) = \gamma(t)\cos(\omega t + \phi_0)$

$s_{OUT}(t) = b(t)\cos(\omega t + \phi_1 + a(t))$

Figure 2-3   Polar modulation operates the transmitter final amplifier as a 3-port, not as a 2-port. The second input controls a signal polar parameter, either (a) signal magnitude, or (b) signal phase. Often for magnitude control, the final RF power stage operates as a multiplier, and the output signal magnitude is set by the control port, not the magnitude of the input signal.

a polar-operating final stage is best described as a power modulator. It is a key component to the output stage of the envelope elimination and restoration (EER) technique [2-2] [2-3] [2-4].

Importantly, the characteristic of being a polar signal process does not change if the output signal does not change when the input signal magnitude varies. When the input signal has some envelope variation through any process $\gamma(t)$

$$s_{IN}(t) = \gamma(t)\cos(\omega t + \phi(t)) \tag{2.8}$$

and the output signal remains

$$s_{OUT}(t) = a(t)\cos(\omega t + \phi(t)), \tag{2.9}$$

where the output magnitude is fully controlled by the control input, this block remains a polar signal process. $\gamma(t)$ may be a constant, but it does not have to be constant as long as the output signal is unchanged by any variation in $\gamma(t)$: $\dfrac{\partial s_{OUT}(t)}{\partial \gamma(t)} = 0$. Chapter 6 shows that it is actually good design practice not to have $\gamma(t)$ as a constant.

There are some who claim that for a signal process to be polar, it is required that the input signal magnitude must be constant. This is not correct. First, we do not define a circuit's operation by what the input signals look like, we define circuit operation by what it does. It is certainly possible to leave the input signal at a fixed and constant magnitude. But this is not necessary, as shown by (2.8) and (2.9). A fixed magnitude signal can easily be put into a linear class A amplifier and work just fine. The necessary conditions for polar modulation are (2.9) and Figure 2-3(a). Details on why these are both necessary and sufficient are the topic of Chapter 6.

2.4   Gain when in compression

Gain is a well-known characteristic of a linear amplifier, defined as the slope of the amplifier voltage transfer function

$$y(x) = ax \qquad (2.10)$$

and is fundamentally defined by

$$gain \equiv \frac{dy}{dx} = a. \qquad (2.11)$$

Other literature and textbooks define gain as the ratio of the output to the input

$$gain \equiv \frac{y}{x} = a. \qquad (2.12)$$

The definitions (2.11) and (2.12) for amplifier gain are very different. Which of these is correct? We are very fortunate that whichever of these methods is used to calculate gain, when the amplifier is operating linearly they both give the same answer. This fortuitous situation has allowed the circuit design community to get away with this ambiguity for nearly a century. Dare I say we all are lazy when discussing gain?

The correct situation is that both of these gain definitions are correct, and that they provide different insights into what the amplifier circuit is doing. We need to use them both. To use them both without ambiguity, we need to have separate names for each gain measure. The names I favor are

$$slope\ gain = g_S \equiv \frac{dy}{dx} \qquad (2.13)$$

and

$$ratiometric\ gain = g_R \equiv \frac{y}{x}. \qquad (2.14)$$

Real amplifiers exhibit output compression where the output signal y becomes less than the level predicted by (2.10), eventually reaching a maximum level $y = y_{max}$, which does not change with further increased input x. This is called output clipping and is shown in Figure 2-4.

Only the basic concepts of gain measurements in amplifier compression are presented here. More details are explored in Section 4.1.5.

2.4.1   Slope gain

The unique characteristic of slope gain g_S is that it goes to zero when the amplifier output clips. This means that variations in slope gain correspond directly to waveform distortion, and more generally to waveform fidelity. When linearizing an amplifier, it is slope gain that must be made constant up to the onset of clipping, when the amplifier slope gain rapidly transitions from its nominal value down to zero. This linearizer characteristic is illustrated in Figure 2-5.

Slope gain is measured from the amplifier's output voltage behavior. When power measurements are used to characterize an amplifier, slope gain is not the result. Instead, it is a ratiometric gain that is being measured.

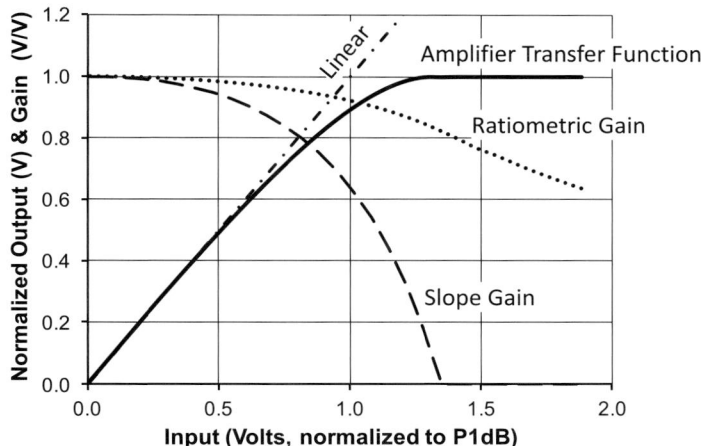

Figure 2-4   Examples of slope gain and ratiometric gain for an amplifier. As this amplifier compresses, the values of the two gain measures increasingly diverge. When the output stage clips, the slope gain goes to zero.

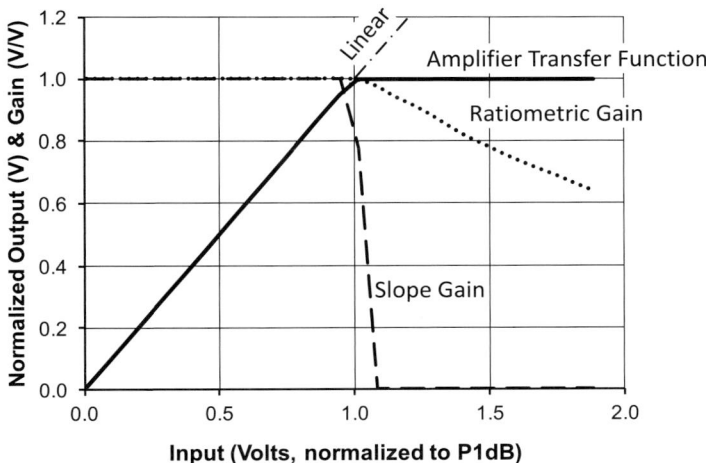

Figure 2-5   Linearizing an amplifier is achieved by keeping the slope gain consistent up to the onset of output clipping. Accuracy of the ratiometric gain is an immediate consequence.

2.4.2   Ratiometric gain

Figure 2-4 shows that ratiometric gain g_R does not go to zero when the amplifier output is distorted. Indeed, ratiometric gain never gets to zero because the output power is not zero, particularly when the amplifier is saturated. It is therefore not immediately obvious from looking at g_R data when the amplifier output waveform becomes distorted from output clipping.

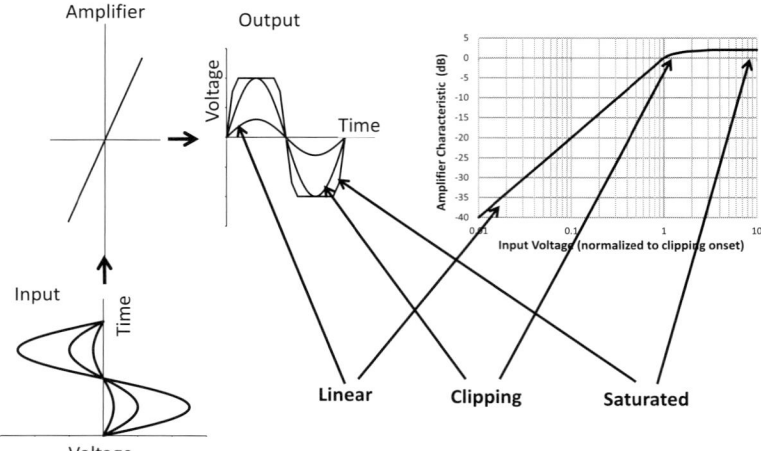

Figure 2-6   Linear, clipped, and power-saturated output waveforms and their corresponding regions along the output power transfer function.

Using g_R data to design an amplifier linearizer is not optimum. This is not the proper gain to be corrected, because it does not directly represent waveform distortion. Unfortunately, the temptation is to do exactly this, because essentially all amplifier gain measurements made in an RF laboratory are done using power measurements, and so are related to g_R data [2-5].

2.4.3   Power transfer function and RF waveforms

One of the effects operating during compression is clipping, which happens if the output waveform tries to exceed the maximum voltage the amplifier can provide. Shown in Figure 2-6, output power still increases while the output waveform magnitude stays fixed at the clipping level and is increasingly distorted by clipping. One cannot then look only at a power transfer curve to determine when waveform distortion is happening. When the output power saturates, as we see in Figure 2-6, the output waveform is essentially already square. The power in the fundamental frequency component of a square wave is $4/\pi$ times higher, about 2 dB, than that of a sinewave with the same peak value. To first order, we can estimate that the onset of waveform distortion from clipping happens at 2 dB below the amplifier saturated power [2-6]. This result holds for a perfectly linearized amplifier, the model of which is often referred to as a "soft" (i.e. finite gain) limiter.

This power increase of the fundamental frequency signal during clipping is certainly real, and will be measured when the amplifier output is filtered to remove all harmonics from the clipping distortion. If such a bandlimiting filter is present, then a linearizer can take some advantage of this "extra" 2 dB of output power that is made available by waveform clipping distortion at the transistor output.

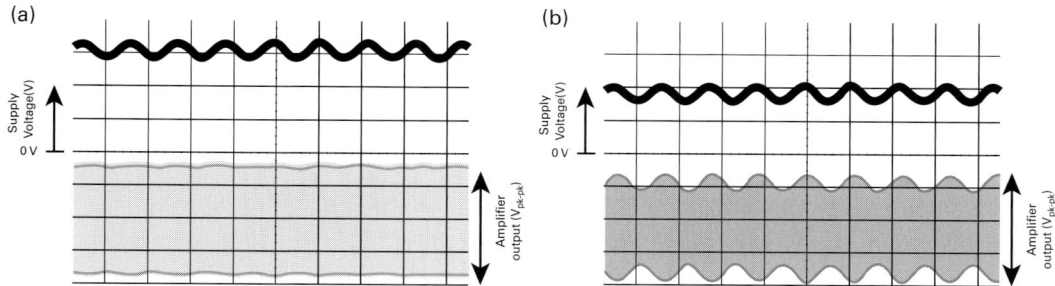

Figure 2-7   Examples of amplifier sensitivity to power supply variations: (a) weak sensitivity; (b) strong sensitivity.

2.5   Power supply rejection

Power supply rejection (PSR) measures the output envelope change from a power supply value change, given no change in the amplifier input power. Variations in the power supply value can come from intentional changes, and from noise in the power supply circuits. Noise on the power supply is considered to be rejected when there is no amplitude modulation of the amplifier output signal due to this noise. This is a desirable situation for three reasons:

- the output signal is completely controlled by the amplifier input signal, and is independent of the supply voltage value;
- design of the power supply is eased by not requiring the details and expense of using low noise design techniques; and
- accuracy of the power supply voltage is not critical, further reducing costs.

The natural test for PSR is to inject a "noise" signal on to the power supply output, and then measure any AM sidebands corresponding to this test signal. For simplicity, the test signal is usually a simple sinewave. Also the amplifier input (and output) is usually an unmodulated RF carrier to make it easier to identify any AM sidebands due to the varying power supply. Results from two PSR measurements are shown in Figure 2-7.

Full details on supply noise rejection, or the lack of it, are provided in Section 4.6.

2.6   Dynamic range

Dynamic range is the difference between the largest output signal and the smallest output signal from any amplifier. In general, the largest useful output signal is set by the onset of output compression. The amount of output compression that can be tolerated is dependent on both the signal modulation used and the spectral confinement requirements on the application. At the bottom end, the minimum output signal power is limited by noise.

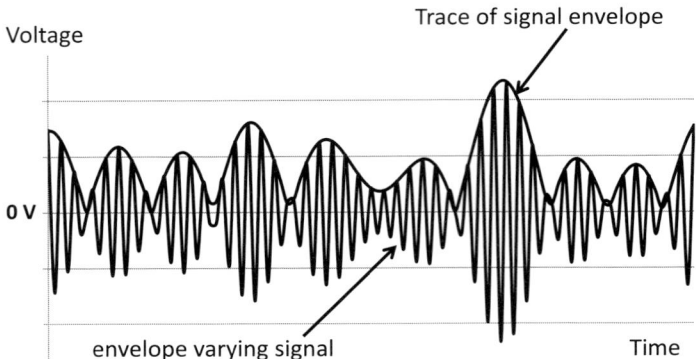

Figure 2-8   Definition of the signal envelope: the locus traced by the modulated RF signal peaks, cycle by cycle.

Dynamic range generally has three measurement types that must be explicitly identified. These are the envelope dynamic range, power control dynamic range, and total dynamic range.

2.6.1   Signal envelope

A signal envelope is the single-sided trace of RF signal voltage peaks over time, as shown in Figure 2-8. This means that the signal envelope is equivalent to the time-varying signal magnitude, a polar coordinate. The signal envelope is not a physically real signal. Only the RF signal itself is physically real. Still, the signal envelope is a very useful concept in the design of systems and transmitters.

Properties of the signal envelope have a dominant impact on transmitter design. When a digital wireless signal has envelope (magnitude) variations when modulated, then there is a corresponding probability with regard to what output power the transmitter is putting out at any particular instant or interval in time. In order to understand these properties, we use three probability measures.

Probability density function (PDF) $P(x)$: This is the probability that the signal envelope (magnitude) has a particular value x, and is shown in instrument measurement selections as the envelope PDF.

Cumulative distribution function (CDF) $P(x<c)$: The first cumulative probability of interest is the CDF, which reports the probability that the envelope value x is less than or equal to a particular value c. The CDF curve is the integral of the PDF characteristic from 0 up to c. This must be a monotonically increasing curve, since the probability that the envelope value is less than zero is, well, zero – and the probability that the envelope is less than the peak envelope value is unity. The CDF curve for a signal provides useful detail on low envelope characteristics and no detail regarding the signal peak characteristics.

Complementary cumulative distribution function (CCDF) $P(x>c)$: very little information is available on the envelope peak characteristics from the CDF. To address this problem, wireless signal engineers use the complementary cumulative density function,

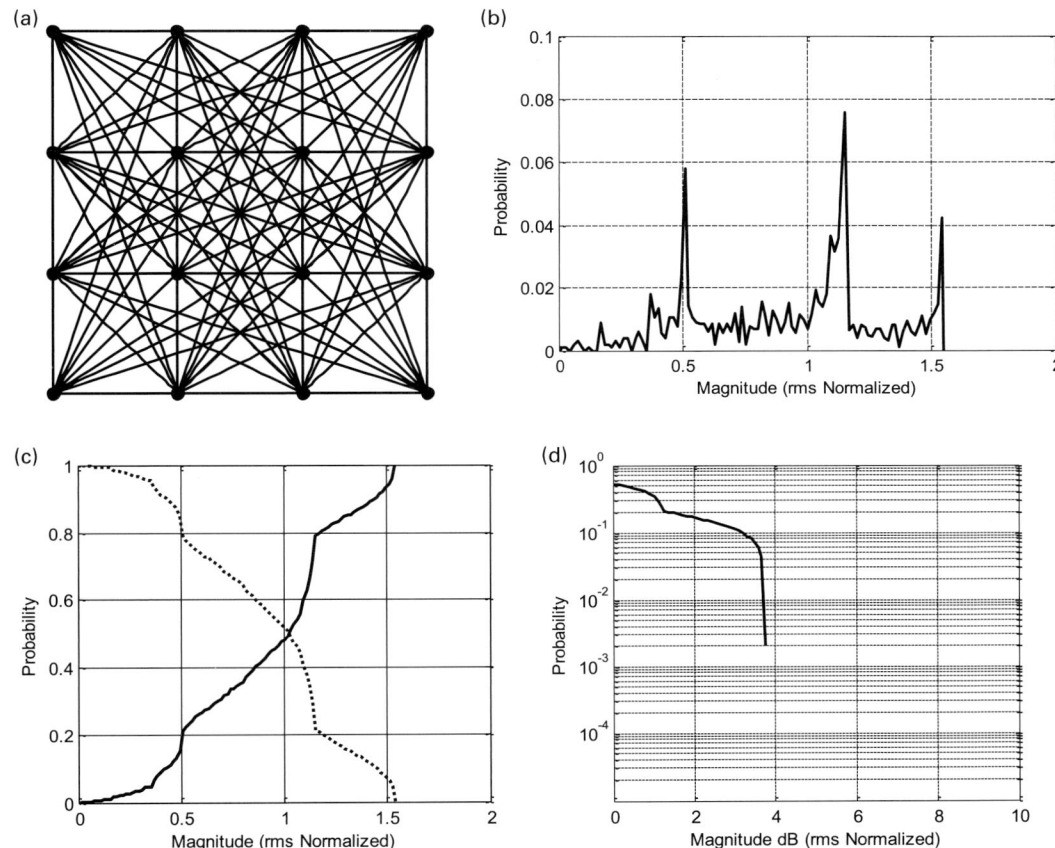

Figure 2-9   Probability measure examples for an envelope-varying signal: (a) vector diagram for a temporal-raised-cosine (TRC) 16 QAM signal; (b) the magnitude PDF; (c) the CDF and CCDF in linear units; and (d) the CCDF of this signal in logarithmic units.

which is simply the CDF subtracted from one (CCDF = 1−CDF). The information in the CCDF, when plotted on logarithmic scales, provides useful resolution on the peak characteristics of the signal at the expense of information resolution at low signal magnitudes.

All of these statistics are derived from the PDF data, which is a histogram of signal magnitudes taken from the signal vector diagram. The signal PDF measurements usually are normalized in one of two ways. They can be normalized to the envelope peak value, or to the envelope root-mean-square (rms) value. The PDF presented in Figure 2-9(b) is normalized to the envelope rms value, and this normalization is maintained in the corresponding CDF and CCDF curves in Figure 2-9(c) and 2-9(d).

From these statistics, we can determine the following properties of the signal envelope:

- peak envelope power (PEP): the highest power the signal may ever have which occurs at the maximum value of the signal envelope;
- rms envelope: the rms value of the signal envelope;

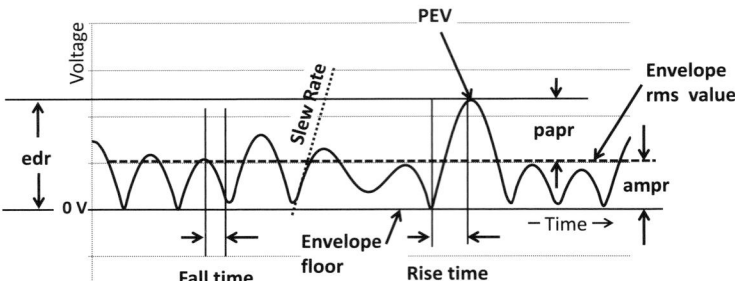

Figure 2-10   Definitions of signal envelope terms, shown on the envelope for the signal in Figure 2-8.

- envelope floor: the minimum value of the signal envelope − must be nonnegative;
- peak-to-average power ratio (PAPR): linear ratio of the PEP over the signal average power; PAPR = $10\log_{10}$(papr);
- average to minimum power ratio (AMPR): linear ratio of the PEP over the signal minimum power AMPR = $10\log_{10}$(ampr);
- envelope dynamic range (edr): voltage ratio of the maximum envelope value to the minimum envelope value. $1 < \text{edr} < \infty$
- rise and fall time: time required for the signal envelope to rise (or fall) from one value to another;
- slew rate: first derivative of the signal envelope, usually the peak value of this derivative − slew rate is usually highest at low envelope values, but not always.

These envelope properties are illustrated in Figure 2-10.

It is advantageous to designers of power amplifiers to work with signals that have finite (nonzero) envelope floor values. This is because the modulated signal's polar characteristics encounter discontinuities at an envelope value of zero. Consider the common implementation of the binary phase-shift keying (BPSK) signal of Figure 2-11(a). The envelope of this signal, shown in Figure 2-11(b), is continuous at all points, but its derivative is not continuous at the places where the envelope is zero. The signal phase, shown in Figure 2-11(c), is discontinuous at this same point, having one value just before the envelope zero point, and being shifted by 180 degrees just past the envelope zero point. These polar discontinuities (remember the physical signal is polar from (2.1) through (2.4)) require the PA to be particularly linear at zero magnitude.

Many commonly used signals have transitions that pass through zero magnitude at the center of any signal vector diagram. Examples shown in Figure 2-12 are QPSK, 64 QAM, and OFDM. Each of these signals shows transitions at the vector diagram center, which means that each of these signals have the characteristics from Figure 2-11. Experience also confirms that each of these signals also requires stringent linearity characteristics from power amplifier circuitry.

The traditional way to avoid these discontinuities is to design the modulation such that there is never a direct path from one constellation point to another

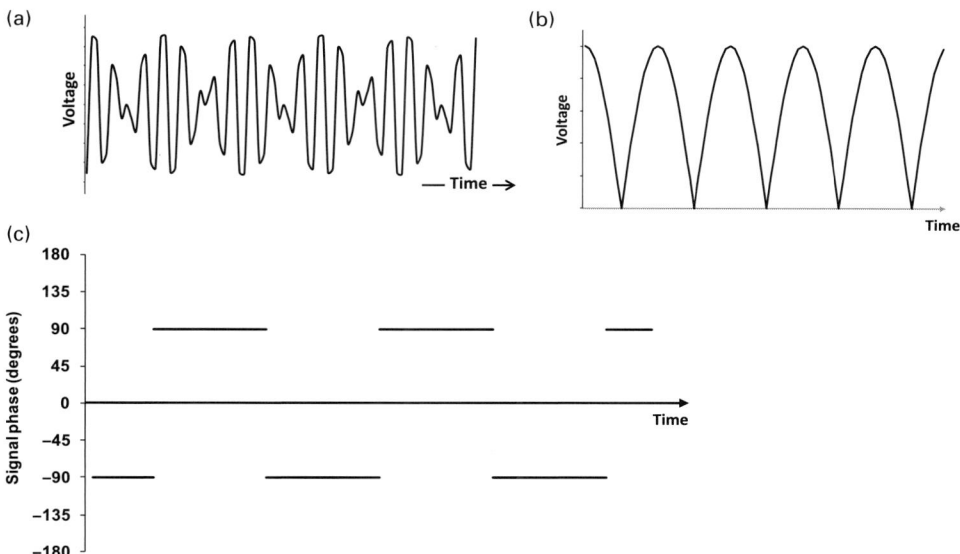

Figure 2-11   Envelope transitions at zero have polar discontinuities: (a) the original signal; (b) envelope for this signal, having an undefined first derivative at the envelope zero points, and (c) the signal phase characteristic, having phase discontinuities at each envelope zero.

that passes through the phasor origin. Usual techniques are illustrated in Figure 2-13. The first two signals use a technique called constellation rotation, where there are actually two constellations used (open circles, filled circles). Any signal transition only happens between one constellation and the other, and never within the same constellation. This eliminates transitions through the signal origin and opens up the center of the vector diagram. Another technique is used in Figure 2-13(c), where the in-phase (I) and quadrature phase (Q) modulation waveforms are offset by one-half of the symbol time. This technique also eliminates the possibility of any transition through the origin.

It is also possible to eliminate these problem-causing origin crossings by using signal processing. The result is the opening of a central region in the vector diagram, leading to the descriptive name of "hole blowing" for this process. This signal process can be done with no impact on the signal PSD at all [2-7] [2-8]. Unfortunately, the necessary changes to the signal transitions are a type of distortion, so the error vector magnitude (EVM) specification must degrade. More details, and very detailed references, are provided in [2-7] [2-8], and Chapter 6.

2.6.2   Power control

The communication range is set by a combination of transmit power and receiver sensitivity. It is a general principle (and also a regulatory rule) that the transmitter

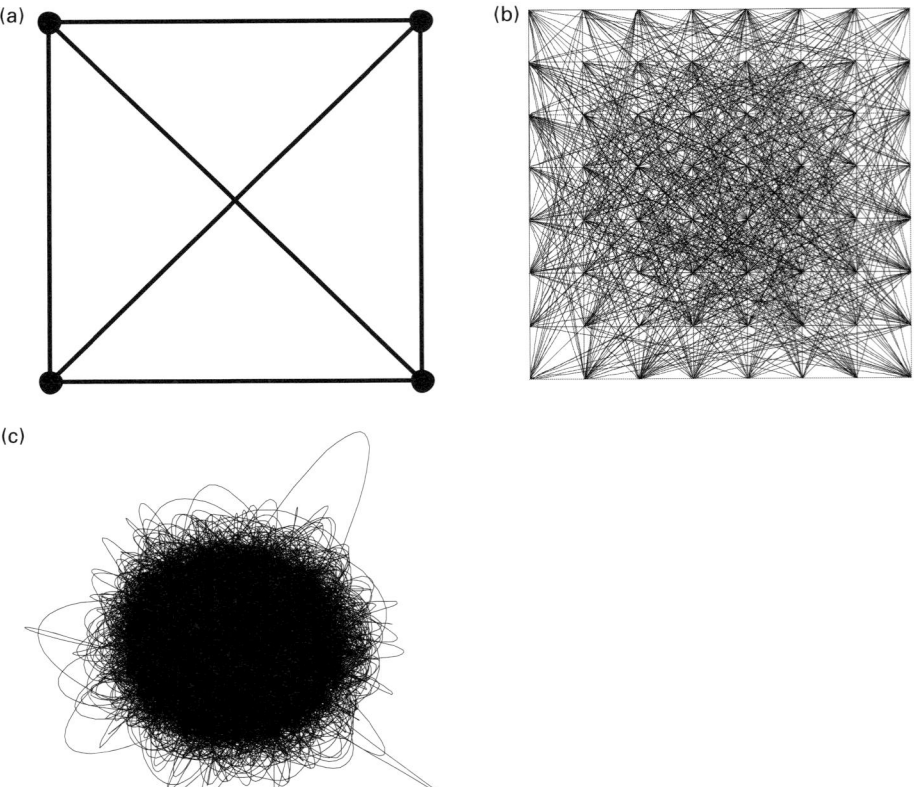

Figure 2-12   Signals with infinite EDR: (a) QPSK; (b) 64 QAM; (c) OFDM. All of these signals are closed in the center of their vector diagrams.

should produce only as much power as is necessary for basic reliability of the wireless communication. The power measure here is the rms power. When evaluating the range and coverage area of any wireless communication link, envelope dynamics are not important. Only the signal average power is important.

It is well understood that as the distance that must be communicated over varies, the transmitter power needs to vary. The relationship between received power P_R and transmitter power P_T for a particular distance d between them is

$$P_R(d) = P_T G_T G_R \left(\frac{\lambda}{4\pi d}\right)^p, \tag{2.15}$$

where G_T is the gain (due to directivity) of the transmit antenna; g_R is the gain (due to directivity) of the receive antenna, and $\lambda = c/f$ is the free-space wavelength of the radio signal. Here the exponent p describes the propagation characteristics of the radio channel: $p = 2$ is the best case of free-space line of sight (LOS)

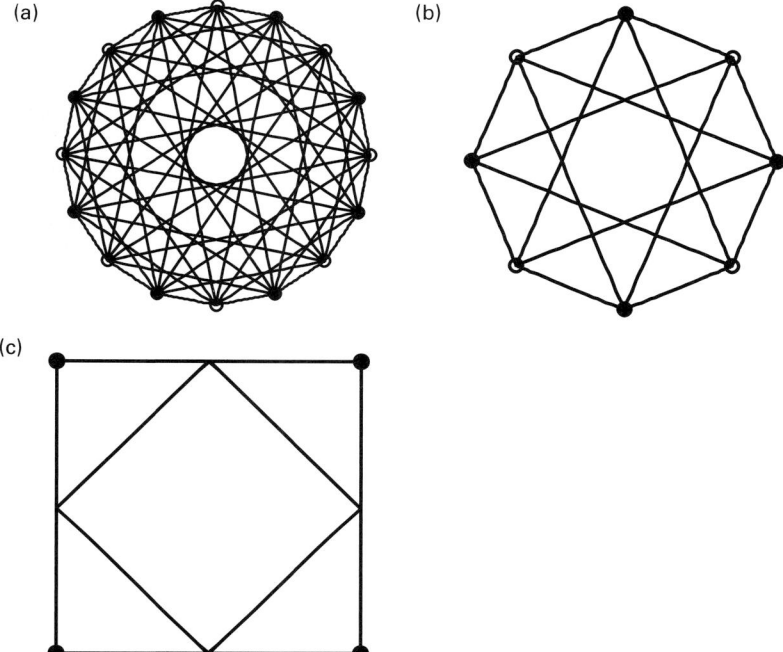

Figure 2-13   Signals with finite envelope floors: (a) 3π/16-8PSK used for EDGE; (b) π/4-DQPSK used for NADC, TETRA, and Bluetooth™ EDR, and (c) O-QPSK with straight-line symbol transitions. All of these signals have open regions in the center of their vector diagrams.

propagation, $p \sim 3$ for occluded non line of sight (NLOS) spaces, $p \sim 4$ for reflected signals, and so on.

To ensure a fixed received power of -90 dBm into a receiver, the transmitter output power must increase if the distance to the receiver increases. Ensuring that the transmitter has the correct output power is achieved through the system's power control mechanism. Figure 2-14 shows a representative case where there is a combined antenna gain of 10 dB at a frequency of 2 GHz. Depending on the value of propagation descriptor p, the required transmitter power increases either relatively slowly ($p = 2$) or very rapidly ($p = 4$) with distance. Very rarely is the value of p a constant value along a terrestrial radio path. The curve in Figure 2-14 labeled "3m break" shows the required transmit power when the propagation behaves with $p = 2$ within 3 meters of the receiver, and, in that example environment p has a value of 4 at distances greater than 3 meters from the receiver.

2.6.3   Total dynamic range

For any transmitter circuitry, the total dynamic range is the sum of the power control dynamic range and the envelope dynamic range. For example, if the envelope dynamic range is 20 dB for the modulation used, and the power control dynamic

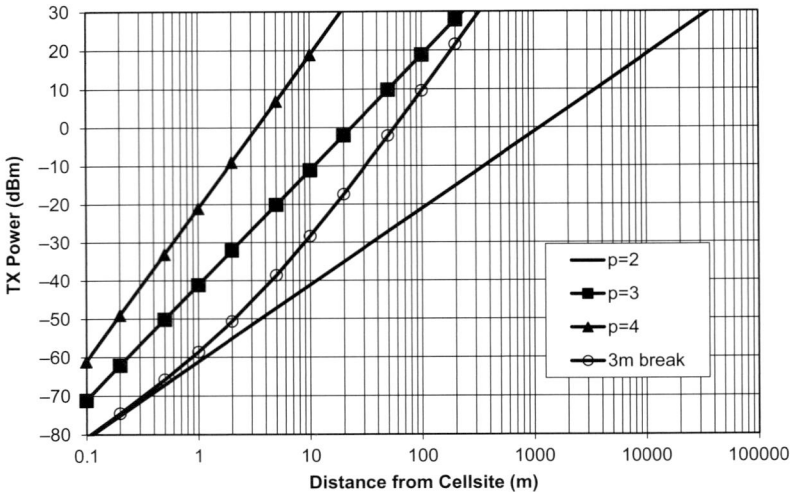

Figure 2-14   Uplink power required to maintain 90 dBm into the base station receiver.

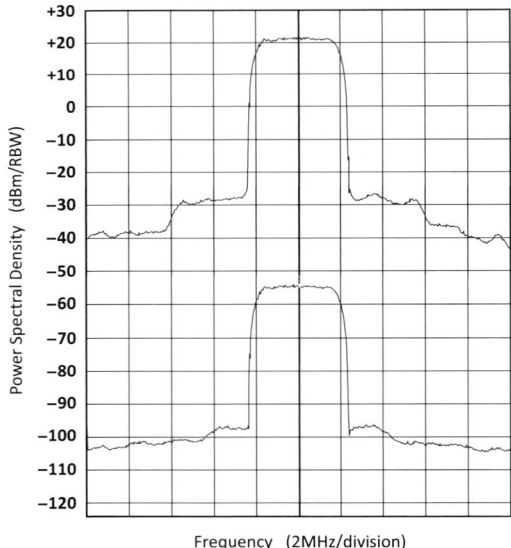

Figure 2-15   A composite overlay of WCDMA PSD measurements at the specified power control maximum and minimum values.

range is 40 dB, the total dynamic range of the transmitter is 20 + 40 = 60 dB. One measure of the transmitter dynamic range of a third-generation cellular mobile is shown in Figure 2-15. The total output signal dynamic range shown exceeds 120 dB, including the average power control dynamic range, plus the envelope variation dynamic range.

2.7   Bandwidth expansion

The signal with absolute minimum bandwidth is a pure sinewave. Any deviation from a pure sinusoid requires the presence of additional frequency components, which is a requirement of the Fourier transform. Such changes from a sinusoid waveform *require* some kind of nonlinear operation. Thus, any nonlinear operation must correspondingly increase the bandwidth of a signal. The converse is also true: if the bandwidth of a signal increases, then a nonlinear process has occurred. This characteristic holds for the envelope as well as for the RF carrier. The bandwidth expansion of interest here is with regard to the envelope.

Consider for example the DSB-SC (double sideband with suppressed carrier) signal with single-tone sinusoid modulation presented in Figure 2-11(a). The RF bandwidth of this modulated signal is twice the modulation frequency, having single-tone upper and lower sidebands centered about a nonexistent carrier. The envelope, shown in Figure 2-11(b), is the track of the magnitude (restricted to positive-sign only) of the RF peaks and is a full-wave rectified sinewave waveform, in this case of the modulating tone. The spectrum of this envelope has many even-order harmonics, all of them required in order to accurately construct the signal zero crossing. Thus, the bandwidth of the envelope is much wider than that of the RF signal itself, even in this very simple case.

Except for a constant envelope signal, the envelope (magnitude) component of any signal is wider in the bandwidth than in the output RF signal. This is a direct result of the nonlinear relationship between the signal in-phase and quadrature components and the signal magnitude

$$A(t) = +\sqrt{I^2(t) + Q^2(t)}. \tag{2.16}$$

The phase-only signal is also subject to bandwidth expansion from the original envelope-varying signal. Determining the signal phase is also a very nonlinear relation with the linear in-phase and quadrature components

$$\phi(t) = \tan^{-1}\left(\frac{Q(t)}{I(t)}\right). \tag{2.17}$$

Spectral regrowth from amplifier nonlinearity is maximized at the phase-only signal, because it is the final result of stripping all envelope variations from the signal. Examples of this modulation component bandwidth expansion are presented in Figure 2-16.

A common rule-of-thumb in the literature is that the envelope signal bandwidth is between five and seven times the intended signal bandwidth. Whether this is true or not depends completely on the actual signal modulation details and also on how much distortion is allowed (e.g. in-band EVM, out-of-band ACLR) at the transmitter output. In reality, all this is saying is that we only care about the envelope signal bandwidth up to a certain point. It is no surprise that this bandwidth is essentially identical to the bandwidth of a distortion correction system that needs to correct for third-order, fifth-

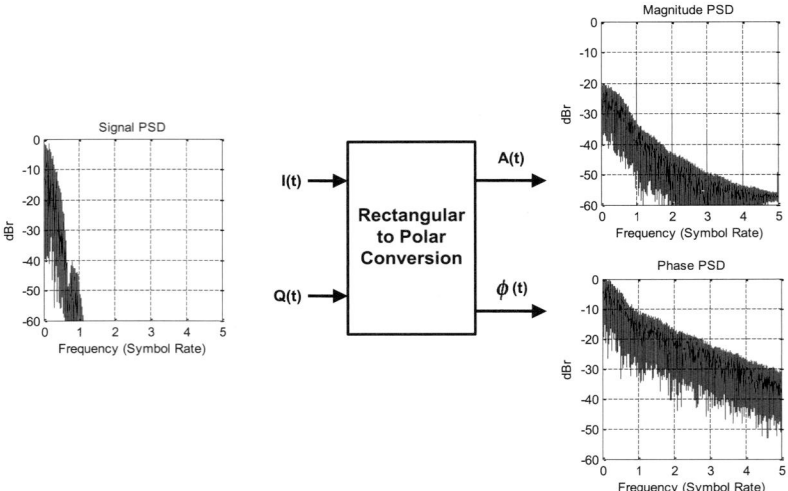

Figure 2-16   One example of modulation component bandwidth expansion from the conversion to polar coordinates.

order, and seventh-order distortion components. Technically, for any zero-crossing signal the true bandwidths of $A(t)$ and $\phi(t)$ are infinite in that mathematically there never is exactly zero signal power spectral density (PSD), no matter how high the evaluation frequency gets.

2.8   References

Note: Patent references are provided for bibliographic use only. Citation of specific patents here is not indicating any view on priority issues.

[2-1] S. Ramo, J. Whinnery, and T. VanDuzer, *Fields and Waves in Communication Electronics*, John Wiley & Sons, New York, 1965.

[2-2] L. R. Kahn, "Single Sideband Transmission by Envelope Elimination and Restoration," *Proceedings of the Institute of Radio Engineers*, vol. 40, no. 7, July 1952, pp. 803–806.

[2-3] F. H. Raab, "Intermodulation Distortion in Kahn-Technique Transmitters," *IEEE Transactions on Microwave Theory and Techniques*, vol. 44, Dec. 1996, pp. 2273–2278.

[2-4] F. H. Raab, P. Asbeck, S. Cripps, P. B. Kenington, Z. B. Popovic, N. Pothecary, J. F. Sevic, and N. O. Sokal, "Power Amplifiers and Transmitters for RF and Microwave," *IEEE Transactions on Microwave Theory and Techniques*, vol. 50, no. 3, March 2002, pp. 814–826.

[2-5] E. McCune, "Gain: Changed Meanings for Compressed Amplifiers," *Proceedings of the 2013 Midwest Symposium on Circuits and Systems (MWSCAS)*, Columbus OH, August 2013.

[2-6] R. Bracewell, *The Fourier Transform and Its Applications*, 2nd edn., McGraw-Hill, New York, 1978.

[2-7] R. Booth, S. Schell, T. Biedka, and P. Liang, "Reduction of Average-to-Minimum Power Ratio in Communication Signals," US Patent 7054385, issued May 30, 2006.

[2-8] R. Booth, S. Schell, T. Biedka, and P. Liang, "Reduction of Average-to-Minimum Power Ratio in Communication Signals," US Patent 7675993, issued March 9, 2010.

3 Dynamic power supply common principles

In any dynamic power supply transmitter (DPST), the performance of the DPS is of critical importance. This is independent of the actual implementation of the RF power amplifier that the DPS is working with. Thus, there are many common principles that apply to the DPS, no matter if the transmitter is being designed for envelope tracking (ET) or direct polar (DP) operation. This chapter explores these common principles. Details on specific DPS circuit issues are the topic of Chapter 9.

3.1 Top principle: PA efficiency visibility to top supply

The DPS is located between the RF PA and the top-level power supply. This effectively makes the DPS the "window" through which the top-level power supply "sees" the RF power amplifier. This relationship is shown in Figure 3-1, where the output voltage and current are referenced to the DPS itself.

It is desired that the power in to the DPS is equal to the power out of the DPS. This makes the ideal DPS a constant power (lossless) device, much like an ideal transformer. Of course, the efficiency of the DPS is not perfect (not unity), so the actual relationship between output power and input power becomes

$$P_S = V_S \cdot I_S = \frac{V_{PA} \cdot I_{PA}}{\eta_{DPS}}. \tag{3.1}$$

The operating conversion efficiency η_{DPS} varies with the actual output voltage of the DPS. This efficiency is not usually a constant. If we want to evaluate the "average" conversion efficiency for the DPS, we need to first know the PDF of the DPS output waveform $p(V_{DPS})$. Then the expectation

$$\overline{\eta}_{DPS} = \int_0^{V_{MAX}} \eta_{DPS}(V_{DPS})p(V_{DPS})dV_{DPS} \tag{3.2}$$

can be evaluated, where

$$0 \le V_{DPS} \le V_{MAX}.$$

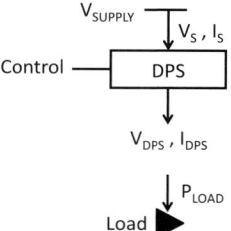

Figure 3-1   The DPS draws the current load I_S from the top-level power supply voltage V_S to provide its output voltage V_{DPS} and current I_{DPS}. Details of this conversion depend strongly on the DPS load impedance and the resulting characteristics of the output current.

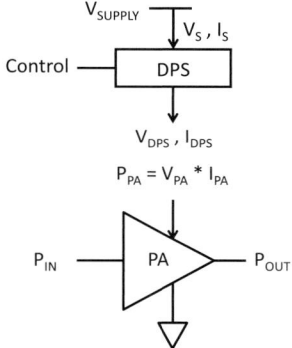

Figure 3-2   General architecture of any DPS transmitter, common to both ET and DP techniques.

Of course, even the result from (3.2) is specialized for a specific signal and power level. This is particularly true if the conversion efficiency varies significantly at different output voltages. Evaluation of the operating efficiency of any DPS is difficult and specific. Nothing is easy.

3.2   General (shared) architecture

All DPS transmitters follow the block diagram in Figure 3-2. The connection between the main power supply and the RF power amplifier is interrupted by insertion of the DPS: the load to the DPS from Figure 3-1 is the transmitter power amplifier. The desired effect is to have the DC to RF energy efficiency of the PA, which is assumed to be very high, to become more visible to the top power supply by reducing the current drawn from that supply. Therefore, to the extent that $V_S > V_{PA}$, we desire $I_S < I_{PA}$.

Another very important aspect of Figure 3-2 is the apparent double labeling of the voltage and current between the DPS and the PA. This is intentional for a very important reason. It is very common for the DPS to be developed by one team, and the PA to be developed by a completely different team. Very often these teams are in separate companies, which makes cooperation among the design teams essentially impossible. Each team must have its own target specifications: the output characteristics of the DPS, or independently the input power characteristics of the PA. When the output of the DPS is of interest, the appropriate labels are V_{DPS} and I_{DPS}. For the PA team, the appropriate labels are V_{PA} and I_{PA}. Only at the late step of connecting these two nodes together do we get $V_{DPS} = V_{PA}$ and $I_{DPS} = I_{PA}$. And even then, when evaluating this connection it is reasonable to refer to that voltage as V_{DPS} because it is the DPS that controls its value. Similarly, once connected the current is reasonably called I_{PA} because it is the PA and its supply input characteristics that sets what current flows from the DPS.

This visibility into the energy efficiency of the PA is seen through the following relationships. We start from the overall efficiency of the PA, which is the ratio of the RF output power over the total DC power into the PA

$$\eta_{PA} = \frac{P_{OUT}}{V_{PA} \cdot I_{PA}}. \tag{3.3}$$

Through the DPS, using (3.1) the power required of the top supply is found to be

$$P_S = \frac{P_{OUT}}{\eta_{DPS} \cdot \eta_{PA}}. \tag{3.4}$$

There are many ways to interpret (3.4). The more apparent interpretation is to note that if $\eta_{DPS} = 1$ (perfect conversion efficiency) then the power P_S from the top power supply is only dependent on the RF output power and the energy efficiency of the PA. This is equivalent to saying that the top power supply has full visibility of the inherent efficiency of the PA. Any reduction in DPS efficiency requires more power from the top power supply, reducing this "visibility."

Addition of the DPS in Figure 3-2 inserts an additional interface into this transmitter design; the connection between the DPS output and the PA power input. For this interface to behave well, all of the following must be well understood: the characteristics of the DPS output impedance, the effective absolute and differential characteristics of the PA power input impedance, the reaction of the PA to the DPS output impedance, and the DPS reaction to the PA as its load. The critical nature of these issues motivates the significant consideration given to these topics in Chapters 5, 6, and 9.

3.3   Power dissipations

Power dissipation – the transformation of electrical energy to heat energy – is an unavoidable consequence of the in-phase presence of voltage and current in any circuit branch. Each such circuit branch is a separate source of power dissipation. Heat flows begin at each location where there is power dissipation. Understanding the consequences of this heat generation and subsequent flow through circuit materials are the focus of this section.

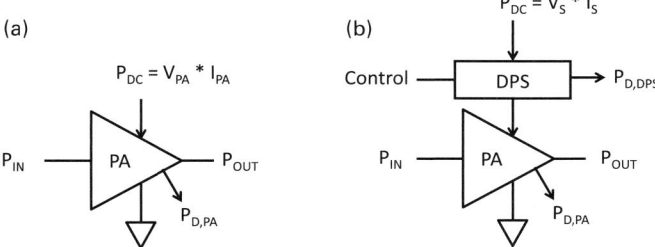

Figure 3-3   Conservation of energy models for two transmitter architectures: (a) conventional PA with stable DC power supply; (b) transmitter with any type of control on the PA power supply.

3.3.1   Conservation of energy (CoE) relationship

Conservation of energy is a fundamental physical principle. The unit of energy measure is joules. In power amplifier analysis, power (watts) is the unit usually measured, which is not a measure of energy but of energy flow: watts = joules/second. One does not see CoE analyses of power amplifiers (using joules) because fortunately there is a way of "cheating" where power numbers can be directly used when considering CoE calculations. For example, when all the measurements are in watts at the input and output of any circuit block we can write

$$P_{IN}\left(\frac{\text{J}}{\text{sec}}\right) + P_{OUT}\left(\frac{\text{J}}{\text{sec}}\right) = 0. \tag{3.5}$$

The seconds in all of these units can be disregarded without changing the relationship among the numbers. This very fortunate result allows the analysis of heat (energy) flows through a transmitter by only – but consistently – using measurements in watts. The result is that the sum of all powers in a circuit is equal to zero, in our case the direct equivalent of the CoE demand that energy (joules) must sum to zero in any closed system.

   The CoE evaluation for Figure 3-3(a), written in the form of output power = input power, is

$$P_{OUT} + P_{D,PA} = P_{DC} + P_{IN}. \tag{3.6}$$

The relationship of interest is the overall energy efficiency of this transmitter

$$\frac{P_{OUT}}{P_{DC}} = 1 - \frac{P_D - P_{IN}}{P_{DC}}, \tag{3.7}$$

which has the form of $(1 - \text{loss terms})$. Having power dissipation P_D as a loss term makes perfect sense. But in (3.7), the input power P_{IN} also appears as part of the loss term. This makes sense in any inverting amplifier (common emitter or common source), which are by far the most widely used circuit designs. Because the input signal and output signal have opposite phases in these circuits, there is no way that the input signal can pass through and add to the output signal (as is possible in common-drain or common-collector amplifiers). Indeed, these inverting amplifiers require significant isolation between the output signal and the input signal in order to maintain circuit stability. Therefore, the entire input signal must be dissipated within the PA and end up in the heat sink.

Dissipation of the input signal, if that is the only power dissipation in the PA, would result, as (3.7) shows, in a perfectly efficient PA from the point of view of DC power supply to RF output power. Any real transistor has resistance, so it *must* dissipate power in this DC to RF conversion. Thus, the requirements for a high efficiency PA cannot simply be to have $P_D = P_{IN}$, even though this solution is the most obvious suggestion from (3.7).

The second design approach to (3.7) that gives a near unity result is to have both P_D and P_{IN} nearly zero. Figure 4-32(a) shows that to get P_D close to zero it is necessary to operate the transistor close to the axes of the characteristic curves, something that is not compatible with inherent circuit linearity unless the load resistance is extremely high. And to get P_{IN} close to zero, the amplifier stage must have high gain, since $P_{IN} = P_{OUT}/\text{gain}$ when the input and output impedances are the same. There are some circuits where it is practical to raise the input impedance of the PA as a technique to reduce P_{IN} and increase overall energy efficiency [3-1].

For the DPS transmitter of Figure 3-3(b), there is also power dissipation from, and signal power input to, the DPS. Writing the equivalent of (3.7) for this architecture gives

$$P_{OUT} + P_{D,DPS} + P_{D,PA} = P_{DC} + P_{IN} + P_{ctl}. \tag{3.8}$$

As before, this is rewritten to determine the overall energy efficiency giving

$$\frac{P_{OUT}}{P_{DC}} = 1 - \frac{(P_{D,DPS} + P_{D,PA}) - (P_{IN} + P_{ctl})}{P_{DC}}. \tag{3.9}$$

Analysis of (3.9) is equivalent to that for (3.7). In this case, the input impedance to the DPS for the control signal is usually very high, keeping the control input power P_{ctl} essentially at zero. The same argument from earlier applies here: maximizing the energy efficiency of a DPS transmitter is equivalent to minimizing power dissipation in both the DPS and the PA, along with increasing the RF gain.

In many cases, the goal is not to absolutely maximize energy efficiency, because achieving any maximum – or reaching any physical limit – nearly always involves higher costs. Thus, the real design goal is to make the overall energy efficiency "good enough." A DPS transmitter therefore allows energy efficiency to be traded off – equivalently reduced power dissipation – between the two circuit functions, the PA and the DPS. In most instances, one of these will be easier to manipulate than the other.

Results (3.7) and (3.9) both show that the goal of maximizing energy efficiency, whether the circuit is the PA or the DPS (or any other circuit for that matter), is equivalent to designing to minimize circuit power dissipation. This view is much more tangible, particularly from a design and measurement perspective, than optimizing the general concept of "efficiency."

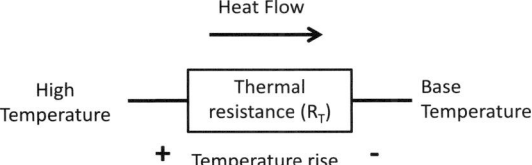

Figure 3-4   Temperature rise is a result of heat (energy) flow through the thermal resistance.

3.3.2   Temperature vs. heat

A short discussion about heat and temperature is in order here, because these are not the same thing at all, and are often confused. Heat is a form of energy. One convenient way to think of this process is that temperature is a response to the flow of heat, which can result in large temperature changes. But it is also common to have large heat flows without any temperature change.[1] Indeed, Ohm's Law is useful here too, because heat flow H and temperature T are related by material thermal resistance R_T according to

$$T = H \cdot R_T. \tag{3.10}$$

We see in (3.10) and Figure 3-4 that by equating current I with heat flow H, voltage V with temperature T, and electrical resistance R with thermal resistance R_T, the math model (3.10) is of exactly the same form as Ohm's Law. For power amplifiers, the heat energy flow follows from any power dissipation in a component (simultaneous voltage and current in any circuit branch). Temperature rise only occurs when there is a nonzero thermal resistance present to interact with this heat flow.

It is widely accepted to write the thermal resistance between nodes a and b of a thermal path as θ_{ab}, which has units degrees/watts. This leads to the general thermal relationship

$$T_a = P_D \cdot \theta_{ab} + T_b, \tag{3.11}$$

where the temperature T_a is the result from the power dissipation P_D causing a heat flow through the thermal resistance θ_{ab} into a base (ambient) temperature T_b. This is of the same form as the Ohm's Law relationship

$$V_2 = I \cdot R + V_1$$

so normal analysis techniques in circuit analysis directly apply to thermal analysis.

3.3.3   Thermal paths vs. signal paths

Signal flows and power-dissipation-based thermal flows in the DPST architecture of Figure 3-2 are illustrated in Figure 3-5. The signal paths in Figure 3-5(a) show that a combination of these signals may occur in the PA, though in some DPST designs we will want to minimize this interaction. Because there is only one output taking into account both inputs, the signals through the DPS and the PA are effectively in series.

---

[1] Consider, for example, the thawing or freezing of water ice.

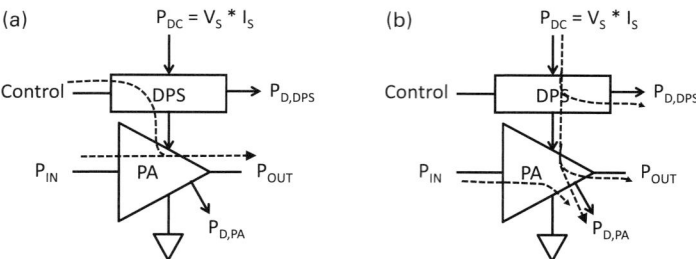

Figure 3-5   Flow of signals and heat in the DPST architecture: (a) signal paths are in series; (b) thermal paths are in parallel.

Thermal paths in the DPST architecture are shown in Figure 3-5(b). Here there are three outputs: power dissipated from the DPS, power dissipated from the PA, and output power into the RF load. The power dissipated from the DPS and from the PA flow into local heat sinks, while the RF output power goes to where it is needed, likely an antenna.

When the DPS and the PA are physically separated, the thermal path from the DPS to its heat sink does not flow through the PA. Similarly, the thermal path from the PA to its heat sink does not flow through the DPS. These paths are in parallel (as long as the circuits are not stacked!). The figure also explicitly shows that the RF input power eventually ends up as dissipated power and therefore also flows to the PA heat sink.

From (3.11), the operating temperature of any location where power is dissipated (T_a) depends on the amount of power being dissipated, the thermal resistance from that location to the heat sink (θ_{ab}), and the temperature of the heat sink (T_b). In contrast to a conventional linear amplifier where all power dissipation occurs in the transistor, in the DPST there are two locations where power is dissipated. When all works well, the sum of these two power dissipations is smaller than the original linear PA power dissipation. Divided into two separate locations, these individual power dissipations are each individually lower than the original single power dissipation. Therefore, if the thermal resistance is the same, then the operating temperature of the transistors will be proportionally lower. This is very good, because long-term reliability is strongly dependent on operating temperature.

One additional degree of flexibility in the DPST architecture is the possibility of having different thermal resistances from the two power dissipation locations. Specifically, because the thermal resistance of silicon is generally lower than the thermal resistance of compound semiconductors, for a PA based on a compound semiconductor, moving power to the DPS (which is almost always made using silicon) has significant temperature lowering benefits.

3.4   DPST node voltages and currents

The relationship among the voltages and currents at the nodes in the DPS transmitter are not simple. The economic value of any transmitter is largely dependent on minimizing the current $I_S(t)$ drawn from the top power supply, given that all other performance specifications are met. But like any other circuit, the current drawn by the load reduces performance further back in the system. In this case, we assume that the antenna load is reasonably well matched, so the RF current is in-phase with the RF voltage, and power is generated in the load, as desired.

From the PA power supply input (the DPS output), the situation is much more complicated. Of course, Ohm's Law applies, so we have

$$I_{PA}(t) = \frac{V_{DPS}(t)}{Z_{PA}(V_{DPS})}, \tag{3.12}$$

where the supply impedance looking from the DPS into the PA (Z_{PA}) is complex (resistive and reactive). Further these resistive and reactive values are generally strong functions of the applied voltage from the DPS. This is a very difficult load for the DPS to deal with. And therefore it is very critical to understand for successful DPST designs.

Reflecting this PA load current to the top supply through the DPS using (3.1) gives this expression for the load current from the top power supply

$$I_S(t) = \frac{V_{DPS}(t)}{V_S \cdot \eta_{DPS}(V_{DPS})} I_{PA}(t). \tag{3.13}$$

Here dependence of the DPS conversion efficiency on output voltage is explicitly shown. Combining (3.13) with (3.12) gives this relation for the top supply load current

$$I_S(t) = \frac{V_{DPS}^2(t)}{V_S \cdot \eta_{DPS}(V_{DPS}) Z_{PA}(V_{DPS})}. \tag{3.14}$$

In (3.14), we see how important it is to know what the supply-input impedance of the PA actually is. This topic is discussed in detail in Chapters 4, 5, and 6.

3.5   Cost and architectures

Nothing comes for free, meaning that complexity generally impacts on cost. All else being equal, simple structures are preferred over complicated ones. History shows that simple structures tend to have the lowest performance. But, if that is good enough for the design at hand, simple is the correct decision.

For best performance, the DPS must be a constant power circuit, as described in (3.1). There are two architectures that (when ideal) perform as constant power circuits: the transformer, and the switch-mode power supply (SMPS). Transformers only work as constant power devices for AC signals, so they are not directly applicable to the DPS problem since the DPS output must generally include DC. This means that efficient DPS designs must generally include some type of SMPS.

Linear regulators use resistive elements to change the output voltage, varying their internal resistance as the output current changes. Current through this resistance causes power dissipation, which we do not want. That said, a dynamic analog voltage regulator

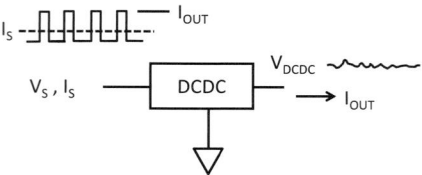

Figure 3-6   DC–DC conversion necessarily has an AC "noise" component on its output.

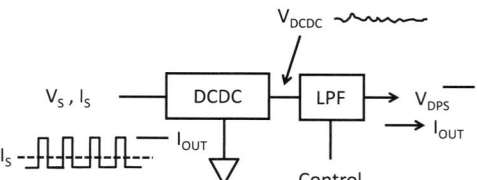

Figure 3-7   Series filtering to manage the DC–DC output noise.

is by far the lowest cost implementation of a DPS. To improve conversion efficiency, the SMPS designs must therefore use reactive elements to perform the needed voltage conversion. SMPS designs based on capacitors as the reactive element are called charge pumps. When inductors are used as the reactive elements, the SMPS is often simply called a DC–DC (direct current in to direct current out) converter. For this discussion, the DC–DC converter will be used in the examples.

A functional description of a buck (voltage lowering) DC–DC converter is provided in Figure 3-6. The output voltage is V_{DCDC} with load current I_{OUT}. The input voltage is V_S with *average* input current I_S. The switching nature of the DC–DC converter connects the internal inductor between the input V_S and the output V_{DCDC} with a duty cycle D based on the energy transfer needs and voltages at that moment. This means that the sourcing power supply V_S must be able to supply the entire output current magnitude for the brief time the internal switch is on. At all other times, the current into the DC–DC converter is zero. At no time is the input current actually equal to the time average of this pulsed characteristic.

This pulsing input current is not completely removed at the output, leaving an output voltage variation called the ripple voltage about the main DC value (V_{DCDC}). This ripple voltage is a problem and cannot remain in most DPS transmitters. There are two main ways to eliminate this DC–DC ripple from the DPS output, called series filtering and shunt filtering.

Series filtering is shown in Figure 3-7. One option is to remove the output ripple with a lowpass filter (LPF). If this LPF is implemented using passive components, it will have a very low bandwidth – typically 1 kHz for common DC–DC switching frequencies about 100x higher. Another option is to use a linear regulator and to remove the DC–DC ripple with the regulator's power-supply rejection (PSR) performance. This latter approach requires that V_{DCDC} be higher than the needed V_{DPS} by an amount sufficient to ensure that the series regulator can remove the ripple with its PSR.

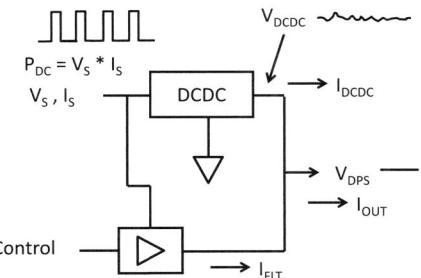

Figure 3-8   Shunt filtering to manage DC–DC output noise.

Shunt filtering is shown in Figure 3-8. This structure adds an amplifier in parallel to the DC–DC converter and the two outputs are connected together. The idea is to have the DC–DC converter provide nearly all of the necessary load current I_{OUT}. The amplifier sources or sinks current from the output node so as to achieve the output voltage and current needed.

For shunt filtering to work, it is necessary that the output voltage of the DC–DC converter can be "pulled" by the shunt amplifier to make up any differences between V_{DCDC} and the needed V_{DPS}. Ohm's Law says that this is possible if the DC–DC converter has a high output impedance: it must be a current source, not a voltage source. Fortunately, this characteristic is practical, because the inductor in the DC–DC converter does act as a current source, as all inductors do. Reducing this output impedance by adding a large output capacitor is not desired at all.

Both the series filter in Figure 3-7 and the shunt filter in Figure 3-8 have important loss mechanisms, on top of the conversion loss in the DC–DC converter shown in (3.1). Beginning with the series filter, this additional loss comes from the linear regulator dropping its input voltage to the needed DPS output voltage in the presence of the full load current

$$(V_{DCDC} - V_{DPS}) \cdot I_{OUT}. \tag{3.15}$$

As long as this voltage drop is small, the conversion efficiency of the linear regulator is actually quite good. If V_{DCDC} ever drops too low, then this regulator cannot recover the difference. Because this structure is a series circuit, all of the input current does eventually flow into the PA.

Loss in the shunt filter structure comes in two forms: one where the amplifier sources current into the output node, and another when the amplifier draws current from the output node. These are described below:

$$\begin{cases} V_{DPS} \cdot I_{FLT} & \text{(sink)} \\ (V_S - V_{DPS}) \cdot I_{FLT} & \text{(source)} \end{cases}. \tag{3.16}$$

When the shunt amplifier draws (sinks) current from the output node, this means that the DC–DC converter is supplying too much current for the immediate load demand. Power dissipation, being the product of in-phase voltage and current, is the DPS output voltage

V_{DPS} times the excess inductor current I_{FLT} that this linear amplifier is required to draw away from the PA node. Additionally, this loss mechanism must further reduce overall efficiency because it represents input current that does not flow into the PA. This amplifier-sink current goes directly to ground and does no useful work.

The second shunt filter loss mechanism in (3.16) occurs when the amplifier must source additional current into the output node, meaning that the DC–DC converter is not providing enough current to meet the immediate load demand of the PA. This current is drawn from the input supply, and the resulting power dissipation is the value of this current times the voltage difference of the DPS output from the input supply value. This additional current from the amplifier does flow through the PA, so its flow does not cause an additional inherent drop in overall efficiency.

3.6   DPS bandwidth

The output from the DPS must respond fast enough to meet the demand of the transmitter. A natural question is how to define the bandwidth requirement that the DPS must have to successfully support its function within the DPST. To answer this question, we first need to understand the actual requirements of the DPS.

If the requirement of the DPST is to provide maximum energy efficiency, then the discussion in Chapter 6 demonstrates that the DPS must provide an accurate match to the envelope waveform. This means that the envelope waveform must be accurate in *all* its respects, both its magnitude values and its phase characteristics. The Fourier transform requires that for complete time-domain waveform fidelity, all aspects of the frequency-domain representation of that waveform must be preserved.

This is a requirement that goes far beyond the traditional −3 dB (half power) definition of circuit bandwidth, simply because the "all aspects of the frequency-domain representation of that waveform must be preserved" requirement is certainly not compatible with the idea of "half power." Considering for now only the simplest circuit, the one-pole RC filter, the frequency response characteristics for this filter are shown in Figure 3-9. Frequency is normalized to the traditional −3 dB bandwidth definition. The magnitude response is plotted in dB, and the phase response is shown in degrees of phase shift.

For applications where the DPS output waveform must be held to precise shaping, then the DPS must accurately pass both the magnitude and the phase characteristics of the envelope waveform. Looking at Figure 3-9, the useful bandwidth is then much less than BW3. Depending on how much phase distortion is actually tolerable, the true useful bandwidth of the DPS may be 10% of BW3. For precision applications, the useful bandwidth will be much less than this, down to as little as 1% of BW3. And to be clear, all of the envelope signal frequency-domain components must be contained within this restricted range.

The bandwidth of the DPS can get very wide indeed. This is one reason why narrower bandwidth DPST implementations, such as ET, are of interest. Nothing comes for free though. As Chapter 5 describes, the price for getting this lower DPS bandwidth requirement is that the overall DPST energy efficiency *must* be reduced.

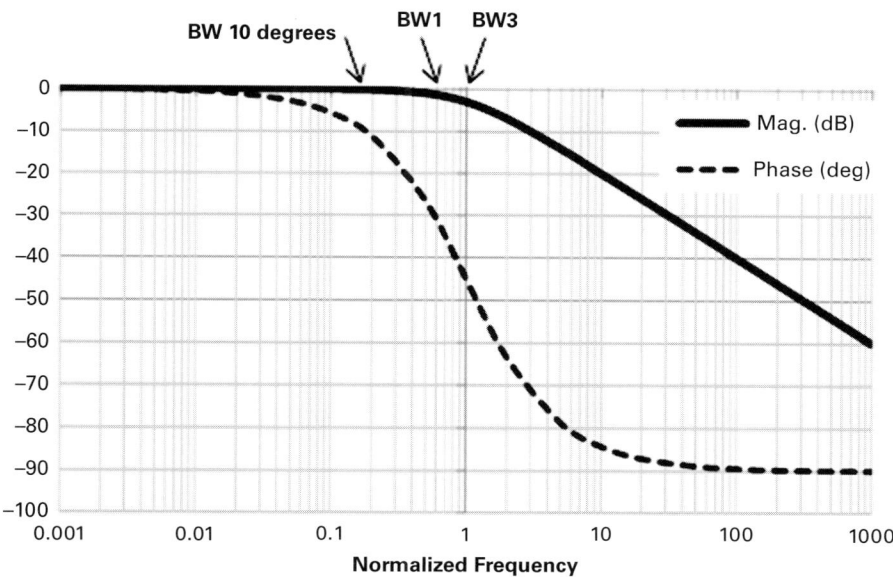

Figure 3-9   DPS bandwidth must be measured as the phase bandwidth to ensure waveform fidelity; magnitude bandwidth alone is not sufficient to ensure output waveform fidelity.

Such a price is hard to tolerate, so some type of phase equalization may become acceptable. This equalization will increase delay through the DPS, which must be accounted for in the final DPST design.

3.7   Reference

Note: Patent references are provided for bibliographic use only. Citation of specific patents here is not indicating any view on priority issues.

[3-1] R. Meck, "RF Power Amplifier Having High Power-Added Efficiency," US Patent 7265618, issued Sept. 4, 2007.

4 Linear power amplifiers

Linear amplifiers are the core of most electronic signal processing, and RF transmitters are no exception. Linear circuitry *always* operates with the transistor as a controlled current source (CCS). We can state both that a linear amplifier operates as a CCS, and the converse that with CCS operation we have a linear amplifier. With a century of history, the design of linear power amplifiers is covered extensively in the literature [4-1] and will not be repeated here.

In this chapter, the points about linear RF power amplifiers that are most important to their application in DPS architectures are described.

4.1 Overview

The entire objective of any linear amplifier is to provide an output signal $y(t)$ that is proportionally scaled from the input signal $x(t)$. The constant of proportionality is called the gain of the amplifier. Mathematically we write this as

$$y(t) = ax(t), \tag{4.1}$$

where the proportionality constant a is the amplifier gain.

In the real world, the relationship in (4.1) is an ideal goal that is never precisely reached. How hard we must work to make our amplifier approximate this ideal performance more closely is dependent on the signal type we are to amplify, and the output performance specifications we need to meet. In general, the greater the signal order is [4-2], which is the number of possible information values that can be transmitted in any signal symbol, the more precise the amplifier linearity performance must be.

It is very important to be clear that the concept of a linear amplifier is a port-based specification, as shown in Figure 4-1. What the actual circuitry is within the amplifier, and how it precisely operates, *does not matter*. It is very common to implement a linear amplifier function using linear circuitry. But this is not necessary at all.

What *is* a requirement, though, is that the input and output signals be essentially sinusoidal in wave shape. All communications signals used are of the *narrowband* type, which simply means that the bandwidth occupied by the signal is much less than the center frequency of the signal. This definition includes any power present around the harmonics of the signal center frequency. Fourier theory clearly states that any signal that is narrowband is also sinusoidal in wave shape. Therefore, if the amplifier block of

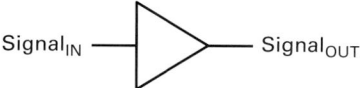

Figure 4-1   Amplifier linearity is only measured at the amplifier ports.

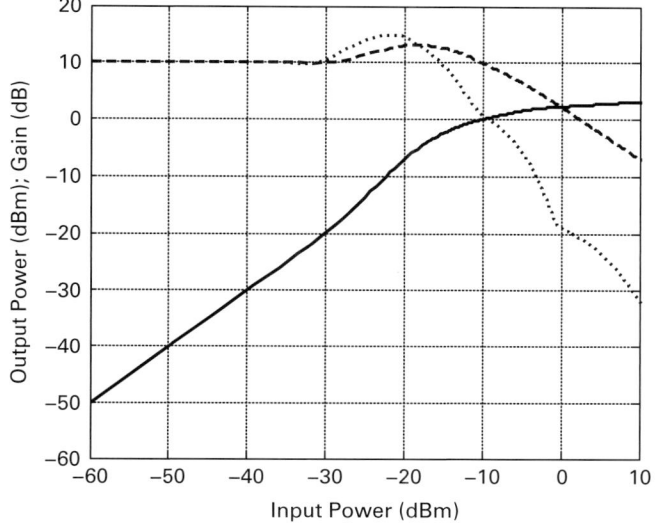

Figure 4-2   Amplifier gain is typically not constant for all values of the input signal, shown here evaluated for an example amplifier transfer function (solid line) with its corresponding slope gain (dotted line) and ratiometric gain (dashed line). See Section 4.1.4 for details on the two gain measures used here.

Figure 4-1 contains nonlinear circuitry, *to be a legitimate amplifier it must also contain circuitry to guarantee that the output signal is narrowband*. This particularly applies to so-called switch-based or switch-mode amplifiers, such as class S [2-4].

Any actual amplifier exhibits a nonconstant gain as the input signal varies. A typical example of this gain variation is presented in Figure 4-2. Here we see both gain expansion, where the gain increases from its very small signal value, and also gain compression, where the gain is below its value at very small signal operation. The two different measures of gain shown in Figure 4-2 are defined in Section 2.2 and detailed in Section 4.1.4.

Both gain expansion and gain compression lead to signal distortion. Gain expansion typically arises from using reduced quiescent current biasing usually adopted for power savings. Gain expansion is not experienced as often when class A biasing is used.

Compression effects
As the output signal from any amplifier approaches the maximum value it can attain, a deviation from linear behavior, called compression, occurs. Compression describes how

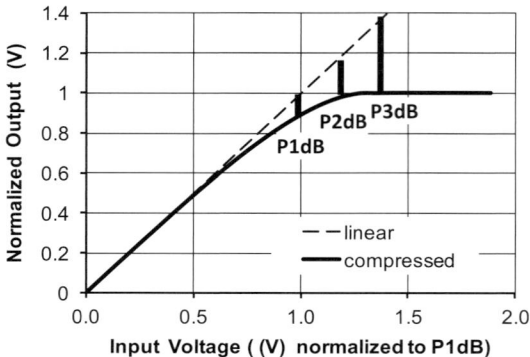

Figure 4-3   Definition of compression points using a simulated amplifier voltage transfer function.

the output from an amplifier does not reach the expected full value if that amplifier remained operating linearly. One measure of compression is the output level where the measured output power is 1 dB below what we expect it to be from the linearity extension from very small signal conditions. This is the power at P1dB, the 1 dB output compression point. We can similarly measure P3dB, the 3 dB output compression point. And, of course, any other output compression value. Three measures of output compression are shown in Figure 4-3. For this particular transfer function, the input values where the actual output is 1 dB, 2 dB, and 3 dB below the ideal linear response (dashed line) are shown. The onset of waveform distortion is much sooner, near to the normalized input voltage of 0.6 V. In Figure 4-3, we note that the 3 dB output compression point (P3dB) is very close to the onset of output clipping. This is not always true for all amplifiers. Each amplifier needs to be individually characterized.

From Section 2.3.3 and Figure 2-6, we note that for a perfectly linearized amplifier, often called a "soft limiter" because of its finite gain, the onset of clipping is essentially 2 dB below the output power saturation. This is a completely different measure from the present discussion surrounding Figure 4-3, because here we have a reverse-looking $(P_{SAT} - 2$ dB$)$ measurement instead of the forward-looking measurements of the output compression points (decibels below ideal linear projection). Therefore, there is no equivalence of the output compression value at power saturation. We can only say that we know that output power saturation is reached when the ratiometric gain is dropping decibel for decibel with further increases of the input power. In this region, the compression point is also increasing decibel for decibel with further increases of the input power.

Sizing the design of any power amplifier is a key decision on any project. The primary requirement is that the power amplifier must be able to provide the absolute maximum output power that the signal envelope can attain when the transmitter is at maximum power. This follows from Figure 2-6, which shows that any amplifier is an output power limited device. The design is driven by PEP, not average signal power.

Setting P_{SAT} = PEP may still not be sufficient for some applications. Section 4.1.5 shows that at P_{SAT} the amplifier output waveform is essentially square, severely clipping off any signal peak from the input sinusoidal waveform. For several modulation types, this signal distortion due to clipping is unacceptable. For these signals, it is necessary to set PEP at a lower output power than P_{SAT}, possibly at P1 dB. Whatever the reduction is, this type of operation is called output back-off (OBO) where (in dB) $OBO = P_{SAT} - PEP_{MAX}$ dB.

4.1.1   Bias classes and their waveforms

For completeness, here is a very brief review of linear amplifier bias class definitions. These include class A, class B, and the various intermediate bias conditions of class AB. Class C is included because it definitely is a controlled current source (CCS) amplifier. Finally a discussion of the two-device push-pull amplifier is included with particular attention to its operation as the device bias transitions from deep class AB through class B and into class C.

Controlled current source (CCS)

CCS amplifiers are the fundamental designs taught in engineering school, and date from the beginning of amplifier design one century ago. The differences among them are set by what bias is applied to the transistor controlling element, as shown in Figure 4-4. Class A requires that current flows through the transistor at all times. All of the other classes require that current flow be interrupted during some fraction of each RF signal cycle.

Class A

Class A amplifiers are the original amplifier circuit. After allowing for the output DC offset, the RF output signal is directly proportional to the input signal. If there is no input signal, a current called the quiescent current continues to flow through the RF transistor. This quiescent current must be greater than the maximum negative excursion of signal current, to ensure that device current will always flow as required for this class. Figure 4-5 illustrates several signal currents for class A operation. Along with the DC

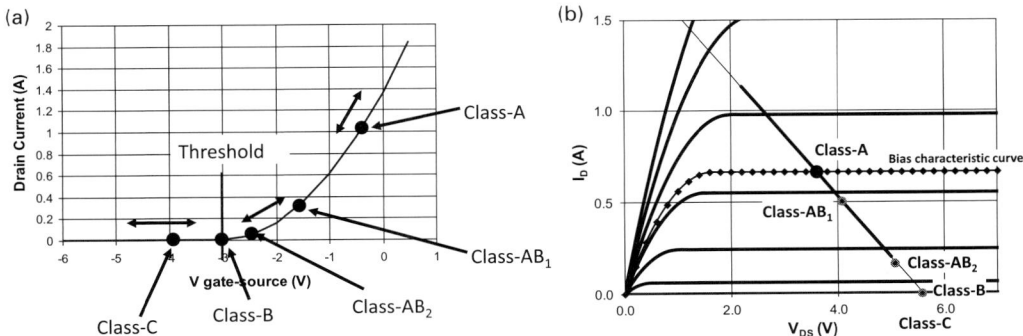

Figure 4-4   Biasing of CCS amplifiers: (a) control characteristic view (depletion FET example); (b) load line view.

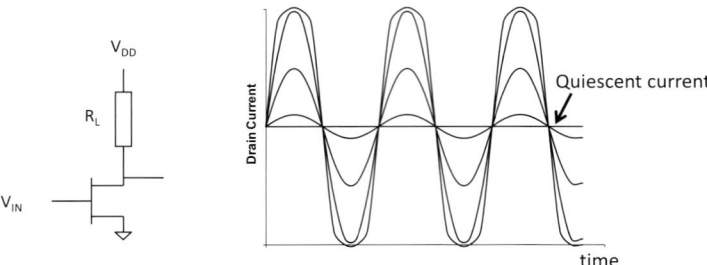

Figure 4-5   Class A operation, showing that current flows through the transistor at all times. When there is no output signal, the quiescent current continues to flow.

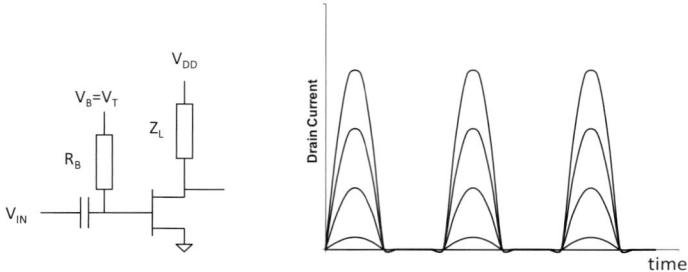

Figure 4-6   Class B single-ended operation, showing that current flows through the transistor only half of the time. When there is no output signal, the quiescent current is ideally zero.

offset from the bias, this causes power dissipation in the RF transistor, and lowers the energy efficiency of this amplifier type. When the bias design is perfect and the transistor is perfectly linear, at its maximum output signal the class A amplifier can approach its theoretic energy efficiency limit of 50%.

One of the most important characteristics of any class A amplifier is that the continuous current flow requirement ensures that the output signal will be sinusoidal as long as the input signal is. This means that no filter is necessary to make the output a sinewave. This is unique to class A, and is not shared by any other CCS amplifier type.

Class B (single-ended)

If the bias point is moved so that it sits exactly on the threshold of the transistor, we get class B operation. Here there is output current only for positive voltage excursions on the input, as shown in Figure 4-6. Here we restrict discussion to single-ended (only one transistor) implementations of class B. In the later section on push-pull amplifiers, we will encounter a very popular class B amplifier architecture that uses two transistors.

Class B is attractive because half of the time there is no current flow in the RF power transistor. This reduces power dissipation in the transistor and improves amplifier energy efficiency to a theoretical maximum of $\pi/4$ (78%) when the transistor turns on at full transconductance immediately (which never happens). It also severely distorts the input

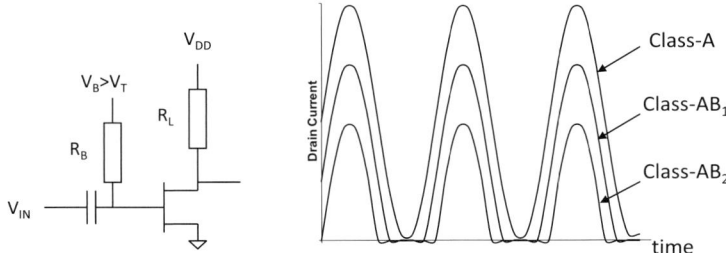

Figure 4-7   Class AB operation, showing that current flows through the transistor more than half of the time. Quiescent current is lower than the class A value but always greater than zero.

waveform with half-wave rectification, so a bandpass filter is required at the amplifier output to restore the output waveform to a sinewave.

Class AB

Figure 4-4 shows that bias can be set anywhere between class A and class B conditions. Amplifiers using this strategy are called class AB. A small shift of the bias from class A toward class B is called class AB_1. When the bias is closer to class B, then the amplifier is said to operate in class AB_2, or also in deep class AB bias. As seen in Figure 4-7, the further the bias point moves toward class B, the greater the clipping of the negative signal peak, which increases waveform distortion.

This signal current clipping at the negative peaks also reduces the power dissipated in the RF transistor, improving PA energy efficiency. Additionally, the quiescent power dissipation is reduced by the lower quiescent current. The waveform is distorted in all cases at the highest output power, so an output bandpass filter is required to restore the output waveform to the necessary sinewave.

These current waveforms are shown near maximum output power. As the input signal decreases in an amplifier biased in class AB, the current eventually returns to continuous operation. When this happens, the amplifier returns to class A operation, by the class A definition of continuous current flow through the amplifier transistor.

Class C

The RF transistor bias can also be set to have the transistor normally in full cutoff. This is called class C and with this bias there is no quiescent current, nor therefore is there any quiescent power dissipation. It is no surprise that this is the most energy efficient of any CCS amplifier. Efficiencies approaching 90% are achieved in commercial products [4-3].

This is still a CCS linear amplifier because when the input signal is large enough to turn on the RF power transistor, the current flow is proportional to input signal variations.[1] Load current flows for less than half of each RF cycle, so the output waveform distortion is very severe, as seen in Figure 4-8. A bandpass filter is required to restore the needed sinusoidal output waveform.

---

[1] After all, class C biasing is used in Doherty architectures for the peaking amplifier [1–1, p. 456].

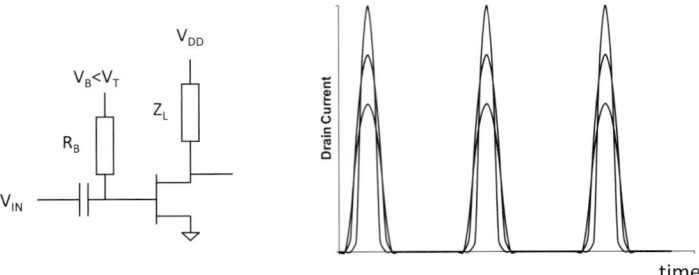

Figure 4-8   Class C operation showing that current flows through the transistor less than half of the time. Quiescent current is always zero.

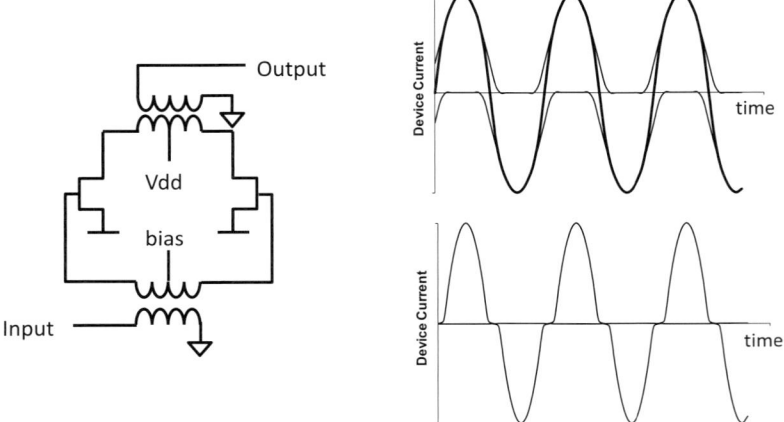

Figure 4-9   Push-pull class B operation, showing the basic circuit on the left. On the top right, if the bias is actually class AB$_2$, then the linearity is good, but power dissipation during crossover gets high. On the bottom right, if the bias is too low (actually class C), then the linearity is bad with crossover distortion.

Note the waveform distortion as the transistor begins to conduct – a consequence of the rapidly varying transistor transconductance (g_m) in this region. This distortion is well-known, so practical class C amplifiers are designed to spend very little time in this very nonlinear region of operation.

These current waveforms are again shown near maximum output power. As the input signal decreases in an amplifier biased in class C, the current eventually goes to zero and remains there. When this happens, the amplifier effectively turns OFF. This is very different than what happens with class AB bias.

Class B push-pull
Shortly after the discovery of the class B operation, because the transistor operates only half of the time, it was realized that if two amplifying devices (at that time they were vacuum tubes) are used, one can operate for the positive half of the input cycle and the other can operate during the negative half. This is shown in Figure 4-9, and was given the

name push-pull. In this way, all of the input signal can be used, recovering the half that is ignored in a single-ended class B operation.

There is some mention in the literature of a "pseudo-differential" amplifier type [4-4]. This structure is shown as having a differential input signal, but the sources (or emitters) of the transistors are not connected together. Instead, both are connected directly to the ground reference. In reality, this is the old push-pull structure from Figure 4-9. The use of the term "differential amplifier" describes a circuit with feedback between transistors by direct connection of their sources (or emitters). When both sources (or emitters) are grounded, this feedback path does not exist. Therefore, *the name "pseudo-differential" is meaningless* and should not be used. Instead, the original name of push-pull remains descriptive and appropriate [1-1, p. 382].

4.1.2   Linearity goals

Three metrics are used in the evaluation of amplifier linearity. Each metric evaluates a different aspect of amplifier performance, giving a purpose to each of them. These metrics are:

- error vector magnitude (EVM),
- adjacent channel power (ACP),
- harmonic distortion (HD).

These linearity metrics are generally independent, which means that having very good performance in one metric is certainly not a guarantee that the other metrics are also good. It is particularly true that ACP can be excellent while the EVM can be very poor.

EVM: in-band signal accuracy

The goal of any signal modulator is to provide the output signal phasor (complex envelope $\rho e^{j\theta}$) that is identical to the desired signal phasor. Any difference between the output signal phasor from the ideal phasor is called the output error vector. This principle is presented in Figure 4-10.

This error vector, like any vector, has a magnitude and phase. As long as the magnitude is small, then the phase of the error vector has little importance. An accurate output will have a small EVM no matter what the phase is of the error vector, so standards committees specify limits only on this EVM.

EVM measures the quality of the signal modulation, showing that the signal phasor is near to where it should be. This is a measure of in-band (within the channel) signal accuracy. EVM is strictly a time-domain measurement. *It is not possible to determine whether EVM is good by only looking at the signal PSD*. This is most easily explained by a single example. It is well known that the signal PSD is defined only by the symbol rate and the filtering used. Indeed, the PSD of QPSK or 256 QAM are identical as long as they have identical symbol times and use the same baseband I and Q filters. If the signal

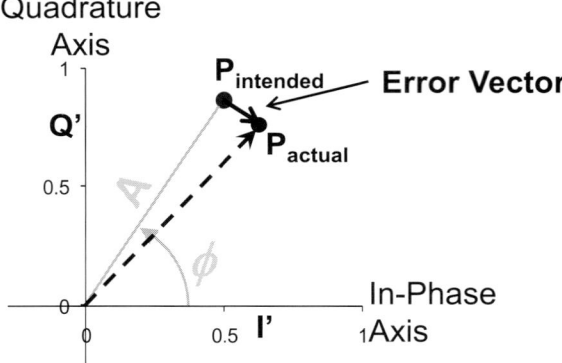

Figure 4-10 Error vector is the phasor difference between the actual and ideal signal phasors. Units: % is most common, though dB is sometimes used.

is actually 16 QAM and it should be 64 QAM, the EVM will be huge, even though there is no discernable difference in the signal spectral measurements.

EVM can be defined in two ways, either focused on the signal phasor only at symbol times or as the accuracy of the entire modulation waveform. These two EVM calculations are discussed in this order. When only the accuracy at the signal sampling times (T_S) is desired, the actual reported EVM value is an averaged and normalized magnitude value across a defined number (L) of consecutive states. This measure is an rms average, normalized to the ideal intended signal

$$\text{EVM}_{rms} = \sqrt{\frac{\sum_{k=1}^{L} |\hat{x}(kT_S) - x(kT_S)|^2}{\sum_{k=1}^{L} |x(kT_S)|^2}}, \tag{4.2}$$

where here we define $x(kT_S)$ to be the ideal signal state (constellation point) intended, and $\hat{x}(kT_S)$ to be the actual modulation achieved. Both of these values are usually sampled once per symbol, at times kT_S. Usually, the result of this averaging is reported as a percentage, since the EVM calculation is normalized to the ideal intended signal. Common requirements for EVM performance are around 5–15%.

In some applications, it is desirable to not only evaluate EVM at the symbol times (constellation points), but at all sample points x_n of the signal waveform. This leads to the "continuous" waveform version of (4.2), which is

$$\text{EVM}_{rms} = \sqrt{\frac{\sum_{n} |\hat{x}_n - x_n|^2}{\sum_{n} |x_n|^2}}, \tag{4.3}$$

where here we define x_n to be the ideal waveform value intended for the nth waveform sample, and $\hat{x}_n$ to be the actual waveform value achieved.

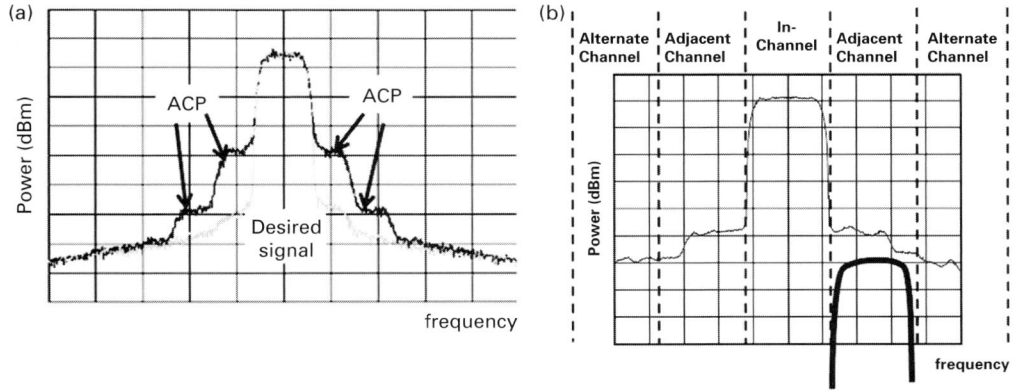

Figure 4-11 Adjacent channel power (ACP) is an undesired spreading of the signal power resulting from nonlinearity in the power amplifier: (a) generation of ACP when a nonlinear amplifier is inserted; (b) the measurement of ACP represents what another similar radio will experience.

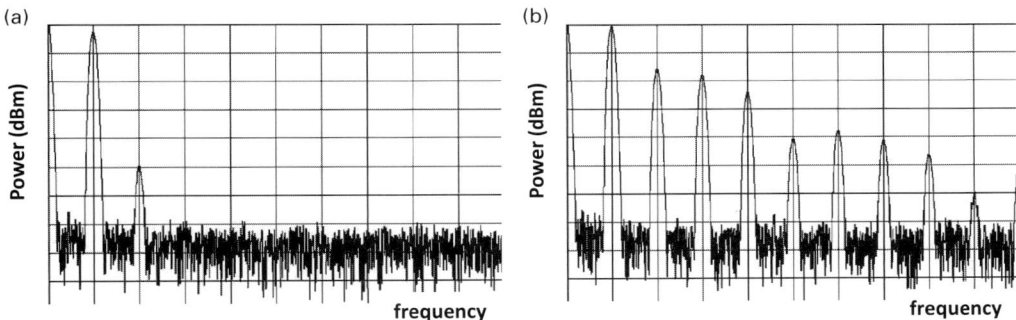

Figure 4-12 Harmonic performance from two amplifiers: (a) a linear amplifier with low efficiency; (b) a strongly nonlinear amplifier with high efficiency.

ACP: out-of-band signal accuracy

For signals that have varying envelopes, distortions in the output envelope result in a broadening of the signal PSD at low to moderate power levels compared to the transmitter output power. Sometimes referred to as *spectral regrowth*, one example is shown in Figure 4-11. Not all systems are equally sensitive to this effect, so specifications vary widely.

Harmonics: wideband waveform accuracy

Harmonics come from distortion of the carrier waveform, which are required from the Fourier transform whenever the power amplifier has any nonlinearity at all. Figure 4-12 shows two signals with very different harmonic contents. In Figure 4-12(a), the harmonic content is low, and only the second harmonic power is measurable on this scale. The situation is very different in Figure 4-12(b), where the amplifier is operating with sufficient nonlinearity (very near to P_{SAT}) such that harmonics up to the tenth order are easily measurable. Fundamental output power is nearly the same for both of these measurements.

As seen in Section 2.1.2, nonlinear operation is required from any amplifier that is inherently energy efficient. When efficient amplifiers are needed, then there will always be a significant number of harmonics present at the transistor output port. Yet from the definition of the linear amplifier at the beginning of Section 4.1, the transmitter design is required to suppress these harmonics before they reach the final transmitter output.

4.1.3   IV curve model (load line)

Amplifier gain is present when the output signal is larger than the input signal. The output signal voltage is generated from the transistor output current flowing through the load resistance R_L. The key performance parameter for the transistor is then its transconductance (g_m), the amount of output current change for a change in the input voltage: $g_m = dI/dV$. From this, we get the device gain of g_m*R_L, which is a slope gain.

A very useful graphic technique to relate these concepts and understand what is actually happening in the circuit is an overlay of the transistor characteristic curves and the load line. One example is shown in Figure 4-13. The endpoints of the load line are readily understood by remembering that if the output is at the power supply voltage, there is no current through the load. If the output is at 0 V, then the current through the load resistance is V_{SUPPLY}/R_L. Whenever the load line crosses a characteristic curve, this is a valid operating point for the circuit.

Device transconductance g_m is not constant for class AB bias, particularly for bipolar devices. The situation is worse for class C bias, where the g_m is zero until the input signal

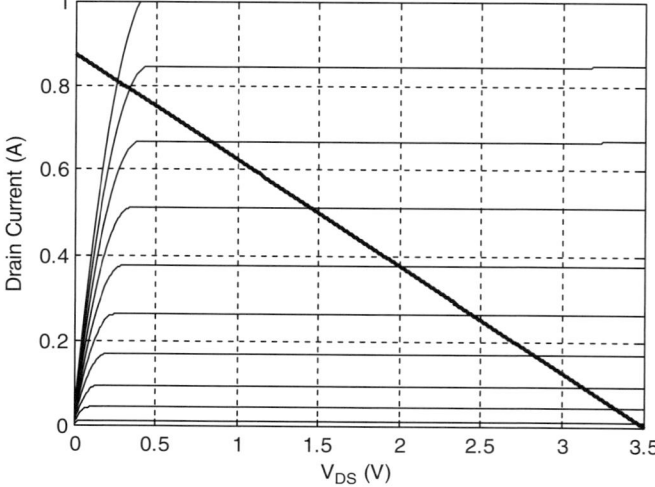

Figure 4-13 Load line overlaid on transistor characteristic curves; intersections are valid operating points.

exceeds the negative bias value and then progresses through rapid changes until significant device current flows. This is the same as saying that the spacing of the characteristic curves in plots such as Figure 4-13 is not uniform for equal steps in the controlling parameter.

4.1.4 Power series models

Amplifier behavior in the presence of nonideal characteristics, generally all lumped together under the single heading of "nonlinearities," is often simply modeled as a power series

$$y(t) = a_1 x(t) + a_2 x^2(t) + a_3 x^3(t) + \cdots + a_k x^k(t) + \cdots, \tag{4.4}$$

where the polynomial coefficients a_k when $k > 1$ are referred to individually as the kth-order distortion coefficients. This power series representation reduces to the ideal characteristic (4.1) when all of the series coefficients of order 2 or higher are zero.

The Taylor series (4.4) is often described as a memoryless model, because there is no past result used in its evaluation. To account for system memory, the more complicated Volterra series is used. Details of this are well beyond the scope of this book [4-5].

4.1.5 Four gain definitions

Here we build more detail into the summary on gain measures from Section 2.2. In that earlier section, two gain measures for a nonlinear amplifier are presented. Here we show that there are actually four different measures of gain in use, all of which give the same result when the underlying amplifier is operating linearly, and all of which give different results when the amplifier operates in compression. Two of these have been recognized earlier [4-8] and the additional power-based gain measurements are newly recognized here.

Taking the first variable term of the power series expansion (4.4) and dropping the explicit display of time dependence, we get a linear voltage transfer function (also seen as (2.10) and (2.11))

$$y(x) = ax, \tag{4.5}$$

$$gain \equiv \frac{dy}{dx} = a. \tag{4.6}$$

Real amplifiers exhibit output compression where the output signal y becomes less than the level predicted by (4.5), eventually reaching a maximum level $y = y_{max}$, which does not change with further increased input x. This is called output clipping and is shown in Figure 4-14. Ignoring memory effects (generally dangerous for amplifier design [4-1] [4-6], but is acceptable for present purposes) this compression phenomenon can be described mathematically with a polynomial model (valid only below clipping)

$$y(x) = \sum_{k=0}^{\infty} a_k x_k = a_0 + a_1 x + D, \tag{4.7}$$

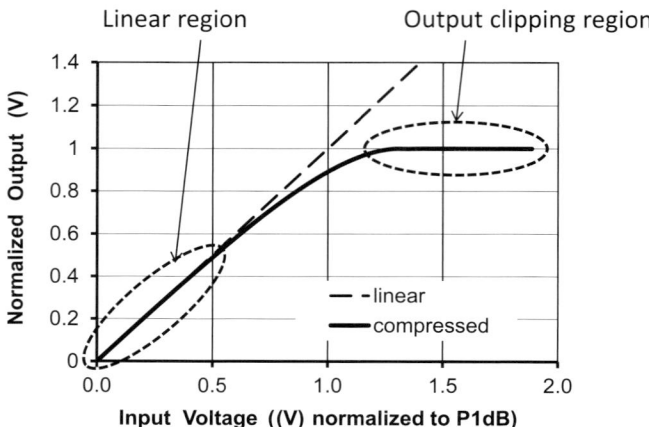

Figure 4-14 Linear and clipping regions of an amplifier transfer characteristic.

where a_0 is the output DC offset, a_1 is the linear coefficient, and D is the distorting term $\sum_{k=2}^{\infty} a_k x^k$. We define x_c as the input level at the onset of output clipping: $y(x > x_c) = y_{max}$. The *slope gain*[2] for the amplifier modeled by (4.7) is found using (4.6)

$$g_S = \frac{d}{dx} \sum_{k=0}^{\infty} a_k x_k = \begin{cases} a_1 + \sum_{k=2}^{\infty} k a_k x^{k-1}, & x \le x_c \\ 0, & x > x_c. \end{cases} \tag{4.8}$$

When the amplifier reaches and enters the clipping region ($x > x_c$), the slope gain (4.8) goes to zero. This means that changes in the input signal no longer effect changes in the output signal. This also means that an input pre-distorting linearizer stops having any effect.

Plots of (4.5) and (4.7) are shown in Figure 4-14, where the polynomial (4.7) is arbitrarily truncated to seventh order. Figure 4-14 compares the polynomial amplifier model (4.7) (solid line) with the ideal amplifier (4.5) (dashed line). Regions where the linear model (4.5) is valid, and regions for clipping, are highlighted.

Figure 4-15 adds a plot for (4.8), showing the normalized slope gain as it drops from its nominal value in the linear region to zero when the amplifier is clipping. This transition is smooth, showing that the amplifier transfer function has a progressive onset of compression, until it hits zero when the amplifier output stage clips at the power supply.

The *ratiometric* gain for this same amplifier is defined by

$$g_R \equiv \frac{y}{x} = \frac{y - 0}{x - 0}. \tag{4.9}$$

Applying (4.7) to (4.9) and setting $a_0 = 0$ (to ignore any DC offset in the transfer function) gives the expression for ratiometric gain as

---

[2] This also can accurately be called *differential* gain, but here we do not want to get confused with the gain of a differential amplifier.

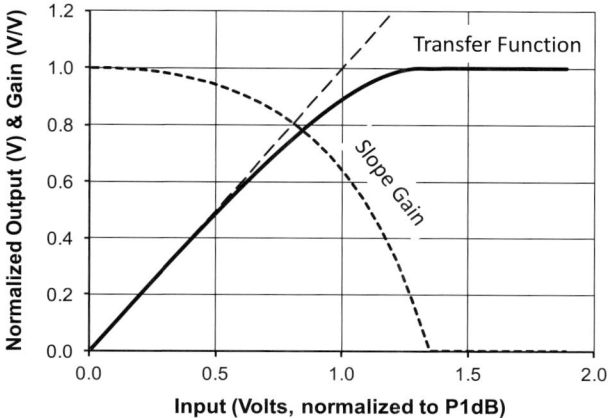

Figure 4-15 Slope gain (narrow dashed line) associated with this compressed transfer function (solid line).

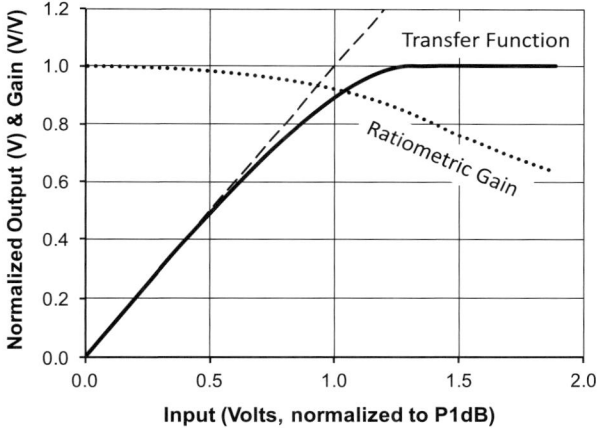

Figure 4-16 Ratiometric gain (dotted line) associated with the transfer function of Figure 4-14.

$$g_R = a_1 + a_2 x + a_3 x^2 + \cdots. \qquad (4.10)$$

Note that (4.10) is valid for all input values (x), and does not carry the domain restrictions of slope gain (4.8). Figure 4-16 shows the ratiometric gain for the transfer function of Figure 4-14. Note that the ratiometric gain never goes to zero, clearly seen from (4.10) for finite input values x and output values y.

Graphical illustration of the relationship between slope gain and ratiometric gain is shown in Figure 4-17. In the (small signal) linear region, where the input x is small and all higher order terms in (4.8) and (4.10) can be neglected, the values of g_S and g_R are identical. When sufficient compression exists, so that the higher order terms in (4.8) and (4.10) can no longer be neglected, the values of g_S and g_R increasingly diverge.

One additional characteristic is unique to ratiometric gain. When the output signal is clipped, as shown in Figure 4-18(a), the output power measured from the amplifier

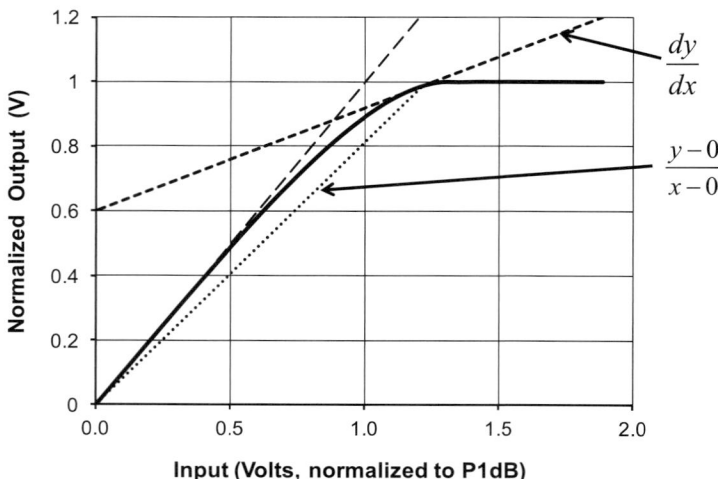

Figure 4-17 Relationship between ratiometric gain and slope gain at a particular point (x, y) along this compressing transfer function.

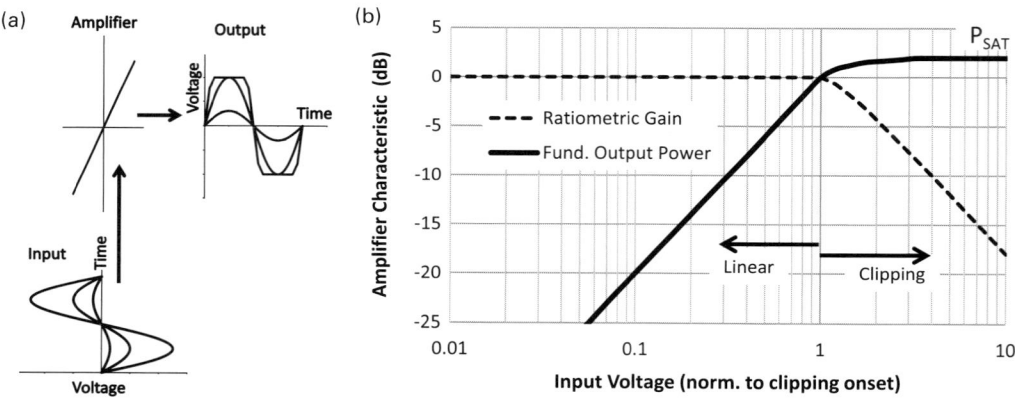

Figure 4-18 When measuring ratiometric gain, the output power increases as the waveform distorts (becoming square) through increasing compression: (a) waveform distortion mechanism from increasing clipping; (b) output power increases and g_R decreases with harder clipping.

continues to increase, even though the signal peak magnitude is held constant by clipping. This output power increase is a consequence of waveform distortion as the sinusoid waveform converts to a square wave with increasing compression. At the fundamental frequency, this apparent output power increase appears as a larger sinusoid magnitude when a spectrum analyzer is used as the measurement equipment. Care is required to remember that we only observe the fundamental frequency component of the output waveform with this measurement method, and obtain no information on wave-form distortion. The magnitude of the Fourier fundamental frequency component exceeds the absolute waveform magnitude of a square wave by $4/\pi$ [4-7]. This power transfer characteristic is shown in Figure 4-18(b).

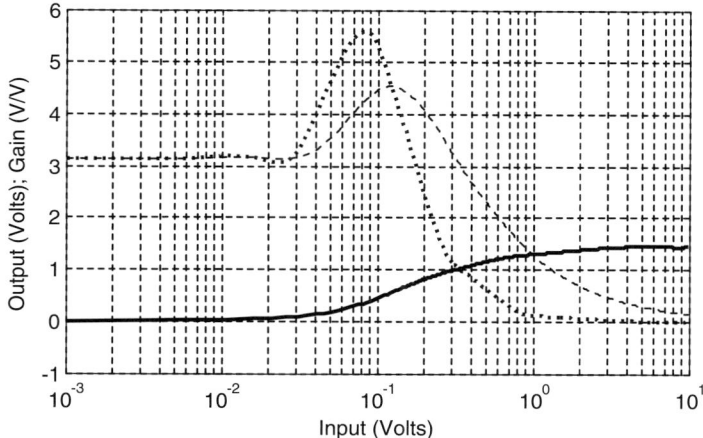

Figure 4-19 Results from Figure 4-2 in linear units: compressing transfer function (solid line), slope gain (dotted line), and ratiometric gain (dashed line).

This additional 2 dB of signal power at the fundamental frequency is physically real, of course, but it is a consequence of waveform distortion and not of any additional signal voltage capability from the amplifier. When only output power measurements are available, we need to interpret them for information on what output signals have a linear (undistorted) waveform and which are distorted. From Figure 4-18(b), we see one optimistic method for achieving this: signals with output power that is more than 2 dB below the saturated output power (P_{SAT}) are more likely not to be clipped.

Applying these two gain measures to the amplifier data from Figure 4-2 provides the curves in Figure 4-19.

The power-slope gain g_{SP} for this amplifier is the slope of the power transfer curve of Figure 4-18(b):

$$g_{SP} \equiv \frac{dp_{out}}{dp_{in}}, \tag{4.11a}$$

$$= \frac{d(y^2/R)}{d(x^2/R)} = a_1^2 = g_R{}^2. \tag{4.11b}$$

This gain measure is sometimes referred to as "dB per dB." When the amplifier goes into compression, the value returned for g_{SP} is in between that of g_S and g_R, shown in Figure 4-20.

As for g_S, the value of g_{SP} also goes to zero. Though when $g_{SP} = 0$, this means that the amplifier output is *power* saturated, and the output waveform is already squared (maximally distorted) by clipping (output voltage limiting).

For completeness of this gain discussion, it is evident that there is another form of ratiometric gain, one derived from the power transfer function – the ratiometric power gain

$$g_{RP} \equiv \frac{p_{out}}{p_{in}} = \frac{y^2}{x^2} = \frac{y^2 - 0}{x^2 - 0}. \tag{4.12}$$

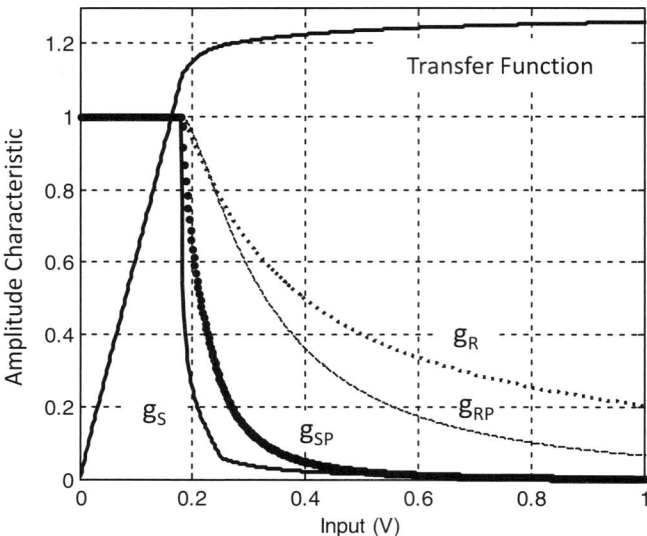

Figure 4-20   Four gain metrics for the same amplifier and its two transfer functions (voltage and power), normalized to identical values when the amplifier is linear to demonstrate their divergence as the amplifier compresses to allow their relative comparison.

Using Ohm's Law, we readily observe that in linear operation $g_{RP} = g_R{}^2$, so no new information is contained in g_{RP}. The relationship between g_{SP} and g_{RP} is graphically the same as that between g_S and g_R; one is the slope of the transfer function at a point of interest, and the other is the slope of a line drawn between the point of interest on the transfer function to the origin. A graphic equivalent to Figure 4-17 can be prepared for g_{SP} and g_{RP}, based instead on the fundamental-tone power transfer function like that in Figures 4-18 and 4-20. As we see in the next paragraph, g_{RP} is widely used in the course of normal lab work. All three of the other gain metrics are also used in lab work, usually, and unfortunately, completely interchangeably, which leaves the ambiguities unresolved. The presentation here clarifies the differences among these gain metrics, and motivates their careful use to avoid perpetuating these ambiguities.

Standard RF engineering practice is to describe the power gain of a stage as the difference in the decibel power measured at the output and input of the stage. Using capital letters for decibel and lower case for linear measures of gain and signal power, this becomes

$$G = P_{out} - P_{in} = 10\log_{10}(p_{out}) - 10\log_{10}(p_{in}) = 10\log_{10}\left(\frac{p_{out}}{p_{in}}\right) = 10\log_{10}(g_{RP})$$

$$(4.13)$$

in full accordance with (4.12). For equal input and output impedances, (4.13) can continue as

$$10\log_{10}\left(\frac{v_{out}^2}{v_{in}^2}\right) = 10\log_{10}(g_R{}^2).$$

$$(4.14)$$

Thus, gain calculated by decibel differences between output power and input power is a *ratiometric power gain* result.

Regarding the other gain measures and their correlation to standard RF laboratory practice, the following list presents common situations:

- Measuring g_m and calculating g_mR_L: transconductance is directly readable from a curve tracer (such as the venerable Tektronix 576), and has units dI/dV_{GS}. Resistor R_L has units V/I, and since the resistor I–V characteristic does not compress, the units dV_{DS}/dI also apply. Multiplying these together provides the net units dV_{DS}/dV_{GS}, which means that this is a **slope gain**.
- Measuring voltage waveforms on an oscilloscope: the common practice of observing the input waveform on one trace and the amplifier output waveform on another trace, and then dividing the trace measurements point-by-point gives units of V/V. This gain measurement is **ratiometric gain**.
- Using a spectrum analyzer and a signal generator: linearity of any RF stage is often checked by stepping the output power from a signal generator in 1 dB steps, and then measuring the output signal power step on a spectrum analyzer to see if it also takes 1 dB steps. This method is a **slope power gain** measurement.
- Using a power meter or spectrum analyzer: this is the measurement described by (4.13), measuring the applied input power and the resulting output power and taking their ratio (equivalently the decibel difference). The result is a **ratiometric power gain** measurement.

All of these gain measures are correct. Their different values represent different measures of the amplifier performance. Of these four measures, the two most important ones are g_S and g_{RP}. Slope gain g_S is the most sensitive to waveform distortion, so it is most important to the design of amplifier linearizers. Ratiometric power gain g_{RP} is by far the most common gain measurement made in RF laboratories. It is not very sensitive to waveform distortion, but it is very easy to measure.

4.1.6   Variable supply behaviors

All amplifiers are actually 3-port circuits, with two inputs and one output. The two inputs are the normal RF input and the power supply, leaving the RF output as the third port

$$P_{OUT} = f(V_S, P_{IN}). \tag{4.15}$$

While we normally ignore the power supply V_S as a circuit input port, with DPS operation this is no longer practical. Allowing the power supply to be a second input can be viewed as allowing the load line to shift horizontally as shown in Figure 4-21.

A performance scan of a multistage amplifier considering both input ports is shown in Figure 4-22. This is called a Booth chart, named after the engineer who first showed this chart and its usefulness in understanding how an amplifier behaves when considered as a 3-port [4-9]. There are seven curves shown in Figure 4-22, for seven unequally spaced values for the power supply input.

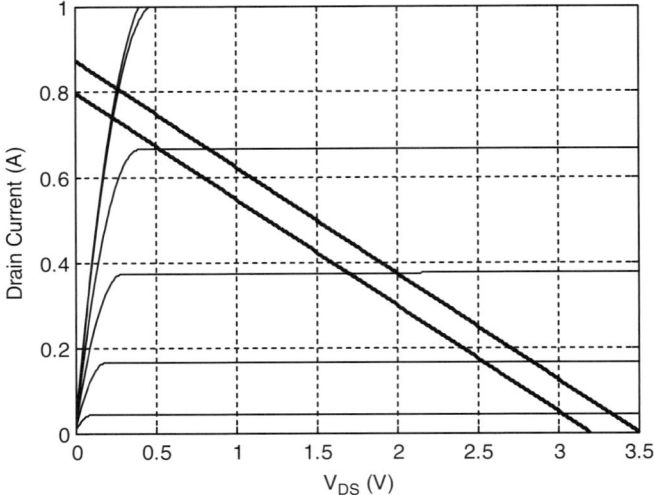

Figure 4-21 Varying the power supply as a second input is equivalent to shifting the load line left and right such that the intersection with the voltage axis is at the present power supply value.

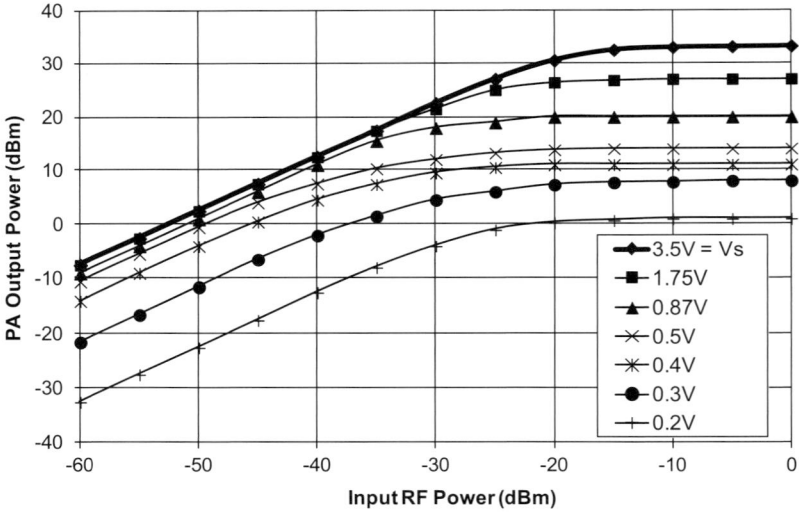

Figure 4-22 Booth chart for a typical amplifier (simulation result).

There are three operating modes evident in Figure 4-22 for this amplifier [4-10], which are:

- where the output power depends on the input power and not on the power supply,
- where the output power depends on the power supply and not on the input power, and
- where the output power depends on both the input power and the power supply.

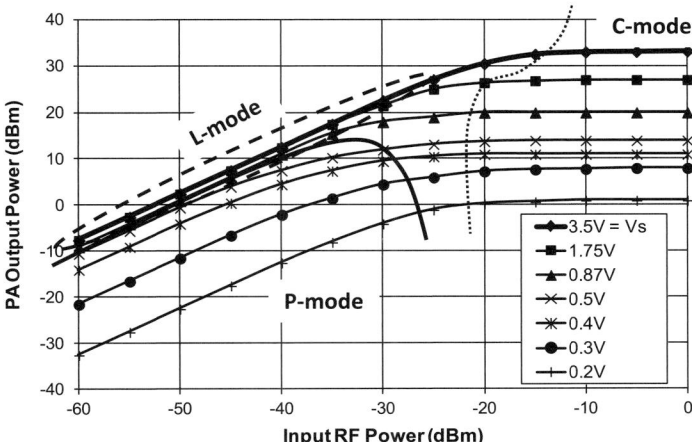

Figure 4-23 Identifying the three operating modes of an amplifier operated as a 3-port, with the power supply as the third port: L-mode (within the dashed ellipse), C-mode (to the right of the dotted line), and P-mode (below the solid line). Between these modes is a transition region.

These modes are identified in Figure 4-23 as: L-mode (where the output power depends on the input power and not on the power supply), C-mode (where the output power depends on the power supply and not on the input power), and P-mode (where the output power depends on both the input power and the power supply). A transition region exists between these three modes where the mode characteristics are shared.

Each of these three operating modes is now individually examined in more detail.

L-mode characteristics

L-mode is the conventional linear operation of a CCS amplifier. As the Booth chart (Figure 4-22) shows, here the RF output power is independent of the value of the power supply for any particular value of RF input power. This is the first characterizing trait for the L-mode.

Because the RF output is independent of the power supply, we know from characteristic transistor curves how the active device is functioning. With this regulated RF output power, the transistor is operating as a current source, where the transistor regulates the current flowing through the load. The traditional load line and operating point analysis shown in Figure 4-24, and the circuit model of Figure 4-25, are appropriate for understanding L-mode operation.

Biasing techniques for the amplifier transistor are also clearly evident from Figure 4-24. For an amplifier with maximum available linear output at a low power supply value, the bias point is shown in the lower left corner of Figure 4-24. If this bias remains unchanged as the power supply is varied, then we have a fixed-bias design, which shifts from class A operation at low supply voltage to a deep class AB when the supply voltage is high. If the linear operation of L-mode is to have maximum extent, then the bias point must change along with the value of the power supply as shown in Figure 4-24 along the line labeled

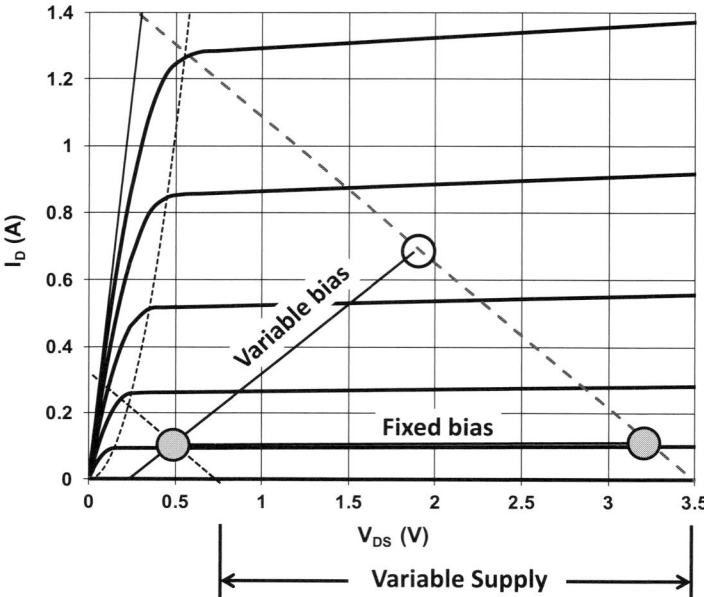

Figure 4-24   Biasing options for the active device for L-mode operation.

"variable bias." This keeps PA operation in class A at all supply values, but not at a constant quiescent current.

Because the transistor (active device) is regulating the current flowing through the load, the actual value of the power supply is not critical at all. The only important characteristic is that the power supply value be high enough so as to not operate the amplifier in compression. This means that L-mode operation has a high tolerance of noise on the power supply. This is the second key characteristic of L-mode operation. Testing for rejection of power supply noise is an easy way to verify that an amplifier is operating in L-mode. This property is symmetrical: an amplifier suppresses noise on its power supply if and only if it is operating in L-mode.

Figure 4-25(b) also shows that the operating points for L-mode are well removed from each of the axes. This means that power dissipation in the transistor is as high as it can be. L-mode can never be energy efficient.

The intersection of the two characteristic curves with the two load lines in Figure 4-25(b) defines a parallelogram. The parallel sides correspond to the regulating behavior of this circuit, showing that it is indeed acting linearly. This seems a trite result at the moment, but this geometric interpretation will become very important for the remaining two modes.

C-mode characteristics

When the amplifier operates in C-mode, the sensitivities and nonsensitivities are swapped from L-mode. Now the output power sensitivity is with the power supply, with no sensitivity to the input power. This operating mode is nonlinear. Therefore, while the load line analysis method still applies, the operating points are not in the

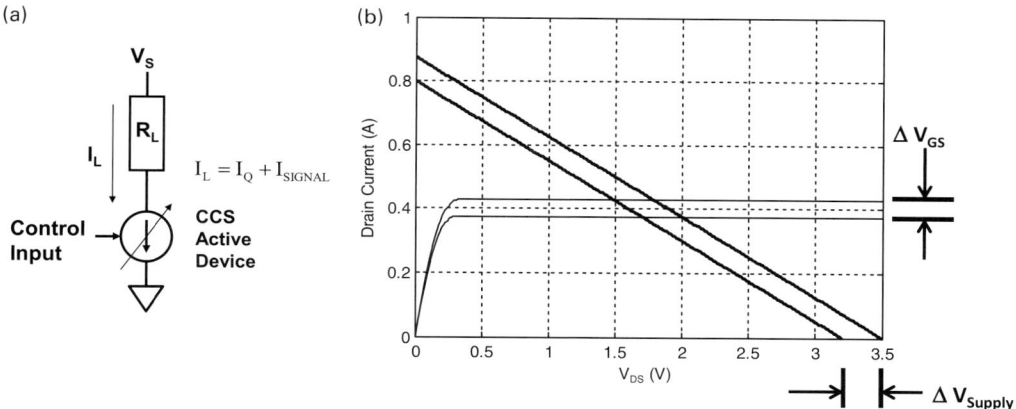

Figure 4-25 L-mode operation: (a) circuit model; (b) load current does not vary with power supply variations, only with input variations, validating the CCS circuit model.

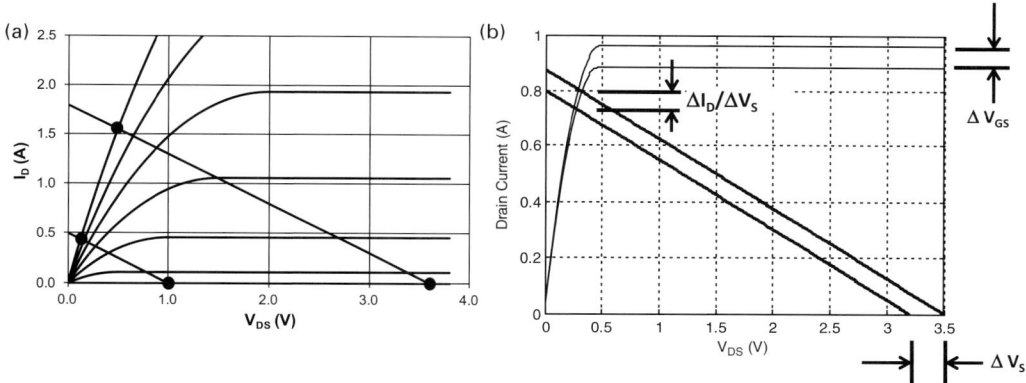

Figure 4-26 C-mode operation is (a) at the endpoints of the load lines, with a minimum amount of time in transition between these points; (b) load current varies widely with power supply variation, and hardly at all with input variations.

middle but rather at each end, as shown in Figure 4-26. The amplifier transistor does not regulate the current flowing through the load. Instead, it only selects when current flows through the load.

Load current only flows at the upper operating points in Figure 4-26. This intersection of the device characteristic curves with the load line is strongly governed by the power supply value and essentially independent of the input signal value, as long as the input signal is sufficiently large. This is consistent with Figure 4-22. The geometric shape outlined by the four operating points in Figure 4-26(a) is a trapezoid. This means that this C-mode circuit acts as a multiplier. It is definitely not a linear amplifier.

A different circuit model than the one from Figure 4-25(a) for L-mode is required. The appropriate circuit model for C-mode operation is shown in Figure 4-27.

Load current is governed by applying Ohm's Law to the series circuit shown in Figure 4-27. For design conditions where the power supply value is significantly larger

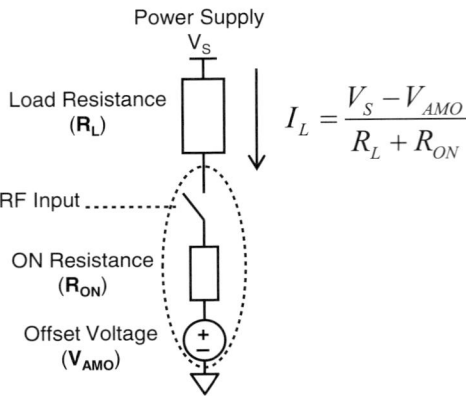

$$I_L = \frac{V_S - V_{AMO}}{R_L + R_{ON}}$$

Figure 4-27 Circuit model for C-mode operation.

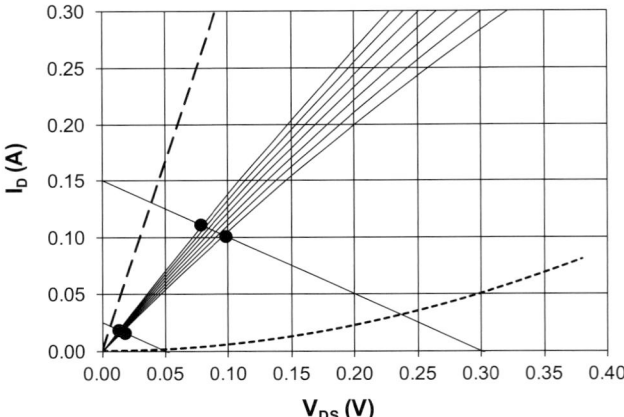

Figure 4-28 Device operation in P-mode, where sensitivity to both power supply and input power is realized with variable resistance device operation.

than any voltage offset present in the transistor, and also where the load resistance is much greater than the transistor ON resistance, then the load current is no longer dependent on transistor characteristics at all, but rather only on the external characteristics of load resistance design and power supply value.

P-mode characteristics

P-mode is so named because the output power is a product of the power supply value and the input RF power. It only appears when the power supply has a very low value. To be sensitive to both input power and power supply variations, it is not possible for the transistor to operate as a current source. Similarly, the transistor cannot operate as a switch. Instead, in P-mode the transistor operates as an input controlled variable resistance, as shown in Figure 4-28.

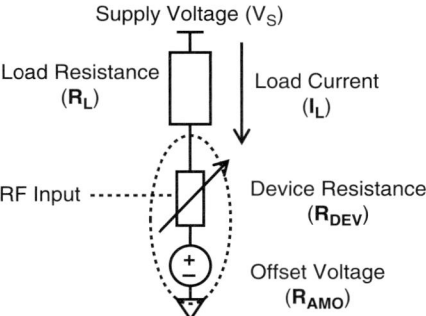

Supply Voltage (V$_S$)

Load Resistance
(**R$_L$**)

Load Current
(**I$_L$**)

RF Input

Device Resistance
(**R$_{DEV}$**)

Offset Voltage
(**R$_{AMO}$**)

Figure 4-29   Circuit model for P-mode operation.

Looking at the geometric figure traced by the four operating points in Figure 4-28, we again observe a trapezoid. As for C-mode, this means that the amplifier operating in P-mode is really a multiplier. In P-mode, the sides of the trapezoid are not along the voltage axis (where the transistor is OFF) or along the minimum channel resistance line. All operation is in the intermediate region. For an FET device, the operation is completely within the "deep triode region," where device operation is as a variable resistance. The corresponding P-mode circuit model is shown in Figure 4-29.

Because the P-mode output envelope depends on both the input envelope and the power supply, P-mode is a major cause of distortion for envelope tracking (ET) transmitters. This forces any power supply variations in ET designs to have a floor value that does not go low enough to enable the P-mode (see Chapter 5). In essence, at low envelope values the ET transmitter must hold the power supply constant and return to two-port, but not necessarily L-mode, operation.

On the other hand, P-mode is an excellent very low-level power control method for direct polar (DP) transmitters. Details of this are discussed in Section 6.7.3.

P-mode for an FET only occurs when it operates within the "deep triode" operating region, which means that the drain voltage is very small compared to the enhancement voltage on the bias. To mathematically illustrate how this works, as shown in Figure 4-28, we begin with the classic triode region FET model

$$I_D = K\left(2(V_{GS} - V_{Th})V_{DS} - V_{DS}^2\right). \tag{4.16}$$

When $V_{DS} \ll (V_{GS} - V_{Th})$, the V_{DS}^2 term can be neglected, leaving the approximation for "deep triode" operation as

$$I_D = 2K(V_{GS} - V_{Th})V_{DS} = 2K(V_{GS}V_{DS} - V_{Th}V_{DS}). \tag{4.17}$$

The $V_{GS}V_{DS}$ term in (4.17) is the multiplication product that the P-mode is named after. When RF is applied to the control port, V_{GS} becomes time-varying. Therefore, I_D is also time-varying and we finally get

$$\begin{aligned} I_D(t) &= 2K(V_{GS}(t)V_{DS}(t) - V_{Th}V_{DS}(t)) \\ &= (2KV_{DS}(t))V_{GS}(t) - 2KV_{Th}V_{DS}(t). \end{aligned} \tag{4.18}$$

The first term of (4.18) is an amplitude modulation of the input signal. If the input signal is already envelope-varying, then this is a major distortion mechanism. This mathematical form also describes a variable gain amplifier, clearly showing why the "gain collapse" phenomenon occurs, because in this operating region the drain current becomes directly proportional to the supply voltage. The second term is a time-varying bias at baseband (envelope) frequencies. It is readily removed with highpass or bandpass filtering.

System-level performance varies widely among these modes, particularly with the presence of power supply noise immunity in L-mode, the complete lack of power supply noise immunity in C-mode, or in the greatly increased signal distortion of P-mode. Understanding that these modes exist, what their properties are, and their manifestations from transistor device characteristics, is important for the timely success for DPS transmitter development projects.

4.2   Linearity/energy efficiency trade-off

The discussion in Section 3.3.1 presents how transmitter engineers "cheat" when doing conservation of energy (CoE) calculations, allowing the use of power measurements instead of energy measurements. A more detailed diagram for evaluating CoE as an equivalence of power out − power in for any power amplifier is shown in Figure 4-30.

Power is dissipated at many places in a power amplifier design. For a typical three-stage design, power is dissipated in each stage (first stage, driver stage, and final stage) along with the bias network plus the input power itself, schematically described by the multiple arrows into the heat sink in Figure 4-30. Mathematically this is

$$P_D = P_{IN} + P_1 + P_{DRIVER} + P_{BIAS} + P_{FINAL}. \tag{4.19}$$

This gives, using (3.7)

$$\frac{P_{OUT}}{P_{DC}} = 1 - \frac{P_1 + P_{DRIVER} + P_{BIAS} + P_{FINAL}}{P_{DC}}. \tag{4.20}$$

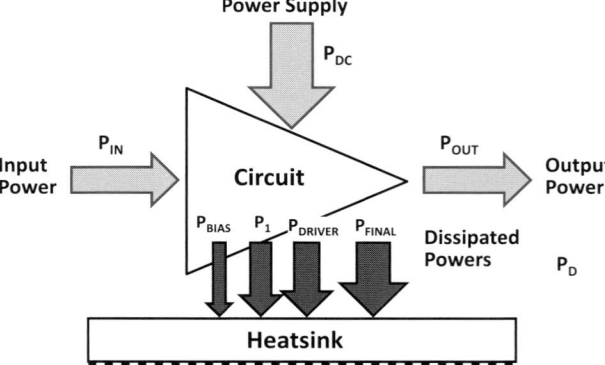

Figure 4-30 CoE as applied to PA efficiency analysis.

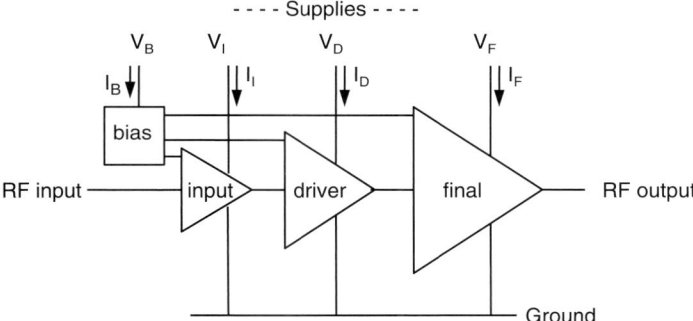

Figure 4-31   Terms needed for various efficiency measures of a multistage power amplifier.

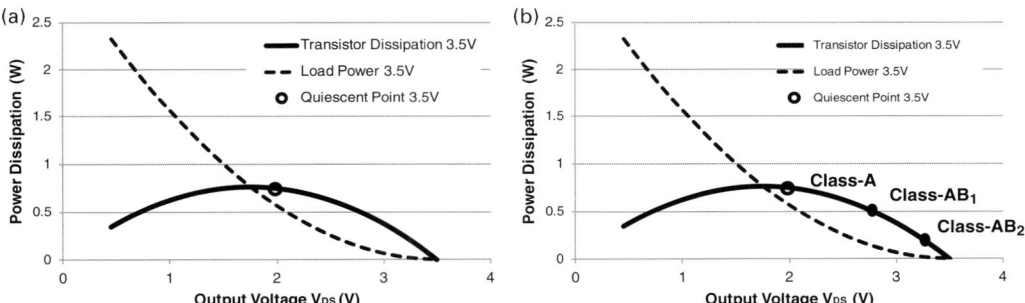

Figure 4-32   Power dissipation in the RF transistor (solid line) and the output load (dashed line) for the transistor and load line from Figure 4-13. The transistor knee voltage limits the lower output voltage available from the amplifier: (a) class A bias point for best inherent linearity; (b) power dissipation reduction with bias into class AB$_1$ and class AB$_2$.

Energy efficiency in this PA is improved by reducing each of these power dissipation sources, within the limits of acceptable linearity degradation.

In a real amplifier, there are usually multiple stages along with a bias circuit as shown in Figure 4-31. When considering the power flows from Figure 4-30, these stages must all be evaluated, individually and in combination.

The power dissipation sharing between the RF power transistor and the output load along the load line is shown in Figure 4-32 for the transistor and load line from Figure 4-13 at constant power supply. When the output voltage is at the power supply, there is no current in either the load or the transistor, so the power dissipation in both is zero. Power dissipation in the load increases quadratically as the load current increases. Power dissipation in the transistor does not: it peaks when the output voltage is one-half of the power supply voltage, and then decreases as the voltage across the transistor continues down. The voltage across the transistor cannot go below the "knee voltage" – the minimum voltage needed to achieve CCS operation at that particular current – for a linear amplifier. The quiescent bias point for this amplifier is at a current that is half of the peak current

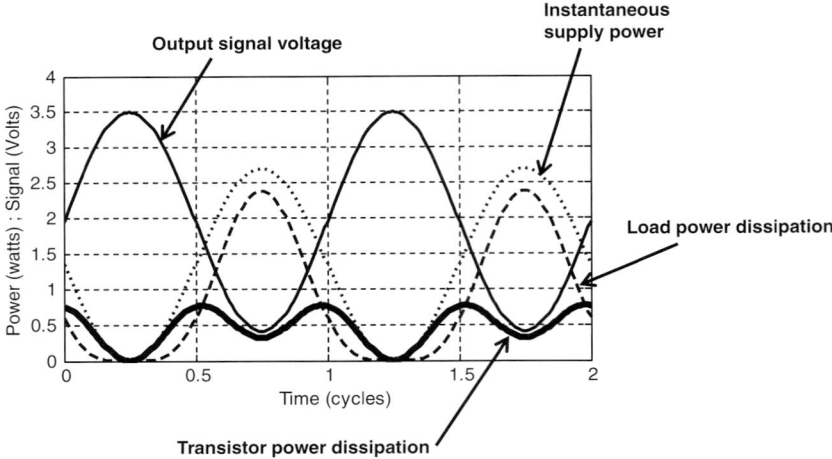

Figure 4-33   Waveforms and instantaneous power dissipation for 2 RF cycles in a class A amplifier.

drawn. It is offset from the center of the power supply, because of the knee voltage required by the transistor.

When there is no output signal, the transistor power dissipation is the value shown at the bias point in Figure 4-32(a), very near to its maximum possible value. In order to reduce this quiescent power dissipation, the intermediate bias conditions of class AB are used, as shown in Figure 4-4 and with the consequences of Figure 4-7. The actual power dissipation results for this biasing are shown in Figure 4-32(b), confirming that quiescent power dissipation does reduce. As the bias is reduced, the operating point shifts toward the cut-off condition, as it must.

Power dissipation consequences of the knee voltage are seen in Figure 4-33. When the transistor pulls maximum current, this voltage bound keeps its power dissipation from going toward zero. This also limits the magnitude of the output signal, at its negative peak. The time-varying power required from the power supply is the sum of the power dissipated in the transistor and the load, which we see is nearly sinusoidal, even though the power actually dissipated in the load is parabolic, as seen here and in Figure 4-32.

Whenever energy efficiency is not perfect – meaning always – something gets hot due to the dissipated power interacting with its associated thermal resistance. This is true no matter what the cause of the power dissipation. Whether it is the transistor, the inter-connect dielectric, the output matching network, or some/all of the above, something *always* gets hot! Two different transmitter design examples are analyzed with their heat flow diagrams in Figure 4-34. The ultimate goal is to minimize the input power while maintaining the output power. The benefit of having an energy-efficient power amplifier is that not only does the input DC power (bought from the utility) go down, but also the heat sink gets smaller, and the power supply gets smaller, both greatly lowering cost. *This* is the reason we care so much about energy efficiency.

This benefit is realized by a reduction in the power required from the utility into the transmitter power supply. Some numbers are shown in Figure 4-34, which follow from the general power conversion relationship

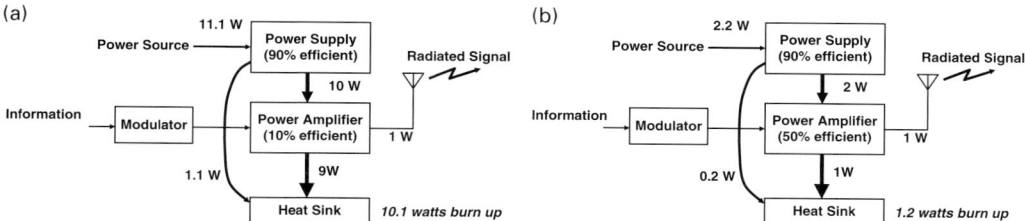

Figure 4-34 Heat flow diagrams for a set of two transmitter types: (a) conventional class AB₁ linear design; (b) deeply compressed design. Both provide the same RF output power, but input utility power drops significantly as the design evolves with higher PA efficiencies. Power from the modulator is here assumed to be zero.

$$P_{S,IN} = \frac{P_{OUT}}{\eta_{PA}\eta_{PS}}.$$

(4.21)

Providing a design opportunity to improve either the PA energy efficiency η_{PA}, or the power supply conversion efficiency η_{PS}, or both, is very important to transmitter design.

4.3 Stability

Stability of any design is imperative for successful manufacturing. It is essential that the designed characteristics remain within specified bounds, even as parts and their environment vary over place and time. In this section, three primary stabilities are pointed out: circuit stability, thermal stability, and manufacturing stability.

4.3.1 Circuit stability

Circuit stability is a vital consideration for any linear amplifier. By definition, an unstable circuit is one where the output changes without a corresponding change at the input. An oscillator certainly meets this definition, where we require an RF output with no RF input. Anything with gain can oscillate, and the *Barkhausen criteria* [4-11] are a useful visualization on the conditions as these are necessary (but not sufficient) for oscillation:

- feedback exists (either intentional or parasitic) and the net gain around the feedback loop is 1 or greater, and
- the net phase shift around this feedback loop is an integer multiple (including zero) of 360 degrees.

For circuit stability (meaning no oscillations), we definitely do not want to satisfy the Barkhausen criteria. Feedback is often unavoidable, and gain >1 is essential for any linear amplifier. The only thing left is to manage phase shifts to stay far away from any integer multiple of 360 degrees.

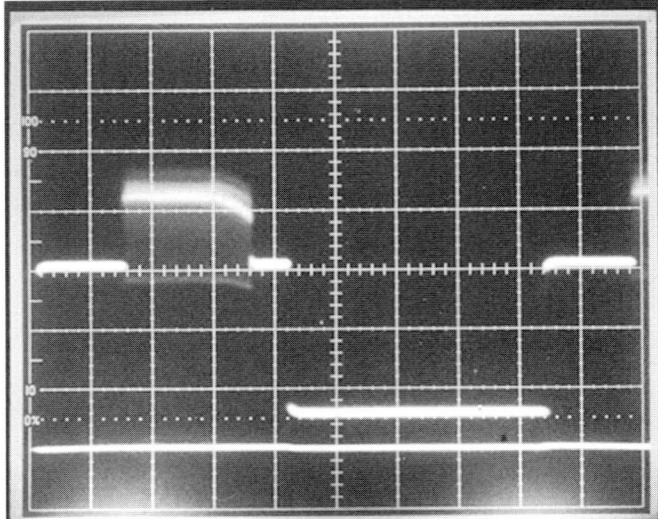

Figure 4-35 Unstable circuit behavior (oscillation) during the bias-high state of this circuit.

There is another mechanism where oscillations can occur, and that is if a resonant circuit exists and the transistor for some reason presents a sufficiently negative resistance to this resonator such that the resonator losses are overcome. This is essentially what is evaluated by the two-port s-parameter stability factor test

$$k = \frac{1 - |s_{11}|^2 - |s_{22}|^2 + (s_{11}s_{22} - s_{12}s_{21})^2}{2|s_{21}||s_{12}|}. \tag{4.22}$$

If this stability factor k evaluates to a value much greater than 1, then stability is good. Below 1, the stability is conditional [4-6]. Don't sneeze. One example of unstable circuit behavior is shown in Figure 4-35.

Stability at low frequencies is a particular problem, because all transistors have their greatest gain at low frequencies. This makes satisfying the Barkhausen criteria easier, and thereby increasing the tendency for low frequency oscillations. It is essentially universal practice to solve this problem by reducing the gain at low frequencies with a large shunt capacitor. There is no need to be concerned with the Barkhausen phasing aspect when the gain is held below unity. Hence, nearly all linear amplifiers have large valued "bypass" capacitors across the power supply.

4.3.2   Thermal stability

For each individual product, as temperature changes it is important that the performance remains consistent. For a transmitter power amplifier, this is particularly difficult because its internal power dissipation also impacts on its operating temperature through thermal resistance (see (3.11)), plus the fact that transistor CCS operation is inherently temperature sensitive.

This is apparent by examining the controlled current mathematics for field effect and bipolar transistors. Starting with the FET CCS model

$$I_D = K(V_{GS} - V_{Th})^2 \; ; \qquad V_{DS} > V_{GS} - V_{Th}, \qquad K = \frac{\mu(\varepsilon_{OX}/t_{OX})}{2} \frac{W}{L}, \quad (4.23)$$

the parameters K and V_{Th} are both temperature-varying. Thermal effects are more directly seen in bipolar transistor operation

$$I_C = I_S\left(e^{V_{BE}/(kT/q)} - 1\right) - \frac{I_S}{\alpha_R}\left(e^{V_{BC}/(kT/q)} - 1\right); \quad I_S = \frac{qAD_n n_i^2}{W_B N_A} \qquad (4.24)$$
$$\approx \beta I_B,$$

where temperature T is explicitly present in the bipolar CCS equation. Temperature compensation within the PA circuitry is therefore necessary to achieve thermal stability.

4.3.3 Manufacturing stability

For each unit manufactured, there are different parts used. All parts differ from each other, even of the same type, so the performance of a design across manufacturing is an exercise in the combination of many random "variables." For each instance of the product, the components are all "known" (not variables any more). The product performance with this set of components must still be within the specifications.

Reviewing the CCS equations in Section 4.3.2, we note that there are terms which depend on manufacturing details. For the FET (4.23), the threshold V_{Th} varies with manufacturing differences, along with all of the terms making up the scale factor K. The bipolar transistor (4.24) is more consistent in its V_{BE} value, but there are many terms within the scale factor I_S that depend on manufacturing details. Therefore, calibration on each manufactured unit is often required.

4.4 Major distortion mechanisms

While linear operation is the intention, this is an ideal only toward which the designer can strive. All circuitry is inherently polar in its operation, as presented in Section 2.1.1. This is evident from the general characteristic that the phase of an input signal does not matter, while the magnitude of the input signal has a very large impact on circuit operation.

This leads to another important set of definitions needed in order to avoid ambiguities. The important terms are *magnitude* and *amplitude*, which are defined as follows:

- amplitude: a signed parameter relating to scaling of a sinusoid signal;
- magnitude: a nonnegative (positive or zero) measure of the peak value of a sinusoid.

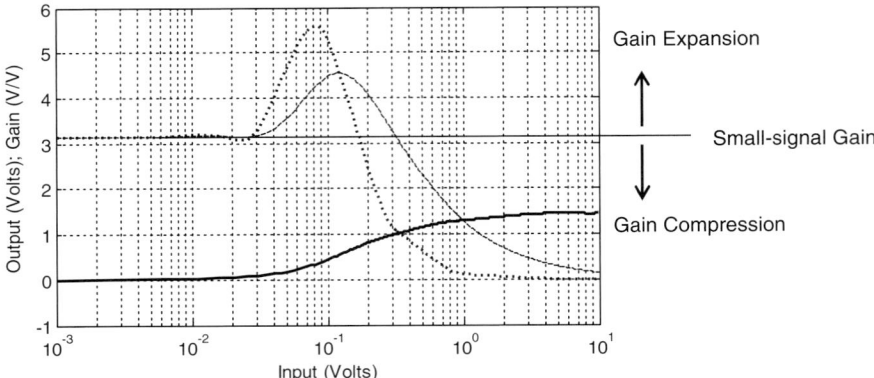

Figure 4-36 AM-AM distortion is any deviation from small signal gain, here illustrated using the amplifier of Figure 4-19.

Whenever polar coordinates are discussed, only magnitude is defined. It is no surprise then that the major distortion mechanisms of the following sections are polar in nature and driven by signal magnitude variations.

4.4.1   AM-AM

AM-AM distortion is a measure of errors on the output signal power due to variations of the input signal magnitude. This distortion is largely driven by output compression, though gain expansion is also a cause of AM-AM distortion. This is illustrated in Figure 4-36 and written as

$$\frac{\Delta P_{OUT}}{\Delta P_{IN}} = g_{SP} \neq \text{constant.} \tag{4.25}$$

Another source of AM-AM distortion is from bias changes that cause transistor trans-conductance changes. Since $g_m R_L$ is the gain, whenever g_m changes then the gain changes and the output is not linear. This effect is particularly prominent when bipolar transistors are used and biased in class AB_2 to reduce quiescent power dissipation.

4.4.2   AM-PM

With linear operation, the phase shift through an amplifier is constant. When compression begins, it is possible that the phase shift also begins to change. This is called AM-PM distortion: a change in amplifier phase response driven by a change in the input signal magnitude

$$\frac{\Delta \phi_{OUT}}{\Delta P_{IN}} \neq 0. \tag{4.26}$$

AM-PM is particularly important for wideband signals. These signals contain many frequency components, and if the frequency components change their phase

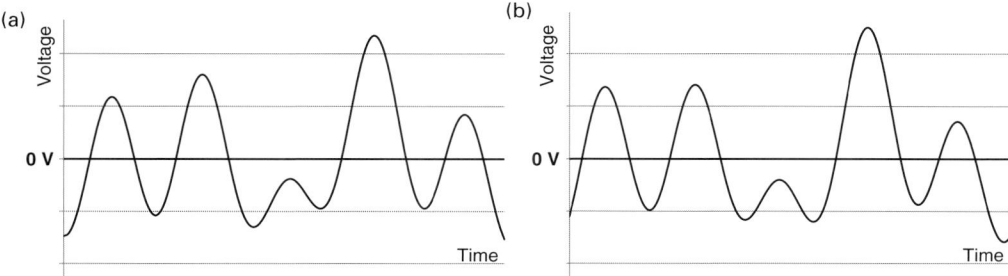

Figure 4-37   Waveform distortion from phase shifts among the frequency components: (a) input waveform; (b) output waveform.

relationships, then the output waveform will be changed. Figure 4-37 shows one example of how a waveform can change when the frequency component magnitudes are unchanged, but small phase shifts are experienced among the various frequency components.

4.5   Gain and linearization principles

Even DPS transmitters must perform as if they are linear amplifiers, as defined in Section 4.1. When the design problem is to linearize an amplifier, we must select one of the gain definitions of Section 4.1.5 to use in the linearizer design. The gain characteristics of Figures 4-15 and 4-16 are different, yet they both describe the same compressed amplifier. Is one more appropriate than the other for rapid success in linearizer design? The answer is yes: use slope gain, since this measure goes to zero when clipping distorts the waveform.

To improve the linearity of an amplifier, it is necessary to delay the onset of compression. One way to do this is shown in Figure 4-38. Here the amplifier transfer function is linearized to stay on the linear response line until the clipping output level is reached, when the amplifier output can no longer increase anyway. The corresponding curves for slope gain and ratiometric gain are also shown. As before, the value of ratiometric gain is always greater than, or equal to, the value of slope gain.

Figure 4-38 shows that slope gain g_S changes rapidly from the nominal amplifier gain down to zero at the onset of clipping. This is immediately indicative of waveform characteristics, showing that there is waveform distortion resulting from clipping. Ratiometric gain is the same for this linearized amplifier until clipping begins, as it must with a linear operation. Then g_R begins to fall slowly, which may be interpreted (incorrectly) as a slow compression effect.

Comparing Figure 4-38 with Figures 4-15 and 4-16, we note that the ratiometric gain curve looks similar whether the amplifier behavior is compression or clipping. Any drop in ratiometric gain therefore only tells us that the amplifier is operating in a nonlinear (distorting) manner. No additional insight is available on whether the

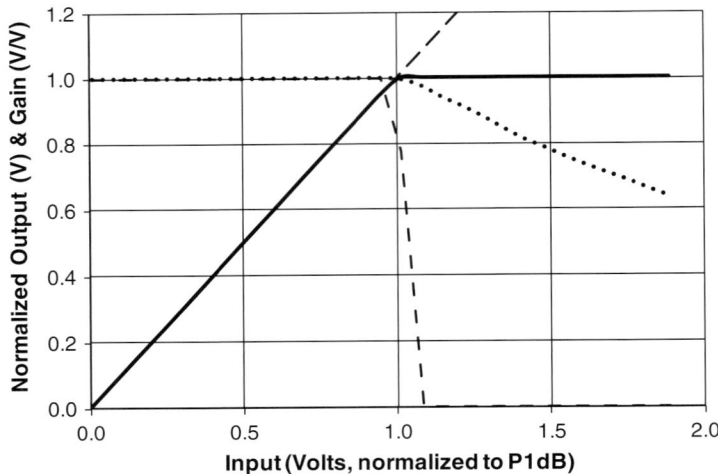

Figure 4-38 Relationship between ratiometric gain (dotted curve) and slope gain (dashed curve) for a linearized amplifier.

distortion mechanism is compression or clipping. On the other hand, slope gain measurements provide very clear information on whether the distorting mechanism is compression or clipping.

4.6   Supply noise suppression

CCS (linear) amplifiers exhibit a very desirable characteristic that any noise on the power supply has a very tiny impact on output noise sidebands. Here we go into physical details on why this is so.

Looking at Figure 4-25(b), it is apparent that when the transistor operates as an ideal current source, if the power supply changes value, the current through the load does not change. The earlier conclusion was that the exact power supply value was not critical. When the power supply has output noise, the equivalent model to that of Figure 4-20 is to have a single load line at the nominal power supply value, and that load line "wiggles" about that mean value along with the output noise. The different action of the load line does not change the earlier conclusion: any noise on the power supply is similarly not a problem because it does not transfer to changes in the load current.

Practical RF power transistors usually do not have this ideal behavior. There is nearly always a finite conductance across the output (controlled) port, shown in Figure 4-39(b) as r_o. The effect of r_o is to have the characteristic curves increase in current as the voltage across the device increases, shown in Figure 4-39(a). Now, when the power supply value changes, there is a change in the device current and an equivalent change in the load current. Isolation is no longer perfect.

Noise on the power supply now couples into the output signal. If the tilt to the characteristic curve is small, this noise coupling into the output signal is weak and likely

(a)

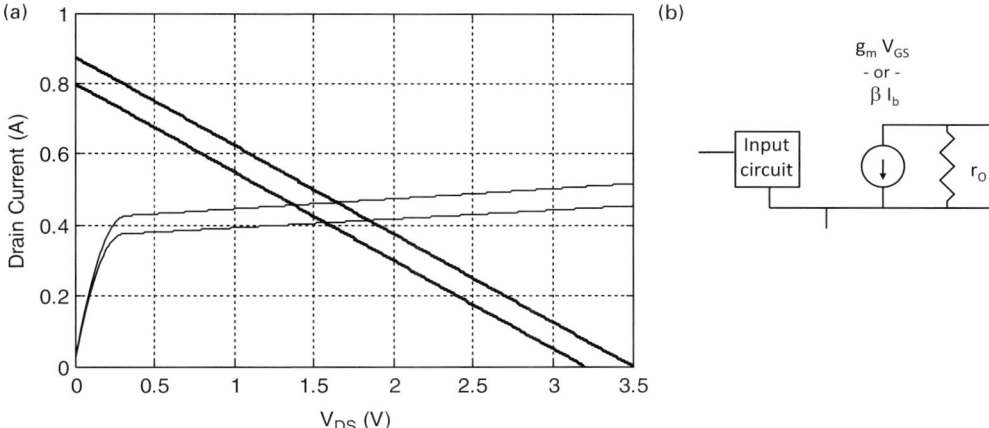

(b)

Figure 4-39 Load current variations as the power supply is changed: (a) transistor with finite output conductance; (b) generic transistor circuit model.

of only minor consequence. The greater the conductance (lower the resistance r_o), the steeper the tilt will be and the stronger the noise coupling. Clearly, the best transistors for linear PA use will have nearly ideal current source behavior.

4.7   References

[4-1] S. C. Cripps, *RF Power Amplifiers for Wireless Communications*, 2nd edn., Artech House, Norwood MA, 2006.

[4-2] E. McCune, *Practical Digital Wireless Signals*, Cambridge University Press, 2010.

[4-3] I. R. Skarbek, "New High-Efficiency 5-kW AM transmitter (with) Unique Class C Amplifier Operates with 90% Efficiency," *Broadcast Transmitter Engineering*, RCA Broadcast News #107, March 1960, available at http://nrcd xas.org/articles/bta5t/.

[4-4] R. Wu, Y.-T Liu, J. Lopez, C. Schecht, Y. Li, and D. Y. C. Lie, "High-Efficiency Silicon-Based Envelope-Tracking Power Amplifier Design with Envelope Shaping for Broadband Wireless Applications," *IEEE Journal of Solid-State Circuits*, vol. 48, no. 9, Sept. 2013, pp. 2030–2040.

[4-5] M. Schetzen, *The Volterra and Wiener Theories of Nonlinear Systems*, rev. edn., Krieger Publishing Company, 2006.

[4-6] D. Schreurs, M. O'Droma, A. Goacher, and M. Gadringer, *RF Power Amplifier Behavioral Modeling*, Cambridge University Press, 2009.

[4-7] R. Bracewell, *The Fourier Transform and its Applications*, 2nd edn., McGraw-Hill, New York, 1978.

[4-8] E. McCune, "Gain: Changed Meanings for Compressed Amplifiers," *Proceedings of the 2013 Midwest Symposium on Circuits and Systems (MWSCAS)*, Columbus OH, August 2013.

[4-9] R. W. Booth, private conversation with the author, Tropian Corp., Cupertino CA, 2001.

[4-10] E. McCune, "Operating Modes of Dynamic Power Supply Transmitter Amplifiers," *Proceedings of the 2014 International Workshop on Integrated Nonlinear Microwave and Millimetre-wave Circuits (INMMiC)*, Leuven, Belgium, April 2014.

[4-11] E. Lindberg, "The Barkhausen Criterion (Observation?)," *Proceedings IEEE Workshop on Nonlinear Dynamics of Electronic Systems (NDES2010)*, Dresden, Germany, pp. 15–18.

5   Envelope tracking principles

Adding a dynamic power supply (DPS) into a conventional linear transmitter (see Figure 5-1) costs money, requiring a very good reason to go through the time, effort, and expense to do this. That motivation is provided by the wireless communication standards committees, who adopt signal modulations with increasing values of peak-to-average power ratios (PAPR) (see Section 2.6.1). If the signal PAPR is 3 dB or less, then there may not be much advantage to adding the DPS. When the signal PAPR exceeds 5 dB, the advantage of incorporating a DPS into a transmitter in order to improve its energy efficiency builds rapidly.

Envelope tracking (ET) is a DPS technique that aims to improve the energy efficiency of a linear power amplifier (PA) by matching the voltage across the RF power transistor to the lowest practical value needed to accurately provide the *present instantaneous* output power. The PA power supply therefore tracks alongside the envelope of the output signal with moderate correlation – hence the name: envelope tracking – *while leaving the power transistor linear operation nominally unchanged* [2-4]. Fundamental to this core principle is that whatever the applied voltage is at the PA, it is not critical to the accuracy of the output signal modulation. Thus, envelope tracking is one of the many techniques used in the century-long quest to build linear power amplifiers and simultaneously to have them be energy efficient.

5.1   History of the technique

Recognizing that the efficiency of an amplifier operating below its maximum output signal power is improved by reducing the applied supply and bias voltages includes a key publication from 1989 [5-1]. This does not attempt to follow the signal envelope, but only the signal power, which is a technique now called average power tracking (APT). A power amplifier incorporating a step-varied power supply controlled by the input signal envelope was first published in 1995 [5-2]. This design switched the supply voltage applied to the RF power transistor between two values, using the lower supply voltage the most, but applying a much higher voltage when the input signal has a large peak value coming. Now, nearly 20 years later, the name for this technique of temporarily switching power to support high peak output power is called class G [5-3].

The first known discussion of ET, where the applied power supply varies continuously along with the input signal envelope to reduce power dissipation in the transistor is [5-4], published in 1999. In this paper, the DPS is a DC–DC boost converter operating at a

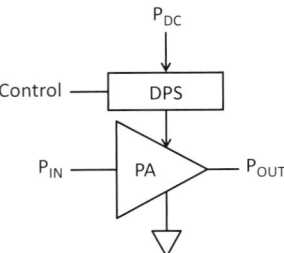

Figure 5-1   Adding a DPS into a conventional linear transmitter.

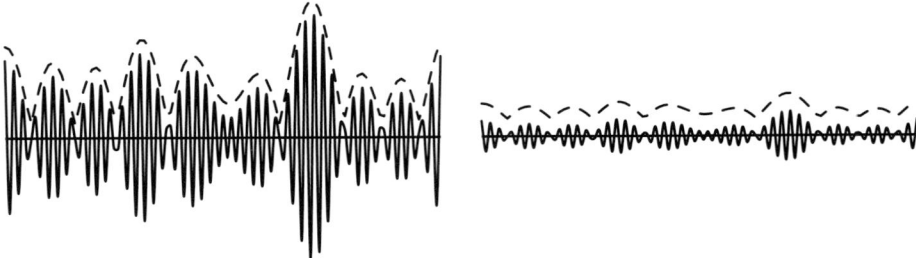

Figure 5-2   Envelope tracking reduces RF PA power dissipation using a variable power supply (dashed curve) that remains sufficiently above the RF waveform peaks to ensure that the RF power amplifier remains in linear operation at all output powers.

switching frequency of 10 MHz. Coupling of ripple at the output of this supply on to the output signal as undesired sidebands, to ensure that the linearity of the RF amplifier maintains its suppression of this power supply noise, receives particular attention. The intent of ET is to implement the operation presented in Figure 5-2. Drawn from a mobile device perspective (using a battery as the power voltage source), the intent is to keep the voltage drop across the RF transistor as low as *practical*.[1] The current through the RF transistor is governed largely by the necessary output power, so transistor current cannot change. Therefore, lowering the supply voltage does reduce power dissipation in the RF transistor.

While reduced power dissipation in the RF power transistor is very useful, it is usually equally important to reduce the current drawn from the primary voltage source (e.g. a battery) to capitalize on the improved energy efficiency of the RF power transistor. In other words, the operating energy efficiency of the RF power transistor must be "visible" to the primary power source. This is the primary objective of a successful DPS design, discussed in detail in Chapter 9.

5.2   Power supply value tolerance

One of the major values of ET is that the actual value of the power supply is not critical. How the amplifier must operate for this to be so is discussed in Section 4.6. For this

---

[1] Quantifying what 'practical' means here is the point of this chapter.

property to hold, the characteristic curves show that the RF power transistor must operate as a controlled current source (CCS) – even if it is a lossy one. Therefore, the ET RF PA operates in L-mode.

5.2.1 Ideal case

When the RF PA is operating linearly, to first-order, the exact value of the power supply has no impact on the current flowing through the load. This is equivalent to saying that the RF power transistor is operating as a current source, regulating the load current. When this is true, then the exact value of the power supply voltage, as long as it is high enough to support the output signal peaks, is immaterial. We can then say that power supply sensitivity (*PSS*) is zero.

$$PSS = \frac{\Delta I_D}{\Delta V_{DS}}, \qquad (5.1a)$$

and

$$PSS = 10 \log_{10}\left(PSS^2\right). \qquad (5.1b)$$

Using (5.1), we now calculate the power supply sensitivity of this amplifier, with the results shown in Figure 5-5, using the same orientation as that used for Figure 5-4, in Figure 5-5(a). As expected, we see that the power supply sensitivity, when the transistor operates as a current source, is zero. Something very different happens at low values of V_{DS} across the transistor. Circuit operation changes markedly. Operation in this region is a very different mode, and is not envelope tracking. This different operation is discussed in Chapter 6.

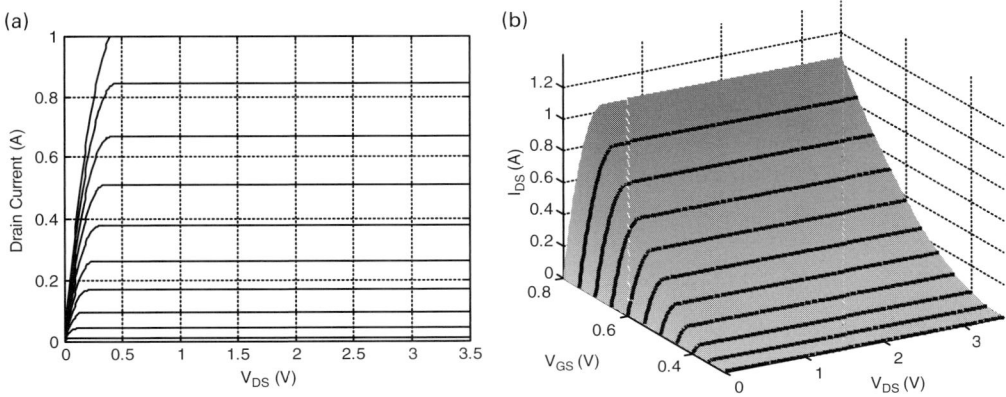

Figure 5-3 Idealized FET characteristic curves for the present discussion: (a) conventional family of curves; (b) three-dimensional plot to illustrate the transconductance variation of this FET.

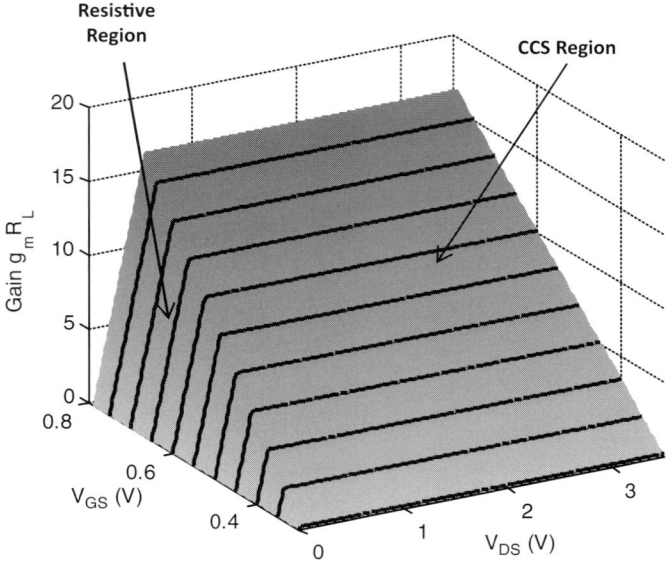

Figure 5-4   Amplifier performance with the transistor described with the curves of Figure 5-3 working into a
load resistance of 4 ohms. Device characteristic curves are overlaid on this surface, spread out
along their corresponding V_{GS} values. Both the CCS and resistive operating regions are identified.

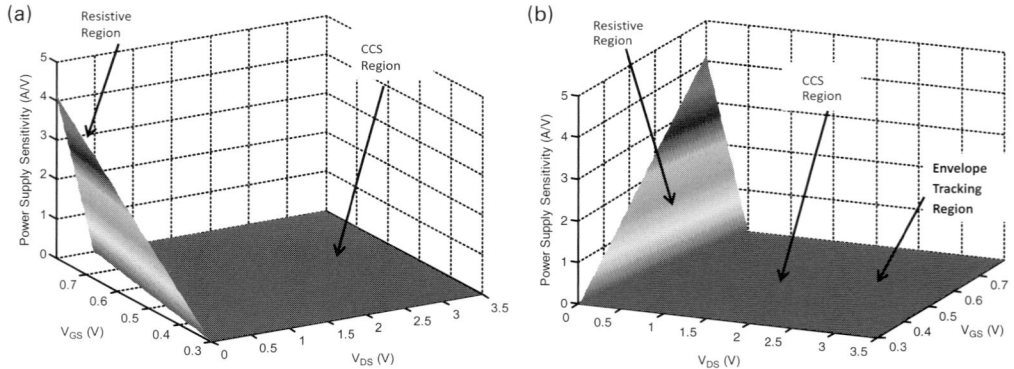

Figure 5-5   Power supply sensitivity for this amplifier: (a) same orientation as Figure 5-4; (b) rotated view to
see the entire region of $PSS = 0$ where the ET operation is valid.

5.2.2   Finite output conductance case

Any practical transistor has a finite output conductance, meaning that the current
flowing through the load is always influenced somewhat by the value of the power
supply. Adding a finite output conductance to this transistor provides the characteristic curves in Figure 5-6. The slopes of the CCS regions for this transistor are very
evident.

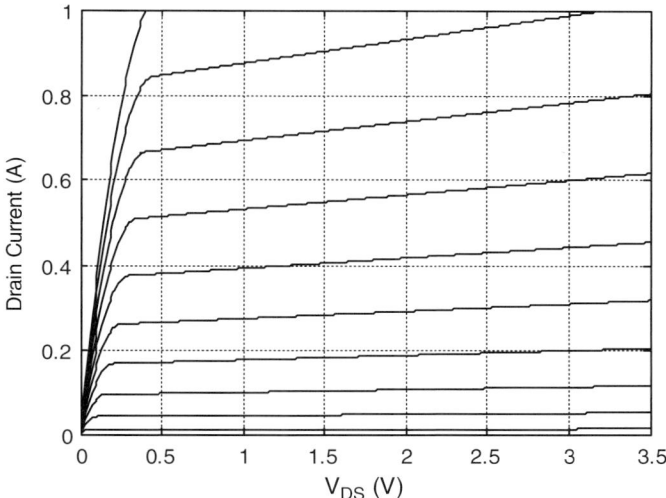

Figure 5-6   Characteristic IV curves for the example FET transistor. Note the presence of finite output conductance, evidenced by slopes in the CCS region.

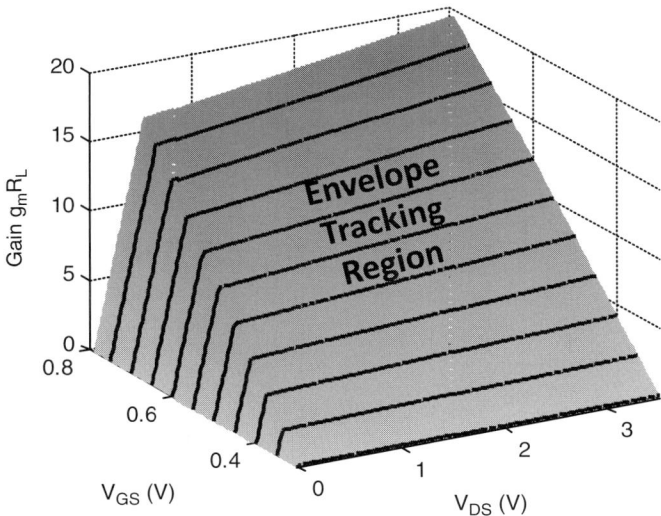

Figure 5-7   Gain surface of an FET-based amplifier, derived from the product of device transconductance and the transistor load resistance. Finite output resistance is included, which twists the CCS region of this surface.

In Figure 5-6, this finite output conductance is the result of channel length modulation (similar in performance to the early effect in bipolar transistors), causing the device current to increase with higher V_{DS}.

Changes to the amplifier operation surface from Figure 5-4 are shown in Figure 5-7. The CCS operation part of this surface is no longer planar, signifying increases to the power supply sensitivity of this amplifier.

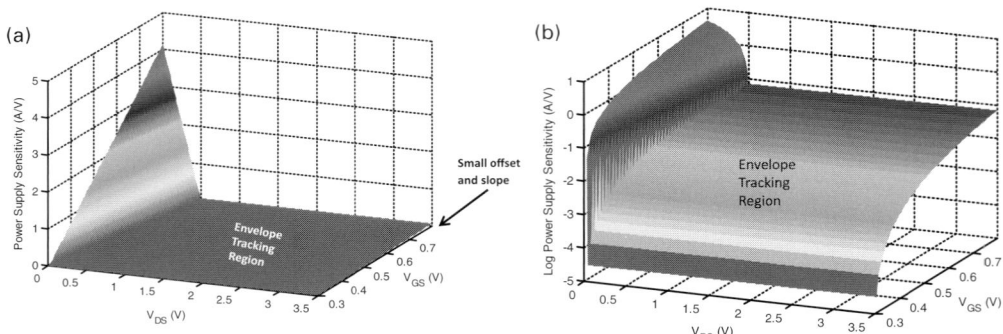

Figure 5-8   Power supply sensitivity of this amplifier across the possible set of device operating conditions (V_{GS}, V_{DS}). Compared with Figure 5-5, the power supply sensitivity is increased due to the finite output resistance: (a) linear plot where a slight tilt to the "floor" is visible at high V_{GS} and high V_{DS}; (b) log plot of the same data to emphasize the increase in PSS in the CCS operation region.

Power supply sensitivity is degraded with this more realistic transistor. In Figure 5-8(a), the "floor" is raised above zero at the larger values of V_{GS} and V_{DS}. On this scale the difference appears minor, so ET is still expected to work reasonably well. Better visibility into actual power supply sensitivity is seen in Figure 5-8(b), where the logarithm of the power supply sensitivity is plotted on the vertical axis. Even at high V_{GS}, the value of pss is still below 0.1. The presence of the other operating mode, with much higher power supply sensitivity, remains very evident.

It is important to be very clear on this point. If the value of pss *ever* becomes nonzero, this means that the value of the power supply does have some influence on the actual output signal envelope. Whether this additional modulation is important or not completely depends on the application in which the ET system is to be used. It is rare to find a RF power transistor that is close enough to an ideal current source that the value of pss can be considered to be effectively zero. This means that more often than not there is dependence of the amplifier output signal on variations in the power supply. Something we must live with.

5.2.3   Measuring the supply variation to sideband conversion

Supply noise to AM sideband conversion is the transfer function through which noise on the power supply converts to distortion of the output signal. Measurement of this transfer function is not easy, though, because of the very large impedance differences present at the circuit points of interest: the power supply impedance is very low, transistor output resistance is (hopefully) very high, and the load is near 50 ohms. Normal calculation of gains and transfer functions in RF cascades, which all assume a consistent 50 ohm impedance environment, do not work here. A different technique is needed.

The best technique to calculate how any variation in the power supply impacts on the output signal is to consider this distortion as an amplitude modulation (AM) process.

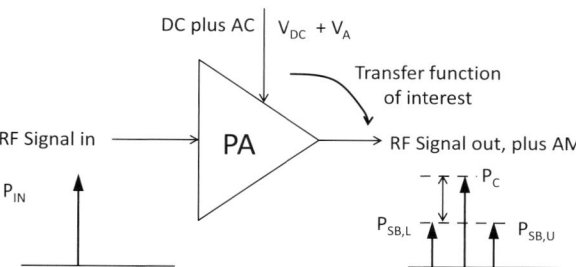

Figure 5-9 The AM process resulting from power supply to RF sideband conversion.

This avoids the need to measure any currents or impedances. Using this method, we only need to measure the power supply voltage variation and the resulting sidebands around the output signal. It is hard to get any simpler.

This process is shown in Figure 5-9. Any pure AM process has symmetrical sidebands. If desired, proving that the measured sidebands are indeed AM can be done by passing the output signal through a limiter, which will remove any AM sidebands but pass through any phase modulation sidebands unchanged. If any spectral asymmetry is measured, this indicates that there are two modulation processes active, both amplitude and phase modulation.

First apply a power supply which has a small sinewave with rms magnitude V_A added to the nominal DC voltage V_{DC}. After the input of an RF signal to the amplifier, if there is any power supply to sideband conversion, two sidebands will appear on both sides of the output signal, separated from the main signal by the frequency of the sinewave of the power supply.

Measure the power (P_C) of the central component RF output signal (from here on this will be called the carrier), and the power (P_{SB}) of each sideband. Knowing that the RF signal is in an $R = 50$ ohm environment, we then determine the voltages of these carrier and sideband signals using

$$A_C = \sqrt{2R} \cdot 10^{\frac{(P_C - 30)}{20}} \tag{5.2}$$

and

$$A_M = \sqrt{8R \cdot 10^{\frac{(P_{SB} - 30)}{10}}} \quad \text{(proof is below).} \tag{5.3}$$

The modulation index of this AM process is calculated using

$$m_{AM} = \frac{A_M}{A_C}. \tag{5.4}$$

A chart of AM sideband levels referenced to the carrier power as a function of this modulation index is provided in Figure 5-10.

The desired transfer function, the *DPS to AM transfer gain*, is

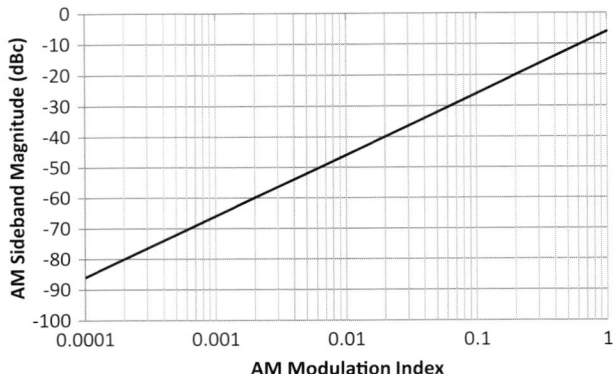

Figure 5-10   Chart of AM sideband levels (dBc) related to the AM modulation index m_{AM} (5.4); actual AM sideband power (P_{SB}) requires knowing the carrier power P_C.

$$g_{DPS\_AM} = \frac{A_M}{V_A}. \tag{5.5}$$

Only two simple measurements are required:

1. AC rms voltage (V_A) on the power supply voltage (independent of the specific power supply voltage) using an oscilloscope, and
2. AM sideband power (either sideband or the arithmetic average power of both sidebands if they are asymmetrical) converted to an equivalent rms voltage (A_M) using (5.3).

Note: A_M is independent of the signal power. There is no real need to ever do the calculation of (5.2) when evaluating g_{DPS_AM}.

There is no need for any additional information, including any information on the PA and its design. This measurement process is completely independent of characteristics specific to the PA, including the impedance of its supply input. There is also no need to know the output RF carrier power, meaning that this transfer function is independent of the output signal. The power supply variation directly translates into amplitude modulation sideband power through (5.5). The modulation index (5.4) does vary with output power, but this is of secondary (if ever) concern.

In general, we expect

$$0 \le g_{DPS\_AM} \le 1 \tag{5.6}$$

because at the PA transistor $0 \le A_M \le V_A$. An output matching network (OMN) that transforms a low impedance R_L at the transistor up to an output impedance R_0 of 50 ohms (or whatever impedance it may be) will result in A_M measuring larger than V_A at the output port of the PA. This is not a problem, because we simply define the measurements to be made at the observable points and go with the results. We then need to change the expected range of the DPS to AM transfer gain to be within

$$0 \le g_{DPS\_AM} \le \sqrt{\frac{R_0}{R_L}}. \tag{5.7}$$

When $A_M \ll V_A$, the PA is suppressing the signal on the power supply. We expect and demand this for envelope tracking. Indeed, if this is not true, then the circuit is not doing envelope tracking. It is doing something else: multiplying the RF signal by the power supply signal V_A. This is a modulator process, not an amplifier process.

For example, we can evaluate the results presented in [5-4] to determine the value of g_{DPS_AM} in that design. While that paper does not speak to their intent behind the word "optimum," the figures show sufficient details. Their Figure 3 shows that the PA supply voltage is 2 V or more above the output envelope for all powers. The photo of Figure 4 is hard to see, though it appears to show that the DC–DC output ripple is just under 1 Vp-p (0.5 V peak), confirmed by the text. The AM sidebands in their Figure 8 correspond to a modulation index of $m_{AM} = 0.002$ shown in Figure 5-10 above. The signal power is close to 0 dBm. The envelope value for this output power is not directly readable from their Figure 3, though this figure shows that the envelope voltage peak for +20 dBm is close to 3 V. This means that the 0 dBm envelope voltage peak is around 3/10. Switching frequency ripple on the output signal envelope is now calculable, and is 0.002*0.3 = 0.0006 V. PSRR of the RF PA is about 20*log(0.0006/0.5) = –58 dB, but since the input numbers are not very well known, all we can conclude is that the PSRR is in the range of 55-60 dB at this output power. This value is rather high, proving that the RF power transistor is operating in L-mode, and that this design is properly referred to as envelope tracking (ET).

Proof of (5.3)
Proving the correctness of (5.3) uses the terms defined in Figure 5-11.

The difference between the measured carrier and the AM sidebands depends on the AM modulation index

$$M_{SB} = 10 \log_{10} (m_{AM}/2)^2 = 10 \log_{10} \left(\frac{A_M}{2A_C}\right)^2 \text{ dBc.} \tag{5.8}$$

The carrier power P_C is related to the signal voltage A_C and impedance R by

$$P_C = 10 \log_{10} \left(\frac{A_C^2}{2R}\right) \text{ dBW,} \tag{5.9}$$

and the sideband power P_{SB} is related to the carrier power by the power difference

$$P_{SB} = P_C + M_{SB}. \tag{5.10}$$

Substituting (5.9) and (5.8) in (5.10) gives

$$P_{SB} = 10 \log_{10} \left(\frac{A_C^2}{2R}\right) + 10 \log_{10} \left(\frac{A_M}{2A_C}\right)^2. \tag{5.11}$$

Expanding and collecting terms yields

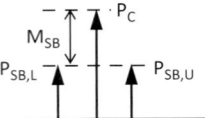

Figure 5-11   Term definitions for the proof of (5.3).

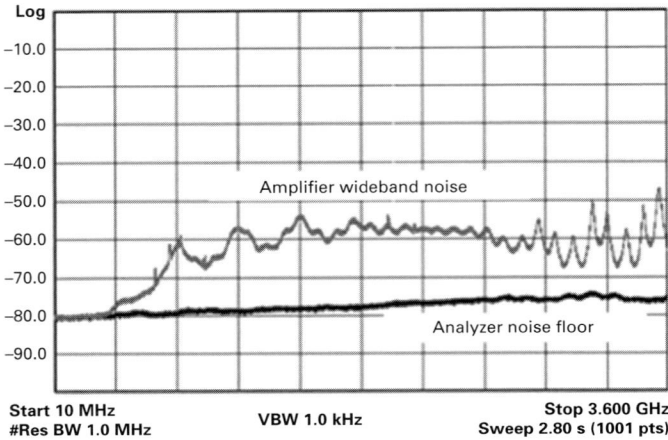

Figure 5-12   Wideband noise output from a linear power amplifier, without an input or output signal.

$$= [10 \log_{10}(A_C{}^2) - (10 \log_{10}(2) + 10 \log_{10}(R))]$$
$$+ [10 \log_{10}(A_M{}^2) - (10 \log_{10}(2^2) + 10 \log_{10}(A_C{}^2))] \tag{5.12}$$

$$= 10 \log_{10}(A_C{}^2) - 10 \log_{10}(2) - 10 \log_{10}(R) + 10 \log_{10}(A_M{}^2)$$
$$- 20 \log_{10}(2) - 10 \log_{10}(A_C{}^2) \tag{5.13}$$

$$= 10 \log_{10}\left(\frac{A_M{}^2}{8R}\right), \tag{5.14}$$

where all terms concerning the carrier signal have canceled out. From (5.14), we note that (5.3) immediately follows.

5.3   Broadband output noise

Noise figure is a concept that applies to *any* linear amplifier, not just receiver input amplifiers. Power amplifiers that operate linearly also have noise figures. The effect of this power amplifier noise figure is the presence of wideband noise on the output, whether or not there is any input signal, such as the wideband noise measurement seen in Figure 5-12.

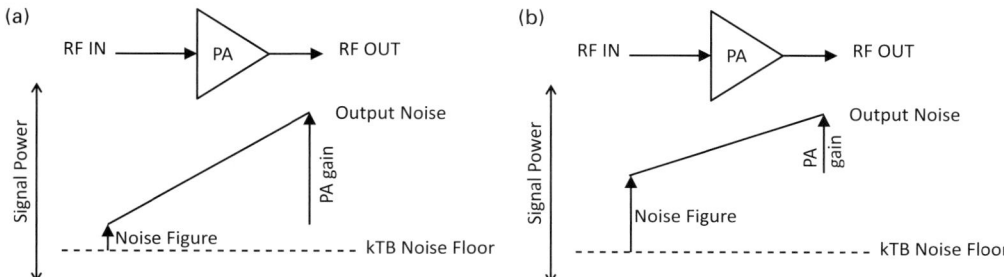

Figure 5-13   Level diagram showing equivalence of (a) low NF, high gain; (b) high NF, low gain.

5.3.1   Noise figure x gain

Noise figure (NF) is effectively another input signal, present within the RF transistor itself. This noise is amplified, like any other input signal, to provide an output signal. The bandwidth of this output signal is unaffected by any tuning on the amplifier input, because this noise input is in the transistor, which follows the entire input circuit. Any tuning on the amplifier output does shape the bandwidth characteristics of the output noise.

It is common to analyze this behavior using a tool called the level diagram. There are two examples shown in Figure 5-13. The case shown in Figure 5-13(a) is of a typical power amplifier with a noise figure of 8 dB and having 30 dB gain. Noise is always measured as a power density (power with respect to bandwidth (dBm/band-width)) with units like dBm/Hz, dBm/MHz, etc. In Figure 5-12, the noise is shown in units of dBm/MHz.

5.3.2   Present whenever gain is active

Because NF is another amplifier input signal, it is always active: whenever there is gain, this noise source causes a noise output through the PA gain process. It is possible to determine the noise figure of the linear power amplifier by measuring properties of this output noise "floor."

The converse is also true: if the output noise changes, then the circuit gain must have changed. Under small signal conditions, any gain changes are unusual in a well-designed amplifier. At the upper end of the amplifier output range, compression begins which does reduce the overall gain. Figure 5-14 shows two output noise plots, one with a small enough input signal that the amplifier remains linear, and one with the signal large enough that the output stage is driven into 5 dB of compression (P5dB). The wideband noise is suppressed by 5 dB in the compressed case. At lower frequencies, intermodula-tion among the noise components acts to increase the low frequency noise floor below the normal amplifier passband.

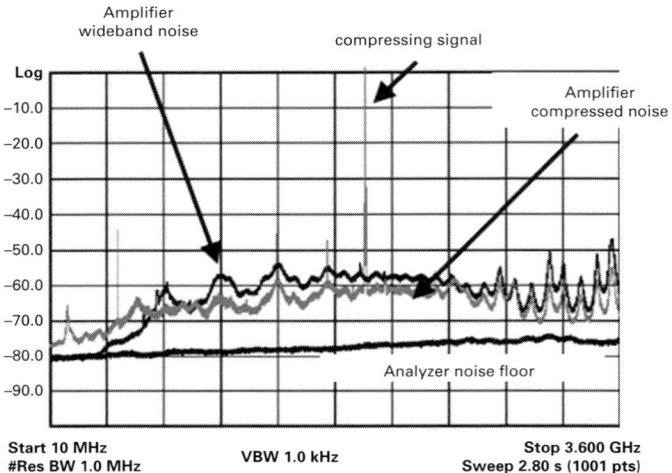

Figure 5-14   Wideband noise measurements from an amplifier with linear operation, and with the output stage compressed by 5 dB.

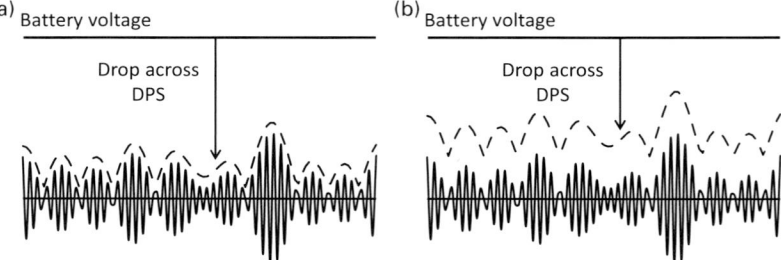

Figure 5-15   ET does not require accuracy from the variable power supply in order to produce an accurate output signal: (a) close spacing; (b) wider spacing.

5.4   Supply accuracy tolerance

Any text that claims that an ET transmitter can achieve gain normalization by adjusting the PA power supply voltage V_{PA} fundamentally does *not* understand what ET is. Any ability in the transmitter to adjust output power by varying the power supply is fundamentally a polar phenomenon (2.9). This is true no matter what the input signal is because of the fundamental properties of the required C-mode operation.

The tolerance of ET to variable power supply values provides much of its relative ease of use. Besides the obvious value of not requiring precision at the DPS output, we have already seen that this feature also provides suppression of any noise that may be present at the DPS output. This tolerance is illustrated in Figure 5-15. The signal output envelope

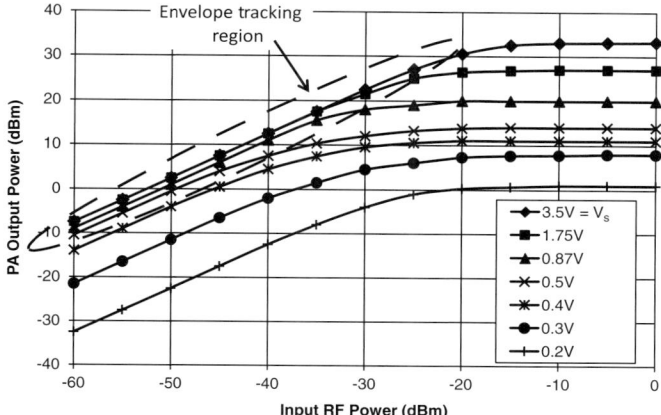

Figure 5-16   ET operation region on the Booth chart is within the dashed ellipse.

(V_{env}) is unchanged, and the supply offset V_{OS} is the difference between the DPS output (V_{PA}) and the output envelope. This V_{OS} leads to additional power dissipation in the RF power transistor.

Similarly, there is no need to hold V_{OS} to a constant value. All that is required is that V_{OS} is sufficiently large to ensure that the RF transistor operates as a CCS, and the PA therefore operates in L-mode, shown in Figure 5-16. This allows the tolerance of significant distortion in the V_{PA} waveform compared to the actual envelope. This makes the control for the DPS easier than strictly following the signal envelope, allowing for filtering of the "envelope waveform" to reduce its bandwidth. Strict timing alignment of the DPS output to the envelope is also not necessary.

5.4.1   Finite transistor output conductance

With finite output conductance in the RF power transistor, the sensitivity to power supply variations never goes to zero. The cause for this is the characteristic curves' slope in the CCS region (device voltage above the knee voltage for the particular input drive level used), like that seen in Figure 5-6. This means that the supply accuracy does become more critical, even with ET operation.

In essence, the ratiometric gain in an amplifier using such a transistor becomes a variable, having dependence on the power supply. This becomes a situation similar to P-mode, where in (4.17) we see dependence on both the input signal and the device voltage. Of course, dependence on power supply variation is much weaker in the present case – but the mechanism is still present.

The message here is:

- for good ET performance, the RF power transistor *must* have low output conductance (high output resistance).

Applying the ET technique to a transistor having finite output conductance *adds* a distortion mechanism. Maybe this is the reason most literature on ET also includes some amount of digital pre-distortion to obtain an output signal of good quality. What is certain is that if the RF transistor does have finite output conductance and the transmitter does not exhibit this additional distortion when the power supply varies, then the mode of operation is *not* envelope tracking.

5.4.2   Supply voltage profile vs. envelope value

To obtain the maximum available energy efficiency in the RF transistor while keeping supply noise suppression, it is necessary to keep V_{OS} to a small value. This leads to the question: what is the minimum V_{OS} value that meets the envelope tracking criteria? This can be readily answered for FETs. To establish a mathematical basis for this profile, we ignore any bias shifts from varying input signal power.

The boundary between the CCS and triode regions of an FET is described by the parabola

$$V_{DS,boundary} = \sqrt{\frac{I_D}{K}}, \tag{5.15}$$

where K is taken from the expression in (4.23). We can call the voltage from (5.15) the knee voltage of the device, which by definition is the minimum device voltage at the controlled port that supports CCS operation. For CCS operation, required for the ET operation, the operating V_{DS} must be greater than $V_{DS,boundary}$. We add an overhead voltage V_{OH} to account for DPS output noise, timing misalignment with the envelope, and all other distortions that could occur. Then we have the minimum V_{OS} that can be accepted for this ET design

$$V_{OS}(I_D) = \sqrt{\frac{I_D}{K}} + V_{OH}. \tag{5.16}$$

Note that (5.16) is *not a constant*. More importantly, it *increases* with output power as represented by an increasing transistor current. If V_{OS} is set to be a constant for an ET implementation, one of two things must occur: either

- the RF device power dissipation is higher than necessary at all output envelope values below PEP, or
- operation of the transmitter is not ET at signal peaks, because the RF transistor leaves CCS operation at large signal values. This is a distortion mechanism where it hurts the most – at peak signal power.

With (5.16) and knowing the PA load resistance R_L, the corresponding V_{DS} values for particular signal output envelope values can be directly established. This calculation

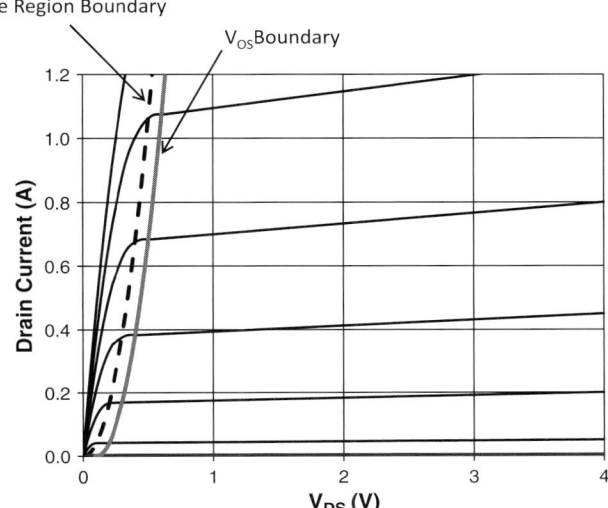

Figure 5-17   Establishing the minimum value for V_{OS} when using an FET for the RF power transistor.

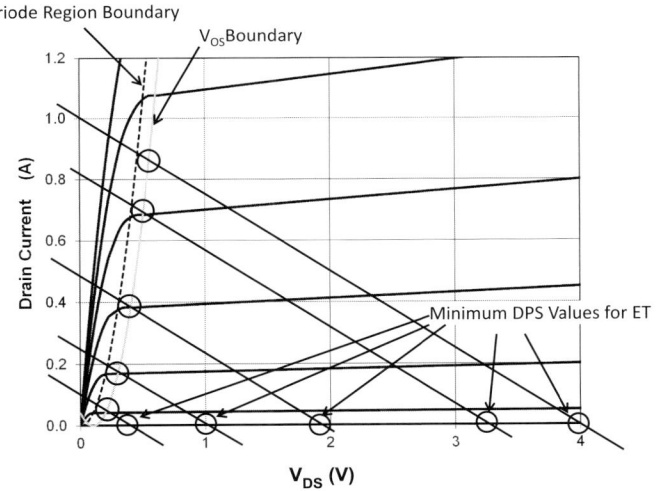

Figure 5-18   Using the minimum value for V_{OS} and the PA load line to determine the optimum DPS output, depending on the output current.

begins with the process shown in Figure 5-18 by identifying the intersection points of the device characteristic curves with the V_{OS} boundary for all values of the transistor control parameter. From these intersection points, the designed load line is projected out and down to the voltage axis. The intersection of these load lines with the voltage axis

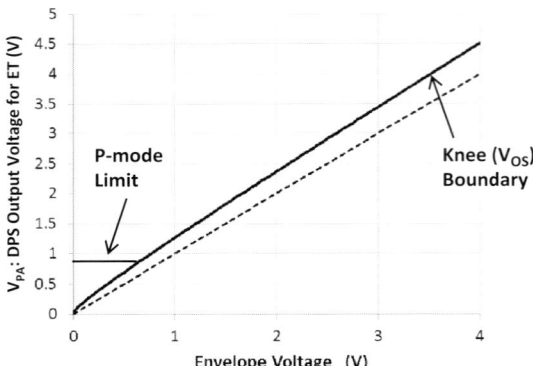

Figure 5-19   The optimum ET profile for the DPS output (solid line) for an FET PA, compared to the desired envelope voltage (dashed line). A typical P-mode avoidance lower limit is also shown, to keep the ET PA away from this strongly distorting region.

correspond to the minimum DPS value V_{DPS} necessary to support L-mode operation at that output signal envelope value, keeping to the ET definition.

The necessary DPS output voltage to guarantee the ET operation when this FET is used for the RF power transistor is, using (5.16) and the definitions in Figure 5-15

$$V_{PA}(V_{env}) = V_{env} + \sqrt{\frac{V_{env}}{R_L K}} + V_{OH}. \tag{5.17}$$

A plot of (5.17) is provided in Figure 5-19.

When a linear amplifier is needed, it is often a good choice to use a bipolar transistor technology, such as a heterojunction bipolar transistor (HBT), for the RF power transistor. Though both FET and bipolar transistors have the required CCS operation needed for ET, bipolar transistors have very different properties in their characteristic curves regarding the boundary for CCS operation. One example, simulated for a GaAs HBT, is shown in Figure 5-20.

One unique characteristic of any bipolar transistor where the forward current gain is greater than the reverse current gain (true for all useful bipolar transistors) is the appearance of an offset near the origin of the characteristic curves. This offset, called $V_{CE,SAT}$, is usually between 0.1 and 0.3 V. For the transistor modeled in Figure 5-20, $V_{CE,SAT}$ is 0.12 V. Once conduction begins, the saturation ON resistance is often very low. Additional voltage across the device (V_X) brings the transistor into CCS operation.

Bipolar transistors unfortunately do not have a simple model describing the onset of CCS operation, such as (5.15) for FETs. Following the model of (5.17), a related expression for the necessary DPS output voltage, this time for a bipolar-based PA, is

$$V_{PA}(V_{env}) = V_{env} + V_{CE,SAT} + V_X + \frac{V_{env}}{R_L} R_{ON} + V_{OH}. \tag{5.18}$$

A plot of (5.18) is provided in Figure 5-21.

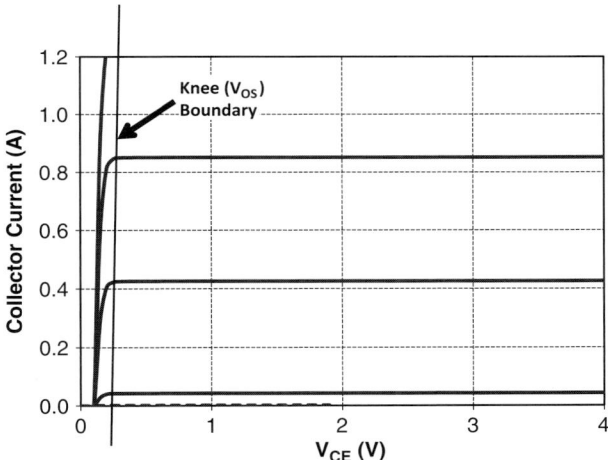

Figure 5-20   Graphically establishing the minimum value for V_{OS} when using a bipolar (BJT, HBT) for the RF power transistor. Each curve represents a uniform step in base current.

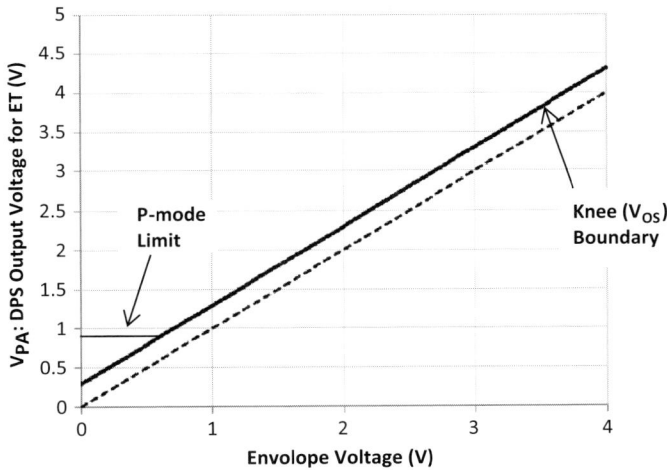

Figure 5-21   The optimum ET profile for the DPS output (solid line) for a HBT (bipolar) PA, compared to the desired envelope voltage (dashed line). A slight increase in V_{OS} is needed as the output envelope increases.

5.4.3   Minimum power supply value

As the applied power supply to the PA is reduced toward zero, Section 4.1.6 shows that the PA enters into P-mode operation. This is a mode of operation that must be avoided in the ET operation because P-mode is a modulator operation, not an amplifier. The question of how low the DPS can go while staying in ET operation must be answered.

Starting with an FET-based RF PA, a close look at the characteristic curves for very low applied voltages is shown in Figure 5-22. The allowable minimum supply voltage

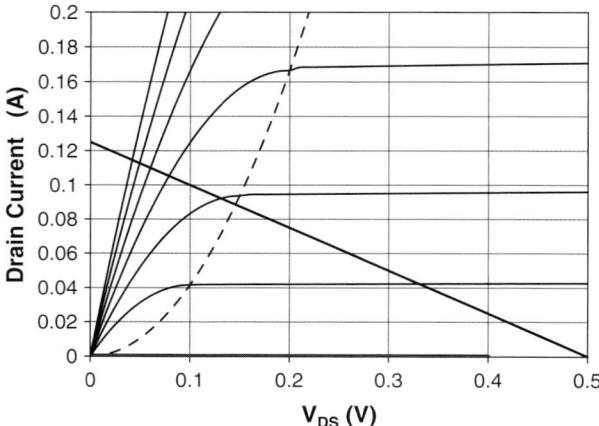

Figure 5-22   Low applied voltage operation of an FET-based power amplifier. The achievable minimum supply voltage depends on the bias (quiescent current) used.

guaranteeing linear (CCS) operation clearly depends on the bias setting, when the power supply is at its minimum value. For the example shown, a bias setting the quiescent current at 40 mA is acceptable for the DPS to go down to 0.5 V. If instead the bias set the quiescent current closer to 100 mA, then the PA would be nonlinear at 0.5 V, and a higher minimum DPS value must be adopted. The primary conclusion is

- the minimum power supply value that can be used for the ET operation is dependent on the quiescent bias of the transistor.

One direct consequence of this is that the acceptable DPS output range for an ET design depends on the bias design of the underlying power amplifier.

A similar situation exists when a bipolar transistor is used for the RF PA. For the example shown in Figure 5-23, a quiescent current of 40 mA would likely not work well, because at signal peaks the PA is transitioning into P-mode operation with its intolerable distortion. If the quiescent bias was closer to 100 mA, operating the DPS at 0.5 V would be completely intolerable, since even quiescently the PA is in P-mode operation. The minimum acceptable DPS value must be much higher than 0.5 V in this particular situation.

More details on the interaction between DPS values and amplifier bias characteristics can be found in Section 5.8.

5.5   DPS time alignment

One of the advantages of the ET operation is that it is tolerant of timing misalignment between the DPS output waveform and the actual signal envelope. This is a direct consequence of operating the PA consistently in L-mode. It is possible to trade off increased tolerance of timing misalignment with increasing V_{OS}, as illustrated in

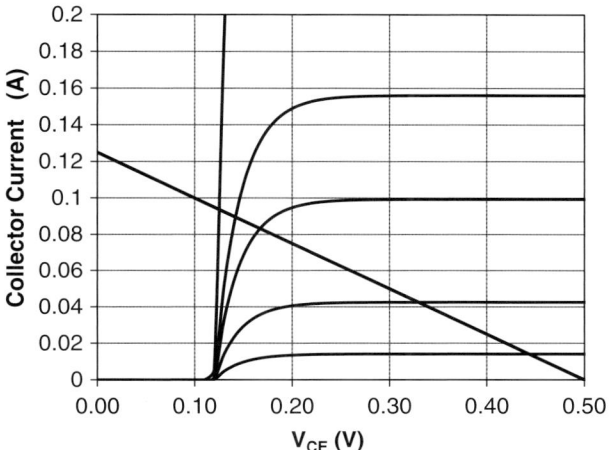

Figure 5-23   Low applied voltage operation of a bipolar HBT-based power amplifier. The achievable minimum supply voltage depends on the bias (quiescent current) used.

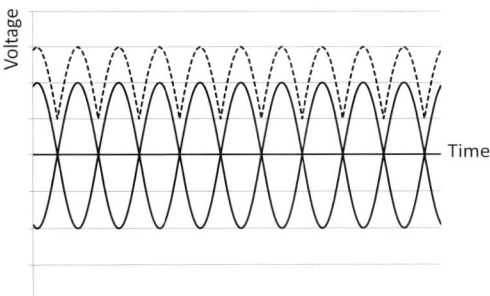

Figure 5-24   Increasing V_{OS}, the difference between the DPS output and the signal envelope, reduces the constraint on time alignment between these two signals.

Figure 5-24. Here a DSB-SC AM signal with single-tone modulation is shown with its envelope (solid lines), and with one possible DPS output waveform (dashed line) with V_{OS} greater than the minimum value (5.16). It is evident in this diagram that the DPS waveform can shift in time to both the left or right while maintaining the ET requirement that the DPS output is always greater than the signal envelope by a value exceeding the minimum acceptable V_{OS} from (5.16).

For the case in Figure 5-24, the amount of time misalignment tolerance increases with an increase of V_{OS} depending on the slope (slew rate) of the envelope. For this signal, the time alignment tolerance depends on the value of V_{OS} as shown in Figure 5-25. Once the value of V_{OS} exceeds the peak envelope value plus the minimum overhead V_{OH}, then the timing can be anything.

Another attraction of the ET technique is that the DPS waveform does not need to match the actual envelope waveform. The particular advantage to this is that any envelope that

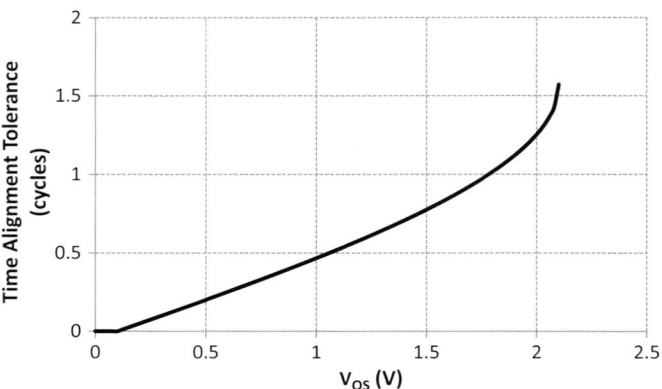

Figure 5-25   DPS waveform time alignment requirements for the DSB-SC conditions of Figure 5-24.

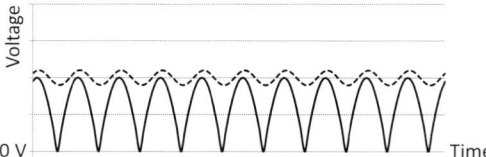

Figure 5-26   Reducing the DPS output signal to minimum bandwidth (a pure sinewave) incurs a variable V_{OS} across the RF power transistor, while making the DPS design much simpler.

gets to zero magnitude has discontinuous derivatives at that zero-magnitude point. This is an infinite curvature (radius = 0) which incurs significant bandwidth expansion as required by the Fourier transform. Instead, the DPS waveform can be reduced to a simple sinusoid, which is the minimum bandwidth signal possible. One example of this is shown in Figure 5-26. Here the DPS bandwidth requirement is minimized, at a cost of a variable value of V_{OS} and its corresponding increase, however slight, in power dissipation in the RF power transistor. Time alignment does not need to be precise, as seen from the time shift already present in the DPS waveform in Figure 5-26.

5.6   Envelope waveform characteristics

Section 2.7 presents the necessity of bandwidth expansion when evaluating the envelope signal compared to the modulated signal. This bandwidth expansion is made worse when the signal envelope goes to zero at any time, and particularly when it goes to zero often. Whether a signal will have this property or not can be implied from reviewing its PDF curve.

Figure 5-27 shows five PDF curves for signals commonly used in the mobile telephone network or wireless data systems. There are two particular characteristics to look for in any signal PDF curve:

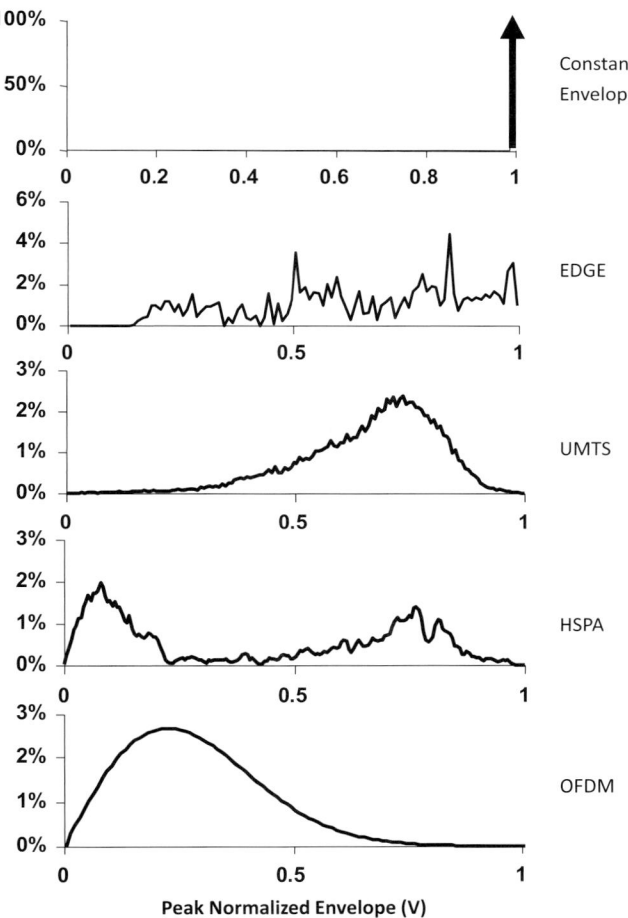

Figure 5-27   Peak normalized envelope PDF curves for five common signals: any constant envelope modulation (e.g. GSM/GPRS, FM/FSK, Bluetooth™ v.1, etc.), enhanced data for GSM evolution (EDGE) ($3\pi/8$-8PSK), universal mobile telephone service (UMTS), high-speed packet access (HSPA), and orthogonal frequency division modulation (OFDM).

- whether the probability of an envelope value at zero magnitude is greater than zero, and if this is true,
- whether the curvature of the PDF curve at zero magnitude is positive or negative.

The constant envelope (CE) and EDGE signals both have finite envelope floors, so the probability of a zero-magnitude envelope value is zero. These signals are relatively easy for PA design. The probability of zero magnitude is nonzero for the remaining signals, UMTS, HSPA, and OFDM. PA design difficulty is influenced by the curvature direction of the PDF characteristic at zero magnitude. If the curvature direction is positive (↻), as it is for the CE, EDGE, and UMTS signals, then the linearity requirements for the PA circuitry are usually not too severe. For signals where the curvature direction is negative (↺) at very low magnitudes, as is seen for the HSPA and OFDM signals, then RF PA

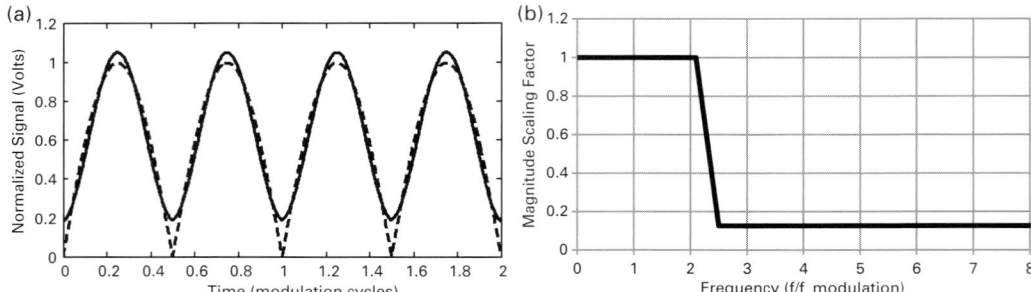

Figure 5-28 The envelope of Figure 5-26 and the effect of reducing higher order signal spectral components using direct frequency-domain manipulation (Fourier transform) techniques: (a) signal envelope (dashed line) and DPS signal with bandwidth reduced by frequency-domain filtering (solid line) according to the frequency-domain filtering function shown in (b).

linearity requirements are more severe because there are a large number of envelope zero events. And the corresponding envelope waveform bandwidth expands to a wider value with so many envelope zero values.

Bandwidth reduction of these envelope waveforms requires making the DPS output waveform continuous in its derivatives, while maintaining the minimum V_{OS} required to keep the PA operating in L-mode. The first things that must be removed are the first derivative discontinuities corresponding to all envelope zero events. This reduces the total dynamic range of the signal.

Curvature (second derivative) of the DPS output waveform is the next significant determinant of DPS waveform bandwidth. One way this bandlimiting can be done, using present digital signal processing techniques, is to suppress the frequency-domain components above the signal bandwidth. One example for the envelope of Figure 5-26 is shown in Figure 5-28. Here the original envelope is shown with a dashed line. After reducing all frequency components above second-order by a factor of 8, the resulting smoother envelope waveform is shown by the solid line. The sharp null is gone, but this waveform is not always greater than the actual signal envelope – a serious problem that can only be remedied by adding a positive offset until there is no crossover. Unlike conventional filtering, no signal delay is inherent in this process. Envelope signal bandwidth is decreased as desired by manipulations made directly in the frequency domain.

Filtering using more conventional methods works too, of course, and this incorporates a delay on the filter output waveform in accordance with the group delay of the filter. Distortion from the nonconstant group delay of analog or IIR (infinite impulse response) digital filters is not a problem for ET, as long as the offset of the filtered waveform remains above the V_{OS} limit. Use of finite impulse response digital filters does not incur such distortions when the coefficients are symmetric, though the delay can get very long when sharp rolloff is used because many coefficients are required.

5.7   Circuit model: CCS

Because any ET design must operate the PA in L-mode, the central circuit model is that of Figure 4-25(a). The transistor regulates the current through the load, any noise on the power supply is suppressed, and the power supply value is not critical.

When the power supply is variable, there are additional complications that must be paid attention to. One of the biggest changes is the inability to keep the conventional large capacitor on the power supply, a topic covered in detail in Sections 5.9 and 7.2. Equally important is recognition that the device parasitic capacitances are themselves usually strong functions of the applied voltages. These variations make the design of matching networks significantly more complicated, because source or load impedances (as appropriate) are not constants.

This also applies to the reverse transfer capacitance, which has additional impact on circuit stability. We do not design oscillators to operate from variable, much less dynamic, power supply values. Yet we do try to design amplifiers to operate this way.

The fundamental conclusion for DPS amplifier design is:

- everything in the model is time-varying, because everything is voltage-varying and the voltage is time-varying.

The mathematics used must be able to handle all of these time value variations. In particular, the use of s-parameters must be very carefully managed, because s-parameters are usually not applied in time-varying forms. Only frequency-varying forms are commonly implemented.

Implied in all of this is that with everything else varying in the circuit models, it is likely that the bias also must vary. This topic is the subject of the next section.

5.8   Bias conditions

Section 5.4.3 points out that the minimum value that a DPS can successfully provide to an RF amplifier depends on the bias applied to the RF power transistor when the DPS gets to low values. Figure 4-24 shows that there are two available design strategies as the power supply value changes: fixed bias with varying class of L-mode amplifier operation, and variable bias that can provide fixed class of operation. Here these are explored in detail.

5.8.1   Easy: variable class with modulation

The easiest design is to set the bias once at design time and then leave it alone. Whatever the eventual power supply value is, the amplifier performs based on that bias. Section 5.4.3 shows that if the DPS is intended to reach a small value in the ET transmitter, then the quiescent bias current must be correspondingly small, even for class A operation. When the DPS output gets to a large value, keeping this same small

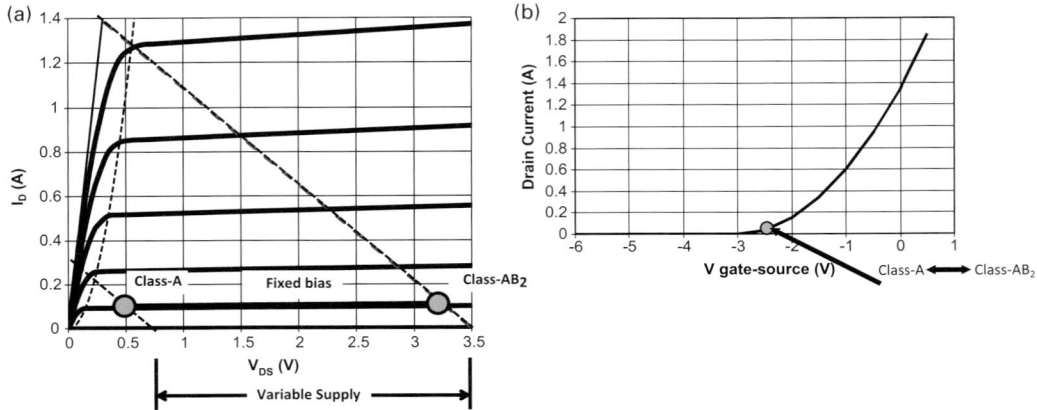

Figure 5-29   Amplifier with fixed bias operates (a) at class A at low power supply values (and low output power) and as class AB$_2$ (deep class AB) at high power supply values (and high output power); (b) the fixed point on this transfer function.

bias current changes the operating class to class AB$_2$, also called deep class AB. This situation is outlined in Figure 5-29.

Operating with class A bias provides good inherent linearity at low output power. Operating in class AB$_2$ at higher output power exposes the output signal to g_m nonlinearity as the transistor goes into and returns from cutoff. This reduces the inherent linearity of the PA – where it hurts the most – at the highest output powers. Inherent linearity at intermediate output power levels varies between these extremes.

It is possible, of course, to operate this amplifier at a fixed bias providing good inherent linearity at maximum output power. Ignoring for the moment that this max-imizes the power dissipation of the RF power transistor (generally a bad thing), for the ET operation this is a serious problem. With a fixed high bias, as the power supply is reduced the signal waveform minima quickly become compressed, operating the ampli-fier in P-mode for that part of the output waveform. Output distortion gets bad very quickly, as described in Section 4.1.6.

5.8.2   Harder: constant class with modulation

Inherent linearity in an amplifier is desirable, because this reduces the need for any linearizer process (and its cost, system complexity, and power consumption) to be included with the amplifier. This requires that the amplifier operating class remains nearly constant at all of the possible DPS voltage values during the ET operation. This is illustrated in Figure 5-30, where linearity is maximized by using class A bias.

In addition, Section 5.4.3 shows that if low values of applied power supply are desired to be used in any ET design, it is necessary to have the bias current reduced to very low levels (for normal power amplifier applications). This is likely to be too low a bias level for higher power operation. This means that variable bias is very likely to be necessary

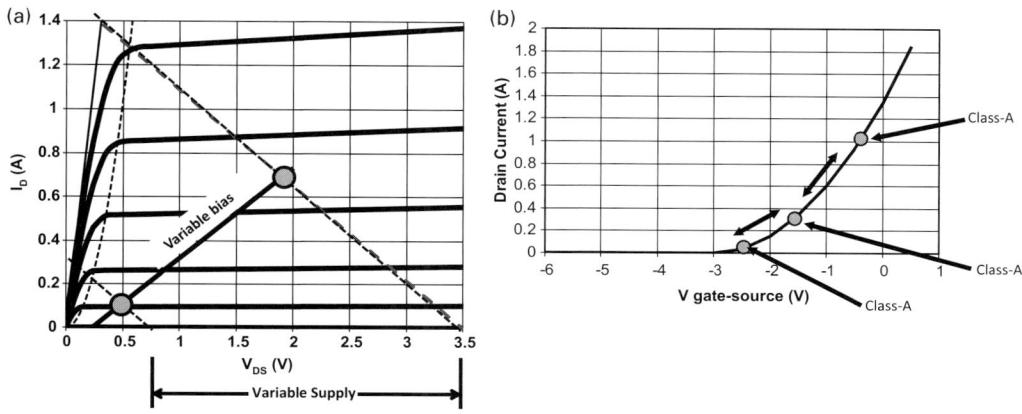

Figure 5-30   A variable bias is required for constant class operation (a) low bias at low power supply values (and low output power) and high bias at high power supply values (and high output power); (b) motion of the bias point along the transfer function to maintain the fixed operating class.

anyway if the ET design intends to improve efficiency at back-off powers by further reduction in the applied power supply.

The major problem with adaptive bias is that it is very dependent on the specific transistor (a manufacturing sensitivity) and on the operating temperature of the RF power transistor. Like most circuit problems, this can be addressed with circuit feedback techniques – often at a cost of efficiency reduction due to the need to sample circuit currents with resistors in the signal path.

5.9   Low frequency stability

Since ET requires the RF power amplifier to always operate linearly, all of the issues required in linear PA design must be addressed. One of the more important design issues is ensuring stability of the high gain transistor at low frequencies. The usual practice to meet this requirement is to connect a large value capacitor at the power supply input to avoid satisfying the Barkhausen gain criterion, as presented in Section 4.3.1.

For envelope tracking, the current that such a capacitor draws when the power supply is a variable input is unacceptable, as discussed in detail in Section 7.2. One approach is to simply reduce the value of this low frequency stability capacitor, which is not a satisfying approach, because the possibility of low frequency instability increases directly as this value decreases.

Recognizing that the important feature for circuit stability is to have a low enough power supply source impedance to keep gain below unity at all frequencies where oscillation may occur, one solution is the DPS itself. If the DPS output impedance is sufficiently low that it provides sufficient stability to the attached PA, then all is fine. Well, sort of. This means that the PA cannot be tested on its own, separate from the DPS. The PA and the DPS become an inseparable pair.

Dynamic operation at envelope frequencies of a high gain RF power amplifier is a difficult problem. This mutual dependence illustrates that this interconnection, the DPS to the PA, is actually now another interface. Being a new interface, it needs to have specifications that allow different design teams at different places, working at different times, to all provide products that will operate as intended when any pair is connected. This situation does not exist today. Key information needed to begin this interface specification activity is included at appropriate places throughout this book.

5.10   Load presented to the DPS by the ET PA

It is important to know what the load is that an envelope tracking PA provides to the DPS. How this impacts on the design of the DPS is discussed in detail in Chapter 9. Here the focus is on the characteristics on how V_{PA} and I_{PA} from Figure 5-1 behave under the ET operation.

Looking at Figure 4-38(b), the load of the envelope tracking PA is dominated by the device's controlled current source (CCS) and its shunt output resistance. A model to evaluate the power supply load characteristics of this linear L-mode amplifier is developed in Figure 5-31.

The DPS ideally sees a constant current load during the ET operation (I_Q) when the output power is low. At high output power, any linear PA draws measurable signal load current (I_{sig}) in addition to its bias (quiescent) current. The load that the PA provides to the DPS can therefore be described by (using voltage definitions from Figure 5-15)

$$R_{PA} = \frac{V_{PA}}{I_{PA}} = \left(\frac{V_{env} + V_{OS}}{I_Q + I_{sig}} \right). \tag{5.19}$$

Knowing that the signal current is related to the envelope voltage and the load resistance by Ohm's Law ($R_L = V_{env}/I_{sig}$), (5.19) is manipulated into the more useful form

$$R_{PA}(V_{env}) = \frac{V_{PA}}{I_{PA}} = R_L \left(\frac{1 + \dfrac{V_{OS}}{V_{env}}}{1 + \dfrac{I_Q}{I_{sig}}} \right) = R_L \left(\frac{1 + \dfrac{V_{OS}}{V_{env}}}{1 + \dfrac{I_Q R_L}{V_{env}}} \right). \tag{5.20}$$

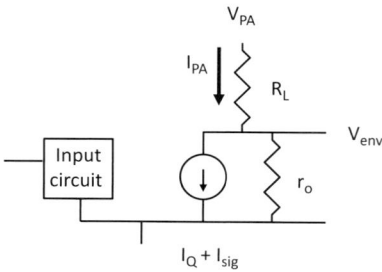

Figure 5-31   First-order model for the power supply load characteristics of a PA used in ET.

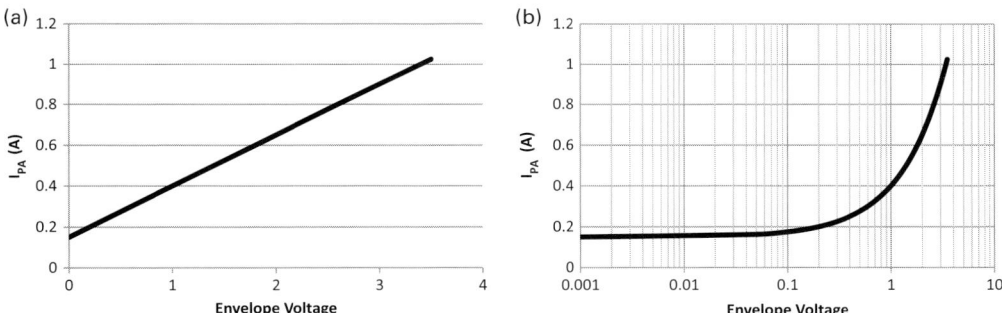

Figure 5-32   Plots of PA supply current (5.21): (a) linear scales; (b) logarithmic with envelope voltage. The logarithmic plot clearly shows both the quiescent and signal currents.

We see from (5.20) that constant resistance behavior occurs as long as the signal current greatly exceeds the quiescent current and the supply voltage offset is a small fraction of the output envelope, which is a common condition for ET at high output power. As the output signal power reduces, it is instructive to look at the PA current component alone, which becomes the expected constant current load with value I_Q

$$I_{PA}(V_{env}) = I_Q + I_{sig} = I_Q + V_{env}\left(\frac{1}{R_L} + \frac{1}{r_o}\right). \tag{5.21}$$

Evaluation of (5.21) provides the curves shown in Figure 5-32, for a load resistance of 4 ohms and the V_{OS} values from (5.16). Two curves are shown, with identical data. Figure 5-32(a) uses linear scaling on both axes, and clearly shows the offset due to I_Q and the slope due to signal current flowing through R_L. The regions where the bias current alone dominates, and how the signal load current then comes to dominate at high output power, are readily understood from the envelope-logarithmic plot in Figure 5-32(b).

In principle, the DPS for an ET transmitter can be designed as either a current source or a voltage source. A current source DPS will work poorly with the current load (5.21) seen under the ET operation. This is true even though the effective load resistance provided to the DPS by the linear CCS power amplifier (5.20) is low, as seen in Figure 5-33.

But hold on – isn't the output impedance of a current source nearly infinite? How can R_{PA} be close to zero at low output power when there is almost no signal current? The answer can be most easily understood from Figure 5-32(a). When the output voltage is very close to zero, the current is still relatively large: the definition of a nearly short circuit. So having all of the curves in Figure 5-33 beginning near the origin makes perfect sense. At higher output signal powers (larger V_{env} values), (5.20) shows that the effective resistance of the ET PA approaches the value of the load resistor R_L. Here, in keeping with all examples so far, the evaluated value of R_L is 4 ohms.

Yet, it is vital that the DPS be a voltage source, not a current source, for successful application in an ET design. The reason is evident from Figure 5-32(b). The load current

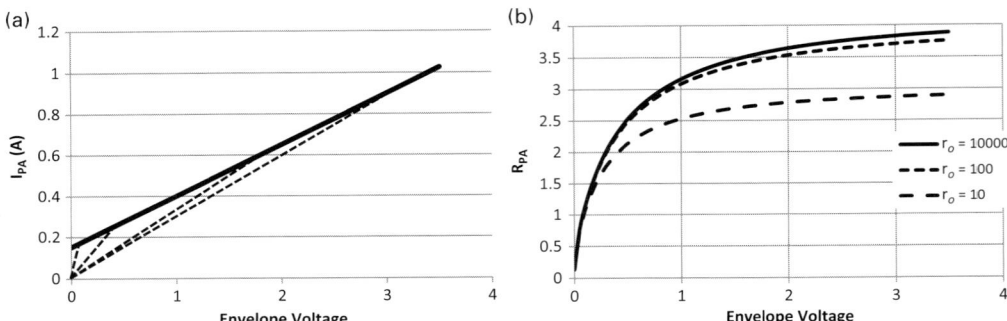

Figure 5-33 Effective resistance of the PA load from (5.20), as experienced by the DPS: (a) copy of Figure 5-32(a) with apparent resistance lines added (dashed lines); (b) effective resistance changes slightly with different output conductance values for the transistor.

begins at a fixed value, the quiescent current, before the signal current in the load becomes significant. Current sources in series is a recipe for disaster. The wide variation in the effective DPS load resistance seen in Figure 5-33 is not easy to manage either. By varying across a ratio of nearly 30:1, the dynamic response of the DPS regulator is difficult to keep stable with a consistent bandwidth (meaning consistent response time). DPS design for ET is not easy.

Not yet considered is the reactive current that the DPS must provide to charge and discharge any bypass capacitors attached to the PA power supply node. At this point, this will remain simply as a mention to be sure that designers will not forget to consider it. For the PA designer working on an amplifier intended for ET applications, the important message is that all capacitance values must be minimized far below conventional bypass values, while maintaining stability across all frequencies. PA design for ET is not easy.

From the simplified circuit in Figure 5-31, we see that the voltage in the load resistor depends on the quiescent (bias) current, the value of the power supply, and the transistor itself. Being only interested in the output RF signal without the bias, we can write the signal envelope as

$$V_{env} = I_{sig}R_L + V_{PA}\left(\frac{R_L}{R_L + r_o}\right). \tag{5.22}$$

Of particular interest is any influence that the varying power supply has on the output signal. This is the second term in (5.22), which can be written most usefully as

$$V_{env}(V_{PA}) = V_{PA}\frac{1}{\left(1 + \dfrac{r_o}{R_L}\right)}. \tag{5.23}$$

The key parameter in (5.23) is the transistor output resistance r_o. If the transistor is ideal, then r_o is infinite and (5.23) evaluates to zero, leaving no influence from the power supply value on the output signal envelope. This is consistent with the discussion in Section 5.2.

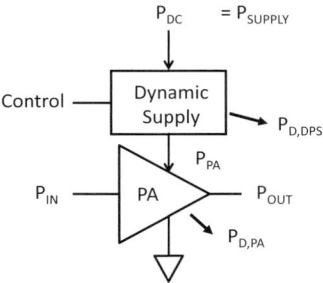

Figure 5-34   Power definitions and dissipations in the architecture of Figure 5-1.

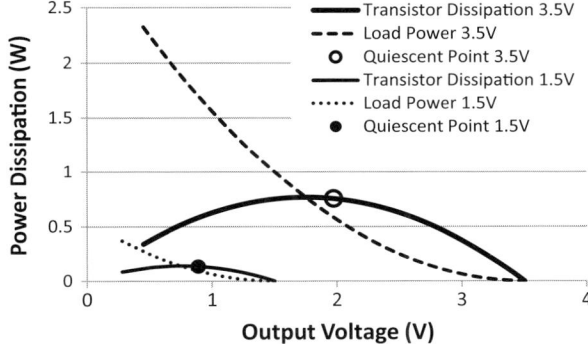

Figure 5-35   Power dissipation behavior in an ET transmitter. When the power supply is reduced at times when the required output power is low, the power dissipation in the RF power transistor drops significantly.

5.11   Energy efficiency effects/power dissipations

The entire reason to even consider using ET can finally come to light, using the tools established in this and prior chapters. We now can look at the effect that using a DPS has on the efficiency of a linear RF power amplifier. The overall situation is shown in Figure 5-34, where the DPS and the PA are electrically connected in series, but the thermal (power dissipation) paths are in parallel. In this section, we only examine the efficiency impact on the PA. Consideration of power dissipation in the DPS is a major topic of Chapter 9.

5.11.1   Referenced at the RF PA

The original linear PA power dissipation problem is seen in Figure 4-32. As the output signal gets smaller, the transistor power dissipation stays close to the quiescent (bias) point. This maximizes the RF transistor power dissipation $P_{D,PA}$ at low output signal conditions, exactly the opposite of what is needed for good energy efficiency.

The ET architecture intentionally has the applied power supply drop as the output signal power falls. The effect of this on device power dissipation is seen in Figure 5-35, where the transistor power dissipation curves (solid lines) and load power (dashed lines)

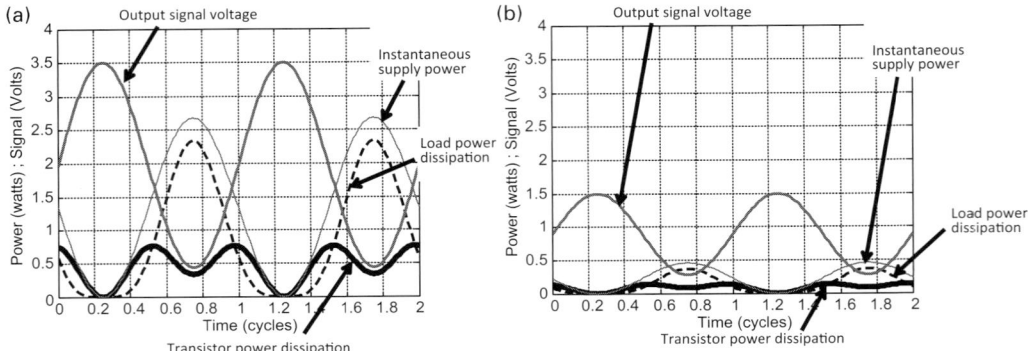

Figure 5-36 RF waveforms for a linear amplifier in the ET operation for the two cases in Figure 5-35: (a) DPS output at 3.5 V; (b) DPS output at 1.5 V.

are shown for two separate values of power supply voltage. Because power is related to the square of voltage for a fixed resistance, both the transistor dissipation and the load power drop significantly at lower power supply values.

The power dissipation curves in Figure 5-35 do not continue to 0 V because of the transistor knee voltage. The impact of this knee voltage is shown with the time waveforms in Figure 5-36.

Calculating the RF PA output energy efficiency

$$\eta_{PA}(V_{env}) = \frac{P_{SIG}}{P_{SUPPLY}} = \frac{P_{SIG}}{P_{SIG} + P_{D,PA}}, \tag{5.24}$$

we get the curves in Figure 5-37. For a conventional linear PA, the power dissipation is floored by quiescent power dissipation, and the energy efficiency falls rapidly with smaller output signals. Adopting a dynamically variable power supply does reduce the transistor power dissipation, keeping the efficiency of the PA much more consistent across output signal levels. In this calculation, perfect dynamic biasing of Figure 5-30 is assumed, keeping the PA operation centered on class A at all values of the DPS.

Trusting any calculation at face value is never a good idea. For this energy efficiency calculation, one way to validate the calculation is to set the knee voltage to zero. For any class A amplifier with its output signal swinging between the power supply and ground, the maximum energy efficiency is well known to be 50%. Doing this validation on the simulator for Figure 5-37 gives the result seen in Figure 5-38. The results do check out, which validates this model.

5.11.2   Referenced at the power supply

To the transmitter user, the PA energy efficiency is only part of the story. Of primary importance is battery life (for a mobile device), which means that the minimum number of electrons must be drawn from the fixed voltage battery. We need to make the energy

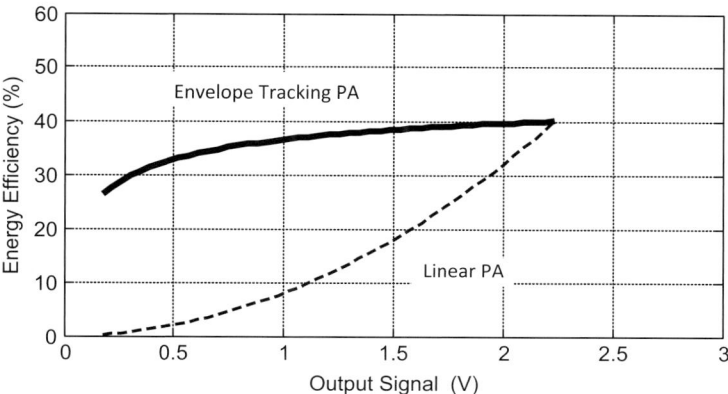

Figure 5-37   PA energy efficiency differences between conventional linear class A fixed-bias operation and ET with perfect dynamic bias (fixed operating class).

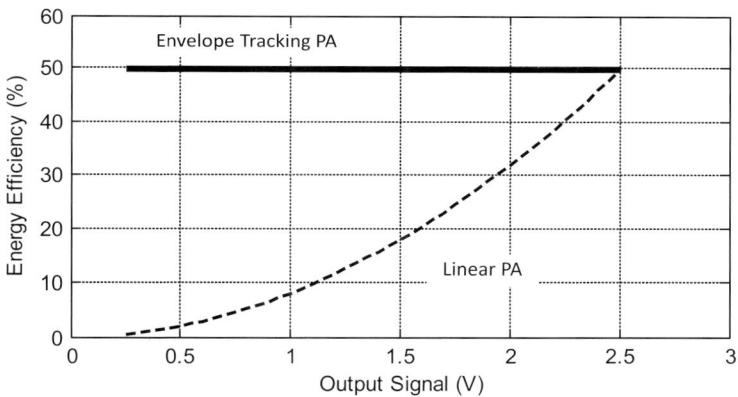

Figure 5-38   Validation of the PA energy efficiency results in Figure 5-37.

efficiency achieved at the RF PA visible to the battery. This puts design constraints on the DPS.

The overall efficiency, referred at the input to the DPS (at $V_{Battery}$ in Figure 5-34), is the product of the PA energy efficiency and the conversion efficiency of the DPS. Thus, the DPS provides a "window" as to how much of the PA energy efficiency is provided to the top-level power supply. Efficiency in the DPS conversion is therefore very important. This is discussed in detail in Chapter 9.

5.11.3   Limit on maximum available efficiency

Assuming that a perfect DPS can be made, the limit to overall transmitter efficiency becomes the efficiency of the power amplifier. For the same output power, (5.24) shows that the only way to improve overall transmitter efficiency is to reduce $P_{D,PA}$, the power dissipation in the RF power transistor. How much the ET technique meets this objective is shown in Figure 5-39.

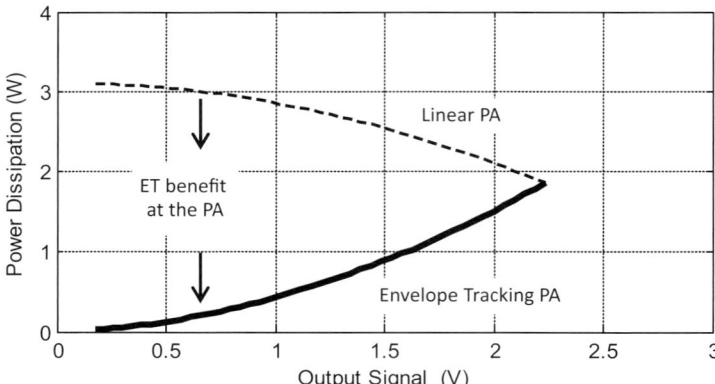

Figure 5-39   PA energy efficiency improvement by reducing power dissipation between conventional linear class A fixed-bias operation and ET with perfect dynamic bias (fixed operating class).

At maximum output power, the power supply is at its maximum and the power dissipation is the same for both transmitters. As the output power is reduced, the power dissipation of the class A amplifier increases, eventually approaching the quiescent power dissipation. Power dissipation falls for the envelope tracking PA. The difference between these curves is a benefit of the ET technique, and provides for the more consistent efficiency seen for ET in Figure 5-37.

5.12   Achieving envelope zero values

Section 5.6 presented the problems with signals that go to zero magnitude in the normal course of their modulation. Such signals are commonly used, so their transmitter must be able to handle them successfully.

For the ET technique, there is no problem with generating an output signal with zero magnitude. This is as simple as letting the input signal magnitude go to zero. The PA in envelope tracking is linear at all times, so zero-in corresponds to zero-out. Because the PA is linear, the wideband output noise presented in Section 5.3 remains at the PA output.

The minimum value of the DPS will be low, but large enough to keep the amplifier running. Depending on the bias conditions described in Section 5.4.3, the DPS output may not get as low as theoretically possible. In any case, the ET amplifier power dissipation will still have a low value, similar to that seen in Figure 5-39.

5.13   TDM burst control

Time division multiplexed (TDM) systems such as GSM and TETRA not only need to control the bandwidth occupancy of the modulated signal, but also the bandwidth characteristics of the effective modulation that occurs when the RF burst comes ON

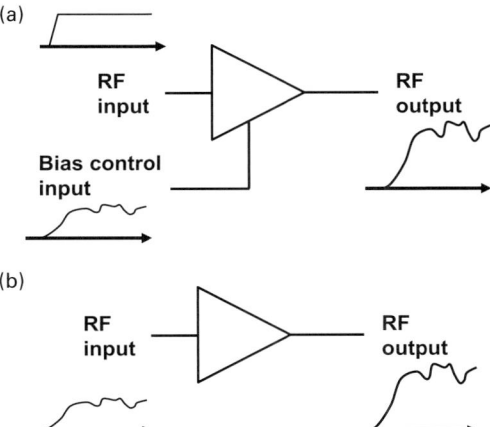

Figure 5-40   Two common methods for burst rise and fall control in a TDM system: (a) PA bias adjustment; (b) input signal shaping.

and then turns OFF. This usually appears in a specification under the heading of transient spectrum.

There are several ways to implement this requirement, two of which are presented in Figure 5-40. One common method in wide use, shown in Figure 5-40(a), is to have the PA initially biased OFF when the transmitter is not operating. Just before the RF output must begin, the modulator applies an RF signal to the PA input. The baseband processor then turns the PA ON in a controlled fashion using the PA bias network. Once the PA is ON, then the input modulation can proceed normally.

A second approach, shown in Figure 5-40(b), is to leave the PA running normally but with no RF input. When the output RF burst is ready, the modulator sends a fully formed signal, including the burst rise and fall profiles, to the PA for it to amplify normally. This approach is simpler (for the PA) than the controlled bias ramping, but it results in more power consumption and an extended presence of wideband noise at the PA output.

5.14   Reverse intermodulation

All antennas are electrically symmetrical. This means that they have exactly the same receive characteristics as transmit characteristics. While this is very useful for antenna characterization (we only need to measure transmit or receive characteristics, whichever is more convenient), this causes a problem in any multiuser operating system. Here the problem is that while a mobile device is transmitting, its antenna is still receiving *all* the electromagnetic signals it encounters. These received signals are sent back to the transmitter, not to the receiver. At any nonlinear element, such as a PA output stage, these *reverse* received signals can interact with the intended signal and cause intermodulation products to be generated.

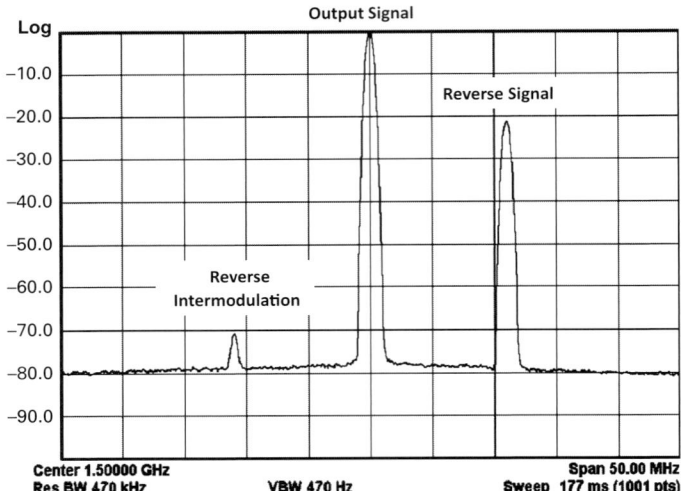

Figure 5-41   Reverse intermodulation example for a linear PA used in an ET system.

One example is shown in Figure 5-41. Here a reverse signal power is received at a frequency above the transmitter output signal. The scaling of the reverse signal is −40 dB from the main transmitter output power, which is the scaling used in the 3GPP specification for this test. Interaction between the transmitter output signal and the reverse signal causes the reverse intermodulation spurious signal, which does appear at the transmitter output and gets radiated.

5.15   Output mismatch

All transmitters operate into some amount of an impedance mismatch at their output. Only engineering test equipment provides a consistently good impedance match to an RF transmitter. The real world is not nearly so nice.

Output mismatch is a serious problem for ET transmitters, because any change in load impedance causes a change in output signal magnitude and power, in direct accordance with Ohm's Law. For an ET design which assumes that the output signal magnitude has a particular value when the input envelope control has the corresponding value for that output magnitude, if there is any output error, the DPS setting can be inappropriate.

This DPS error is particularly troubling if the output impedance goes above its design value, causing the PA output magnitude to increase. When an ET system is designed to have the highest efficiency it possibly can, while remaining envelope tracking, the difference (V_{OS}) between the output magnitude and the DPS is already minimized. Should the output magnitude increase due to output mismatch, the DPS voltage becomes insufficient and the PA output will begin clipping the output signal. Such distortions are rarely acceptable.

There are two solutions to this problem. One is to increase the PA headroom by raising the DPS output voltage enough so that the PA will not clip if there is an output mismatch toward a higher impedance. The consequence of this is a reduction in energy efficiency under all nominal operating conditions. Of course, this also is not desirable – but it is simple. How this is done is discussed in Section 7.10.

A second approach is to sense the output magnitude and feed that back to the DPS circuitry. If the output magnitude is where it should be, then nothing needs to change. But if there is an output mismatch, then the DPS can adapt to either higher output magnitude or lower output. This approach inserts a feedback control system into the transmitter, significantly increasing the complexity of the DPS, and likely reducing the bandwidth of the DPS from the feedback dynamics; very likely also increasing the cost of the DPS. None of this is desirable either. But nothing comes for free.

5.16   Envelope tracking property summary

A quick summary of the properties of envelope tracking is presented in Table 5-1 [5-5]. The key properties are linear PA operation and power supply noise rejection. The desired benefit is a very significant increase in the energy efficiency of the power amplifier. All of the usual techniques used to design linear amplifiers apply, with added complexity when the bias and supply operating points now move around dynamically.

Table 5-1 Property summary of envelope tracking

Characteristic	Envelope tracking (ET)
Output signal envelope control	Input RF signal
Input signal modulation	Precise from specification
DPS accuracy	Low, when above a minimum
Output power = 0?	No problem
Power supply noise	Rejected
Circuit stability	Critical and difficult
Thermal stability	Requires compensation
Energy efficiency	High
RF PA transistor operation	CCS
Device output matching	Complex conjugate
Wideband noise, dominant source(s)	PA noise figure and gain
Input signal magnitude accuracy	Critical
Load current control	Regulated by the RF transistor
Envelope time alignment	Tight, if emphasizing efficiency

5.17   References

[5-1] B. Geller, P. Assai, B. Gupta, and P. Kline, "A Technique for the Maintenance of FET Power Amplifier Efficiency Under Backoff," *Digest of the 1989 IEEE Microwave Theory and Techniques Symposium*, pp. 949–952.

[5-2] C. Buoli, A. Abbiatti, and D. Riccardi, "Microwave Power Amplifier with 'Envelope Controlled' Drain Power Supply," *Proceedings of the 25th European Microwave Conference*, Sept. 1995, pp. 31–35.

[5-3] J. S. Walling, S. S. Taylor, and D. J. Allstot, "A class G Supply Modulator and Class E PA in 130 nm CMOS," *IEEE Journal of Solid-State Circuits*, vol. 44, no. 9, Sept. 2009.

[5-4] G. Hanington, P. Chen, P. M. Asbeck, and L. E. Larson, "High Efficiency Power Amplifier Using Dynamic Power-Supply Voltage for CDMA Applications," *IEEE Transactions on Microwave Theory and Techniques*, vol. 47, no. 8, Aug. 1999, pp. 1471–1476.

[5-5] E. McCune, "Envelope Tracking or Polar – Which Is It?," *IEEE Microwave Magazine*, June 2012.

6   Polar transmitter principles

A polar transmitter by definition operates in accordance with one or more of the principles shown in Figure 2-3. By far the most common implementation is direct control of output signal magnitude from (2.8) and (2.9) and illustrated in Figure 2-3(a). The implementation shown in Figure 2-3(b) is actually a phase modulator, and is usually also implemented ahead of the stage controlling output magnitude.

6.1   History of the technique

Polar modulation, defined as a signal process implementing Figure 2-3(a), has been in use for nearly a century. Early in the implementation of AM broadcasting, it was discovered that the energy efficiency of class A transmitter power amplifiers was unacceptably low. Many alternatives were attempted [6-1], including constant current modulation, variable gain (controlled grid) modulation, and plate modulation.

The technique that won this competition is plate modulation.

6.1.1   Plate modulation (predates 1920)

The controlled port of a vacuum tube (valve) is called the "plate." It was found before 1920 that the output power from an amplifier can be made to vary if the power supply applied to the amplifier varied. This output signal variation was roughly proportional to the power supply value, as long as the amplifier operated in compression [1-1]. The proportionality was not perfect, but for the modulation requirements of AM broadcasting it was good enough. Later it was learned that when class C bias was used, along with very large drive signals to achieve sufficient output compression, the operation of a plate modulated transmitter could achieve efficiency values greater than 90% at AM band frequencies [4-3]. Plate modulation of class C biased power amplifiers dominated AM broadcast transmitter designs for more than 60 years, making it a very successful technology.

Plate modulation, as used for AM broadcasting, is a single coordinate polar signal process. Only signal magnitude is varied according to the information signal. There is no corresponding modulation of the phase coordinate. Indeed, there is no phase modulation at all for conventional AM broadcasting. The development of RF signal modulations that

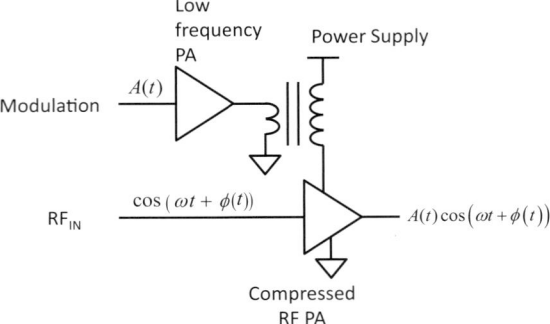

Figure 6-1   Plate modulation varies the voltage on the anode (plate) of the high power amplifier tube by adding the envelope signal to the DC power using a transformer. All of the sideband power for the output signal must be supplied by this transformer, so the envelope power is 50% of the DC power.

use both phase and magnitude coordinates arrived in the early 1950s with the advent of RF use for the bandwidth-efficient single sideband AM (SSB-AM) signal.[1]

6.1.2   EER by Kahn in 1952–1957

Modulations that are solely amplitude modulation (AM) or solely phase modulation (PM) must have two sidebands, one above the carrier frequency and the other below it with mirror symmetry. The information contained in each sideband is duplicated, so these modulations are not as bandwidth efficient as possible. When one of the sidebands is removed (to improve bandwidth efficiency), the remaining single sideband (SSB) signal of necessity has both AM and PM components. Amplifiers that successfully bring SSB signals to full transmitter power, particularly SSB signals with suppressed carrier (SSB-SC), have more constraints than needed for the original AM signal. The most important tighter constraint is the need for accurate linearity.

Linear amplifiers have the highest power dissipation, and therefore the lowest energy efficiency, as described in Section 4.1.6 regarding L-mode operation. Getting around this problem for high power SSB-SC transmitters led Kahn [6-2] to propose the envelope elimination and restoration (EER) technique. Envelope elimination and restoration is an *amplifier* technique that meets the requirements of Figure 2-1, where the internal circuitry operates nonlinearly but the input and output signals are proportional to each other as required from any linear amplifier. The general structure of the EER technique is shown in Figure 6-2.

The EER amplifier internal circuitry first separates the polar modulation coordinates of the input signal. Once the input envelope is demodulated from the input signal, the phase path removes the envelope by passing the input signal through a limiter. This now constant envelope (and much wider bandwidth) signal is applied to the input of a fully

---

[1] While SSB originally was envisioned in 1915 [Carson] and was used at low carrier frequencies in the 1920s on telephone cables, it did not come to use in radio communications until the early 1950s.

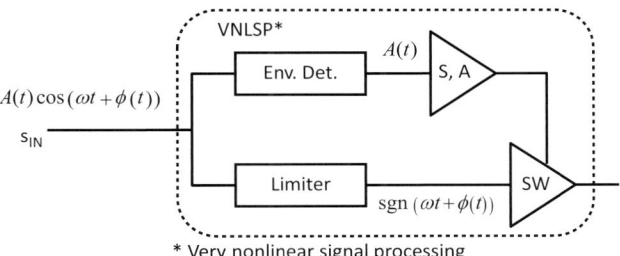

Figure 6-2 The Kahn EER amplifier technique is a true amplifier, while it is comprised of many extremely nonlinear signal processing steps.

compressed power amplifier, so high RF power can be generated with very good efficiency. The envelope signal goes through a separate path and is amplified to the level necessary to provide all the envelope sidebands and power for the output signal. These two paths combine at the power amplifier using plate modulation, which reconstructs the output signal as a scaled replica of the input signal. The overall process is much more energy efficient than any linear CCS circuitry can achieve. It is a linear amplifier operation that internally contains a polar power stage.

This is brilliant in theory. EER never was implemented for its originally intended purpose, which means that there are sufficient problems in practical circuitry such that other techniques are more cost effective and have adequate performance. A short list of the major problems of EER implementation includes: finite gain in the limiter, limited dynamic range in the envelope demodulator, delay matching of the phase and magnitude paths, and getting the compressed power amplifier to accurately resolve zero-magnitude output signals when they occur. All of these issues still hold for any polar transmitter, and they are discussed within this chapter in the appropriate sections.

6.1.3   Resurgence in 1990s

In the 1980s, digital modulation techniques supplanted plate modulation for AM broadcast transmitters [6-3]. Digital modulation techniques also began to supplant analog techniques for communications in public services (police, fire, etc.) and for personal communications (e.g. cellular radiotelephone). All of these signals found that conventional linear amplifiers performed well enough for their applications such that there was no need to adopt unconventional transmitter techniques.

As the new century came closer, foresighted people began to realize that the energy efficiency problem addressed in earlier decades by plate modulation was going to reappear and be important to solve [6-4] [6-5] [6-6] [6-7]. The problem was not just the need for improved linearity of digital modulations adopted for increased bandwidth efficiency and/or data rates, but more from the increased PAPR and the inherent energy efficiency penalty of such signals in linear amplifiers. While the energy efficiency resulting from using digital signals degraded as the PAPR increased from 0 dB to 3.5 dB, it was apparent that when the PAPR exceeded 5 dB the energy efficiency penalty of conventional linear amplifiers became economically less tolerable. Doherty amplifier architectures [6-8] [4-1] extended the usefulness of linear-type amplifier designs to PAPR values up to 6 to 8 dB. Above that, something else is needed.

This "something else" is polar technology using dynamic power supplies, which in the literature is often erroneously called envelope tracking (ET). A careful read of most papers published at the time of this writing show that the circuit operation is in line with the conditions described in the remainder of this chapter, and not in accordance with the conditions detailed in Chapter 5. There is also interest in the general heading of class S, but such designs to date do not meet the operating requirement of Figure 2-1. Specifically, the problem is wideband noise and signal processing aliases that require a bandpass filter to remove them before the main class S output signal can be used. Until this output bandpass filter design problem is successfully solved, class S is not a viable alternative.

6.2   Magnitude control mechanism

As in Chapters 4 and 5, here we also begin with device characteristic curves and amplifier load lines. Figure 6-3 shows the same FET characteristic curves from earlier chapters, this time with contours added for constant power dissipation in the FET. One contour is present for every 100 mW from 100 mW to 1000 mW. The load line is drawn for 4 ohms and a power supply value of 3.12 V. An open circle is shown for the maximum CCS current that can be drawn with this load and power supply, which is also intersected by the 300 mW power dissipation contour.

Power dissipation in the transistor is shown by the solid line in Figure 6-4, which corresponds to tracing the intersection of the load line with the power dissipation contours in Figure 6-3. Power dissipation in the transistor peaks near the middle of the

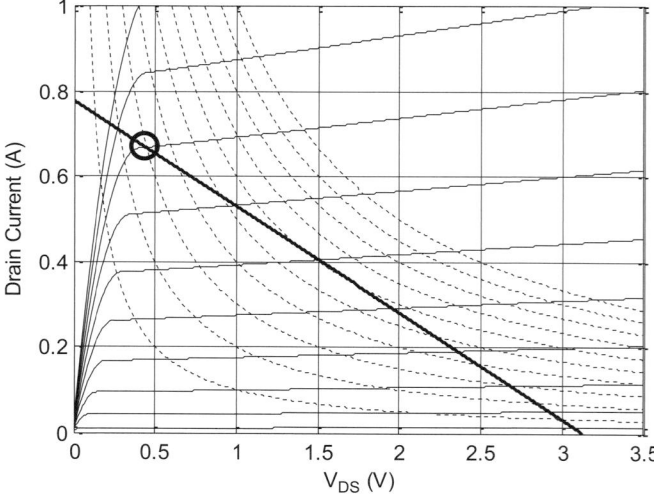

Figure 6-3   Example FET characteristic curves with load line and constant power dissipation contours (dashed lines) for (0.1, 0.2, ..., 1.0) watts. Maximum CCS (linear) current for a power supply of 3.1 V is shown in the open circle.

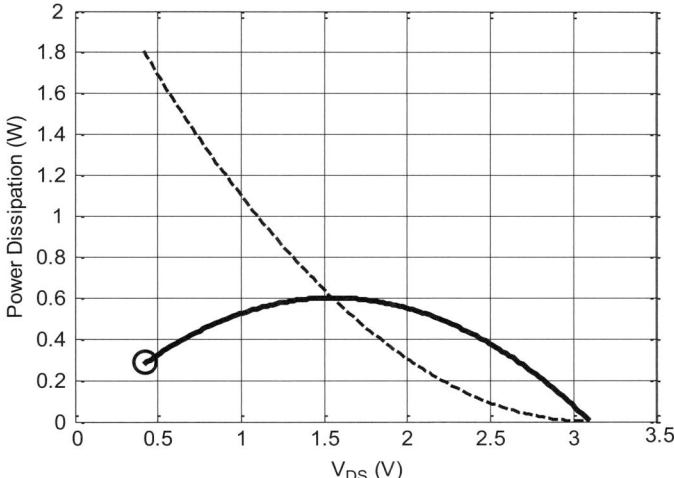

Figure 6-4   Power dissipation (solid line) along the load line from Figure 6-3. The peak dissipation is near 600 mW, dropping to 300 mW at the peak current (open circle) and to zero at peak voltage. Power dissipation in the load resistance is shown by the dashed line.

load line, corresponding to the class A quiescent bias power dissipation. Power transferred to the load is maximized at the maximum current point. Instantaneous output power transfer efficiency is seen in the relative values of these two curves at any particular value of V_{DS}. If the dashed line (output power) is below the solid line, then instantaneous efficiency is below 50%. Instantaneous efficiency is at 50% when the two curves cross. The desired condition is to have instantaneous efficiency above 50%, which happens when the dashed line is above the solid line. This difference is largest at the maximum current point, which occurs at the lowest available value for V_{DS}.

When transmitter efficiency must be improved, it is necessary to reduce power dissipation in the transistor for the same output power in the load. This requires characteristic curves to be present at lower power dissipations than the operating point in Figure 6-3. There are several options for this, as seen in Figure 6-5. Keeping the same load current, represented by the horizontal line through the operating condition of Figure 6-3, we can obtain the lower open circle, which has a power dissipation below 150 mW, less than half of that for the operating condition from the endpoint of Figure 6-4. This produces a significant efficiency increase.

Figure 6-5 provides another option that provides reduced power dissipation, shown by the upper open circle. Here the load current is larger than before, signifying larger output power. Power dissipation is below 200 mW, an improvement of nearly 40%. More output power with less loss in power dissipation is a double win for energy efficiency.

In order to obtain these two higher efficiency conditions, it is necessary to increase the input signal to the FET gate – this step alone gets to the upper open circle. In order to return to the original output power, it is also necessary to reduce the power supply voltage as shown in Figure 6-6. The new power supply voltage is near to 2.9 V.

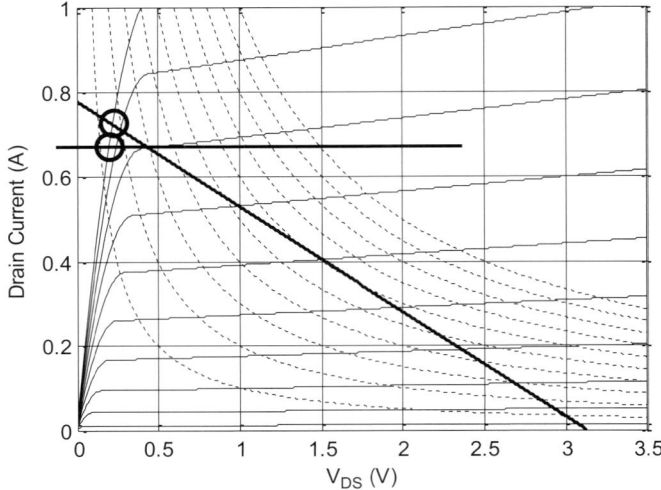

Figure 6-5   Reduced FET power dissipation from the operating condition in Figure 6-3 can take two forms shown here. The upper open circle is realized by simply increasing the input drive level to the FET gate. Load current increases while power dissipation falls below 200 mW. For an identical load current, shown by the horizontal line through the original operating point, the power dissipation can fall closer to 100 mW.

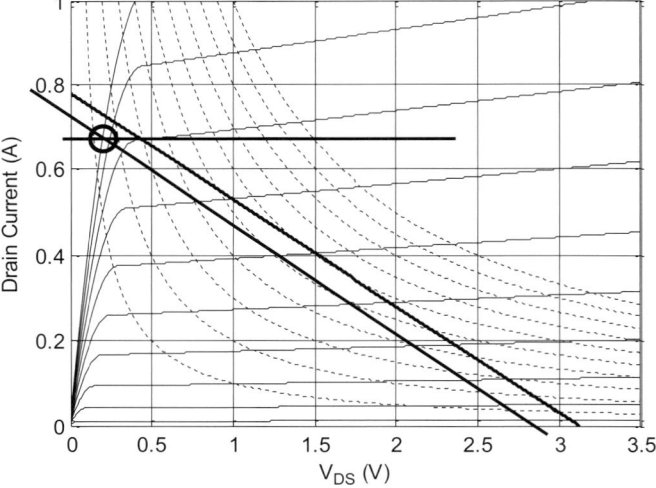

Figure 6-6   Reduced FET power dissipation from the operating condition in Figure 6-3 with an identical load current requires both an increase in the gate drive and a reduction in the power supply value.

Something else has also happened from the transition to these higher efficiency operating conditions. The slope of the characteristic curves at these points is greatly increased. Small changes in the power supply now correspond to large changes in device (and load) current. Looking back at Figure 5-5, this circuit power supply sensitivity has

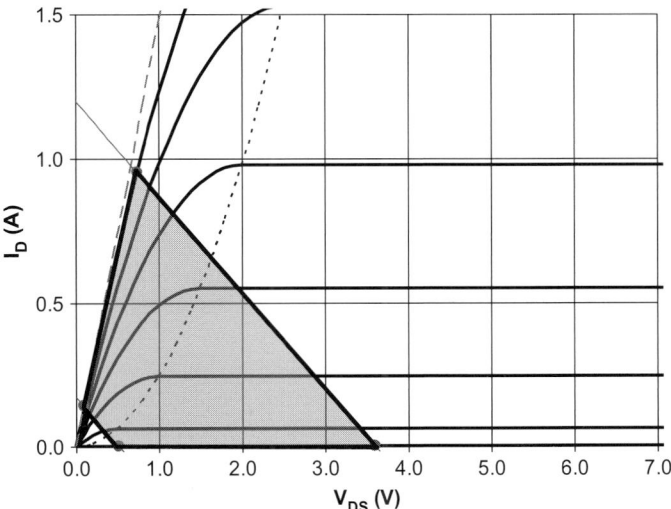

Figure 6-7   Trapezoid shape on the characteristic curves that implements the multiplication of (2.9).

transitioned off of the flat region of low supply sensitivity on to the highly sloped curves showing large sensitivity to power supply value. This circuit now operates very differently from the ET architecture detailed in Chapter 5.

Looking back at Figure 4-26, this new operating condition is recognized as C-mode. With this small drop in power supply (here to 2.9 V from 3.1 V) and slight increase in gate voltage, the amplifier has transitioned completely away from L-mode. Now in C-mode, it is expected from Figure 4-23 that the power supply value will set the output power. This is exactly what is needed for direct magnitude control of the polar operation, and the power supply itself is the third port from Figure 2-3(a). It is also seen that when an ET design is modified to provide improved energy efficiency, the circuit operation must shift from L-mode to C-mode, leaving ET behind and shifting to polar operation.

The gain of this stage is now an interesting question to consider. From Section 4.1.6, C-mode operation requires that the circuit operate with $g_S = 0$, and also g_{SP} very nearly zero. Ratiometric gains g_R and g_{RP} are well above zero, because there is plenty of output signal power above the smaller input signal. Looking at Figure 4-20 confirms that the input drive level must be increased above that for linear operation for this condition to exist.

There is another physical relationship to examine regarding this direct magnitude control. The multiplication (2.9) is effected through the trapezoidal shape seen in Figure 6-7. When the power supply changes value, the load line moves with it. The current through the amplifier follows the corresponding ON operating points in the transistor resistive region. The OFF operating points stay on the voltage axis. To the extent that the ON part of the characteristic curves follows a straight line as the power supply varies, the current correspondingly varies in direct proportion. This effects the desired multiplication.

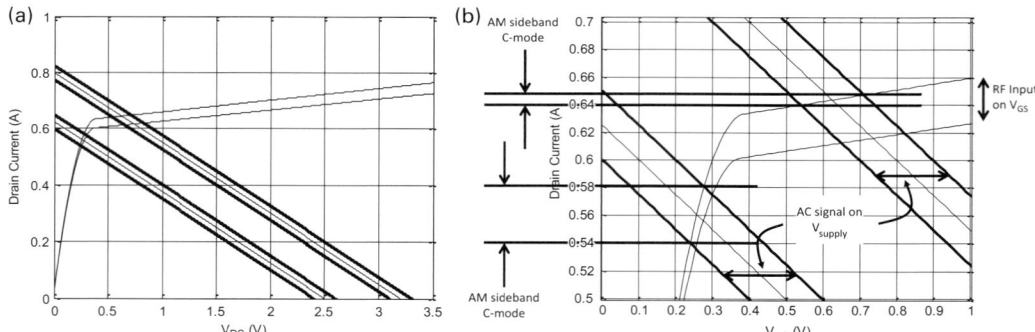

Figure 6-8   Test plan for measuring the L-mode to C-mode transition: (a) full view of operating conditions; (b) close view of AM sideband changes as the transistor operating mode shifts.

6.2.1   Measuring the supply-variation-to-sideband transfer function

Measuring the effect of direct magnitude control through the power supply is readily done by adding a small sinewave on to the power supply voltage. One useful test is to add a 100 mVpp sinewave on to the DPS output, keep the input RF drive constant, and vary the power supply value. The plan for this test is shown in Figure 6-8.

One typical result of this test is presented in Figure 6-9. Suppression of modulation output sidebands due to the AC signal on the power supply during L-mode operation is seen in Figure 6-9(a). At the lower power supply value from Figure 6-8, the RF PA transistor shifts into C-mode, and the sideband magnitude increases as the sensitivity to power supply variation increases. This is seen in Figure 6-9(b).

Measurement results from a sweep across this transition are shown in Figure 6-10 for the amplifier whose characteristics are shown in Figure 5-10. At supply voltages below 1 V, the AM sideband has constant magnitude, while the output power varies according to the square law (6.1). This is classic behavior of C-mode operation (5.13). As the power supply increases, the output power increases but deviates from the square law, characteristic of the transition to L-mode operation. When the power supply approaches the design value of 3.3 V, the output power stays near the same value, and the AM sideband power drops by slightly more than 10 dB, signifying L-mode operation. The AM sideband does not drop completely away, telling us that the PA transistor has some finite output conductance that appears as a finite slope to the characteristic curves in the CCS region. The long transition region implies that the transistor transition between saturated and CCS operation is broad, which is also consistent with the curves in Figure 5-10.

6.2.2   Stage dynamic range

Ohm's Law says that power = V^2/R. For a polar transmitter, this is written as

$$P(V_{DPS}) = \alpha \frac{V_{DPS}^2}{R_{PA}},$$ (6.1)

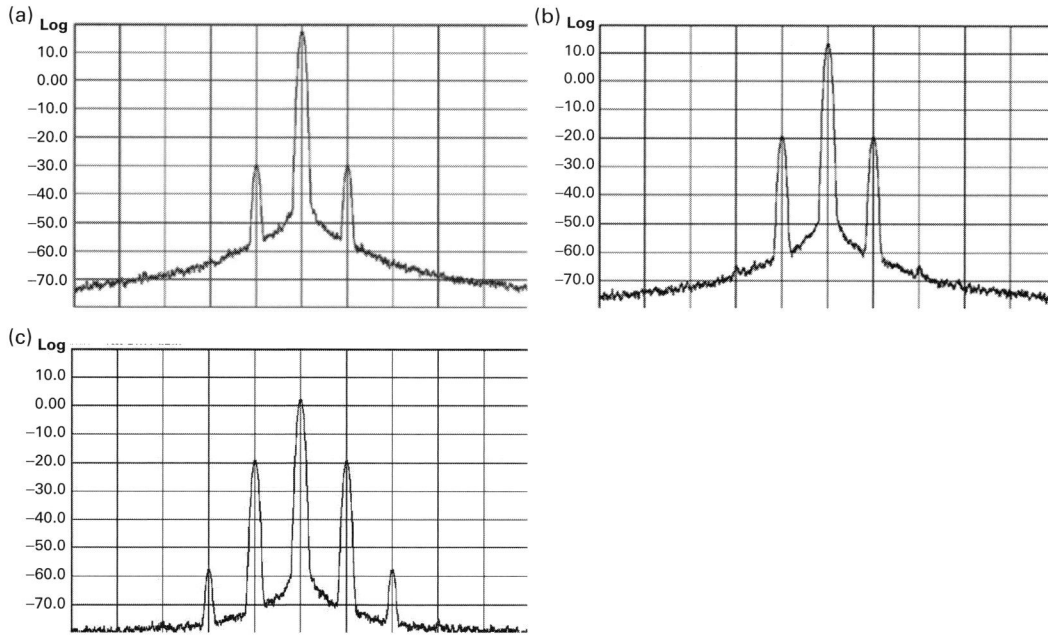

Figure 6-9   Sideband measurements across the transition from L-mode into C-mode operation: (a) L-mode suppression of 100 mVpp on the power supply; (b) full sidelobe magnitude resulting from C-mode operation; (c) sidelobe magnitude is unchanged with additional changes in the power supply.

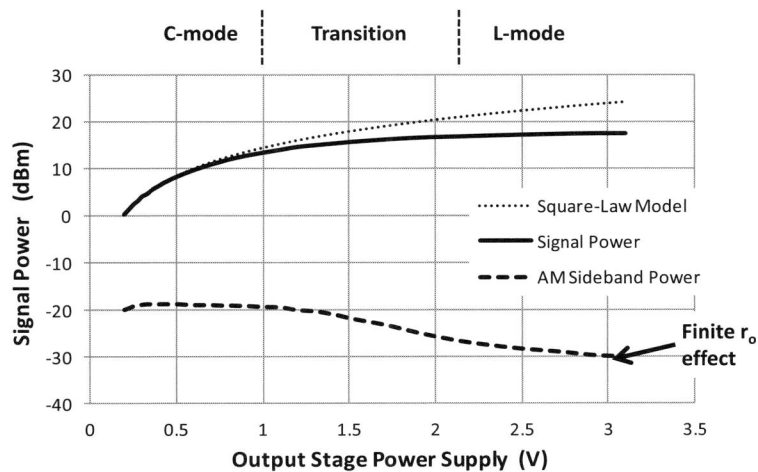

Figure 6-10   Sideband magnitude variation from varying the power supply across the L-mode to C-mode transition (vs. AC magnitude, to resolve the transition region size). PA for this measurement is from Figure 5-10.

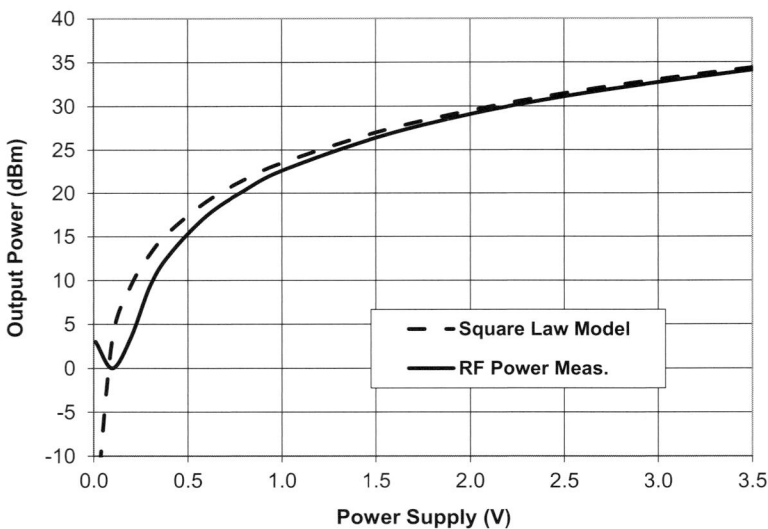

Figure 6-11   Example stage dynamic range measurement.

where R_{PA} is the effective load resistance at the transistor and α is a scaling factor that absorbs the efficiency of the PA and otherwise fits this model to the measured data. *Stage dynamic range* is defined by how well this model actually applies to any particular PA. One example of why this matters is shown in Figure 6-11.

From these data, we see that the fit to the model of (6.1) is at its best at the highest output powers, and steadily degrades as the output power reduces along with the applied power supply.

The cause of this problem is primarily due to leakage of the input signal to the output port through the RF power transistor's reverse transfer capacitance C_μ. A detailed discussion of the shape of this output power vs. supply voltage curve is provided in Section 6.10.2.

6.2.3   Transistor transconductance reduction

Transconductance is the change in device current resulting from a change in input voltage. For an FET in triode operation, appropriate for C-mode, the transconductance is the derivative of the drain current with respect to gate-source voltage. Using the FET model of (4.23), this gives

$$g_m = 2KV_{DS} \quad \text{(triode operation)}, \tag{6.2}$$

which is much less than the transconductance in CCS operation

$$g_m = 2K(V_{GS} - V_{Th}) \quad \text{(CCS operation).} \tag{6.3}$$

In C-mode, we always have $V_{DS} < V_{GS} - V_{Th}$, so the device g_m is reduced. Examining the characteristic curves for the C-mode operating condition in Figure 6-12 clearly

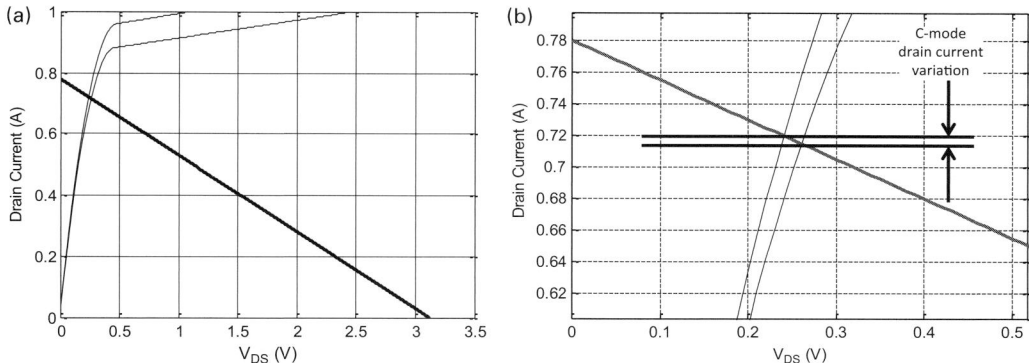

Figure 6-12 Effective device transconductance drops in C-mode operation because the device current changes very little with increases in gate voltage: (a) drain current hardly changes with this varying input; (b) close-in view.

shows why this happens. When the transistor drive is sufficiently large, the drain current becomes effectively constant because the transistor tries to draw more current than the load resistance and power supply value can provide.

6.2.4 DPS output noise requirements

When the power supply value directly controls the output signal magnitude, then any noise on the power supply voltage also directly modulates the magnitude (amplitude) of the output signal. This is completely unavoidable: with C-mode (polar) operation there is not, nor can there ever be, any rejection of noise on the power supply.

The methods of Section 5.2.3 are readily modified to evaluate these resulting sidebands from power supply noise. Like any noise calculation, we now consider noise *densities* instead of integrated noise power. Unlike most noise calculations, the impedance is not the same at the input and the output of the DPS. With input impedance very high and output impedance very low, usual power-based analysis methods are not applicable. Fortunately, we do not need them to get our needed answer. We only need to consider the actual voltages, and can let the currents be whatever Ohm's Law will require in the actual circuits.

Any sideband is measured with a spectrum analyzer with regard to power within the spectrum analyzer resolution bandwidth (RBW). This is obvious for sidebands at discrete frequencies, and it also holds true for distributed sidebands of the power spectral density (PSD). In this latter case, the sideband power P_{sb} has units W/Hz RBW (watts within the RBW in hertz), where the RBW is defined to be the equivalent noise bandwidth (ENB) of the RBW filter [4-2]. For everything to balance, the units on the corresponding voltage waveform must be V/√Hz. Then the voltage density to power density relationship is

$$P_{sb}(A_M) = \frac{(\delta \cdot A_M)^2}{R_0} \quad \text{W/RBW Hz}, \tag{6.4}$$

Table 6-1 Scale factors for measurements other than root-mean-square (rms)

Measurement type	Signal waveform	δ value
V_{rms}	any	1
V_{peak}	sinusoid	$1/\sqrt{2}$
$V_{pk\text{-}pk}$	sinusoid	$1/\sqrt{8}$
$V_{pk\text{-}pk}$	Gaussian noise	$1/6.1$

where the additional parameter δ has been added to allow for different ways of measuring the voltage waveform characteristics. This is practically important because the measurement of noise waveforms is not necessarily straightforward. In particular, it often is more convenient to measure a peak-to-peak value instead of an rms value. Values for δ are provided in Table 6-1.

The interest here is in determining the voltage density corresponding to noise sideband power density at the output of a polar (C-mode) PA. Solving (6.4) for the voltage density and then using (5.5), we get

$$A_M = \frac{1}{\delta} \sqrt{P_{sb} \cdot R_0} = V_{DPS} g_{DPS\_AM}, \tag{6.5}$$

where we know that V_A from (5.5) is here the DPS output voltage AC component V_{DPS}. This shows that the value of the DPS to AM transfer gain for the PA being used, *for the operating conditions that PA is under at that time*, must be known. Only then can the limits on V_{DPS} be determined for a given output noise density requirement P_{sb}

$$V_{DPS} = \frac{\sqrt{P_{sb} \cdot R_0}}{\delta \cdot g_{DPS\_AM}} \quad \text{V}/\sqrt{\text{Hz}}. \tag{6.6}$$

6.3   Broadband output noise characteristics

All amplifiers have wideband noise present on their output. Section 5.3 shows that linear amplifiers have this noise present at all times, even when there is no output signal. This "linear" noise is a result of the PA noise figure, duly amplified by the amplifier gain at each frequency.

A power circuit operating in C-mode also has wideband output noise, but its characteristics are *very* different from that of linear CCS amplifiers. Here we explore this difference.

6.3.1   Phase noise dominates

Any phase modulation passes unchanged through a C-mode stage, which holds true also for phase noise on the input signal. One example of this is shown in Figure 6-13. The graphic in Figure 6-13(a) provides an overlay of the phase noise on the input signal and

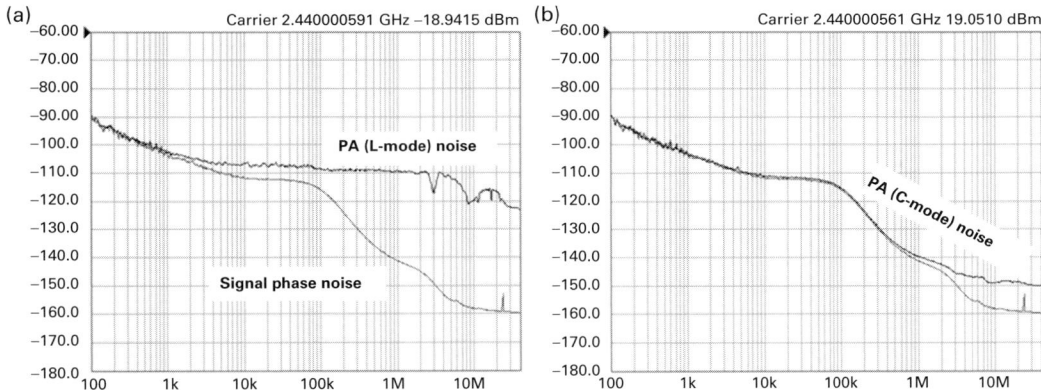

Figure 6-13   Wideband noise at the output of an amplifier depends on the operating mode of the amplifier: (a) output noise is flat when the amplifier operates in L-mode; (b) phase noise passes directly through a C-mode stage, implying that amplitude noise is removed at the PA output.

the noise at the output of a linear PA (operating in L-mode). When the input signal is increased, so that the PA transitions to C-mode, the PA output noise closely follows the input signal phase noise as seen in Figure 6-13(b).

6.3.2   Noise figure effect is suppressed

Wideband noise from the PA is reduced in C-mode operation because when the transistor is fully ON or OFF, not only is the gain reduced, the noise figure action is also suppressed. This action is seen by comparing the two PA noise curves seen in Figure 6-13. Output power is increased, and this combination greatly increases the output signal to noise ratio of a C-mode power stage.

AM from DPS noise is a critical design parameter when any RF stage operates in C-mode. This leads to a very different design consideration from that used in ET transmitters. In polar operation, the actual DPS output value has increased importance.

6.4   Supply accuracy requirements

In contrast to the flexibility afforded by the ET regarding the precise value of the power supply voltage, in polar operation the power supply value directly sets the output power. For accurate output power, particularly in a time-varying envelope, the power supply value must be precisely set. Any error is a signal envelope distortion. Indeed, in polar operation the responsibility for an accurate output signal is transferred from the RF power transistor to the DPS.

This also means that there is no rejection of noise on the power supply. *None.* To the C-mode stage, any noise on the DPS output is the same as any variation intended to control the signal envelope. DPS design to control output noise is a critical activity for any polar transmitter. The converse is also true: if there is no suppression of any noise on the power supply by the transmitter power amplifier, then the transmitter is operating with polar modulation in accordance with (2.9).

6.5   DPS time alignment

With direct control of output signal magnitude, the signal phasor is correct only when accurately matched to the corresponding phase shift

$$s(t) = \rho(t)e^{j\phi(t)} = [\alpha \cdot a(t)]e^{j\phi(t)}. \tag{6.7}$$

If the phase shift is not correctly matched to the magnitude modulation, the signal will be rotated from the correct position at each time value. Because the phase modulation is not constant, this erroneous rotation is not constant and the actual signal phasor is significantly distorted. Similarly, if the magnitude signal is not properly aligned with the phase modulation, the phasor angle will be correct but its magnitude will be in error. The error magnitude will not be constant, and again the signal modulation will become significantly distorted.

Modulation signal paths are inherently not balanced in a polar transmitter, unlike the natural matching usually present in a quadrature transmitter.[2] All else being equal, the delay of any circuit path is inversely related to the bandwidth of that path (lowpass or bandpass). The path with the narrower circuit bandwidth will generally have the longer delay. Time alignment is achieved by inserting a controlled delay in the path with the wider bandwidth. More often than not, the phase modulation path has the greater circuit bandwidth. Time alignment in these cases is implemented with delays (usually digital) in the phase modulation path.

An example of how a modulation phasor path can be distorted due to time misalignment is provided in Figure 6-14. This particular example is from an EDGE modulator. The principle, and net effect, is universal and significant distortions can definitely occur if not carefully managed.

Modulation errors do not only show up in EVM measurements. Any change to the time-domain waveform may also have an effect on the frequency-domain characteristics, the signal PSD. This does NOT have to happen though. It is definitely possible to have a huge EVM and a perfect PSD. It is harder to have a perfect EVM and a distorted PSD.

The amount of signal damage due to time misalignment strongly depends on the signal itself. In general, the amount of tolerable time misalignment is related to the reciprocal of the signal's occupied bandwidth: wider-band signals require tighter time alignment. Another factor is the distortion tolerance of the signal: a signal that has a 15% distortion specification will tolerate more time misalignment than a signal specified to have a waveform distortion below 5%. For these reasons, there is not a general rule that

---

[2] To be sure, there is a time alignment requirement in quadrature modulators that is every bit as critical. Fortunately the inherent balance of the $I(t)$ and $Q(t)$ signal path lengths and bandwidths makes meeting this time alignment relatively easy.

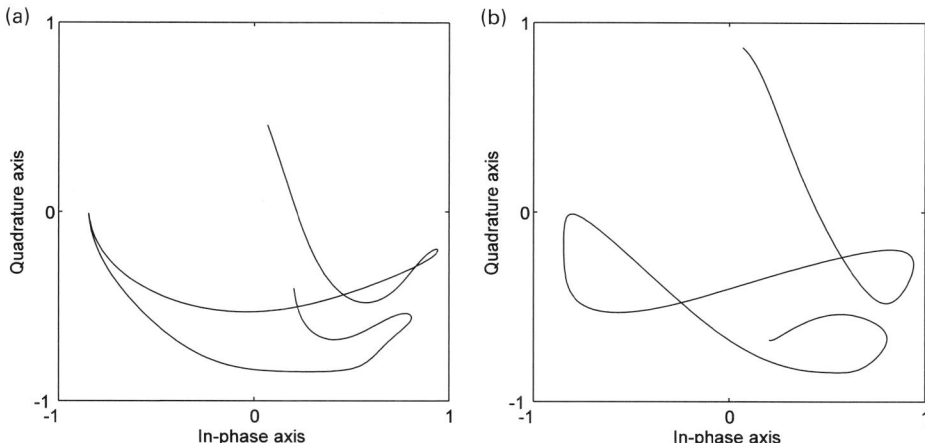

Figure 6-14 Signal trace (modulation path) distortion due to time misalignment: (a) the correct signal; (b) an example of the resulting distortion for a time misalignment of one-half symbol time.

is easily derived to determine the tolerance of time alignment error for any particular signal. This tolerance limit must be individually evaluated for each signal.

6.6 Signal waveform characteristics

With a polar transmitter, the output signal is *always* operating at P_{SAT}. Output signal envelope variations are generated by varying the value of P_{SAT} using changes in the power supply value. By definition then, there is no output back-off (OBO), since all output power is provided at the then set P_{SAT}. This means that whether the signal is envelope-varying or not, the PEP of the envelope-varying signal matches the output power when the signal is a constant envelope type.

There is also no distortion around this envelope-varying signal peak, which is very different than driving a linear L-mode amplifier such that the output $PEP = P_{SAT}$. This means that for a particular transistor size, operating in C-mode allows a higher PEP without envelope distortion. With identical PAPR values, the polar transmitter therefore also provides greater average power, a direct consequence of the higher peak power. For any particular transistor size, more output power is possible for a low distortion envelope-varying signal from the polar transmitter. By definition, there can be no OBO, so there is no problem having an envelope-varying signal PEP be equal to a constant envelope signal operating at P_{SAT}, which is shown in Figure 6-15.

Conventional L-mode design requires the highest PEP (= P_{MAX}) to be below the amplifier saturation power by the OBO

$$P_{SAT} - OBO \geq P_{MAX} + PAPR. \tag{6.8}$$

This situation is shown in Figure 6-16. The design criteria for sizing a polar final stage is to set $P_{SAT} = PEP_{MAX}$. The highest envelope power at the highest average output power

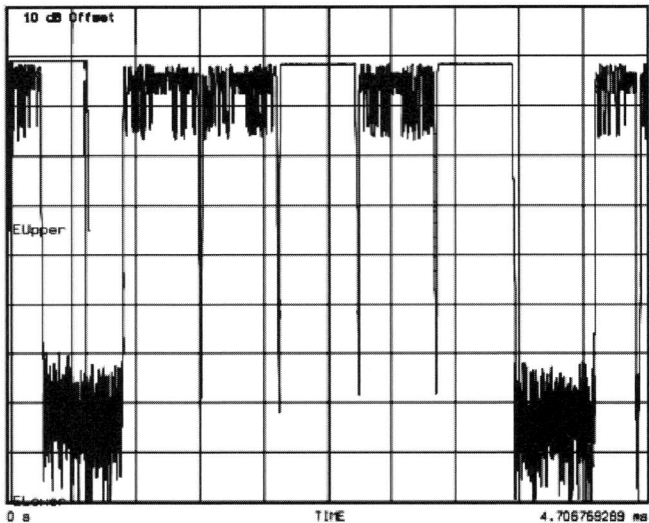

Figure 6-15   No OBO when an envelope-varying signal PEP matches the P_{SAT} for a constant envelope signal.

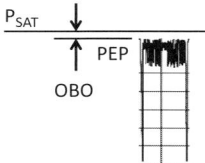

Figure 6-16   Increase in average output power, or decrease in transistor size, available when using polar modulation.

(P_{MAX}) must be provided by the final RF power stage at power saturation. Any other instantaneous signal output power will be below this. Mathematically, we write this polar design condition as

$$P_{SAT} \geq P_{MAX} + PAPR, \tag{6.9}$$

which is identical to (6.8) with $OBO = 0$. This means that for a particular transmitter PEP requirement, a polar transmitter can use a smaller transistor with $P_{SAT} = PEP$, instead of $P_{SAT} = PEP + OBO$ as shown in Figure 6-16. Alternatively, a polar transmitter can provide larger output signal power because, for the same P_{SAT} as the linear amplifier, it can reach $PEP = P_{SAT}$ instead of the lower $PEP = P_{SAT} - OBO$.

Bandwidth expansion

Section 2.7 introduces the expansion of the modulation component bandwidth when using polar components, a consequence of the nonlinear basis transformations (2.15) and (2.16). At the polar output stage, the modulation component signal spectra (e.g. Figure 2-16) are convolved, because these phase and magnitude modulation components are multiplied in the time domain (2.9), meaning that a successful convolution needs accurate interaction among all of the frequency components in each component PSD – both magnitude and phase.

The amount of bandwidth expansion varies with the signal type, from zero for a constant envelope signal such as GMSK to 10x or more for a signal that has a large percentage of zero-magnitude events. Much of this bandwidth expansion comes from rapid changes in the envelope waveform, the phase waveform, or both. This is a direct consequence of the Fourier transform: the greater the waveform curvature (equivalently, the number of waveform derivatives that are not negligible in the modulation component waveform), the wider the signal component bandwidth must be. This is a direct relationship: once we have particular waveforms, then the bandwidth characteristics are set. Any changes made to the frequency characteristics also force changes to the corresponding waveform.

An immediate consequence of this direct relationship is that there is no "rule of thumb" on how much bandwidth expansion happens in all polar decompositions. For any signal that has smooth magnitude and phase waveform characteristics, then there will not be much bandwidth expansion. Similarly, for any signal that has envelope zero crossings, then the envelope waveform has corresponding values of zero and infinite curvature (undefined derivatives at that point), and the phase waveform has direct discontinuities at these same times. Both of these characteristics cause dramatic bandwidth expansion, possibly to 100x out to –60 dB in their spectral characteristics.

In a practical sense, all circuitry cannot support infinite curvatures and mathematical discontinuities, so there is always a finite bandwidth expansion limit. As circuit technologies improve over time, their ability to support tighter curvatures and react rapidly to waveform discontinuities also improves, making the bandwidth expansion problem more interesting. This encourages the selection of signals that do not have these bandwidth expanding characteristics at the system design time, when achieving polar transmitter energy efficiencies is an important system goal.

When wideband signals having these zero-crossing properties are applied to a conventional linear amplifier that is then over-driven in order to improve energy efficiency, nonlinear artifacts appear (spectral spreading) that can readily include seventh-order and ninth-order products. Any linearizer design to correct for these nonlinearities must itself have sufficient bandwidth to measure these distortion products accurately, and also to cancel them precisely. In other words, the bandwidth needed in a precision linearizer and the bandwidths needed by polar modulation circuitry to manage modulation component bandwidth expansion are really not very different. The circuit bandwidth requirements need to be compared fairly at equivalent transmitter energy efficiency performance. Work to date implies that from this comparison there is essentially no difference in circuit bandwidth requirements between polar and linearizer transmitter designs.

It is true that bandwidth expansion is reduced when the signal envelope (magnitude) does not go to zero. When the signal envelope has a finite "floor," both the magnitude and phase components are continuous at all times. With signal bandlimiting filtering, all of the time derivatives of these modulation components are also continuous. One technique to achieve this is presented in Section 6.13.5, with others presented in additional references included in that section.

6.7   Circuit model: "switch"

Physical reality has charge (coulombs (Q)), time (seconds (sec)), mass (kilograms (kg)), and so on. Everything else we derive for our convenience, including energy (joule J = (kg m^2)/sec^2), voltage (volts = J/Q), current (amperes = Q/sec), frequency (hertz = 1/sec), and so on. Embarking on switch-based RF design is best done keeping these fundamentals in mind because the conventional use of circuit steady state in design does not apply.

6.7.1   Port "impedances"

What is the impedance (Z = V/A = (J sec)/(Q^2)) of a switch? This is an unanswerable question because there is no steady state value. When the transistor is ON, then the impedance is very low and largely resistive. When the transistor is OFF, the impedance is usually mostly capacitive. The impedances measured in conventional CCS operation modes (and reported in data sheets) never apply.

This leaves a big question on how an output matching network (OMN) should be designed for a polar operated RF power stage. The usual procedure is to design a network that, at the design frequency, has input and output impedances that are complex conjugates of the impedances that the network will be attached to. This design is simplified when one side is connected to 50 ohms resistive.

For a polar stage, there is no definable impedance with which to do this design. Instead, experience shows that the objective for the OMN is to translate the output load to a resistance at the transistor, which allows it to develop the needed power at the available power supply voltage. For a polar RF power stage, the OMN is not there to tune an impedance at the transistor to the output, instead it starts from the output impedance and converts that to a load *resistance* at the transistor. In the most general design, nothing is tuned. Transistor output reactance is charged and discharged by the switching action of the transistor conductance.

6.7.2   Restricted ability to use s-parameters

The first sentence in an s-parameter design text usually begins with: 'In a nominally linear system' . . . 'then the rest of the text follows'. In this switch-based polar regime, all circuit linearity is suppressed. Therefore, the fundamental assumption upon which all s-parameter theory and practice is based is not valid. The only possible result is that any use of s-parameter techniques with respect to the transistor in the design of polar transmitters is invalid.

That being said, s-parameters are always valid for a network of passive circuit elements. We can and definitely do use them for the design of the OMN described in Section 6.7.1. We cannot use them for design of the switching power stage itself.

The newer X-parameters are very different. These are not simply an extension of s-parameters for a linear "Taylor series" environment into the memory-effect Volterra series environment. Rather, the X-parameters are a measurement-based behavioral modeling method that is designed to support the extreme nonlinearity of switching-based RF circuitry. These are new enough that methods for their application are not yet widespread or well developed for application in wideband transmitter designs. Yet the promise provided by X-parameters is very interesting because, as they result from multidimensional interpolation between measurements from a nonlinear vector network analyzer, they can provide for both fast and accurate design simulations.

6.7.3   Switch-based amplifier classes: D, E, F

There are three amplifier classes that generally assume switching operation for the RF power transistor(s). These are classes D, E, and F [4-1]. In all of these amplifier types, the high energy efficiency that they are famous for is only achieved when they are at their peak, saturated power. This means that there is no circuit linearity remaining at this high efficiency condition. Indeed, the class F set of amplifiers requires that harmonics be present at the transistor output for the wideband output network to manipulate into the intended waveforms. The transistor must operate in C-mode.

What this also means is that when the input drive level is reduced sufficiently so that the amplifier is no longer operating in C-mode, then it is also no longer operating in class E or class F. As the transistor operation transitions from C-mode to L-mode, the switching required by class E and the nonlinearities for generating harmonics required by class F both cease, and the bias on the drive into the transistor defines its L-mode operation within the conventional CCS classes A, AB, B, and C.

6.7.4   P-mode: not a switch, but not linear (L-mode) either

Operation in P-mode at first look appears to be a linear operation, although it has a very restricted upper limit to its available output power as seen in Figure 4-23. It is true that the output power varies decibel by decibel with the input power, just as in L-mode. But to get this result we must add a condition: the power supply must be constant. Any variation in the power supply changes the transfer gain of the amplifier, in accordance with (4.17). P-mode is a one-quadrant analog multiplier. Any analog multiplier is not called a linear amplifier; and this also applies to P-mode. Consistency in use of terms is imperative to avoid ambiguities.

P-mode is every bit as sensitive to noise on the power supply, as is C-mode. And importantly, L-mode is not. We must use descriptions that accurately describe how the circuitry operates, and make clear what the properties of its operation are. P-mode is therefore best described as an analog multiplier. Equivalently, it is accurately described as a variable gain amplifier controlled by the power supply input.

6.7.5   Different design rules

New design criteria are clearly needed for RF circuit design, based on these switching circuits. Using the model in Figure 4-27, the signal voltage across the load resistance is

$$V_{env} = I_L R_L = R_L \frac{V_S - V_{AMO}}{R_L + R_{ON}} = \frac{V_S - V_{AMO}}{1 + R_{ON}/R_L}. \qquad (6.10)$$

Our first new design criterion is to ensure that

$$R_{ON} \ll R_L. \qquad (6.11)$$

Satisfying (6.11) allows the second term in the denominator of (6.10) to be neglected. Meeting this condition is achieved by proper transistor selection, following the appropriate selection of the load resistance R_L at the transistor using conventional techniques [4-1]. The value of R_{ON} changes with transistor temperature, and from transistor to transistor. Inherent thermal stability, and performance consistency across manufactured units, are both improved by decreasing the value of R_{ON} from the value of R_L. Practical limits on how low R_{ON} can be must be evaluated case by case for every design.

With a proper selection of R_{ON} by transistor selection, the second new design condition is to ensure that

$$V_{AMO} \ll V_S. \qquad (6.12)$$

There are two ways to satisfy (6.12). First, the system design must ensure this condition is met no matter what the value of V_{AMO}. With typical values of V_{AMO} within 0.1 V to 0.25 V, the minimum value of the power supply must be around 2 V. All variations of the power supply for envelope control are restricted to be above this minimum. This approach therefore tends to be practical only for higher power infrastructure circuits, due to the restricted supply voltages available in battery operated mobile radios.

The second option is to add an additional condition on the transistor selection, further requiring that the transistor exhibit a very low value for V_{AMO}. In general, this means that the transistor type be an FET type because of the $V_{ce,sat}$ characteristic of bipolar transistors, as seen in Section 5.4.3. If the transistor exhibits $V_{AMO} = 0$, then all restrictions on the possible power supply values are removed.

When these two new design criteria are met, the output envelope becomes

$$V_{env} = I_L R_L = R_L \frac{V_S - V_{AMO}}{R_L + R_{ON}} \qquad (6.13)$$
$$\approx V_S.$$

Equivalently, the load current is

$$I_L \approx \frac{V_S}{R_L}, \qquad (6.14)$$

which is now independent (to first order) from *all* transistor performance details. We expect from (6.14) that a properly designed C-mode circuit – meaning that both of these two new design conditions remain valid at all times – should provide innate stability across both temperature variations and manufacturing variations of the transistor. Indeed, both of these stabilities are realized from these designs. More details are provided in Section 6.15.

The approximation at the end of (6.13) is actually very profound. When the design conditions (6.11) and (6.12) are both valid, we get from (6.13)

$$V_{env} = V_S \qquad (6.15)$$

which shows that the signal envelope is directly controlled by the applied power supply. This is a clear manifestation of (2.9) proving that this implements polar modulation.

6.8 Bias conditions

In the original plate modulation transmitters, the power amplifier was nearly always biased in class C [1-1]. With this bias, there is no quiescent current if the input RF signal is removed. It is also well documented [1-1] that the RF drive into these transmitters must be sufficiently large to ensure that the power amplifier operates deep into compression. This is consistent with the definition of C-mode.

This section evaluates the bias of a modern C-mode circuit, and evaluates whether the older bias selection is appropriate in newer designs – or whether a different strategy is better.

6.8.1 CCS modes: class A, B, C

Conventional bias design is certainly possible for C-mode amplifiers. Bias in class A or class AB would therefore imply that quiescent current would flow in these stages. However, when the operation is in C-mode, it is required that the input drive level is large enough to fully compress the transistor. This means that when the input signal is not sufficiently large to fully compress the transistor, the transistor leaves C-mode as described by the Booth chart.

We are motivated then to learn the best bias condition to make it easiest to enter C-mode and stay there. One way to do this is shown in Figure 6-17, where a fixed magnitude input sinusoid is added to a variable bias and used to drive the amplifier. The family of output current waveforms, one for each of the bias points identified in Figure 6-17(a), is shown in Figure 6-17(b). Evaluating the slope gain g_S for each of these waveforms provides the curve in Figure 6-17(c), showing that the best transistor bias to achieve large drive and quick access to C-mode is class A, not the class C bias historically used in traditional transmitters.

6.8.2 Class PFS

As the magnitude of the input signal increases, eventually the transistor spends most of its time either cut-off (when the input signal is below the transistor threshold) or as a low value resistor (when the drive signal is large enough that the transistor "wants" to draw more current than the load makes available). In this condition, the actual value of the bias voltage or current becomes inconsequential, as we see in Figure 6-18 for

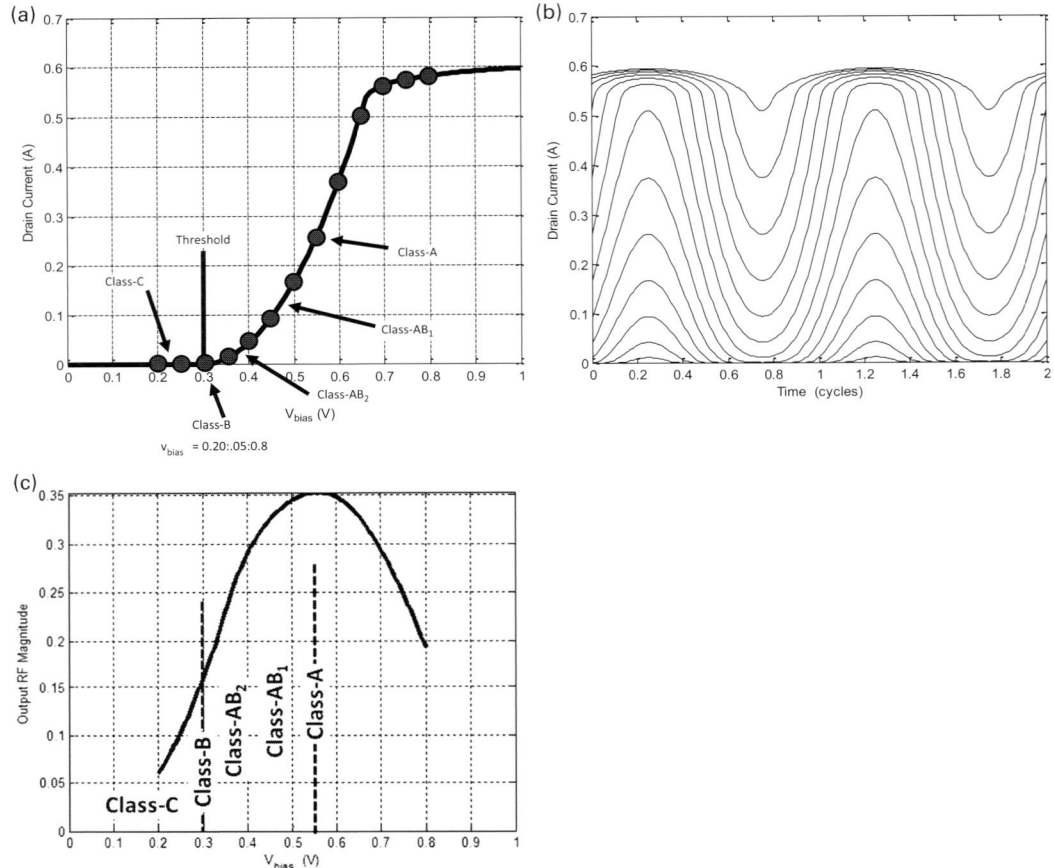

Figure 6-17   For a particular magnitude of input drive, output power varies when the bias is changed: (a) drain current vs. input voltage V_{GS} and the corresponding values as bias class changes; (b) waveforms corresponding to the dots in part (a); (c) output RF signal magnitude varies with bias class and peaks at class A.

large (sinusoidal) drive signal magnitudes into an FET. Whether the bias is set for class C, class B, class A, or anything in between, the output power is effectively the same. This operating condition is colloquially termed class PFS, for "pretty 'fully' saturated."

6.8.3   Dynamic bias with envelope variations

When the power supply is varied for a C-mode stage, as the power supply increases, there is progressively more drive signal necessary to enable the transistor to draw the desired amount of current. This is a consequence of the finite transconductance of any transistor, $g_m = \Delta I / \Delta V_{in}$. The signal impact is that the duty cycle of the transistor current changes along with the power supply variation. One example of this behavior is shown in Figure 6-19, where the transistor used is an FET.

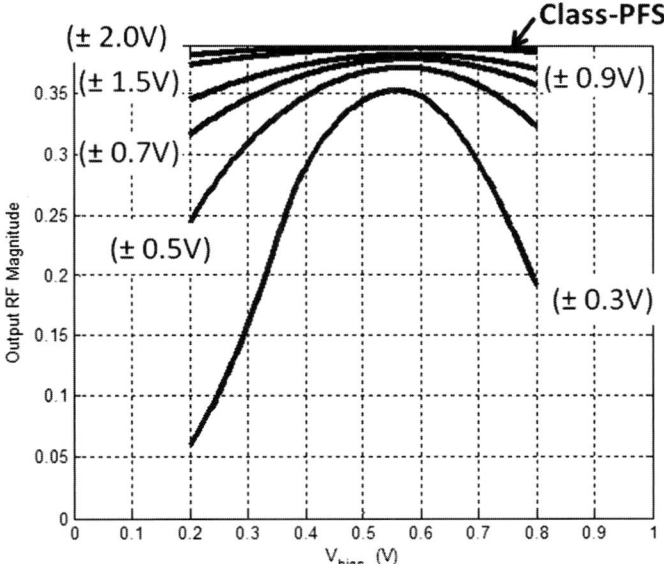

Figure 6-18   When the input drive signal is increased, the variation of output power with changes in the bias diminishes. Eventually, the actual bias does not matter, and the operating condition reaches class PFS.

The distortion in Figure 6-19(c) is clearly a performance problem for a polar operated C-mode RF power stage. The cause of this problem is shown in Figure 6-20. In any switching waveform, the power present in the fundamental frequency spectral component depends on the duty cycle of the waveform. Narrow or wide duty cycles both have low RF power levels. There is a peak at 50% waveform duty cycle, where the RF output power is maximum for the given pulse magnitude. The 1 dB width about this RF power maximum ranges between a 35% and 65% duty cycle.

Two important points are shown in Figure 6-20. First, to maximize the RF output power from C-mode operation, the switching duty cycle must remain close to 50%. This intuitively makes sense. Second, any change in the C-mode switching duty cycle causes unintended variations in the output signal power, always toward the output power being too low. This distortion mechanism is special to a C-mode power stage. In principle though, this is similar to the fixed-bias problem for the ET operation from Figure 5-29: leaving a fixed bias on the transistor, whether it operates as a linear or switching amplifier, results in envelope distortions when DPS operation is used.

For ET in Section 5.8, the linearity recovery solution is to track bias from the applied supply in such a way to keep the amplifier always operating in class A. Here for C-mode operation, linearity is suppressed, which provides two different approaches to solve the linearity problem shown in Figure 6-19(c). One option is to operate the C-mode transistor in class PFS, where the actual applied bias no longer matters. A second option is to also adopt a form of variable bias that adjusts the input signal swing as the applied

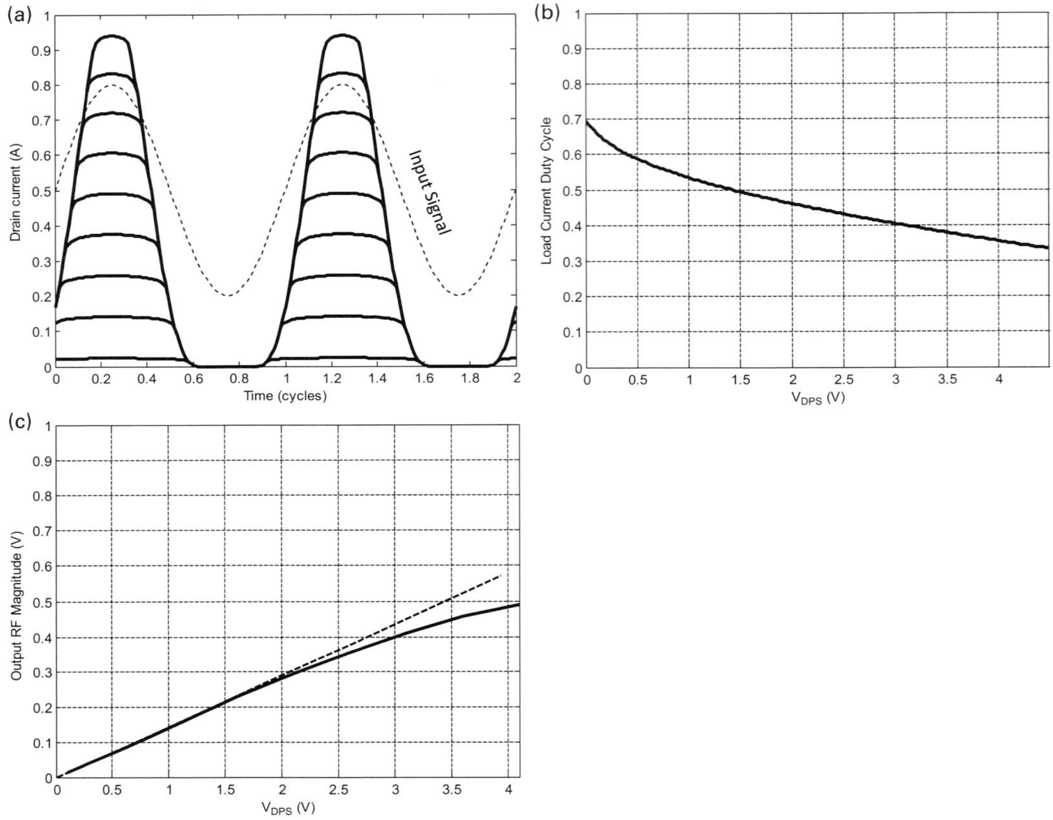

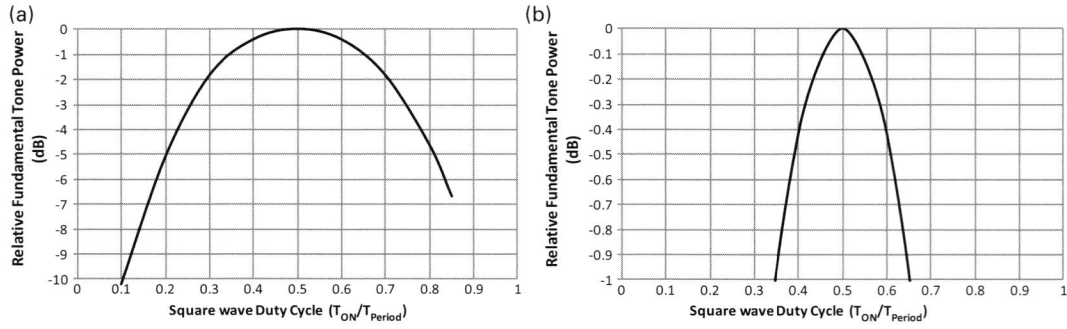

Figure 6-19   In C-mode operation, the duty cycle of output current pulses changes as the power supply varies when the drive bias and magnitude remain fixed: (a) transistor current (solid line) compared to the input signal (dashed line); (b) duty cycle of the transistor current as the power supply is varied; (c) RF output magnitude as the power supply varies (solid line) compared to perfect linearity (dashed line).

Figure 6-20   Output envelope distortion occurs if the RF switching duty cycle varies even when the magnitude of the pulses stays constant: (a) duty cycle region of 10 dB output power variation; (b) the more restricted duty cycle region of 1 dB output power variation.

supply varies, such that the transistor duty cycle is kept very close to 50%. Both of these approaches are detailed in Section 8.1.1.

6.8.4   Drive with rectangular waveforms

Because the objective of C-mode operation is to operate the RF power transistor as a switch, there actually is no need to have a sinusoidal input signal. The input signal can be first passed through a limiter, as was shown for the EER technique in Figure 6-2. Or the RF transistor itself can act as its own input limiter, which is what occurs when C-mode operation is adopted. In fact, any C-mode operation has a self-limiting action at the transistor input because of the C-mode property that the RF amplifier output power is not sensitive to the actual input signal magnitude, and only sensitive to the value of the applied supply from the DPS (Section 4.1.6).

Circuit application of the rectangular input signal drive is discussed in Section 8.1.1. Additional advantages to using this technique for the C-mode stage drive are discussed in Section 6.11.

6.8.5   Differences between bipolar and FET operation

Even though the majority of examples used in this chapter are based on field-effect transistors, it is certainly possible to use bipolar transistors in polar transmitters [6-9]. While there are many similarities in how these two transistor types behave, there are also a number of operational differences between FET and bipolar transistors that must be managed in order to succeed in C-mode polar operation.

At frequencies used for most wireless communication, the bipolar technology used to generate watt-level power is the heterojunction bipolar transistor (HBT), as opposed to the conventional bipolar junction transistor (BJT). For present purposes, the style of the bipolar transistor does not matter. The discussion here applies to all of these types.

To the device characteristic curves, the major difference from FET characteristics is the presence of the $V_{ce,sat}$ voltage offset seen in Figure 5-23. This offset is a direct consequence of making the bipolar transistor asymmetrical, in that the forward current gain β_F is much greater than the reverse current gain β_R. All collector current must flow across this voltage drop, so there is an additional power dissipation of $P_{D,VSAT} = I_C * V_{ce,sat}$. Fortunately, the value of $V_{ce,sat}$ is typically a small 100–300 mV, so this additional power dissipation is relatively small.

There is also an offset seen in the polar power control characteristic from (6.1) and Figure 6-11. This offset is the definition of the curve-fit term V_{AMO}; adding this voltage offset to (6.1) gives

$$P(V_{DPS}) = \alpha \frac{(V_{DPS} - V_{AMO})^2}{R_{PA}}. \tag{6.16}$$

There is no direct theoretical link yet between the physical parameter $V_{ce,sat}$ from the bipolar characteristic curves and the curve-fit model term V_{AMO}. We can easily speculate that they are closely related. But as of this writing there is no proof or description of any

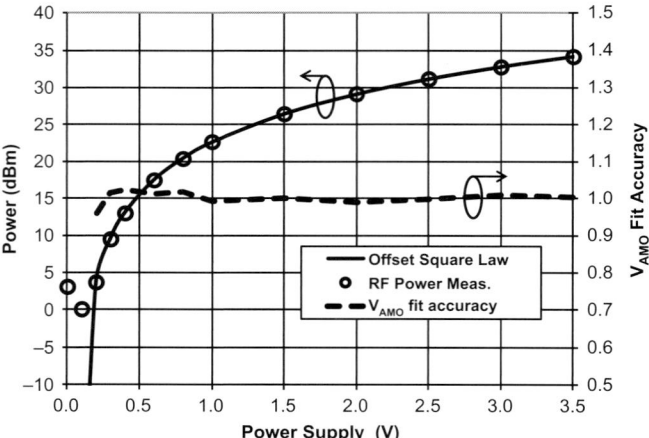

Figure 6-21   Modeling the polar (power supply) controlled output power when a bipolar transistor is used shows a best fit when the applied voltage is modified by an voltage offset parameter V_{AMO}. Here the transistor used is a GaAs HBT.

relationship. Again, to the present discussion this is of no consequence. We simply accept the empirical validity of (6.16) given the excellence of its fit to the measured data seen in Figure 6-21.

Any bipolar transistor is a current controlled device, which means that whenever there is collector current, there also must be base current. This base current flows through the base-emitter junction, which has a forward voltage drop V_{BE}. This in-phase relationship is a resistance, resulting in an additional power dissipation of $V_{BE}*I_b$. When the transistor current gain β is large at RF, then this base current is small. Yet it is another additional power dissipation mechanism present in bipolar transistors that is not present in FET devices.

Exponential V_{BE} characteristic – high g_m

When viewed as a voltage instead of a current, the input to any bipolar transistor is the base-emitter voltage (V_{BE}). The relationship between the collector current and V_{BE} is exponential as noted in (4.23) and seen in Figure 6-22 for both silicon and compound semiconductor bipolar transistors.

Many more details on designing with bipolar transistors for DPST polar transmitters are provided in Section 8.1.3.

Capacitors tend to hold their voltage constant, so a reduction in the RF voltage on one side of a capacitance corresponds to a smaller RF voltage at the other side. This holds for the reverse transfer capacitance C_μ as well. Having small RF input voltage resulting from a high g_m, coupling of the input signal to the output is reduced.

Other effects of device capacitances are less desirable. In bipolar transistors, at higher frequencies the presence of input capacitance C_{BE} (Figure 6-23(b)) shunts some of the input signal away from the transistor action represented by r_π. To get the same output current, increased input current becomes necessary. To the circuit designer, this appears

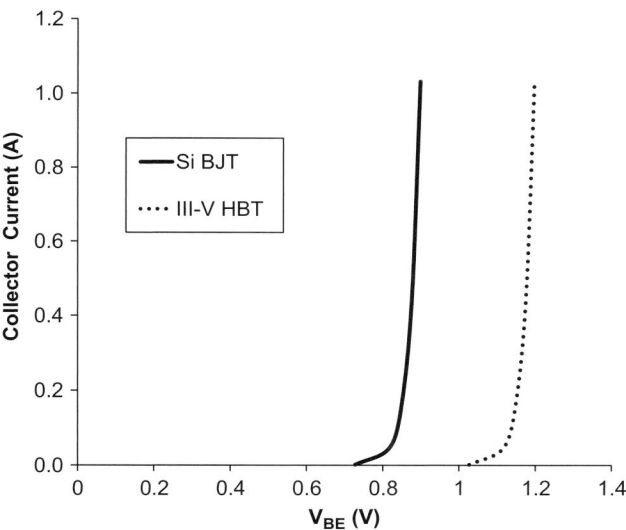

Figure 6-22 The forward voltage drop of the base-emitter junction changes very little as the collector current changes over a very wide range. This corresponds to a high transconductance value for bipolar transistors, usually much higher than typical for FETs.

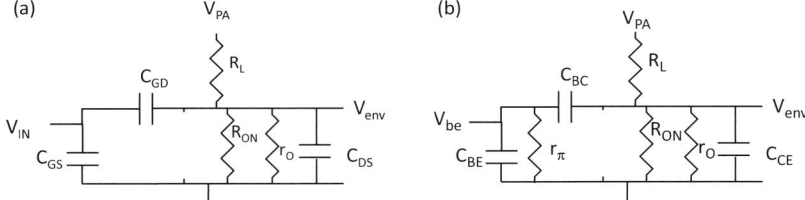

Figure 6-23 Input capacitances for FET and bipolar transistors are a problem for different reasons: (a) FET input capacitance C_{GS} requires charge transfer to change the gate voltage, requiring a transient current flow from the driver; (b) bipolar input capacitance C_{BE} shunts the input signal away from the current gain mechanism (r_π), effectively reducing the transistor gain with increases in frequency.

as frequency dependence in the transistor beta. PA driver designs must provide this increased current, even though it does no useful work.

The input capacitance of FETs causes problems as well. Besides drawing increased current at higher frequencies, in many technologies the capacitance C_{GS} (Figure 6-23(a)) tends to vary significantly – maybe 3:1 or more – as the input voltage varies. This makes design of input matching networks a significant challenge.

6.9   Load presented to the DPS

Following (2.9), when a polar transmitter puts out no signal, then the power supply to the PA also goes to zero. This means that there is no possible quiescent current I_Q for a polar transmitter (at least for the final RF power stage). There is also no power supply overhead value V_{OS} as is required for ET. Applying $V_{OS} = 0$ and $I_Q = 0$ to (5.20), we find that the polar PA provides a load on the DPS described by

$$R_{PA}(V_{env}) = \frac{V_{PA}}{I_{PA}} = R_L \left. \left(\frac{1 + \dfrac{V_{OS}}{V_{env}}}{1 + \dfrac{I_Q}{I_{sig}}} \right) \right|_{V_{OS}=0, I_Q=0}$$

$$= R_L. \tag{6.17}$$

This is a very simple and significant result. The polar PA always appears as a resistive load to the DPS, as long as it continues to operate in C-mode for all values of the DPS. Polar DPST designs exist that provide a constant resistive load to the DPS all the way down to 10 mV at the DPS output. This is very different from the DPS load provided by the L-mode envelope tracking PA (5.20), where at low DPS voltages the PA appears to be a short circuit. The ET PA is likely though to never operate at a DPS value much below 1 V because of the severe output signal distortions that happen with P-mode operation.

6.9.1   Single stage polar operation

Following the same procedure used in analyzing the envelope tracking PA in Section 5.10, the corresponding model for the polar PA current is shown in Figure 6-24.

Using the model in Figure 6-24, PA current drawn from the DPS is directly written, using Ohm's Law, to this updated version of (5.21)

$$I_{PA}(V_{DPS}) = \frac{V_{DPS} - V_{AMO}}{R_L + R_{ON} \| r_o}. \tag{6.18}$$

The three resistances in the denominator of (6.18) have widely varying values in a typical circuit. Transistor output resistance r_o is usually several orders of magnitude larger than the transistor ON resistance R_{ON}. Therefore, r_o can almost always be safely neglected, though this must always be validated. Good C-mode design practice, as described in Section 6.7.5, calls for the design condition that R_{ON} be negligibly smaller than the load resistance R_L. For most practical designs, (6.18) can then be simplified to the form

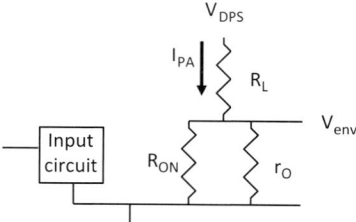

Figure 6-24   First-order model for the power supply load characteristics of a PA used in polar modulation.

$$I_{PA}(V_{DPS}) \approx \frac{V_{DPS} - V_{AMO}}{R_L}. \tag{6.19}$$

The only transistor specific parameter remaining in (6.19) is the voltage offset V_{AMO}. This is determined from a curve-fit procedure described in Section 8.7.6. If data from the particular transistor being determined evaluate to $V_{AMO} = 0$, then we get the further simplification

$$I_{PA}(V_{DPS}) \approx \frac{V_{DPS}}{R_L} \tag{6.20}$$

where all transistor specific parameters are insignificant to the actual stage operation. Consequences of this result are presented in Section 6.15.

6.9.2   Multiple stage polar operation

It is certainly possible to operate more than just the final stage in C-mode. If the driver stage is also operated in C-mode, then the power supply applied at the driver stage is used to control the amount of RF applied at the final stage. In particular, this can in principle be used to control the drive magnitude [6-10] [6-11]. We shall see that there are bad surprises in store when this design is not done carefully.

To illustrate the behavior surprise that lurks within a multiple stage polar design, we need a model of how a two stage amplifier cascade works. The model is shown in Figure 6-25, where the port voltages and currents are defined at the power supply inputs of each stage. Best performance is obtained with a separate DPS supplying each stage. There is a strong economic motivation to use only one DPS to power all of the PA stages together. If the multistage PA is designed for this technique, then all should be OK. Most multistage PAs though are not designed for this type of use, and a very bizarre behavior sometimes happens in these situations.

If a single DPS supplies both stages in Figure 6-25 when the DPS voltage is small, then the output signal from the driver is very likely to be too small to meet the C-mode drive requirements of the final stage. When the C-mode operating conditions are not met, the model of Section 6.7.5 does not apply and the final stage reverts to a CCS class of operation that the remaining bias does define. As the DPS value increases, eventually the signal magnitude from the driver stage does drive the final stage to operate in C-mode. C-mode operating currents are lower than the corresponding CCS bias currents for the same output power, meaning that the current drawn by the final stage can actually go

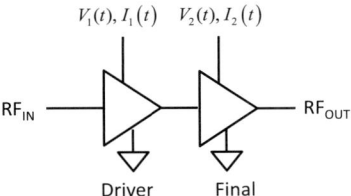

Figure 6-25   Cascade model of two amplifier stages.

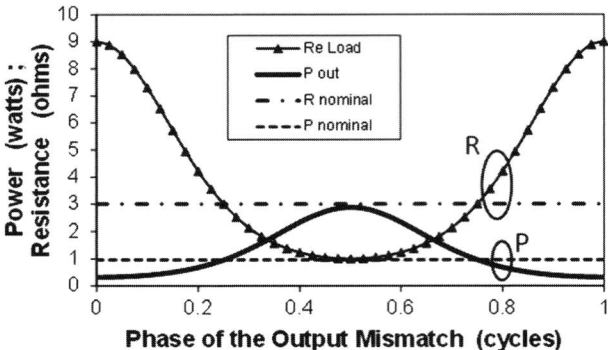

Figure 6-26   In the presence of an output load impedance mismatch, the actual output power from a C-mode PA varies with the phase of the impedance mismatch. Here the mismatch ratio is 3:1.

down with an increase in V_{DPS}. The PA can present a *negative* dynamic resistance characteristic to the DPS! Most DPS circuits will oscillate when presented with this kind of load. Circuit details for this undesirable operating characteristic are presented in Section 12.2.3.

6.9.3   Impact from output mismatch

For a C-mode circuit, the output power is described by (6.1). The value of R_{PA} is nominally equal to R_L from (6.13). When the value of R_L changes because the PA output load impedance (Z_{OUT}) varies away from its nominal design value

$$P(V_{DPS}) = \alpha \frac{V_{DPS}^2}{R_{PA}(Z_{OUT})}, \qquad (6.21)$$

the output power must also vary from its design nominal value. One example is shown in Figure 6-26, where the PA is designed to provide 1 W into a 3 ohm load resistance (the nominal case). With an output impedance mismatch of 3 to 1, as the phase of the output mismatch moves around a complete cycle the actual load resistances swings from a minimum of $R_L/3$ to a maximum of $3R_L$. Output power goes down for higher load resistance values, and up for lower values of R_L. More details are provided in Section 8.8.

The biggest problem occurs when R_L is around its minimum. Following the new design rules from Section 6.7.5, is R_{ON} still negligible compared to this reduced value of R_L? This leads to a further refinement to the design conditions of Section 6.7.5:

- to tolerate impedance mismatch at the output without distorting the output signal, divide R_{ON} from its nominal design value by the maximum voltage standing wave ratio (VSWR) expected.

For example, if the transmitter must tolerate a 3:1 VSWR without increasing output signal distortion, then the actual design value of R_{ON} must be one-third of the design nominal value. The RF switching transistor gets bigger, because nothing comes for free. But the load current does not change because the design conditions of Section 6.7.5 are still met. And now the tolerance of load VSWR without signal distortion becomes innate to the polar transmitter.

System specific requirements will define if this power variation is a problem or not. If it does become necessary to correct this power mismatch the discussion in Section 9.13 presents a method to manage an automatic correction.

6.10   Energy efficiency effects/power dissipations

The primary purpose for interest in polar operation of a DPS transmitter is to achieve the best energy efficiency possible. This means that the output power is maximized at the same time that the internal power dissipation is minimized. Besides the obvious need to operate in C-mode, special attention needs to be paid to the transition time of the output current from OFF to ON or back again. Duty cycle of the switching action also contributes to power dissipation and impacts on the achieved efficiency.

6.10.1   Referenced at the RF PA

The power dissipation characteristics of C-mode operation are quite different from those seen in Section 5.11 for L-mode operation. This difference is seen in Figure 6-27, particularly in comparison to the waveforms in Figure 5-36. Output current is controlled by changes in the input voltage for this FET example. The FET transitions between OFF and ON along the load line, just as if the transistor operates linearly. The same peak power dissipation seen in Figure 5-36 is experienced here too. But now, that high power dissipation happens for a much shorter amount of time, during the output waveform transition. This is visible in both parts of Figure 6-27.

Though the power dissipation profile for the ET operation is largely sinusoidal (see Figure 5-36), for C-mode operation this profile, shown as P_{FET} in

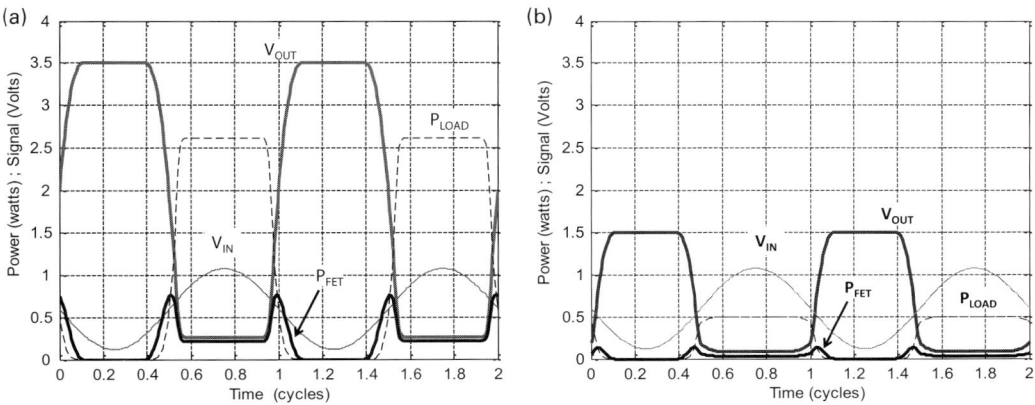

Figure 6-27   Power dissipation profiles for an FET operating in C-mode with the same DPS values from Figure 5-36 and a constant RF input, resulting in varying amounts of input overdrive: (a) 3.5 V supply; (b) 1.5 V supply.

Figure 6-27(a), is nominally rectangular with "ears" at each transition. The peak values of these "ears" are identical to the peak value of the L-mode power dissipation in Figure 5-36. The total heat energy put into the transistor to drive the temperature rise is controlled by the area of this power dissipation curve. With this C-mode operation, the time spent at higher power dissipation is reduced from that necessary for ET operation. This reduces the heat, and improves efficiency. Ideally, we want the time of these transitions – the duration of these "ears" – to be of zero duration. This of course is not possible. But it certainly can be reduced, and parameters affecting these transition times are discussed in Section 8.4.

When the power dissipation is minimized, efficiency can be increased further by increasing the output power under identical operating conditions. The peak-to-peak value of the RF output waveform here is larger than that seen in Figure 5-36, benefiting from both the elimination of the V_{OS} headroom necessary to ensure L-mode operation and from the distorted current waveform by operating at $g_S = 0$. This provides the desired condition of increased output power along with reduced power dissipation. Therefore, for the same value of the DPS the polar PA output power is always greater than the ET output power, the PA power dissipation is always lower, and the energy efficiency increases accordingly.

6.10.2   Overall power dissipation

In the distributed circuit that a polar transmitter uses, power dissipation occurs at each step as shown in Figure 6-28. All of the power dissipations add up to the total dissipated power according to

$$P_{DISS} = P_{OUT} \left(\frac{1 - \eta_{PS}}{\eta_{PA} \eta_{DPS} \eta_{PS}} + \frac{1 - \eta_{DPS}}{\eta_{PA} \eta_{DPS}} + \frac{1 - \eta_{PA}}{\eta_{PA}} \right). \tag{6.22}$$

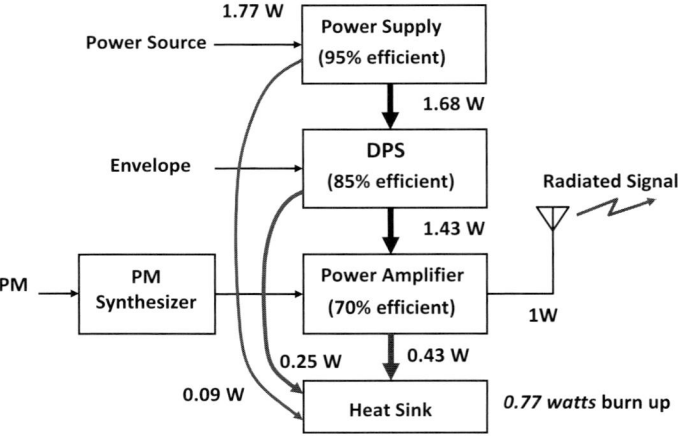

Figure 6-28   Power dissipation distribution among the circuits in a polar transmitter.

Dissipated power is proportional to the necessary output power in accordance with the collection of individual circuit efficiencies active in the transmitter. In (6.22), a separate top-level power supply is accounted for, as might be seen in higher power infrastructure equipment. For mobile devices that connect the battery directly to the DPS, this power supply block is effectively 100% efficient ($\eta_{PS} = 1$) and the first term in (6.22) goes to zero. To keep the total power dissipation small, and therefore the overall energy efficiency high, it is important that the efficiency of each individual circuit is as high as physically possible.

Starting from (5.24), we get for this architecture

$$\eta_{PA}(V_{env}) = \frac{P_{SIG}}{P_{SUPPLY}} = \frac{P_{SIG}}{P_{SIG} + P_{D,PA} + P_{D,DPS}}. \tag{6.23}$$

With this information, it is now possible to compare the efficiency performance of polar transmitters with envelope tracking and conventional linear operation, expanding on the information in Figures 5-37 and 5-39. With $V_{OS} = 0$, since the polar output signal envelope is directly controlled by the power supply and linear operation of the transistor is suppressed, the output signal from the FET has two major differences. First, the waveform is "square" and the magnitude of the fundamental frequency component is greater than the magnitude of the square wave by up to a factor of $4/\pi$ (depending on how square the waveform actually is). Second, the magnitude of this square wave is greater than the maximum magnitude of the output sinewave from the ET or linear PA because $V_{OS} = 0$ for the polar output signal. These combine in the polar architecture to provide greater output signal power for the same transistor size and power supply value than either linear or ET architectures can provide. This is seen in the top curve of Figure 6-29(a) showing available polar PA efficiency using the same transistor from the ET design in Figure 5-37. Corresponding power dissipation of the polar transmitter is added to the ET results from Figure 5-39 and shown in Figure 6-29(b).

Power dissipation in the polar (switching) RF power transistor is lower than either ET or linear amplifiers can provide because the voltage across the transistor is below the knee voltage when it is ON. This power dissipation is globally minimized when the transistor achieves this lowest power dissipation condition in the shortest possible time. Because the transition between ON and OFF states must proceed along the load line, the available speed of the transistor (related to f_T or f_{MAX}) has an influence that can be significant. This aspect is discussed in detail in Section 8.4.

6.10.3   Temperature rise

Temperature rise in any circuit is the direct result of heat flow – power dissipation moving to the heat sink through the thermal resistance as shown in Figure 3-4. To understand the behavior of temperature in any circuit, we need to understand both how power is dissipated while the circuit operates, along with the thermal resistances of the materials where this power dissipation occurs.

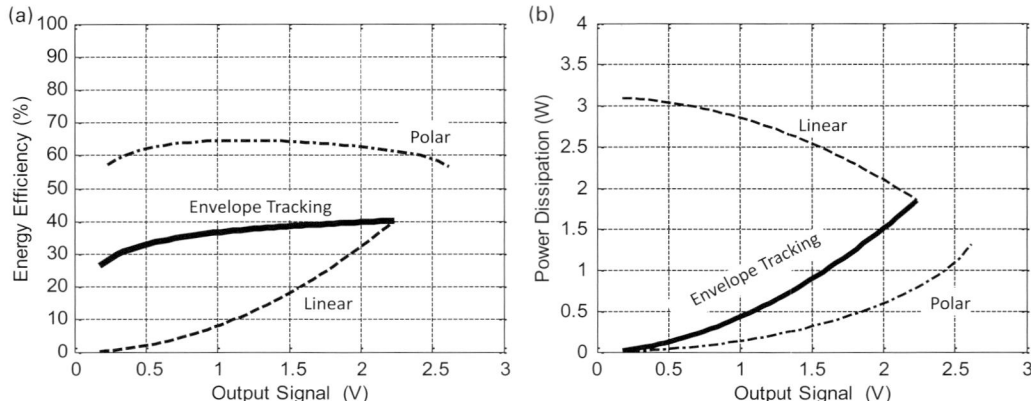

Figure 6-29   Available energy efficiency and RF transistor power dissipations comparing polar operation with earlier results for linear and ET design: (a) adding polar (dash-dot line) to the data in Figure 5-37; (b) polar power dissipation added to the data in Figure 5-39.

In a polar RF power stage, dissipation is zero when the transistor is OFF, then occurs during the ON state and along both transitions between these states. This dissipation is directly related to the output signal envelope, and thus to the value of the DPS, since in any polar operation the envelope is directly controlled by the DPS value. An interesting side effect of this mechanism exists.

For a polar DPS transmitter, this characteristic implies that the power dissipation and corresponding temperature rise should fall when an envelope-varying signal is being generated, compared to a CE signal at the same PEP. This means that unlike the conventional linear amplifier experience where power dissipation increases when an envelope-varying signal is used in a transmitter, here we get the opposite behavior. The amplifier actually gets cooler. Experiments verify that this is indeed true. One case is presented in Figure 6-30. Here the polar DPST operates at PEP = +34.1 dBm (= 2.5 W) with the GPRS TDM signal. When the GMSK CE modulation is used, the temperature rise increases as the number of active slots in the eight-slot frame is increased. When all eight slots are active, the measured temperature rise above ambient at the PA package is 18 °C.

When the EDGE 8-PSK modulation is used, the PEP is identical but the average power drops to +30.9 dBm (= 1.2 W) due to the PAPR of 3.2 dB for that modulation. As predicted, the temperature rise also falls with the envelope-varying signal. If the entire power dissipation came from the output stage, the expected drop in temperature rise would be proportional to the signal average power drop of –PAPR = –3.2 dB or to 48% of the value for the GMSK CE signal. The actual temperature drop is to a rise of +10.5 °C, 57% of the GMSK value. This shows that there are additional power dissipations in this PA, but the final stage is dominant. What is interesting though is that from an ambient temperature of 22 °C, the PA rises to 32.5 °C. This is lower than the normal human body temperature of 37 °C. Therefore by touching this PA operating at full power, heat flows from your finger into the PA. Such a transmitter is inherently safe.

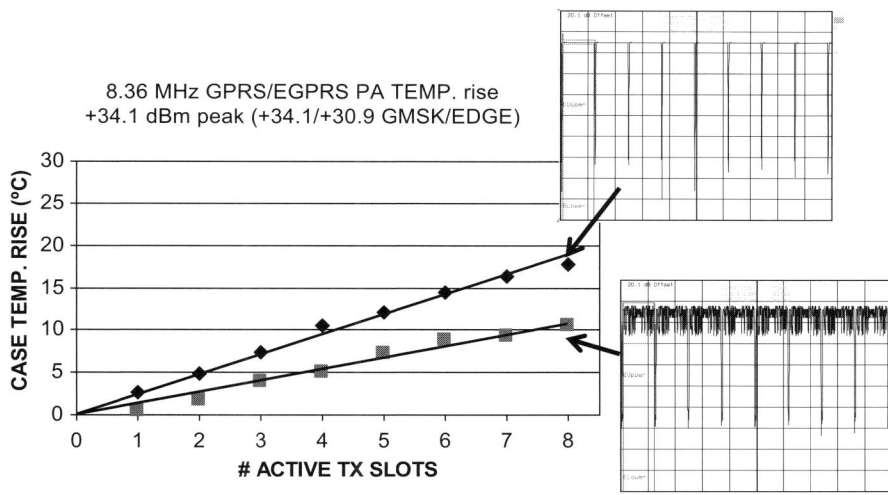

8.36 MHz GPRS/EGPRS PA TEMP. rise
+34.1 dBm peak (+34.1/+30.9 GMSK/EDGE)

Figure 6-30   GMSK and EDGE measured temperature rise of a polar transmitter, with identical PEP values.

6.11   Cross modulation

The output signal of a C-mode stage operating with polar modulation does not exactly implement the multiplication of (2.9). Errors are always present. Whether these errors cause problems, or not, depends on two things:

1. what signal type the transmitter is generating, and
2. how tight the system specifications are on that signal.

For example, the $\pi/4$-DQPSK signal for Bluetooth EDR is not nearly as tightly specified as the nearly identical modulation used in the TETRA system. At signal selection extremes, GMSK is very tolerant of polar distortion mechanisms, while OFDM is not tolerant of any distortion mechanisms, either linear or polar.

Major modulation distortion mechanisms of a C-mode stage are introduced below.

6.11.1   DPS-AM distortion

Linear amplifiers have a distortion mechanism called AM-AM, which refers to distortion on the output signal envelope caused by PA output compression that is encountered by envelope variation peaks on the input signal: AM on the input signal can drive AM distortion at the output if the amplifier compresses. Conventional AM-AM effects are generally very small at low output power, and get progressively worse as the output power increases.

The opposite is encountered in a polar transmitter, where distortion of the output magnitude is seen at lower magnitudes and not at the signal peaks. To begin, the conventional AM-AM mechanism is not possible in any C-mode stage because the slope gain is zero. All input envelope variations are removed at the output by C-mode

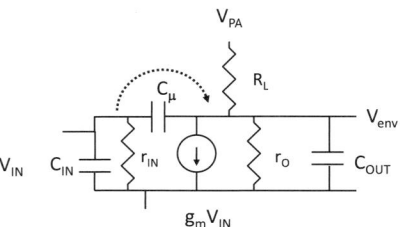

Figure 6-31   Leakage path through C_μ of the drive signal to the output.

Figure 6-32   Drive signal leakage through C_μ adds vectorally to the ideal signal producing the observed output (vector diagram).

operation: the RF transistor acts as its own RF limiter. Instead, the distortion mechanism here is an inaccuracy in the translation of the power supply value to output envelope (see the stage dynamic range discussion in Section 6.2.2), a process called DPS-AM distortion. In Figure 6-11, the translation of the power supply value to output envelope is accurate at the signal peaks and gets progressively worse as the output power gets lower. The reason this occurs is the subject of this section. And it is important to realize that this distortion mechanism is separate and distinct from the magnitude distortion mechanism from Section 6.8.3.

DPS-AM distortion is a result of limited isolation in the RF power transistor between the output and the drive signal at the input. Expanding the circuit model of Figure 6-24, we now add the reverse transfer capacitance C_μ to represent this coupling path.

Experiments show that this leakage signal is vectorally added to the intended polar output signal to produce the observed output signal

$$s(t) = \alpha V_{DPS}(t)\cos(\omega(t) + \phi(t)) + \kappa V_{DRV}(t), \tag{6.24}$$

where the envelope modulation scale factor α is a scalar and the driver signal isolation is defined by complex scale factor κ. A graphic of this mechanism is shown in Figure 6-32, where the first term in (6.24) is labeled the ideal output, the second term in (6.24) is labeled driver leakage, and the combination of these two phasors is shown as the observed output.

Vector summation (6.24) leads to a parametric model. The simplest version takes as one parameter the output signal at peak power and defines that as the normalized output model with unit magnitude and zero phase. The second parameter is the observed leakage signal from the driver at the output measured with respect to the peak output signal. In between these two endpoints, vector summation is done as the ideal output varies from unit magnitude down to zero. The result of this is shown in Figure 6-33 as the dashed line labeled "model." Also included in this figure are the expected output in the absence of any leakage called "square law," and measured data shown as open circles. Both models are normalized to the peak value of the data. This particular polar power stage (using a GaAs HBT) shows output power higher than expected for square-law

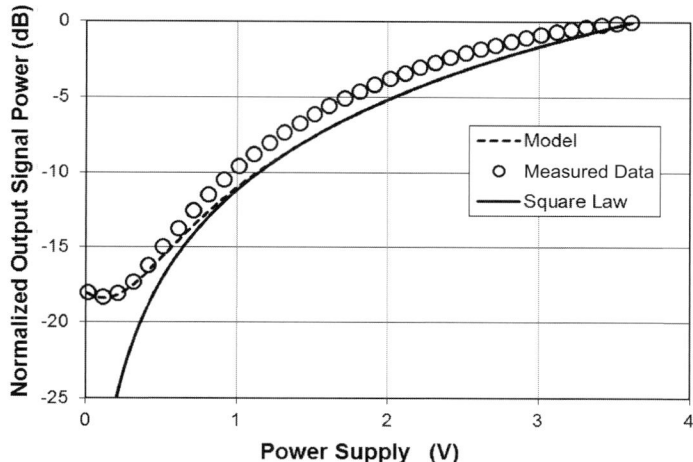

Figure 6-33   Modeled envelope distortion (dashed line) compared to the data (circles) and an ideal square-law response (solid line). Measured data above the ideal square-law line mean that this PA is not operating in C-mode at the highest output power, where the square-law scaling is done.

control. The model fit at minimum power through the output magnitude dip though is very good. This figure also shows that the output power apparently exceeds the square-law bound. This of course is not possible, so the actual behavior is that this PA is dropping out of C-mode at the higher output powers. The square-law model can be rescaled to accurately represent the region of proper C-mode operation, which will then show how the output drops from C-mode.

The fit of this model to the data allows the following parametric model to be used. Knowing only the maximum output power, and setting its phase as the model reference of 0 degrees, and the output leakage magnitude and phase when $V_{DPS} = 0$, the entire leakage curve can be drawn. A magnitude minimum occurs when the phase value of the leakage signal is greater than 90 degrees from the output reference phase defined earlier.

Figure 6-34 presents two sets of envelope magnitude curves using this parametric model for a leakage signal that is phase shifted 140 degrees from the peak output power. When the leakage is large, the distortion is significant at all values of controlling power supply. Observed errors in magnitude always begin toward the magnitude being too low, before turning around and eventually only showing the leakage signal magnitude. It is no surprise that output control accuracy is best when the leakage magnitude is small. It is interesting to compare these results with the data in Figure 6-11.

6.11.2   DPS-PM mechanism

Polar modulation (2.9) assumes that there is complete independence between the magnitude and phase modulation parameters. Experience shows that there is independence from the phase modulation to the magnitude modulation. Experience also shows that the other direction is definitely not independent; magnitude modulation does have an effect on the output signal phase. For a polar transmitter, this effect is most accurately

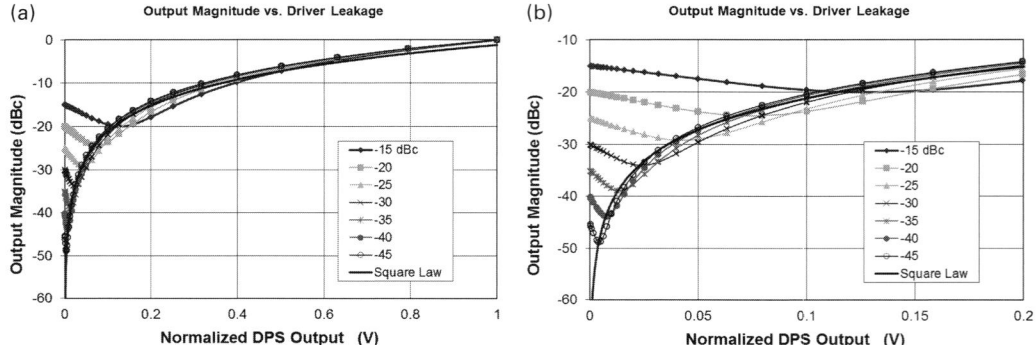

Figure 6-34   Modeled envelope distortion using the parametric model (6.24): (a) sweeping across leakage magnitudes from −15 dBc to −45 dBc; (b) the same data looking only at the bottom 20% of magnitudes showing that the envelope distortion always begins toward lower magnitudes for a properly modulated polar power stage.

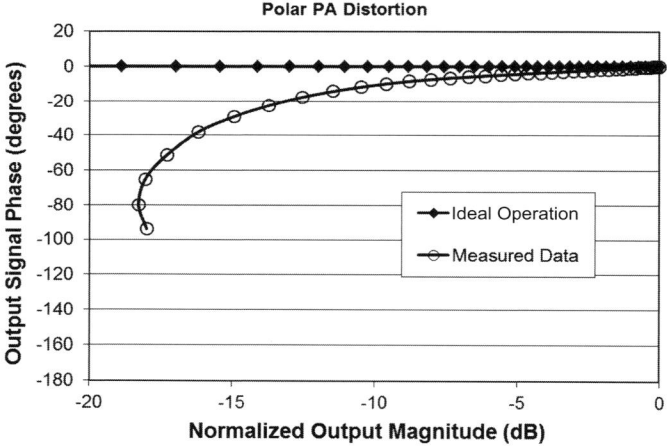

Figure 6-35   Ideal vs. actual output phase shift as the power supply is varied.

called DPS-PM distortion. It is important to keep this effect mentally separate from the commonly used AM-PM distortion measurement and characterization of linear amplifiers, where the PM is an effect from the amplifier input signal magnitude varying. This traditional measurement is not at all descriptive of the active mechanism here.

Unintentional phase shift through any C-mode power stage is also dominated by leakage of the switch driver signal to the output, as shown in Figure 6-35. Because the ideal output phase is nearly inverted (close to 180 degrees phase shift) from the input drive signal phase, the leaked input signal will partially cancel the main output at low magnitudes. This is the effect seen in the data in Figure 6-33. This magnitude variation also comes with a phase rotation, from the phase of the intended output signal to the phase of the leaking drive input signal. This phase rotation is seen in Figure 6-35.

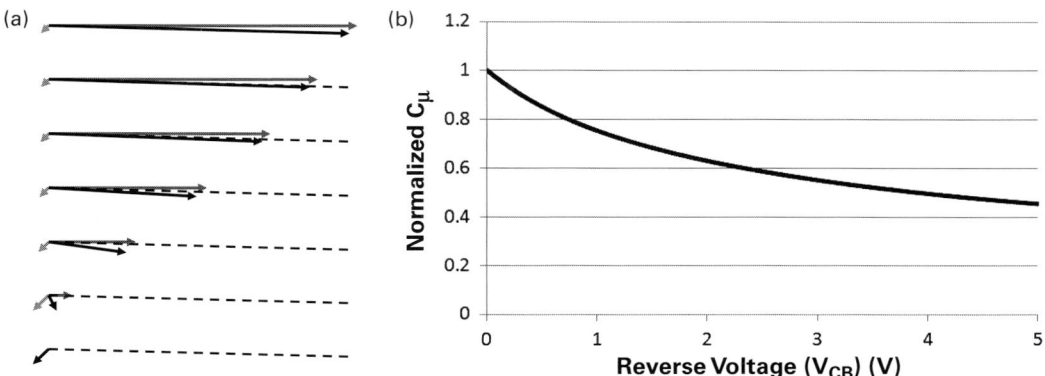

Figure 6-36   Unintentional phase-shift mechanism in a C-mode power stage: (a) vector summation of drive signal leakage with the switching output; (b) typical voltage variation of reverse transfer capacitance C_μ for a bipolar transistor (normalized to maximum value).

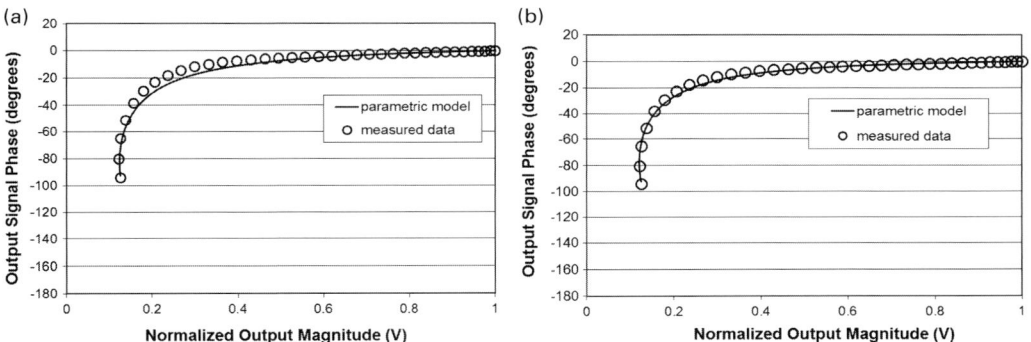

Figure 6-37   AM-PM distortion parametric model validation: (a) constant value leakage model fits the data reasonably well; (b) accounting for voltage variable capacitance improves the model to an excellent fit.

The mechanism for this phase shift is a simple vector summation of the leakage signal with the polar output signal, as shown in Figure 6-36(a). At large output power, the observed output is dominated by the intended polar signal, shown by the top vector. When the DPS voltage goes to zero, this intended signal also goes to zero. The only signal at the output now is the leaked driver input, with its magnitude and phase set by the leakage path isolation, and its phase shift, shown by the bottom phasor. In between these endpoints, the observed signal transitions its phase, with most of the rotation occurring at low output magnitudes, in accordance with the data shown in Figure 6-36.

Isolation between the input drive and the output signal does vary as the applied power supply voltage changes. The cause is voltage variability in the reverse transfer capacitance, where the value decreases as the DPS value increases. One example shape, typical for a GaAs HBT, is shown in Figure 6-36(b).

Assuming a vector summation parametric model based on this input drive, signal leakage leads to the results of Figure 6-37. Data are shown in the open circles, and the peak power normalized model is shown as a solid line. When the leakage is assumed to be constant at all values of the DPS, the model fit is seen in Figure 6-37(a). The fit is good, but pessimistic because it shows slightly more phase shift at intermediate power levels than is measured. Adding consideration of the voltage variation in the reverse transfer capacitance, we get the result in Figure 6-37(b). Now the fit is even better. This validates the simple model of vector summation of input drive signal leakage with ideal polar control of the C-mode output signal, preferably including the voltage variation of the reverse transfer capacitance.

6.11.3   Desired input magnitude variations

Leakage is reduced with lower input signals into the C-mode stage. As the output signal gets smaller, the discussion in Section 6.8.3 shows that dynamic variation of amplifier bias and drive signal magnitude is desirable to inherently linearize control of the output envelope. In this section, we see that the same effect holds for inherently correcting phase distortion from a switching transistor due to leakage of the drive input signal.

Reason would suggest that if leakage from the drive signal dominates distortion mechanisms on the output signal, then managing the drive signal is an appropriate technique to minimize this distortion. The curves in Figure 6-38 show that this strategy does work. The need for a minimum drive signal magnitude to keep the power stage in C-mode, as seen in the Booth chart of Figure 4-22, limits the ability to completely compensate for both of these distortion mechanisms. If the dynamic ranges of magnitude and phase control still need to be extended further, an open-loop modulation pre-distortion on the RF path can be applied. Operation in C-mode shows excellent stability and repeatability for any particular transistor type [6-7], allowing this simple open-loop control for fine signal correction.

These results demonstrate that it is actually *bad design practice* to hold the drive signal constant when doing polar modulation. It is output signal accuracy that counts. Providing this accuracy by minimizing inherent distortion mechanisms through direct

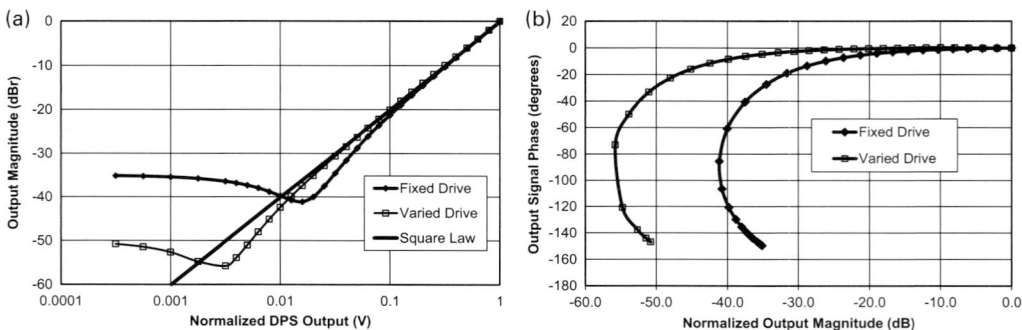

Figure 6-38   Varying the input drive level minimizes both (a) output magnitude error; (b) output phase distortion.

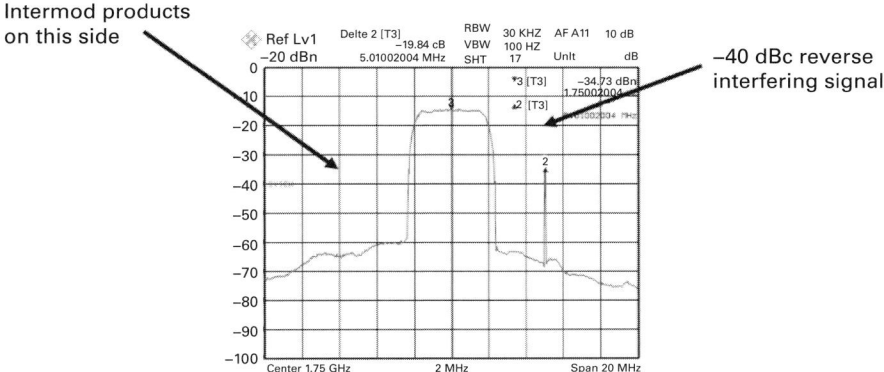

Intermod products
on this side

−40 dBc reverse
interfering signal

Figure 6-39   Reverse intermodulation performance of a C-mode polar power stage is much better than conventionally experienced with linear amplifiers.

control of the modulation process is better design practice than simply accepting the distortions and applying external linearization later.

6.11.4   Reverse intermodulation

Reverse intermodulation is a specification of increasing importance when many radios using similar frequencies are present in the same vicinity. This situation is very common for mobile phone use at events, in restaurants, and so on. This effect is due to the transmitted signal from nearby mobile phones not only going to the base station antenna, but also into your antenna. This sends a relatively strong signal back into your transmitter, where it will interact with the intended transmitter signal you are sending. Because all transmitters have nonlinearities present at the transmitter output, mixing and intermodulation products will be generated. These undesired signals are, of course, present at the transmitter output port. So, along with the desired transmitter output, they will head to the antenna and be radiated into the network.

For a C-mode transmitter, the nonlinearities are maximized (and linearity is minimized) by the power transistor switching action. We expect that reverse intermodulation performance from a polar transmitter to be very bad. Surprisingly the opposite is true. Testing shows that reverse intermodulation of a polar transmitter is actually very good, with a measurement example presented in Figure 6-39. Circuit views of this performance are presented in Section 8.9.2.

6.12   RF output power control

The key feature of any polar transmitter is that the power supply directly controls the output envelope magnitude. Unlike for linear amplifiers or ET, where the amplifier input signal controls the output envelope, polar operation moves responsibility for output signal accuracy from the RF power transistor and places it on to the accuracy of the power supply.

6.12.1 High output powers

When the polar (C-mode) RF power stage is operated in the ways shown in Section 6.8.3 that provide linear translation between the DPS value and RF envelope, then power control is readily implemented by simply scaling the DPS values used to implement the signal envelope in accordance with Ohm's Law. Taking the envelope $a(t)$ from (2.9) with its rms value of a_{rms}, for that envelope output power we can rewrite (6.1) as

$$P_0 = \frac{a_{rms}^2}{R_{PA}}. \tag{6.25}$$

Scaling this envelope voltage waveform for other output powers with the linear parameter γ gets the relationship

$$P(\gamma) = \frac{(\gamma a_{rms})^2}{R_{PA}} = \gamma^2 P_0 \tag{6.26}$$

which is no surprise because power is proportional to voltage squared.

As the output power decreases linearly, the envelope voltage decreases quadratically. Very quickly the important DPS values drop below 100 mV. As long as the isolation provided by the transistor discussed in Section 6.11 remains sufficient to realize control to this dynamic range, such low DPS values are not a problem. Often, however, the driver magnitude management techniques discussed in Sections 6.8.3 and 6.11 are both needed to achieve an envelope control dynamic range exceeding 40 dB.

6.12.2 Low output powers

When the DPS value drops below 1 V, the design requirements from Section 6.7.5 must be carefully checked. This is particularly true when a bipolar transistor type is being used, because the value of V_{AMO} may not be negligible at low DPS values. If, for example, the value of V_{AMO} is evaluated to be 0.11 V, then this will produce a 1 dB error when the power supply voltage falls below 1 V ($= 0.11/(1 - 0.89)$). At DPS values lower than this, some correction for the presence of V_{AMO} is required, which increases the complexity of such a transmitter. Another option is to select a transistor that has $V_{AMO} = 0$, which completely avoids this design issue.

At power levels where the driver signal leakage exceeds the needed envelope floor, then signal quality, measured by both EVM and ACLR, degrades rapidly. The only way to improve this situation is to reduce the magnitude of the drive signal [6-12]. The constraints allowing this to happen while maintaining C-mode operation are presented in Sections 6.8.3 and 6.11. Reducing the drive level in the C-mode stage to no more than is truly necessary for C-mode operation is good design practice in all cases.

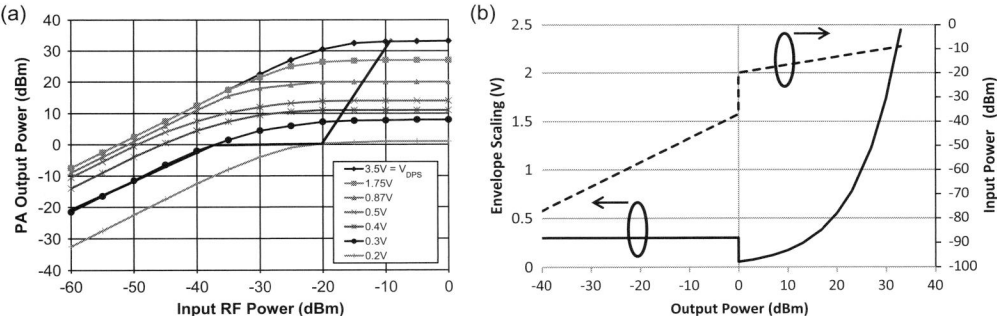

Figure 6-40   Power control strategy for large output dynamic range: (a) trajectory on the Booth chart; (b) envelope scaling and input RF power across the output power dynamic range.

Eventually, if further reduction of output power level must be realized, C-mode operation must be abandoned for either L-mode or P-mode operation. At very low power supply values, L-mode operation is not possible, which only leaves P-mode.

6.12.3   Very low output powers

P-mode is an excellent mechanism to use with a polar transmitter to achieve output power control to nanowatt levels. Using P-mode, the output envelope is applied using the DPS as is usual with C-mode operation. The phase modulated RF signal now *must* be constant magnitude for any particular output power in order to maintain envelope modulation accuracy on the output signal. By varying the (now constant envelope) RF input power, the output power is controllable to very low values with excellent signal quality [6-13].

One way to visualize the transition from C-mode power control to P-mode power control is shown in Figure 6-40. Building on the Booth chart from Figure 4-22, in Figure 6-40(a) within the C-mode region the black solid line shows the selected drive changes for various output powers. The output power itself is controlled only by scaling the DPS output (envelope) signal, in accordance with (6.25). This is the parabolic shaped solid line in Figure 6-40(b).

At 1 mW (0 dBm) power, this particular amplifier no longer easily supports C-mode operation. Looking again at Figure 6-40(a), we see that a transition to P-mode is possible by reducing the RF input power from −20 dBm to −37 dBm. The DPS value must increase to that of this next curve in the Booth chart. Power control then proceeds downward by decreasing the magnitude of the input signal while holding the power supply scaling constant. This is specifically shown in Figure 6-40(b), where the transition from C-mode to P-mode is set at 0 dBm output power. Below this output power, the power supply scaling (solid line) is constant, while the input power (dashed line) slopes down decibel for decibel. It is important to note that the "fixed" power supply here simply means that the envelope variation from (6.25) still happens to set the envelope variation, now with the scaling factor γ set to the constant $\gamma = \gamma_P$ to maintain the same method of envelope control used in C-mode, maintaining polar operation from (2.9). In accordance with (4.17), this polar envelope modulation works as long as the signal applied on V_{GS} is held at a constant magnitude for each output power setting. In P-mode, the input power is generally much less than the output power, so the signal leakage concern from C-mode operation is not an issue. The entire envelope dynamic range at each output power level is set by the DPS.

By converting modes in this way, control of transmitter output power splits when P-mode is active. Here the envelope $a(t)$ is controlled by the DPS, keeping this a polar operation, but the average output power (γ) is controlled by the signal magnitude into the RF transistor [6-14] [6-15] [6-16]. This is very different from when the polar transmitter operates in C-mode, where both envelope and average power level are controlled together through the DPS action on the C-mode RF power stage (6.22).

At the boundary between operating with power control provided with C-mode operation or with P-mode operation, there is a calibration required. It is fortunate that there is a natural internal standard available to do the calibration against, using the inherent precision of C-mode. With a local power detection circuit available, the boundary output power is first produced using C-mode operation and then measured with the local detector. Then the mode shift to P-mode is performed, set to put out the same output power. This power is also measured, and the measurement results are compared [6-17] [6-18]. Any difference is due to inaccuracy in the P-mode operation, which is internally adjusted as necessary to bring the measurements into balance.

One interesting way to illustrate this very high dynamic range power control technique in action is to use the EDGE signal, as shown in Figure 6-41. We use the EDGE signal because it has a finite envelope floor, meaning that if the signal power is scaled down

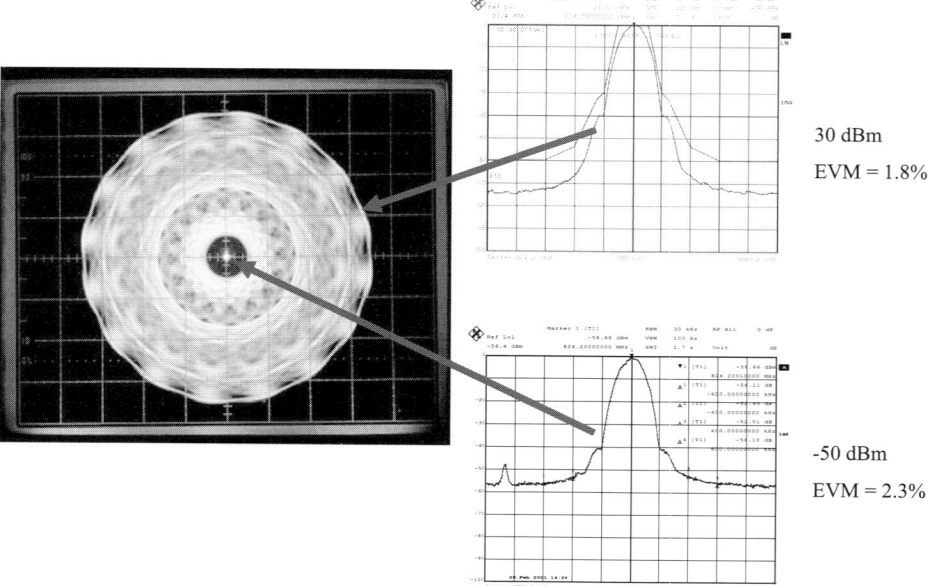

30 dBm

EVM = 1.8%

-50 dBm

EVM = 2.3%

Figure 6-41   Power control using P-mode for the EDGE signal: 80 dB change in output power with essentially the same signal accuracy.

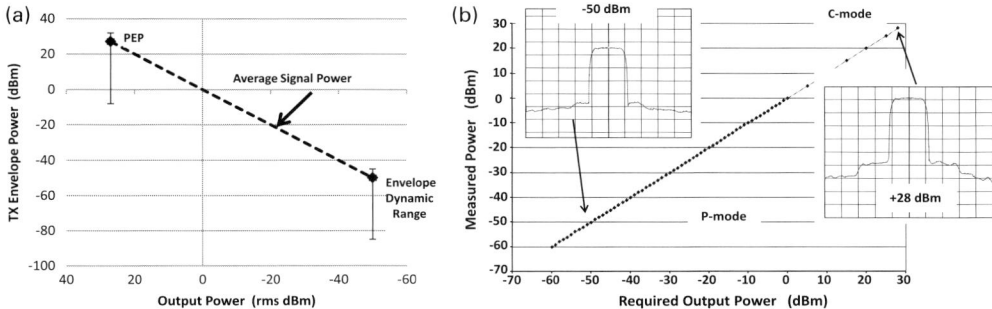

Figure 6-42   WCDMA power control and envelope dynamic range can sum to greater than 100 dB in a polar transmitter: (a) conceptual diagram; (b) implementation within a WCDMA mobile transmitter.

such that the new signal peak power is below the earlier minimum envelope power, an overlay of the two vector diagrams will have the low power signal visible within the central "hole" of the high power signal. The multiple exposure photograph in Figure 6-41 shows exactly this, with the high power (1 W) C-mode signal dominating the measurement, and the low power (10 nW) P-mode signal shrunk to a small dot at its center.

Spectrum analyzer and signal analyzer instruments are used to investigate the modulated signal quality at these widely separated power levels. In both spectral measurements from Figure 6-41, the top 50 dB of the signal power spectral densities are exactly the same. The noise floor is much higher (relatively!) in the lower power signal measurement. This reduces the signal analyzer's ability to accurately determine signal modulation accuracy, and largely accounts for the slight increase in EVM reported.

The available power control dynamic range using all of these techniques can easily exceed 100 dB as shown in Figure 6-42. Whether this power control is used for signal power or envelope variations at a particular power is not material to this transmitter design. To a polar transmitter, envelope value is envelope value whatever the underlying cause might be.

6.12.4   Automatic low battery compensation (ALBC)

For battery powered transmitters, a critical concern is what happens as the battery drains, its output voltage lowers, and its source impedance increases. Operation of the wireless system continues on. If the system requires peak output power from the transmitter and the battery is unable to provide that power, the envelope modulation process will not reach the necessary PEP. This clips the output signal and results in significant distortion on the RF output signal.

One way to manage this problem is to have the local microcontroller monitor the battery voltage, and to manage commands to the transmitter about output power settings when the battery voltage gets too low. This certainly works, but it requires additional routines in the product software with all the necessary integration tests and data flow certifications. If the transmitter could handle this autonomously, then there is no impact

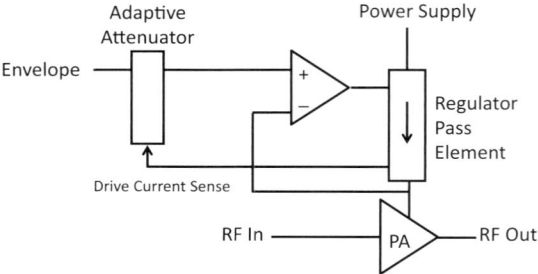

Figure 6-43   Sensing the pending compression of the DPS output pull-up is achieved by noting the need for increased drive into the output regulator before the output stage itself begins to compress from low supply voltage.

on product software and such a transmitter can more easily be adopted for use under multiple product platforms.

Such an autonomous peak power management technique is called automatic low battery compensation (ALBC) [6-19] and has well-proven performance for maintaining output signal precision as battery energy drops, and is replenished by recharging, during transmitter operation. The key concept is shown in Figure 6-43, where circuitry is added to the DPS output circuitry that senses when the output transistors are beginning to experience insufficient voltage headroom, but still have enough gain to ensure that the supply voltage to the PA is still accurate. This sense circuit then notifies a fine-step attenuator on the envelope waveform input to step the envelope scaling down a small amount, just until the voltage headroom within the DPS recovers. The RF output power correspondingly falls slightly, in accordance with the lower voltage available for the transmitter to operate. Full output power returns when the supply (battery) voltage recovers to nominal levels.

A detailed description of the operation of ALBC circuitry and operation is provided in Section 9.12.

6.13   Handling a zero output (IQ origin crossing)

One of the major difficulties with any polar transmitter is its difficulty to generate a zero-magnitude output signal. There are two ways that this problem is readily evident, first using the Booth chart shown in Figure 6-44(a). Looking at the C-mode section of these curves, even with a low power supply value the RF output signal is far from zero magnitude. This conclusion is supported by the data in Figure 6-44(b) where the dashed line shows that at an applied power supply of 0 V, the output shows leakage of the drive signal across the switching transistor to the output. The mechanism for this effect is drive signal leakage across the C-mode transistor which is presented in Section 6.11.

Ideally, transistor isolation of its input signal can improve to infinity, and this problem would be solved. Such a result is impossible with actual transistors, so other techniques are needed to get the output signal to behave as the system signal designers require.

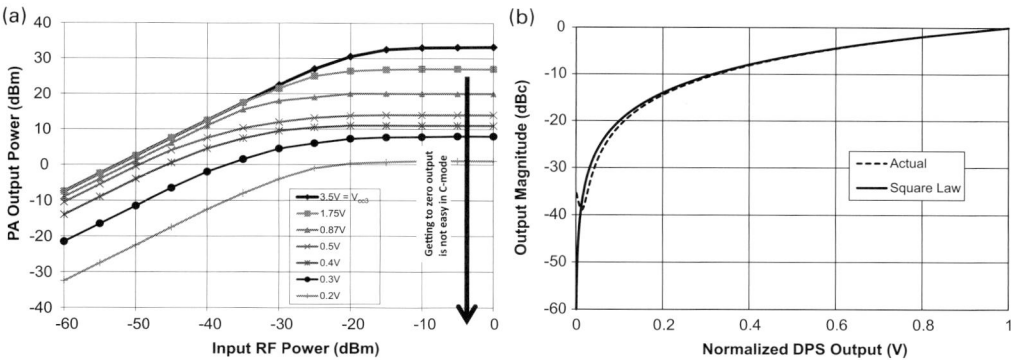

Figure 6-44   C-mode does not readily support the output signal going to zero magnitude: (a) Booth chart view; (b) power curve view.

Techniques to solve this problem do exist, and in the following sections six different techniques designed to mitigate this problem are introduced.

All of these techniques are evaluated equally by using the two-tone signal shown in Figure 6-45. The envelope is shown as a bold curve bounding the signal waveform peaks in Figure 6-45(a). The all-important details of envelope zero performance are shown in Figure 6-45(b). This figure will be used to evaluate the performance of the following envelope zero techniques. The first comparison (see Figure 6-45(c)), shows the output waveform distortion near the envelope zero due to leakage from the driver signal which is 35 dB below peak output power with a relative phase shift of −155 degrees (refer to the model in Section 6.11.2).

6.13.1   Forced zero output

Probably the most obvious technique to get the output of a polar transmitter to a zero value is to turn the output stage off and short the RF signal to ground for the required very short amount of time. There are also artifacts of the digital signal processing that can miss the envelope zero event entirely, keeping it from getting to the polar output stage [6-20]. This is a particular problem because the zero-magnitude event has discontinuous derivatives, which result in significant bandwidth expansion as required by the Fourier transform. An envelope zero event makes it extremely difficult to meet the Nyquist criterion for guaranteed complete waveform reconstruction.

Fortunately, the DSP can determine if an envelope zero event must occur. If so, then the output stage can be powered to zero and the remaining leakage signal at the output can be briefly shunted to ground. This approach is outlined in Figure 6-46. After the envelope zero event, the switch is released and the signal modulation continues on normally.

This shunt switch must remain in a truly OFF state for any output signal that the transmitter may generate. This is especially important for negative swinging output amplitude values. Any current flow, including currents into the switch device capacitances, is a distortion mechanism on the output signal. It is important to remember that

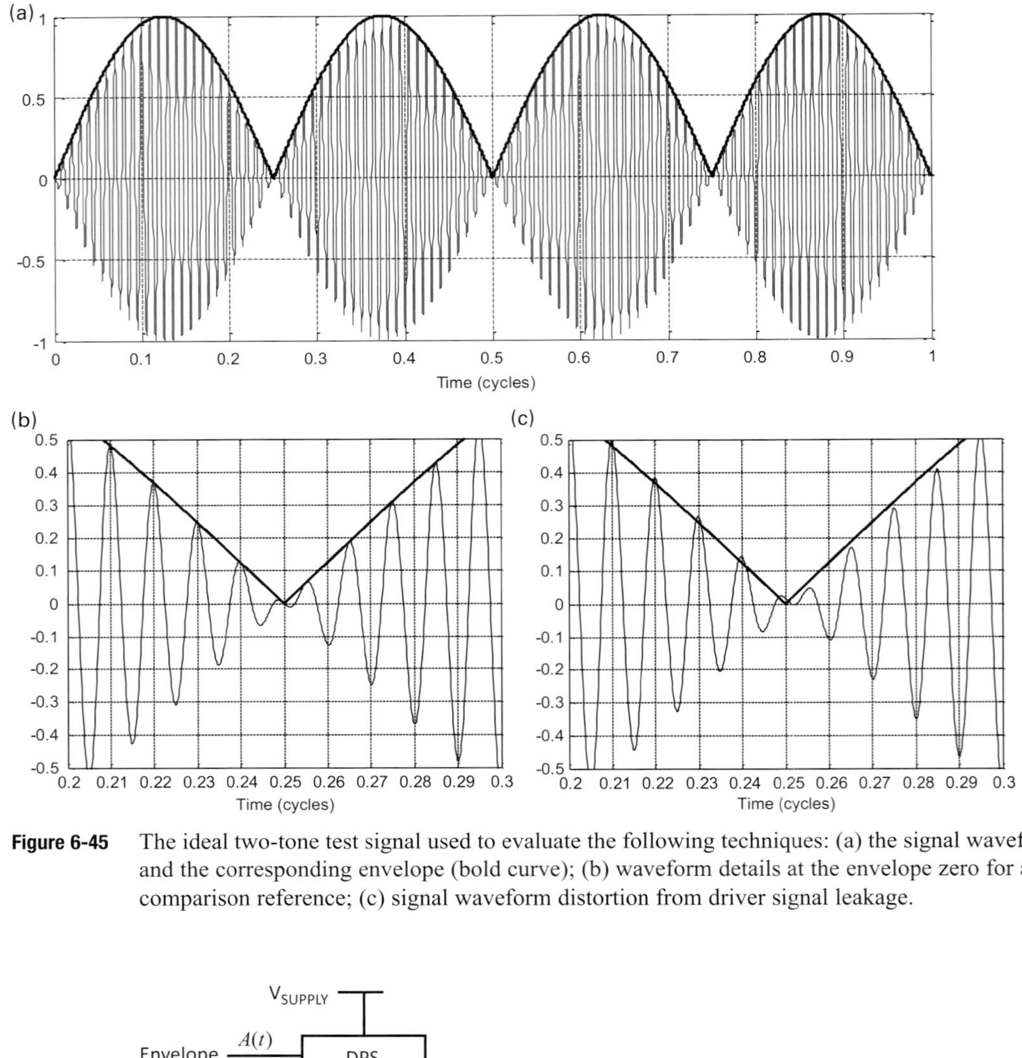

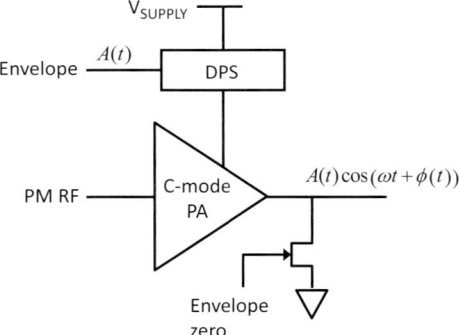

Figure 6-45 The ideal two-tone test signal used to evaluate the following techniques: (a) the signal waveform and the corresponding envelope (bold curve); (b) waveform details at the envelope zero for a comparison reference; (c) signal waveform distortion from driver signal leakage.

Figure 6-46 Using a shunt switch to force the output signal to a zero value, here shown at the output of the polar transmitter.

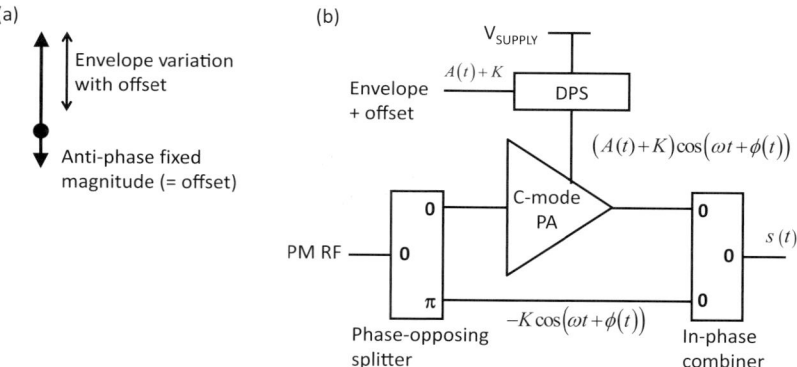

Figure 6-47   Block diagram of the opposing phase offset technique.

transistor parasitic capacitances are usually nonlinear with applied voltage, and there-fore nonlinear distortion effects are likely if the output signal swings become large.

6.13.2   Opposing phase offset

An alternative approach is to add a signal into the output stage of the transmitter which is larger than the leakage signal from the driver through the output transistor [6-21]. This technique is presented in Figure 6-47. Here the injected signal is provided in phase opposition to the nominal output signal. The polar envelope modulator must account for this signal by adding an offset to the envelope which is equal to the magnitude of the added signal. Envelope modulation otherwise proceeds normally. An output combiner adds these two signals together to get the final output. When the minimum envelope combines with the injected signal, the magnitudes are equal and the phases in opposition, providing the desired envelope zero at the output without needing to generate a zero-magnitude signal.

Signals demonstrating the operation of this technique are shown in Figure 6-48. The first illustration, in Figure 6-48(a), shows the component parts of this technique. Here the offset envelope and the resulting polar modulated signal from it are seen exceeding the desired output envelope (long-dashed curve), along with a small magnitude signal in phase opposition to the modulated signal. Leakage from the driver is present in the modulated signal. After the in-phase combiner, the actual output signal is shown in Figure 6-48(b). The desired envelope is restored, and the envelope zero event is now seen. Closer inspection of the envelope zero event is seen in Figure 6-48(c), where the signal waveform is essentially identical to the ideal wave shape shown in Figure 6-45(b).

This technique requires that all of the phase modulation in the output signal must be present in the injected signal. This can be a problem for the phase inversion that occurs at most zero-crossing events. If an isolating combiner is used, there is also power lost in the isolating resistor that reduces overall energy efficiency. Of course, this injected signal is small in magnitude, so the combiner does not necessarily need to be balanced.

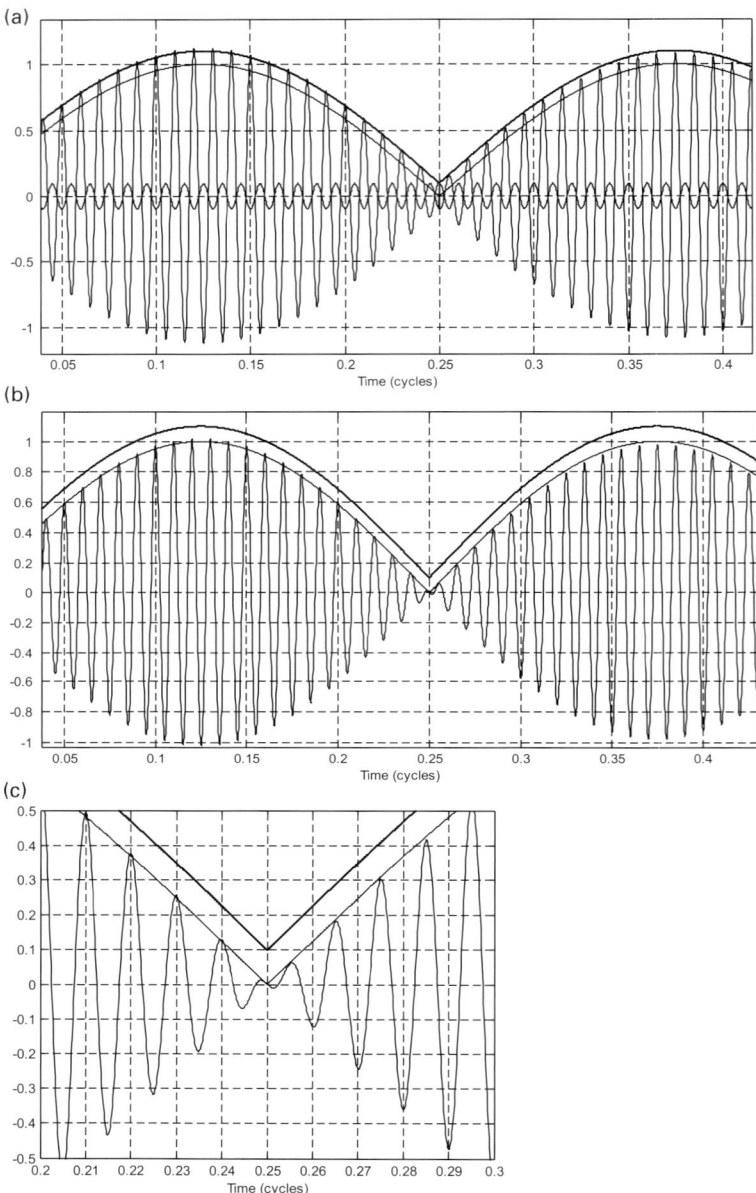

Figure 6-48   Generating an opposite phase signal to avoid distortion from the driver signal leakage: (a) component signals in this technique; (b) the output signal from the combiner; (c) close-up view of the envelope zero event demonstrating the accuracy of this technique.

6.13.3   Opposing phase summing

Avoiding the need to generate a phase reversal in any modulated signal is desirable, and the opposing phase summing technique shown in Figure 6-49 is one way to do this

(a)   (b)

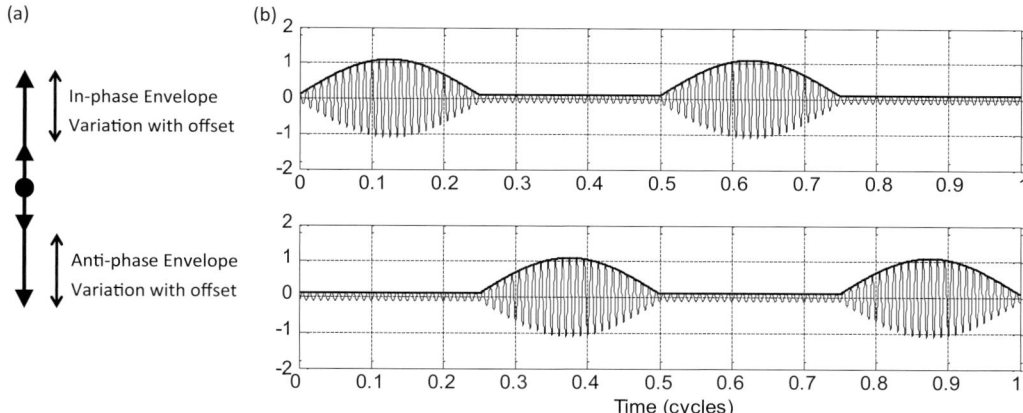

In-phase Envelope
Variation with offset

Anti-phase Envelope
Variation with offset

Time (cycles)

Figure 6-49 Principles of the opposing phase summing technique: (a) phasor diagram; (b) time waveforms of each signal.

[6-22]. Here the idea is to have two phase-modulated carriers, each in phase opposition to the other. Because of this, the actual phase modulation applied to either component carrier only needs to span 180 degrees, not the full 360 degrees.

This technique is an expansion on the opposing phase offset technique of the prior section. Here, whenever one signal is being modulated, the other is held constant at a small level that exceeds the leakage from the modulated stage. When a signal zero crossing occurs, the paths change roles, i.e. which is modulated and which is the offset injection.

Each envelope modulator must provide an offset above the desired envelope, as is required in the opposing phase offset technique. The block diagram for the opposing phase summing technique is seen in Figure 6-50(a). The two paths are now balanced architecturally, even though only one is providing the majority of the output signal power at any particular time. Envelope control signals into the DPS for each path are different from each other. This is readily seen in Figure 6-50(b), where the individual envelope control signals are shown by the solid curves. The envelope control signal for the phase-inverted path is shown here with its sign inverted for clarity. The intended output envelope is shown by the long-dashed curve, as previously. The presence of an envelope zero event is apparent.

Figure 6-50(c) provides a close-up view of this envelope zero event. The controlling envelope signals are quite different from other techniques, but the accuracy of the output waveform does match the ideal from Figure 6-45(b). This technique does meet its objective: low magnitude signal distortion from the transistor driver signal leakage is essentially eliminated.

As before, when an isolating combiner is used the difference in the input signal magnitudes does cause power to be dissipated in the isolating resistor. This presents a loss mechanism that is not desirable. The baseline injection signals are small, but this time it is harder to avoid using a balanced combiner because the output power can come from either input. Use of a nonisolating combiner is possible because of the constant phase-opposition relationship of the input signals.

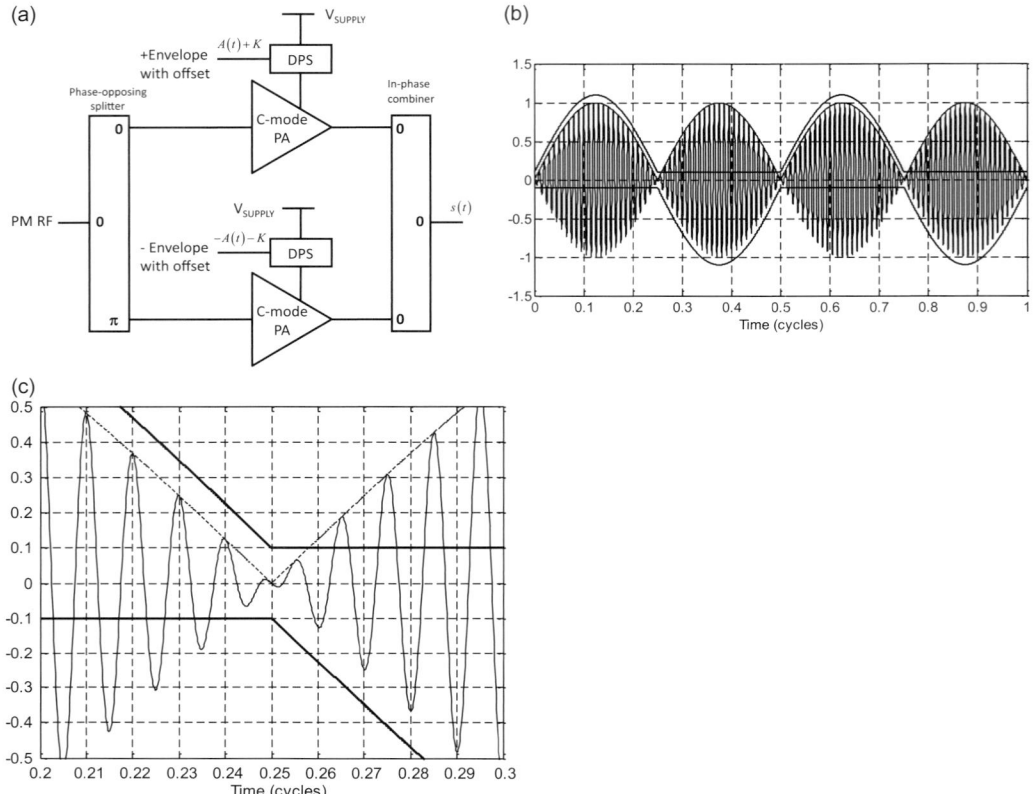

Figure 6-50   Opposing phase summer: (a) block diagram; (b) output signal; (c) close-up view of the accuracy at the envelope zero event.

6.13.4   High-level polar/low-level LINC (HLP/LLL)

Using the phase-opposition techniques of Sections 6.13.2 and 6.13.3, there always is power generated in the polar stage that never gets to the output. This loss can be avoided, even in an isolating combiner, if the magnitude and phase of both input signals are the same. Of course, the problem of zero magnitude remains, so a conversion is done at very low output magnitudes from a matched polar modulation to conventional LINC outphasing modulation. At these low output powers, the power lost in the isolating combiner is small and is considered negligible.

The principle of this technique is illustrated in Figure 6-51(a), best considered in combination with the block diagram in Figure 6-51(b). Each path of this architecture operates with an identical phase whenever the output signal magnitude is high. Inputs to each envelope modulator are also identical to provide well matched signals into the output combiner. This operates the output combiner in its minimum-loss condition, maximizing the possible energy efficiency.

When the output signal magnitude gets small enough that the driver signal leakage begins to be a problem, this phase-matched and polar envelope modulated operation

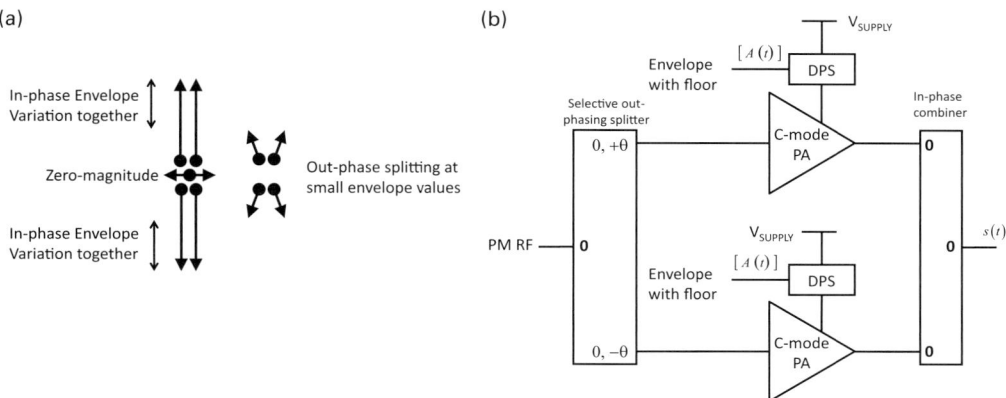

Figure 6-51   Principles of the high-level polar/low-level LINC technique: (a) phasor description; (b) simplified block diagram.

converts to a conventional outphasing approach and the signal envelope control stops going down [6-23]. Any envelope zero event is realized by the two RF signals having exact opposite phases at their reduced magnitudes, as shown in Figure 6-51(a). Output signal phase reversals only occur at the output of the combiner, and never in the phase modulators. On the other side, when the output signal magnitude exceeds the outphasing threshold the two RF signals return to being in-phase and all further magnitude variation returns to using polar modulation of both RF paths. Because the envelope modulation signals are identical for both RF paths, it is possible with careful design (to ensure sufficient isolation) to use only one DPS to feed both polar modulation stages. And at the combiner it is practical to use the easy isolating combiner of the LINC technique to avoid all of the impedance variations and linearity issues of the Chireix combiner, because the power dissipated in the combiner with this HLP/LLL technique is inherently small.

Example waveforms for this technique are shown in Figure 6-52. The two RF signals are shown in Figure 6-52(a), where the envelope control signal never goes to zero, avoiding the distortion and dynamic range limitation from the driver signal leakage. Any phase reversal in the output signal does not need to be generated in the phase modulators, avoiding the bandwidth expansion mechanism in the RF circuitry. Also, unlike the prior techniques the signal envelope generated is the exact envelope needed at the higher output powers, eliminating the combiner loss mechanism.

Adding these two signals in a combiner yields the output waveform shown in Figure 6-52(b). The higher magnitude in-phase signals add directly with minimum loss, even using an isolating combiner. At lower magnitudes, the outphasing technique is used and the output signal envelope can go to zero. Output phase reversals at the envelope zero events therefore also occur only at the output of the combiner. The quality of the envelope zero event is seen in the close-up plot of Figure 6-52(c).

Advantages of this technique include low bandwidth expansion on both AM and PM, low combiner loss, and the ability to use a simple isolating combiner for easy impedance control at each PA output. Large power dissipation in the combiner from a traditional LINC system is completely avoided.

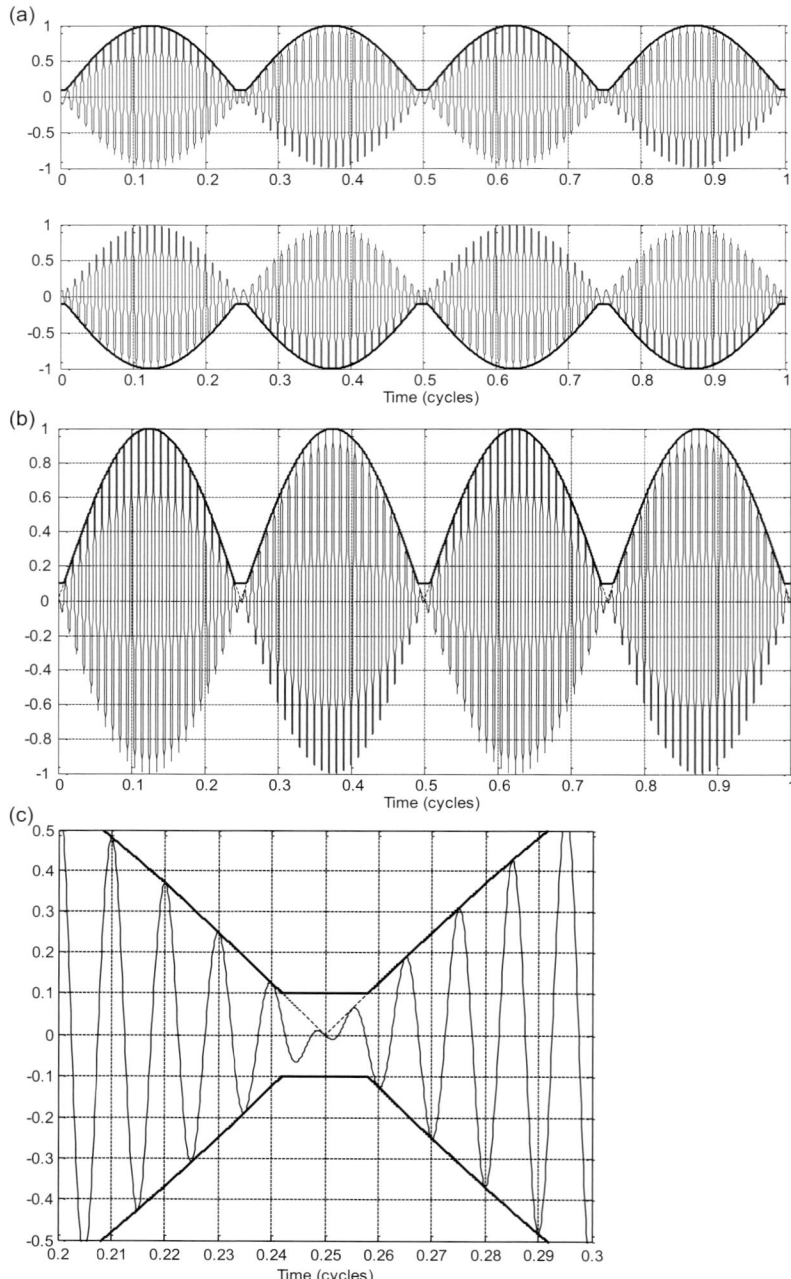

Figure 6-52   Losses in the combiner are reduced, and dynamic range is increased, by combining polar power control at high levels with outphasing/LINC at lower levels: (a) individual RF path signals with their controlling envelope waveforms; (b) output from the combiner showing both envelope control and realization of envelope zero events; (c) close-up view of an envelope zero event.

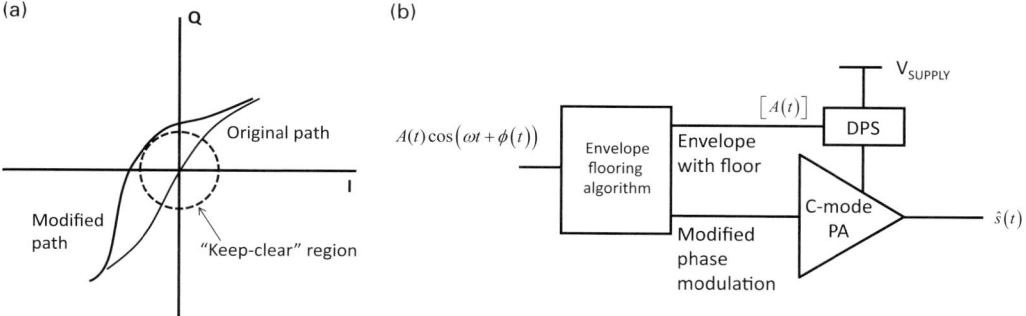

Figure 6-53   Principles of the envelope flooring polar technique: (a) example of a re-routed signal path to be outside the exclusion region around the signal plane origin; (b) block diagram.

6.13.5   Envelope flooring

One strategy to manage the problem of envelope zero events is to simply not allow them to happen. This technique is called envelope flooring, and entails changing the routing of signal modulation paths in the vector diagram such that no path approaches closer to the IQ plane origin than a selected amount [2-7] [2-8]. When the selected exclusion zone, sometimes called the "keep-clear region" or the "white hole," is large enough to avoid problems from the drive signal leakage, then no additional RF path is needed in the transmitter. This keeps cost lower than for any other technique presented in this section.

One example is seen in Figure 6-53(a). Here an original modulation path goes through the origin, clearly violating the keep-clear region. The envelope flooring algorithm adds offsets to the modulation waveform with sufficient magnitude and appropriate phase to temporarily modify the signal path to something else as the exclusion zone is approached and passed. This is equivalent to the following set of steps:

- identify when a modulation path will pass within the exclusion zone,
- determine the time-bounded additional modulation signal needed to avoid the exclusion zone,
- add this time-bounded signal to the original signal.

This signal addition is a linear process, even though the determination of what the added signal pieces need to be is not necessarily a linear process. This signal selection and addition happens in the DPS ahead of the polar transmitter, so only the original RF path is needed as seen in Figure 6-53(b).

More details on this technique are provided in references [6-24] through [6-27]. Here, the general principles of this technique are summarized. Keeping with the two-tone signal example used throughout this section, the example signal and the signal that needs to be added to keep the output signal magnitude no lower than 10% of peak magnitude are presented in Figure 6-54(a). Particularly, note the much smaller scale of the

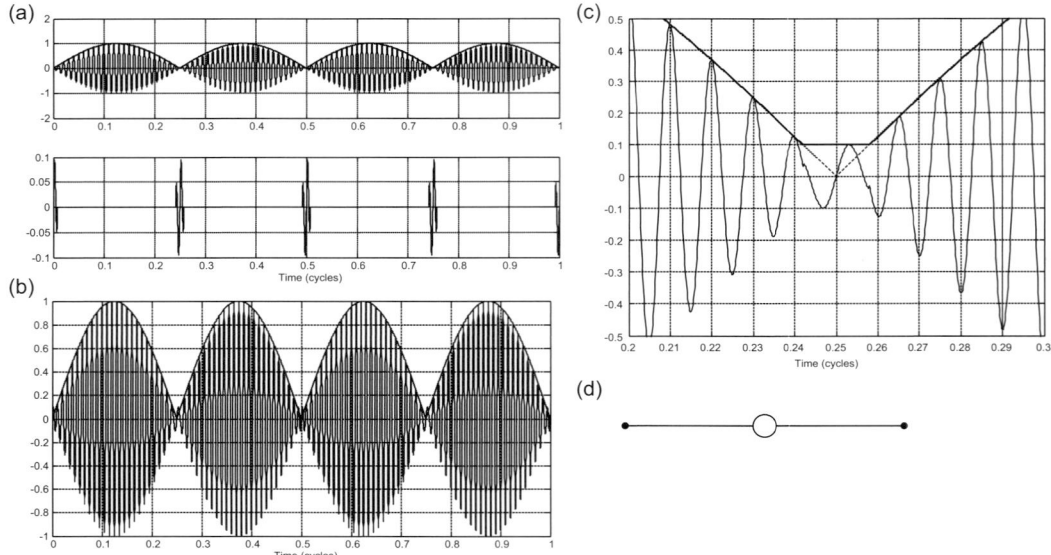

Figure 6-54   Ensuring that the signal envelope does not go to zero using the envelope flooring technique: (a) two-tone signal and the necessary signal to keep the minimum magnitude no lower than 10% of peak magnitude while rotating the phase; (b) final output signal; (c) close-up view to see the phase-rotating minimum envelope interval; (d) envelope flooring vector diagram for this signal.

lower added signal. The result of adding these two signals together is presented in Figure 6-54(b), with a close-up of the envelope flooring result in Figure 6-54(c).

This output signal is not a match to the ideal signal. Because the modulation waveform is intentionally changed, using the envelope flooring technique definitely has an effect on signal EVM. Yet it is definitely possible to ensure that any implementation has no effect at all on the signal PSD. This is completely consistent with the discussion in Section 4.1.2, where it was shown that EVM can be horrible even while signal PSD can be ideal. Envelope flooring achieves an ideal spectrum shape by treating each of these additional added signals as an additional signal symbol that is shaped by filtering with the same pulse used for bandlimiting the signal. The exclusion zone can be 60% of the peak signal magnitude causing EVM to be 90% or higher, while the signal PSD shows no change at all.

It is worth mentioning that if an envelope minimum can be removed by selectively adding these pulses, it is equally possible to remove peaks by selectively adding pulses with the same technique [6-28] [6-29]. Following symmetry arguments, we can call this establishing an envelope ceiling – but reduction of signal peaks is already well known as crest-factor reduction (CFR). Again, this CFR technique can be applied in the described way and have absolutely no impact on the signal PSD. But the output signal EVM will be greatly impacted. In general, it is found that the EVM impact of this form of CFR is much higher than the EVM impact of envelope flooring, mainly due to the size of the added pulses for CFR being much larger.

Envelope flooring, and its CFR equivalent, both act to reduce the bandwidth of the polar magnitude and phase-modulation components. This is a direct result of waveform continuity, and derivative continuity, from removal of the envelope zero events and suppression of signal peaks. The larger the exclusion zone, the greater the modulation component bandwidth limiting process – and the greater the damage to signal EVM.

6.13.6   Flooring and filling

Damage to signal EVM from envelope flooring is due to the waveform distortion resulting from the erstwhile unneeded pulses that are added. Fortunately, the pulses that are added are fully known to the signal processor. If they are saved, they can be later removed from the distorted signal to restore the original EVM, while still allowing the primary signal power to be provided by the polar modulator [6-30].

In essence, this envelope flooring and filling technique inserts known distortions into the signal so that it can be generated very efficiently at high power using polar modulation. Once the high power signal is generated, the distortion is removed from the output signal using the architecture in Figure 6-55. The flooring signal is zero, much if not most of the time, so it is actually appropriate to scale it to match the output signal using an L-mode amplifier. This signal is small, having a peak envelope equal to the size of the exclusion zone, so this amplifier is much smaller than the main polar power stage.

Continuing from the envelope flooring signal from Figure 6-54(c), the saved and inverted signal used to floor the envelope is shown in Figure 6-56(a) as a long-dashed line. When combined with the distorted polar output signal, the result is that of Figure 6-56(b). This closely matches the ideal two-tone test waveform, showing

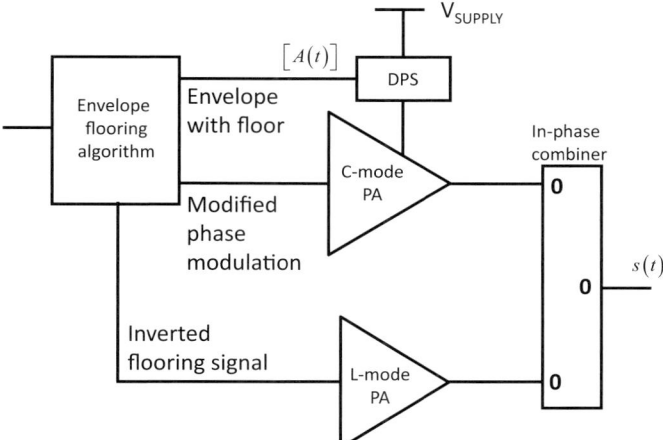

Figure 6-55   Block diagram of the envelope flooring and filling polar technique.

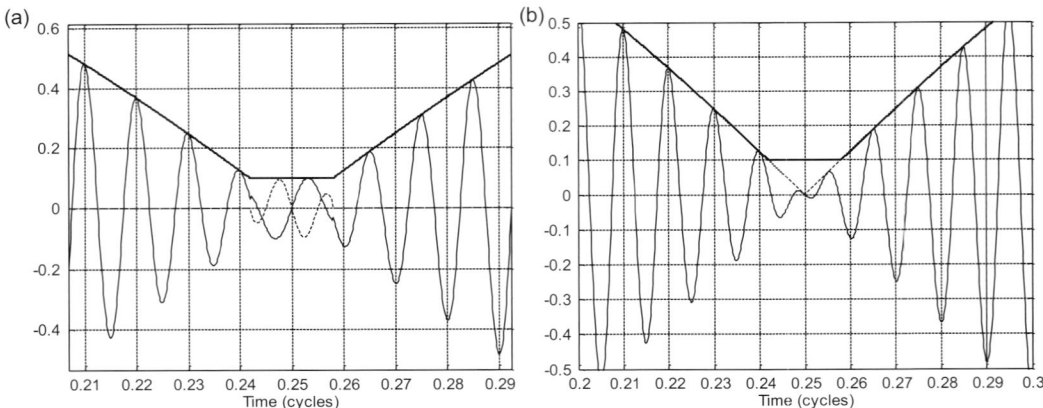

Figure 6-56   EVM distortion from envelope flooring can be corrected by subtracting away the correction signal from the polar PA output: (a) envelope-floored signal (solid line) and inverted flooring signal (long-dashed line); (b) resulting output after the small signal subtraction restores signal accuracy.

that the exclusion zone is no longer present, and the EVM is restored to its input value.

The remaining two techniques described here are not strictly envelope zero polar techniques, but are additional techniques that work within a polar transmitter to simplify transmitter design under important circumstances.

6.13.7   Load impedance manipulation combining outphasing with polar (LIMOP)

The outphasing technique, originally described by Chireix, uses rapid phase splitting to implement envelope variation from the vector summation of two constant envelope signals [6-31]. Energy efficiency is achieved from using a nonisolating combiner that changes the impedance seen by the two amplifiers, as the relative phase angle between them increases. Thus, the Ohm's Law relationship

$$P = \frac{V^2}{Z} \tag{6.27}$$

provides lower power not by varying voltage from the PA, but by increasing the load impedance $Z = R + jX$. Managing the reactive component of Z is one of the problems inherent with implementing Chireix outphasing architecture.

What if, instead of using the outphasing to vary output signal magnitude from fixed RF signal magnitudes, the roles are reversed? In other words, outphasing and polar can be combined where the outphasing angle sets the load impedance experienced by the RF power stages and then is fixed, and output envelope variation is implemented with conventional polar modulation using a DPS [6-32]. This architecture is illustrated in Figure 6-57. Power ranging now is set by selection of the outphasing angle, and tuning of the non-isolating combiner's reactive load impedance remains fixed for

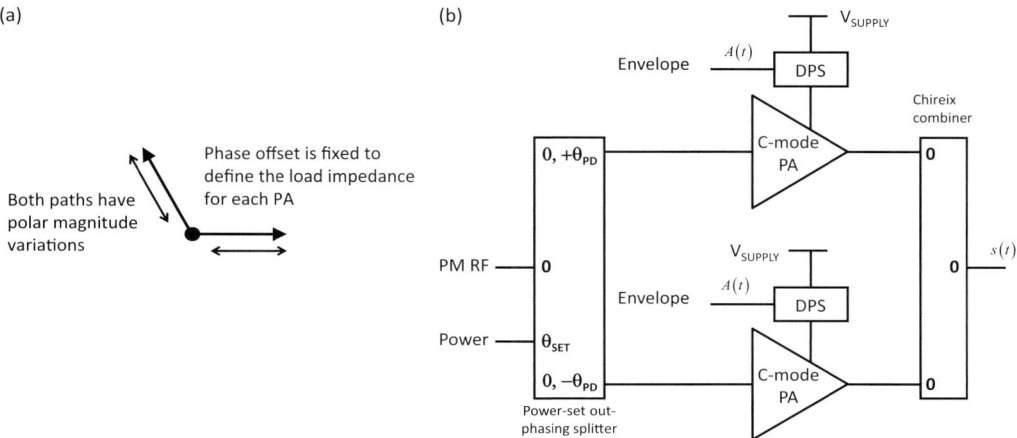

Figure 6-57 Principles of the load impedance manipulation using outphasing polar (LIMOP) technique: (a) phasor diagram; (b) block diagram.

that particular angle, now simply a power range. For signals with a finite envelope floor, this is sufficient to implement a fairly wide power control dynamic range. If the envelope must have a zero-magnitude event, it remains possible to increase the outphasing angle (and retune the combiner) only at that particular time.

6.13.8   Stop the AM

As output power within a communication system is lowered, the demands on output signal quality are also reduced as operation gets closer to the thermal noise floor. For signals with particular characteristics, it becomes possible to avoid all problems with envelope modulation by simply stopping the envelope modulator [6-33]. This is, in essence, the ultimate in envelope flooring in that the signal becomes a constant envelope. Stopping part of the transmitter circuitry when not needed saves on power consumption in both the transmitter and the baseband.

The important signal characteristic needed to be able to take advantage of this technique is to be using a signal constellation that has all points at the same magnitude from the IQ signal plane origin. This, of course, is obvious for FSK, CPM, and PSK signals, it is also true for many HPSK signals used in the UMTS system. With no information contained in the signal envelope variations, going through the effort to generate signal envelope variations is not worth it if not absolutely necessary to meet other specifications. It is always good to save the power and reduce operating costs.

Figure 6-58 shows the ACLR specifications for the WCDMA cellular signal, including the signal ACLR when envelope variations are removed, which for the appropriate HPSK modulations is −15 dB. The specifications show that when the output power is below −35 dBm, then this constant envelope signal is compliant to the specification.

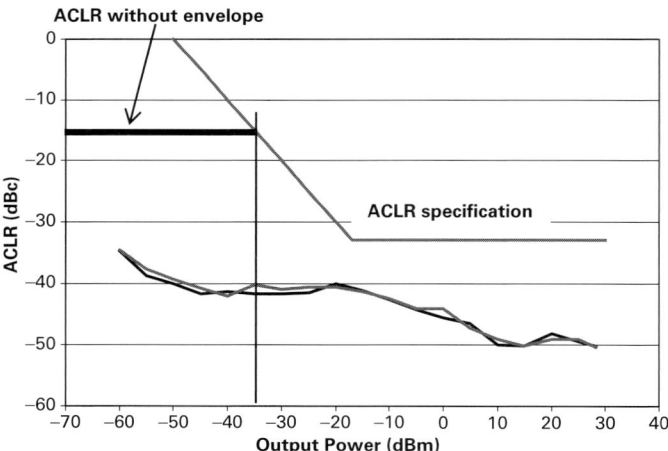

Figure 6-58   Envelope variation can be stopped for constant modulus constellation signals at low output power. UMTS specification for ACLR vs. output power is met when output power is −35 dBm or less.

Such low output powers are used in a dense urban environment when the distance to the closest cell site is very short.

6.14   TDM burst control

For any time division multiplexed (TDM) system, spectral characteristics of the RF bursts turning on and off must be constrained to be close to the bandwidth occupancy of the signal's normal modulation. Further, the tests for transient spectrum behavior are normally performed with peak detection, which is far harder to satisfy than the averaging detector used for modulation spectrum measurements. Constraining this transient spectrum is a difficult problem.

It is well known that as long as a family of signals has the same general modulation format (e.g. QAM) with identical symbol times and using identical filter responses, then the signal power spectral densities are identical. This means that the PSD is effectively independent of the actual constellation: 4 QAM or 256 QAM exhibit the same signal spectra.

This property is useful to ensure that spectral content of a signal burst rise and fall power transient is identical to the spectral content of the modulated waveform [6-34] [6-35]. This also eliminates the usual trial and error in designing TDM rise and fall profiles that are simultaneously fast and spectrally compact [6-36]. Measurements of using this technique are presented in Figure 6-59 for EDGE bursts. Speed of the rise and fall times is evident.

Demonstration of how well this works is shown in Figure 6-60. An unpulsed EDGE modulated signal spectral measurement is in Figure 6-60(a). Adding the specified TDM bursting to this signal and changing the detector to peak as called for in the 3 GPP

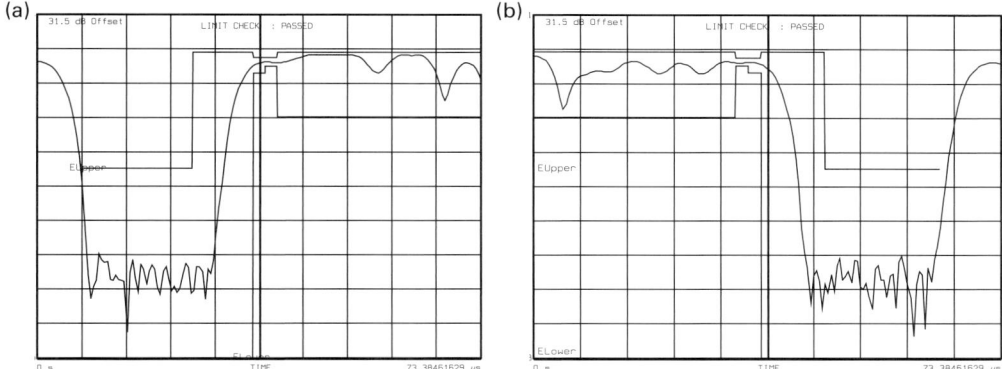

Figure 6-59 Pre-pending and post-pending symbols while using the same bandlimiting filter ensures that spectral performance of transient events matches the spectral characteristics of normal modulation: (a) rise time detail; (b) fall time detail.

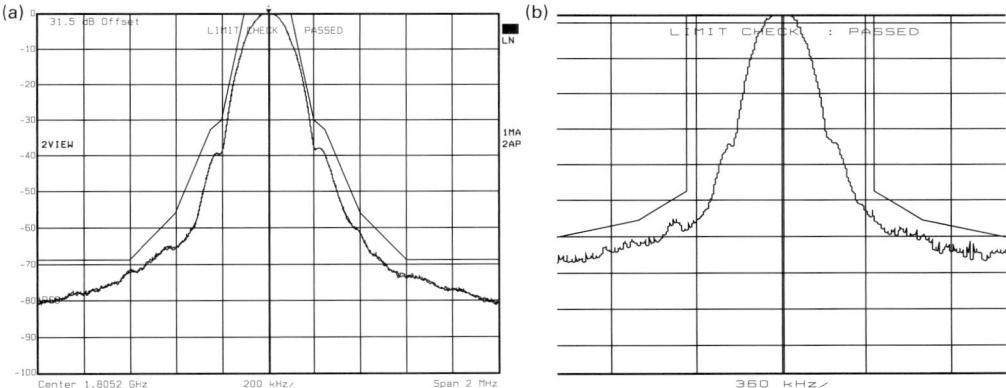

Figure 6-60 Matching of modulation and transient spectrum measurements for the EDGE TDM system: (a) modulation PSD measurement using an rms detector; (b) transient spectrum measurement using a peak detector.

specification leads to the signal measured in Figure 6-60(b). The top 50 dB is identical in both measurements, and both signals easily meet specifications.

6.15 Stability performance

In Section 4.3, three stability measures are discussed that are important to production wireless transmitters. Here these same measures are revisited in the light of C-mode operation, different from the L-mode operation used in Section 4.3. Two additional stability measures are added in Sections 6.15.2 and 6.15.5.

6.15.1   Circuit stability

Circuit stability from Section 4.3.1 is based on the magnitude and phase of the slope gain of the amplifier. Here with C-mode operation, the slope gain is zero. How does this affect the stability evaluations from Section 4.3.1?

The fact that in C-mode operation the slope gain goes essentially to zero has a profound effect on the circuit stability of a polar power stage. With no slope gain (nor any other gain) when the transistor is OFF, and a slope gain very close to zero when the transistor is ON, during these states the first Barkhausen criterion [4-12] is not satisfied and the C-mode stage cannot oscillate. During the transitions OFF to ON and ON to OFF, there is sufficient slope gain to support oscillation if the second Barkhausen criterion is satisfied – and only if the stage remains in this transition state long enough for oscillations from circuit instability to begin and build up.

If the transition conditions of Section 6.10 (Figure 6-27) are met to achieve available C-mode efficiency, there is not sufficient time with high slope gain to allow oscillations to build up. It can be claimed then that a properly operating C-mode polar transmitter is unconditionally stable with regard to circuit oscillations. The converse can also be claimed, which is actually a much stronger case:

- if a C-mode polar transmitter design has any circuit stability problems, then it is not actually operating in C-mode and not truly a polar transmitter.

6.15.2   Inherent low frequency stability

Any linear amplifier can oscillate at frequencies much lower than the operating signal frequency if the bias circuitry is not properly designed, or if the low frequency bypassing is not sufficient to drop the circuit gain at frequencies where the transistor gain gets high. Particularly in the latter case, the oscillation behavior is often referred to as *squegging*.

With C-mode operation, the ON state condition of nearly zero transconductance holds true at all frequencies. This means that when properly designed, a C-mode power stage is also immune from squegging or any other low frequency oscillation. These circuits are truly unconditionally stable at *all* frequencies when properly designed in accordance with the conditions of Section 6.7.5.

6.15.3   Thermal stability

The load current for the circuit operation described by (6.10) is more precisely written as

$$I_L = \frac{V_S - V_{AMO}(m, T)}{R_L + R_{ON}(m, T)}, \tag{6.28}$$

which explicitly shows that the device voltage offset and channel ON resistance both depend on the specific device (m) and its operating temperature (T). Following the

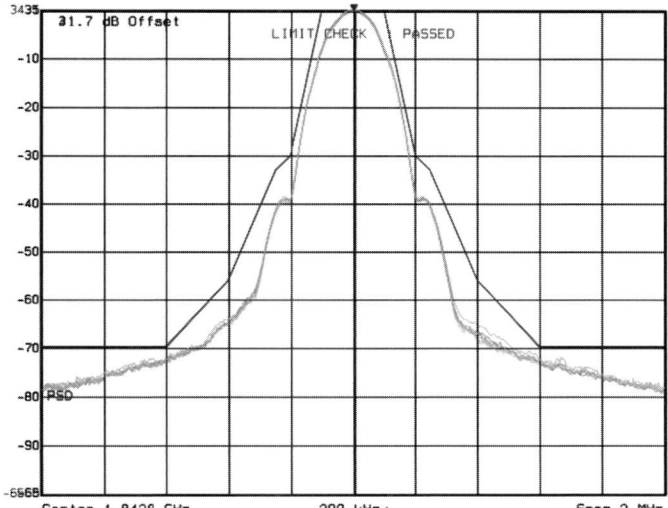

Figure 6-61   Overlay of nine spectrum measurements from a 20 W 1980 MHz polar (C-mode) transmitter across the thermal steps of {+30 °C, +40 °C, +50 °C, +60 °C, +70 °C, +60 °C, +50 °C, +40 °C, +30 °C} showing essentially complete overlap.

C-mode design rules from Section 6.7.5 directly predicts that any drift of transistor characteristics with temperature are negligible with respect to the output signal. Temperature stability of a polar C-mode circuit should be inherent when properly designed. This result is true, as shown in Figure 6-61 where an overlay of nine different spectral measurements from a 20 W EDGE transmitter is shown.

The results in Figure 6-61 are in complete accordance with the prediction of (6.28). Even though the transistor ON resistance is drifting with temperature as it must, as long as this ON resistance remains negligible with respect to the load resistance that the transistor is operating at, then to first order this drift has no measurable impact. An additional benefit is that the condition for high transmitter energy efficiency is also that from (6.11) the load resistance greatly exceeds the value of the transistor ON resistance. Optimizing a polar transmitter for either maximum energy efficiency or minimum temperature drift naturally optimizes for the other parameter.

6.15.4   Manufacturing stability

The more detailed model (6.28) points out that manufacturing variations result in variations in both channel ON resistance and voltage offset, if they exist. Just as the C-mode circuit exhibits essentially no output power drift when the transistor drifts over temperature, if transistor ON resistance varies from one device to another, there is no real difference to the analysis. Drift of transistor ON resistance, whether by temperature or device variation, analyzes identically: as long as the actual transistor ON resistance is negligible compared to the load resistance, the drift of transmitter output power is hard to measure. This aspect of manufacturing stability is illustrated in Figure 6-62.

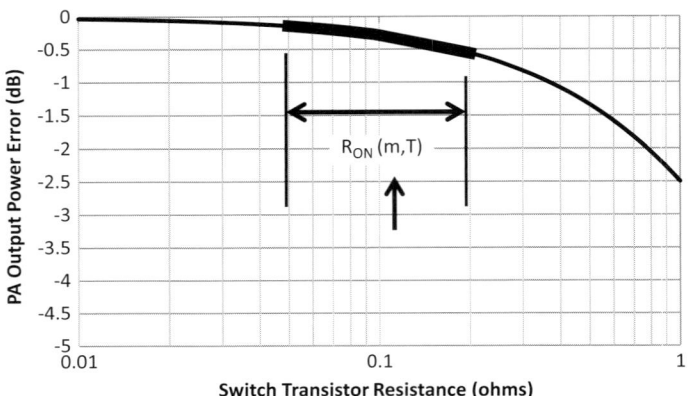

Figure 6-62 Output power variation from a nominally 0.1 ohm ON resistance transistor working into a 3 ohm load shows less than a 0.5 dB total variation even when the transistor varies by an octave in each direction.

This example case in Figure 6-62 is a design having a transistor with nominal ON resistance of 0.1 ohms, working with a load resistance of 3 ohms. The curve along the entire length of the figure describes the output power variation into the load for transistor ON resistances across 0.01 ohms to 1 ohm, normalized to an ideal transistor having 0 ohms ON resistance. As the transistor ON resistance increases, in accordance with the discussion in Section 6.7.5, the output power begins to drop when the ON resistance is no longer negligible in (6.10).

The central section of the curve between transistor resistances of 0.05 to 0.2 ohms shows the expected output power variation as this transistor, or any set of transistors, exhibits ON resistance values across 2 octaves of variation. This figure shows that this output power variation is within a total range of 0.5 dB. This is well within the allowable power variation range of most transmitter specifications.

When temperature is added into the output power accuracy mix across a set of polar transmitters, a set of histograms like that in Figure 6-63 results. The variance for each of these sample sets is well within 0.5 dB at each tested temperature. All samples at all temperatures stayed within ± 1 dB, with sample variance well below that. It is important to note that this stability is inherent in C-mode operation, needing neither temperature compensation nor manufacturing calibration in this design.

6.15.5 Operating stability (ageing)

In the ultimate accounting, stability of circuit performance over extended periods of time is what matters to product consistency in its service life. Since we cannot wait for an entire service life to see how consistent they are over time, there are tests available to speed up the ageing process. These all involve warming up the design and then doing tests at the elevated temperature. And of these tests, among the most stringent is high temperature operating life (HTOL).

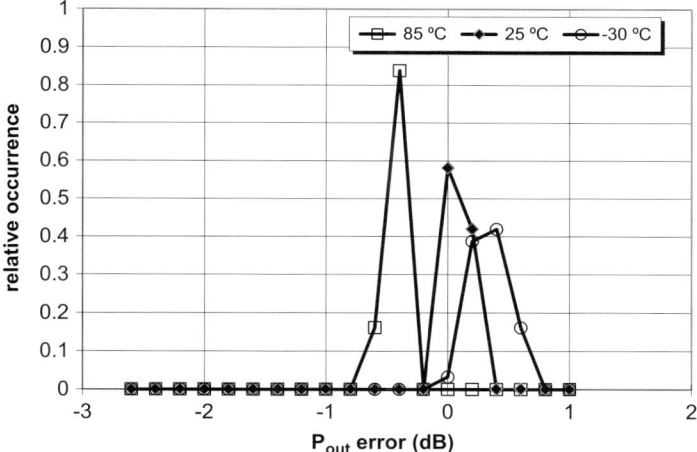

Figure 6-63   Histogram of output power measurements from 1980 MHz polar (C-mode) handset transmitters, showing a very small variance across units along with a slight drift with temperature.

Among the most stringent HTOL tests is when the device is brought up to +85 °C (185 °F) and operated at full power, meaning at highest power dissipation, and held there with periodic performance tests for at least 1000 hours. That is nearly 42 days of nonstop high temperature high dissipation operation. This is proven to be a very good stress test for validating design stability because the device internal temperatures (T_a in (3.11)) can get very high.

Figure 6-64 presents two charts containing results from an HTOL test on a GSM/GPRS/EDGE polar transmitter. Performance is extremely stable, as predicted in Section 6.7.5. This is particularly true for transmitter power seen in Figure 6-64(a), where the sample variance is very small at each power level (PLEV). The modulation accuracy test in Figure 6-64(b) shows similar performance consistency. Variance of the EVM measurements at the lowest power level increased as the lower transmit signal approached the measurement noise floor.

6.16   Limiters

Section 6.8.4 presents the energy efficiency advantage, and DPS-AM transfer linearity, achieved when driving a C-mode stage with rectangular waveforms. One natural approach to getting a rectangular RF waveform is to use a limiter, just as Kahn did for the EER structure shown in Figure 6-2. While this works to first order, using a limiter for this function has some higher order effects that do impact the signal passing through it.

One important aspect of limiter performance is the consistency of the output phase for a fixed input phase as the signal envelope varies. One measurement of an integrated limiter for its transfer phase characteristic is shown in Figure 6-65(a). In general, we do not care what the absolute phase shift is. But we care greatly

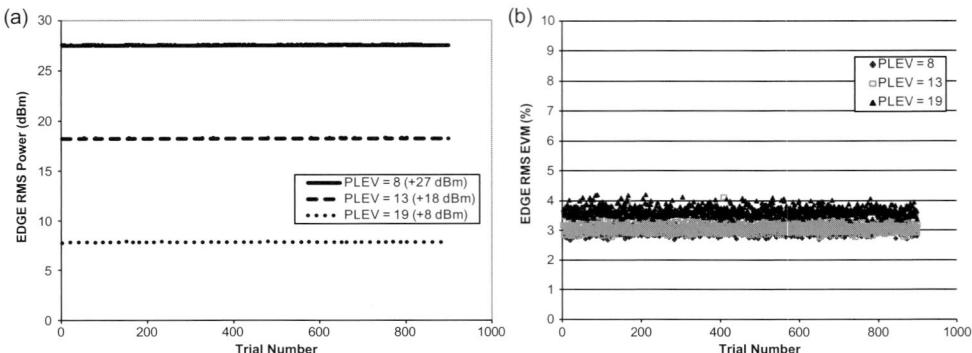

Figure 6-64   Accelerated ageing measurements for output power and EDGE EVM from 1000 hours of operating at +85 °C, taken every hour across the duration of this test. No drift over time is measured, corresponding to the very low power dissipation of the C-mode transmitter: (a) output power consistency; (b) EVM consistency.

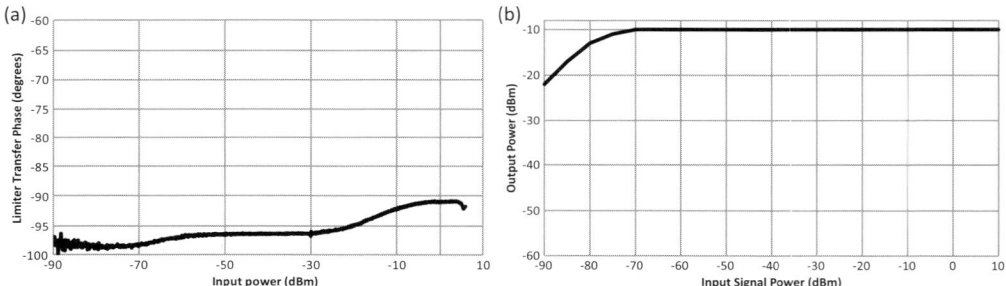

Figure 6-65   Limiter test results from a silicon complementary bipolar MMIC RF limiter: (a) transfer phase; (b) output power flatness.

about that phase not changing as the limiter operates. In this figure, we see that the absolute phase shift is about −95 degrees. Variation from this "center" spans about 4 degrees in both directions. Phase accuracy therefore is 8 degrees peak to peak for this limiter. Output flatness measurements for this same limiter are shown in Figure 6-65(b).

These measurements show that limiters are not ideal, and in important ways. When an envelope-varying signal is applied into this limiter, the envelope will be successfully removed except near the zero crossings. But the signal phase will experience some rotation even while the envelope is removed. These are two reasons that true EER presented in Section 6.1.2 never succeeded in market production. Design and operation of a really good limiter is not at all easy.

It is important to recognize that the RF limiter in a C-mode circuit is usually not a separate circuit at all. By definition, C-mode operation includes self-limiting because the output signal power is independent of input signal power variations. There is no need, nor point, for the external limiter called for in the EER technique of Figure 6-2. RF limiting is an essential feature of the RF transistor operation in C-mode.

Table 6-2 Property summary comparison of ET and polar modulation

Characteristic	Envelope tracking (ET)	Polar modulation
Output signal envelope control	Input RF signal	Dynamic power supply (DPS)
Input signal modulation	Precise from specification	Only PM is needed
Dynamic power supply accuracy	Low, when above a minimum	Critical always
Output power = 0?	No problem	Difficult
Power supply noise	Rejected	Passed through directly
Circuit stability	Critical and difficult	Unconditional
Thermal stability	Requires compensation	Unconditional
Energy efficiency	High	Higher (best possible for PA)
RF PA transistor operation	CCS (L-mode)	Switching (C-mode)
Device output matching	Complex conjugate	Real line
Wideband noise, dominant source(s)	PA noise figure and gain	Phase noise (RF input), power supply
Input signal magnitude accuracy	Critical	Anywhere within C-mode range
Load current control	Regulated by the RF transistor	DPS and load impedance accuracy
Envelope time alignment	Tight, if emphasizing efficiency	Tighter

6.17   Summary table (cumulative)

The summary of circuit operation characteristics provided in Table 5-1 for envelope tracking DPST operation is expanded in Table 6-2 to include the corresponding properties for polar modulation [5-5]. A review of this table shows that the properties of the ET operation and polar modulation are often in opposition. These are very different types of transmitter operation with very different design requirements and likewise very different demands on how the power transistors must perform. These transmitter modes are not duals of each other, and are most definitely not two different names for the same thing. These transmitter operating modes must be treated as the separate but related modes, and carefully described and analyzed as is proper and appropriate.

6.18   References

Note: Patent references are provided for bibliographic use only. Citation of specific patents here is not indicating any view on priority issues.

[6-1] R. A. Heising, "Modulation Methods," *Proceedings of the Institute of Radio Engineers*, vol. 50, no. 5, May 1962, pp. 896–901.

[6-2] L. R. Kahn, "Single-Sideband Transmission by Envelope Elimination and Restoration," *Proceedings of the Institute of Radio Engineers*, vol. 40, no. 7, July 1952, pp. 803–806.

[6-3] H. Swanson, "Digital AM Transmitters," *IEEE Translations on Broadcasting*, vol. 35, no. 2, Feb. 1989, pp. 131–133.

[6-4]   J. Staudinger *et al.*, "800 MHz Power Amplifier Using Envelope Following Technique," *Proceedings of the 1999 IEEE Radio and Wireless Conference (RAWCON'99)*, Denver, CO, Aug. 1–4, 1999.

[6-5]   W. Camp, J. Schlang, G. Gore, R. Boesch, and D. Arpaia, "Circuit and Method for I/Q Modulation with Independent, High Efficiency Amplitude Modulation," US Patent 6194963, issued Feb. 27, 2001.

[6-6]   W. B. Sander, S. V. Schell, and B. L. Sander, "Polar Modulator for Multi-mode Cell Phones," *Proceedings of the 2003 IEEE Custom Integrated Circuits Conference (CICC)*, San Jose, Sept. 2003.

[6-7]   E. McCune, "High-Efficiency Multi-mode, Multi-band Terminal Power Amplifiers," *IEEE Microwave Magazine*, March 2005.

[6-8]   W. H. Doherty, "A New High Efficiency Power Amplifier for Modulated Waves," *Proceedings of the Institute of Radio Engineers*, vol. 24, no. 9, Sept. 1936, pp. 1163–1182.

[6-9]   E. McCune, "Polar Modulation and Bipolar RF Power Devices," *Proceedings of the IEEE Bipolar Circuits and Technology Meeting (BCTM)*, Santa Barbara, Oct. 2005.

[6-10] S. Schell, W. Sander, R. Meck, and R. Bayruns, "Power Control and Modulation of Switched-mode Power Amplifiers with One or More Stages," US Patent 6734724, issued May 11, 2004.

[6-11] S. Schell, W. Sander, R. Meck, and R. Bayruns, "Power Control and Modulation of Switched-mode Power Amplifiers with One or More Stages," US Patent 7042282, issued May 9, 2006.

[6-12] E. McCune, "Variable Bias Control for Switch-mode RF Amplifier," US Patent 6323731, issued Nov. 27, 2001.

[6-13] W. Sander, R. Meck, and E. McCune, "Communication Signal Amplifiers Having Independent Power Control and Amplifier Modulation," US Patent 7010276, issued March 7, 2006.

[6-14] W. Sander, R. Meck, and E. McCune, "Communication Signal Amplifiers Having Independent Power Control and Amplifier Modulation," US Patent 7035604, issued April 25, 2006.

[6-15] W. Sander, R. Meck, and E. McCune, "Communication Signal Amplifiers Having Independent Power Control and Amplifier Modulation," US Patent 7444125, issued Oct. 28, 2008.

[6-16] W. Sander, R. Meck, and E. McCune, "Communication Signal Amplifiers Having Independent Power Control and Amplifier Modulation," US Patent 7515885, issued April 7, 2009.

[6-17] D. Flowers, "Mode Shift Calibration in Power Amplifiers," US Patent 8000663, issued Aug. 16, 2011.

[6-18] E. McCune, G. Do, and W. Lee, "Transmission Power Controller," US Patent 8064855, issued Nov. 22, 2011.

[6-19] W. Sander, "Saturation Prevention and Amplifier Distortion Reduction," US Patent 6528975, issued March 4, 2003.

[6-20] W. Lee, "Methods and Apparatus for Reconstructing Amplitude Modulation Signals in Polar Modulation Transmitters," US Patent 7991336, issued Aug. 2, 2011.

[6-21] S. Schell, W. Sander, and E. McCune, "Method and System of Amplitude Modulation Using Dual/Split Channel Unequal Amplification," US Patent 6751265, issued June 15, 2004.

[6-22] J. Sevic, W. Sander, and S. Schell, "Differential RF/Microwave Power Amplifier Using Independent Synchronized Polar Modulators," US Patent 6653896, issued Nov. 25, 2003.

[6-23] W. Sander, "Efficient Precise RF Modulation Using Multiple Amplifier Stages," US Patent 6690233, issued Feb. 10, 2004.

[6-24] H. Wang, P. Liang, R. Booth, S. Schell, and T. Biedka, "Methods and Apparatus for Conditioning Communications Signals Based on Detection of High-Frequency Events in Polar Domain," US Patent 8331490, issued Dec. 11, 2012.

[6-25] P. Liang and R. Strandberg, "Apparatus and Method for Conditioning a Modulated Signal in a Communication Device," US Patent 7675995, issued March 9, 2010.

[6-26] P. Liang, K. Takinami, and T. Matsuura, "Transmitter Utilizing a Duty Cycle Envelope Reduction and Restoration Modulator," US Patent 8131234, issued March 6, 2012.

[6-27] H. Wang and P. Liang, "Methods and Apparatus for Reducing High-frequency Events in Polar Domain Signals," US Patent 8300729, issued Oct. 30, 2012.

[6-28] S. Schell and R. Booth, "Method and Apparatus for Reducing Peak-to-RMS Amplitude Ratio in Communication Signals," US Patent 7639098, issued Dec. 29, 2009.

[6-29] S. Schell and R. Booth, "Method and Apparatus for Reducing Peak-to-RMS Amplitude Ratio in Communication Signals," US Patent 8050352, issued Nov. 1, 2011.

[6-30] M. Mow and R. Booth, "Signal Enhancement in RF Transmitters Employing Non-linear Filtering," US Patent 8010063, issued Aug. 30, 2011.

[6-31] H. Chireix, "High Power Outphasing Modulation," *Proceedings of the Institute of Radio Engineers*, vol. 23, no. 11, Nov. 1935, pp. 1370–1392.

[6-32] K. Takinami and P. Liang, "Adaptive Impedance Converter Adaptively Controls Load Impedance," US Patent 8140030, issued March 20, 2012.

[6-33] T. Biedka, P. Liang, and G. Do, "Selective Envelope Modulation Enabling Reduced Current Consumption," US Patent 7688157, issued March 30, 2010.

[6-34] S. Schell, "High-Quality Power Ramping in a Communications Transmitter," US Patent 6983025, issued Jan. 3, 2006.

[6-35] S. Schell, "High-Quality Power Ramping in a Communications Transmitter," US Patent 7227909, issued June 5, 2007.

[6-36] B. Sander, W. Sander, and S. Schell, "Multi-mode Communications Transmitter," US Patent 7158494, issued Jan. 2, 2007.

Part II

DPST circuit issues

7 Special linear PA circuit considerations for ET

Any RF amplifier that is intended to be used in the ET operation needs to be designed differently from conventional linear amplifiers. Any ET amplifier must remain in linear operation at all times while the power supply varies in the 3-port DPST application. This places new constraints on device selection and bias strategies. And methods that are commonly used to ensure circuit stability in particular must completely change.

7.1 Core principle: power supply value independence

The core principle of ET from Chapter 5 is that the exact value of the power supply does not impact the output signal from the amplifier. This is equivalent to saying that any noise on the power supply is suppressed/rejected from modulating the output signal with an ET transmitter. Mathematically, we describe this operation of the 3-port amplifier (4.15) as

$$\frac{\partial P_{OUT}}{\partial V_{SUPPLY}} = 0 \qquad (7.1)$$

and

$$\frac{\partial P_{OUT}}{\partial P_{IN}} = g_{SP}, \qquad (7.2)$$

which we recognize from (4.11a) is the slope power gain g_{SP}. For linear amplifier operation, required for ET, this is also equal to the most common RF gain measurement, ratiometric power gain g_{RP}.

How well a particular amplifier meets the ET core principle is most readily seen from the Booth chart (Section 4.1.6) for that amplifier. One example of a measured Booth chart is shown in Figure 7-1. Output power is recorded across an input power sweep of −60 dBm up to +10 dBm for a particular value of power supply voltage, and then repeated at different values of the power supply. In this example, the power supply is varied only on the PA final stage.

This PA exhibits all three modes (L, C, and P) of operation. Power supply values are stepped by 0.1 V between each of these curves. The lowest power supply values operate this amplifier in P-mode, so these must be avoided in the ET operation, setting a power supply value floor.

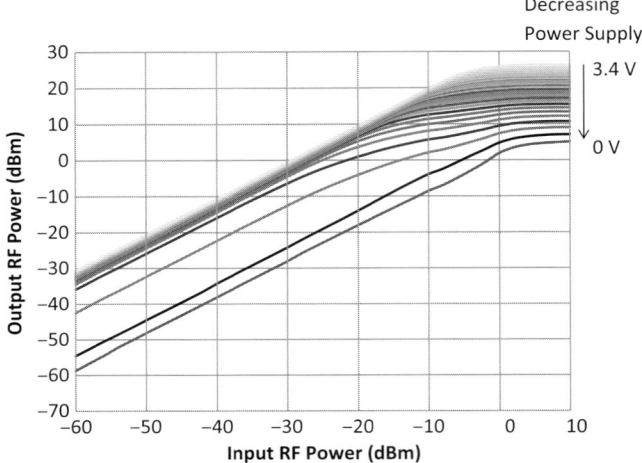

Figure 7-1   Measured Booth chart for an example SiGe HBT PA, where the power supply is stepped every 0.1 V from 0 V to 3.4 V.

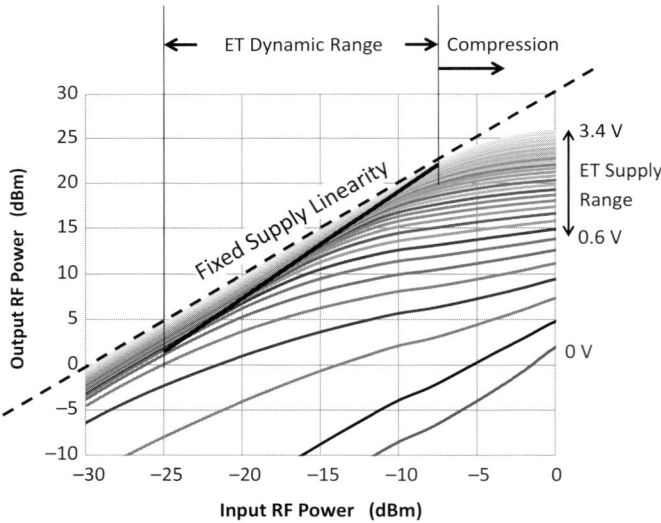

Figure 7-2   Close view of the upper L-mode region for this example PA. Characteristics of the ET operation for this PA are highlighted.

A closer view of the upper end of the L-mode region for this PA is shown in Figure 7-2. The distortion of the compression region is a problem for linear operation, so this region must be avoided. At power supply values below 0.6 V, this amplifier never gets into L-mode, so this sets the minimum ET supply value. The ET dynamic range is set at the

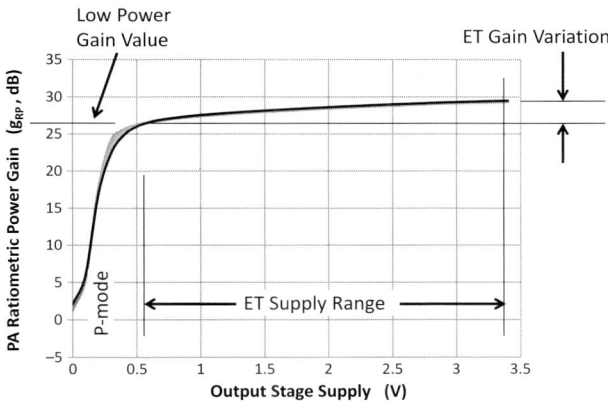

Figure 7-3   Ratiometric power gain for this PA as the power supply is varied across its entire range.

upper end by the onset of output compression for the maximum supply voltage, and at the lower end by where the minimum power supply value enters P-mode. From the Booth chart in Figure 7-2, for this amplifier the ET dynamic range spans from −25 dBm of input power to −8 dBm, a span of 17 dB, with a corresponding span of the supply voltage from 0.6 V to 3.4 V. At lower input power, the power supply must remain fixed at 0.6 V.

Looking at the corresponding output power dynamic range, at the lowest ET RF input and power supply, the output power is +2 dBm. At the upper end, the output power is +22 dBm. This is an output dynamic range of 20 dB, 3 dB greater than the input dynamic range. With this amplifier, applying the ET operation causes 3 dB of gain expansion across the ET range. The straight line connecting these end points is shown in Figure 7-2, which is obviously tilted from the dashed line, showing the gain characteristic for fixed supply operation.

Another useful figure that shows that this must happen is in Figure 7-3. This chart shows that for this amplifier there is always an influence of the power supply value on the output power. The ideal ET operation is not possible, because the ET core principle (7.1) is not strictly met. But the primary motivation for using DPS operation, improved PA energy efficiency, does still occur. This is seen in Figure 7-4, where the curves for output efficiency across input power variation for each power supply value are overlaid. The effective PA efficiency trajectory for this amplifier can be traced by looking at Figure 7-2 and noting what the PA input power is when a particular power supply curve intersects with the ET linearity profile. The intersection of that input power with the efficiency curve corresponding to that same power supply value gives the resulting PA output efficiency for that ET operating condition. Repeating this process through the ET operating range gives the curve overlaid on the output efficiency profiles in Figure 7-4.

Two things are noticed about the ET PA efficiency profile in Figure 7-4. First is that it does provide an improvement above the efficiency provided by fixed supply operation, often double or even triple what the efficiency would be otherwise. More striking though

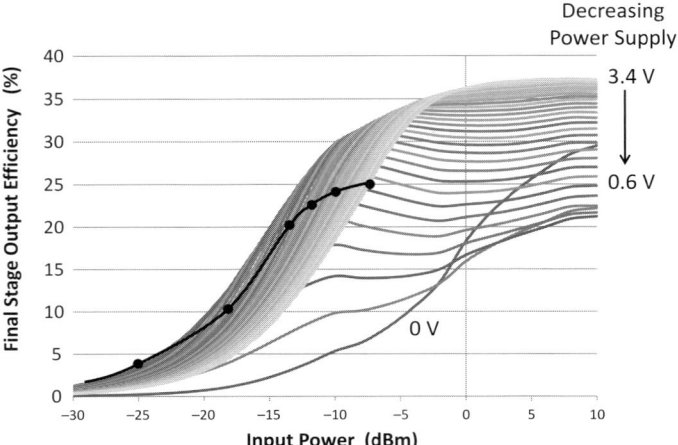

Figure 7-4   Energy efficiency of this example PA as the power supply is varied.

is that this efficiency profile is still below the available efficiency from this PA. This is a direct result of the ET core principle: when we need to have the RF output signal (largely) independent of the actual power supply value, then as seen in Section 5.4.2 and Figure 6-3 there must be sufficient voltage headroom on the power supply to ensure that the transistor operates as a controlled current source (CCS) so that it regulates the signal current flowing through the load. This required voltage overhead comes along with a drop in realizable energy efficiency. You cannot win over Ohm's Law.

Any operation of this PA that achieves higher efficiency *requires* that the ET core principle of RF output independence from the power supply value (7.1) be violated. Achieving higher efficiency directly means that the transistor dynamic operating point must get closer to the I-axis on its IV curves, forcing C-mode operation and resulting in polar modulation. In other words, optimizing any envelope tracking DPST for improved efficiency must result in polar modulation (Figure 6-5). There is no option, and this is completely independent of whether the input signal into the PA is envelope-varying or not, as shown in (2.9) and that surrounding discussion.

Relating this ET amplifier behavior to the transistor IV characteristics is a very important step for understanding what is actually going on. The measured characteristic curves for the PA used in this example are shown in Figure 7-5. There definitely is a tilt to the CCS region characteristics, meaning that the curves are further apart when higher voltages are applied than when the power supply is lower. This explains the gain variation measured in Figure 7.3. This also explains why in the ET operation the gain expansion exists, as seen in Figure 7-2.

In Figure 7-5, the "knee" voltage, the voltage offset from the current axis where CCS operation begins (horizontal curve characteristic) for any of these characteristic curves, is nearly twice the value of the voltage offset due to the ON resistance (nearly vertical curve characteristic) of this transistor. This is the cause for the difference in energy efficiency between the ET and polar operation, seen in Figure 7-4.

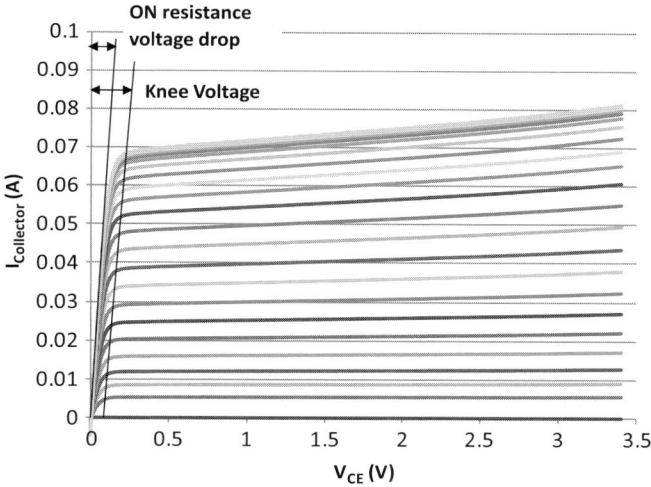

Figure 7-5   Output transistor characteristic curves for this example PA (a GaAs HBT), where each curve is set by stepping the PA bias control input pin by 0.05 V across 1.0 to 3.0 V.

A survey of important transistor characteristics for ET transmitter use is provided in Chapter 10. Many more details, along with the ability to compare characteristics among many different RF power transistor technologies, are there.

The linearity required of an envelope-tracked amplifier needs the transistor characteristic curves to be parallel and equally spaced for uniform changes in the controlling parameter. This combination ensures that the transistor will have constant gain ($g_m R_L$ for an FET, βR_L for a bipolar). This cannot happen in the transistor's resistive operating region, but it can happen in the CCS region. Note though that this does not require these parallel characteristics to have zero slope, just that the slopes must be equal. Characteristic curves of a transistor behaving in this way are seen in Figure 7-6.

There are two additional considerations necessary to achieve the energy efficiency advantages of ET. One is to provide constant gain and to jointly maintain that gain value across considerable variation in the power supply. Achieving this requires that the CCS region characteristic curve slopes must be small enough that the intersection of the "barely on" must at best intersect with the voltage axis at a small voltage value. Ideally, of course, this means that the slope should be zero and there will never be an intersection with the voltage axis: the transistor acts as an ideal current source. Otherwise the minimum available power supply voltage for an amplifier using that transistor must stay above the highest intersection voltage. We conclude that even though the fixed supply gain and linearity are essentially identical for the transistors in Figure 7-5 and Figure 7-6, an amplifier using the transistor in Figure 7-5 is potentially useful for ET use, but an envelope-tracked amplifier based on the transistor of Figure 7-6 will be much more difficult to realize.

The second consideration for the best energy efficiency performance in ET use is to have the CCS region extend close to the current axis, allowing linear amplifier operation

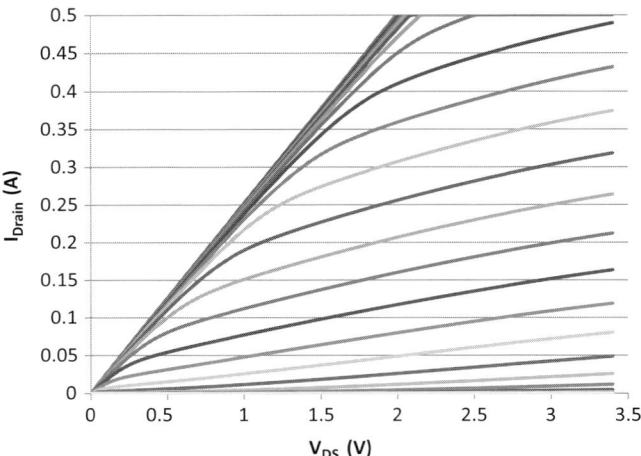

Figure 7-6   Transistor that provides good amplifier linearity without having "flat" CCS characteristics, here a GaAs pHEMT (depletion mode device). V_{GS} is stepped by 0.05 V from −1.5 V up to 0 V.

and minimizing transistor power dissipation. The device voltage where the CCS region ends is called the knee voltage (Section 5.4.2) and it generally is far from being a constant across all transistor operating conditions. The transistor knee voltage profile is actually a more accurate description than defining a fixed value for knee voltage. The two transistors in Figures 7-5 and 7-6 have very different knee voltage profiles. It is desirable to use a transistor that exhibits a low value knee voltage profile, which is yet another reason why the transistor in Figure 7-5 is more attractive than the transistor of Figure 7-6 for ET use.

7.2   Low frequency stability

RF power transistors have a very large amount of gain at low frequencies, here meaning less than 10 MHz. This is the frequency range that bias networks operate in, so for the stability of amplifier biasing it is very important that the amplifier does not oscillate at these (or any) frequencies.

Oscillation becomes possible whenever the two Barkhausen criteria are satisfied (Section 4.3.1). Usual practice is to set the gain of the circuit at these low frequencies to be less than unity, and therefore removing the need to worry about phasing of circuit signal paths.

7.2.1   Problems with usual practice

Getting the amplifier gain to be less than unity is usually achieved by ensuring that the transistor load impedance is less than 0.1 ohm at the bias frequencies, by placing a large capacitor across the power supply input. Figure 7-7 presents a chart of capacitive

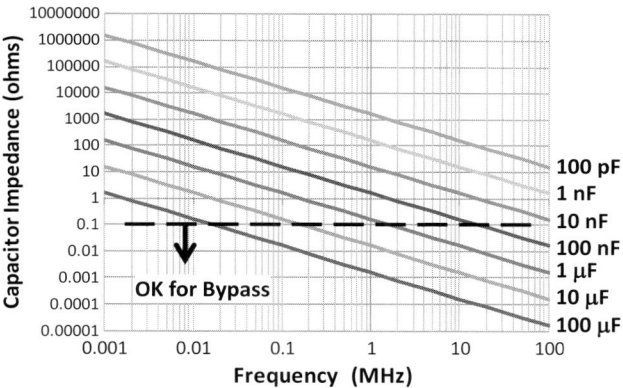

Figure 7-7   Bypass capacitor values for effective low frequency stability.

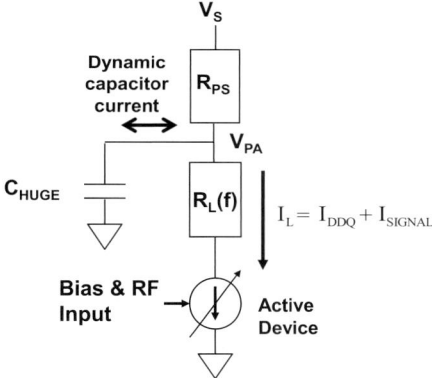

Figure 7-8   Conventional high value capacitor bypass for low frequency stability of an RF power amplifier. Voltage V_{PA} is held essentially constant by capacitor C_{HUGE}, ensuring that low frequency gain is well below unity to not satisfy the Barkhausen gain criterion.

impedance across the bias network frequencies along with how seven different capacitor values perform in this application. A dashed line shows that the threshold for satisfying the Barkhausen gain criterion when the device gain is 10 (20 dB) is to have a net load resistance of $1/10 = 0.1$ ohm. For transistors having higher gain, this value of effective low frequency load resistance must be lowered accordingly.

This usual technique is shown in the block diagram of Figure 7-8. For sufficient bypassing at 0.1 MHz, we need to apply 10s of microfarads across the supply line. This is indeed a common practice. By holding V_{PA} constant, instantaneous signal current for the amplifier flows from the capacitor, and only average current flows (uniformly) from the power supply.

But with DPS applications, this capacitor is a real problem because of the current required to change the voltage across this capacitor in the short time periods of envelope variations. The reason this is a problem for ET comes from the physics of a capacitor relating the current (I_C) flowing in or out based on the voltage slew rate (SR)

$$I_C = C \frac{\Delta V}{\Delta t}$$

$$= C \cdot \text{SR}.$$

(7.3)

When the power supply is dynamically changing, there may be significant slew rate, particularly for wideband signal modulations. For example, if the power supply must change 3 V in 100 nsec (a typical value for wideband modulations) to follow a large envelope rise then SR = 3×10^7 V/sec (equivalently, 30 V/μsec). If there is a 10 μF capacitor on this supply line providing low frequency stability, then the current necessary to change its voltage as required is

$$I_C = \left(10^{-5}\right)\left(3 \times 10^7\right) = 300 \text{ A}.$$

It is nearly certain that the DPS cannot deliver (source or sink) 300 amps for 100 nsec across a 3 V change. Even if it could, Figure 7-8 shows that this huge current is flowing from the supply into the capacitor, and not through the load. This causes a very large drop in overall efficiency from power supply input to RF output.

Although there is no need for more problems before deciding to take a different approach, nevertheless there are two more problems worth mentioning. One is that having large capacitance at the power supply output phase shifts the supply voltage and current waveforms, which can cause supply stability problems. Also, for the commonly used ceramic chip capacitors, there is a large variation of capacitor value with changes in the voltage on the capacitor. Capacitance vs. voltage performance of a typical 10 microfarad ceramic chip capacitor is shown in Figure 7-9, showing that it loses more than half of its capacitance value as the voltage increases to 6 V. At higher voltages, this capacitor stores less than half of the energy that its marked value would lead a designer to expect.

To achieve the high energy efficiencies desired from the ET operation, current from the power supply must flow only through the RF generating circuitry of the PA. Lower capacitor current requires smaller capacitor values, as shown in Figure 7-10. Comparing this figure with Figure 7-7 shows that there is no single capacitor value that

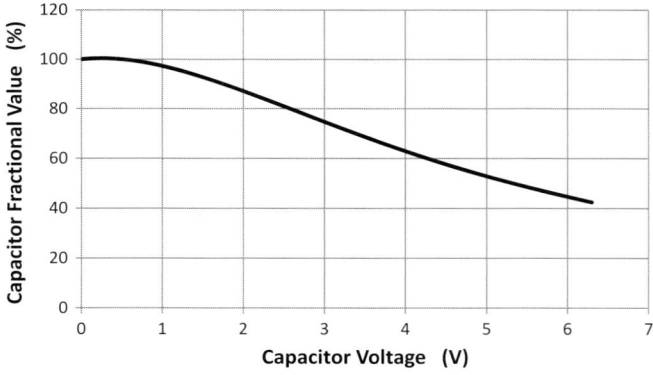

Figure 7-9   Value change of a typical 10 microfarad ceramic chip capacitor across applied voltages.

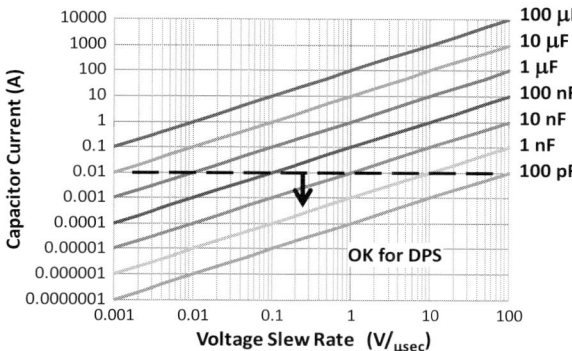

Figure 7-10 Capacitor dynamic current across slew rate values for the same values in Figure 7-8.

simultaneously has low enough impedance to provide low frequency stability and a small enough dynamic current so as not to provide significant degradation to overall PA energy efficiency.

Having large supply currents flowing through low frequency bypass capacitors is not acceptable for amplifiers used in ET transmitters. Another method is needed.

7.2.2   DPS output impedance requirements

Recognizing that the important feature for circuit stability is to have a low enough power supply source impedance to keep gain below unity at all frequencies where oscillation may occur, one solution is to get this low impedance from the DPS itself. From the amplifier's point of view, a very low supply impedance magnitude across the bias frequencies provides low frequency stability, whether it is capacitive or not. All that matters is that any low frequency signal current does not develop voltage at the power supply node.

How low the DPS output impedance must be depends on the low frequency gain of the transistor. Specifically, the output impedance must be less than the reciprocal of the low frequency gain at all frequencies of interest. It is possible to design a feedback regulator to have sub-ohm output impedance near DC. Keeping the output impedance very low at higher frequencies is much more of a challenge. This is discussed in much more detail in Section 9.3.3.

This brings up an interesting situation: the RF power amplifier must have a wideband low-impedance power supply in order to be stable. The power amplifier is therefore not necessarily stable when operated on its own. Design of the entire transmitter becomes a coupled affair, particularly when the dynamic behavior of the RF PA must be characterized.

What becomes clear is that for DPS transmitters there is a new interface that must be managed, which is the connection between the DPS and the PA power input. This is not a conventional RF interface because impedance matching makes no sense. The PA supply-port input impedance should be relatively high so that it does not draw much current. The DPS output impedance must be very low for reasons discussed here.

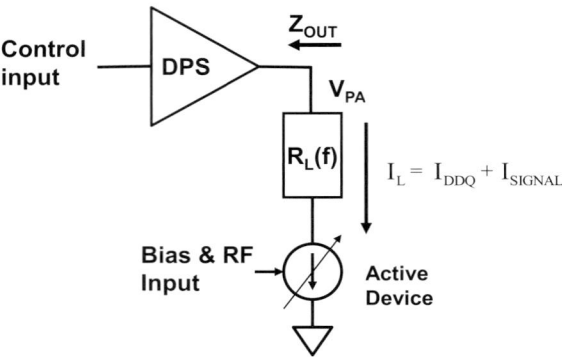

Figure 7-11   Managing low frequency stability with the low output impedance of a DPS to effect the needed "bypass."

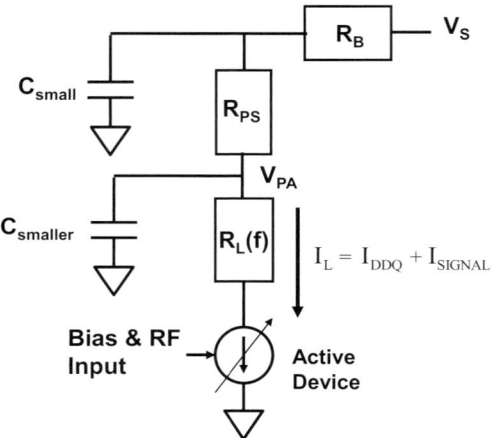

Figure 7-12   Managing low frequency stability of an RF power amplifier without large capacitors or a low impedance power supply by incorporating a low impedance phasing network.

7.2.3   Alternative method to eliminate low frequency oscillation

One effective option to ensure independent PA stability and have small capacitor values is to focus on the Barkhausen phasing criterion, and to make sure that it is never satisfied no matter what the amplifier slope gain value is. This technique ensures that the phase shift at all frequencies, even the low frequencies, never meets the integer multiple of 360 degrees needed to support oscillation. Inserting low impedance phasing networks at each power supply input pin achieves this goal [7-1]. The block diagram of this technique is shown in Figure 7-12. The value of the low frequency bypass capacitor C_{small} is reduced by factors exceeding 100,000x, meaning that a design originally needing 10 microfarads for low frequency stability now only needs 100 picofarads for that same stability along with the insertion of the small series resistance (R_{PS}) in the range of 2 to 10 ohms. Reactive current from the DPS reduces by the same amount. Compared to the above example, the 300 amp capacitor dynamic current falls to 3 milliamps at the same DPS slew rate.

There is an additional feature that comes with the structure in Figure 7-12, and that is an increase in RF isolation along the power supply connections. Having only the large bypass capacitor of Figure 7-8, if any RF signal is present on the power supply side of the load impedance, then nothing is there to stop it from propagating along the power supply interconnect to other circuits. With the series impedance of Figure 7-12, even though it is small, additional attenuation is provided on any RF signal present on the power supply side of the load impedance. This improves interstage isolation at RF.

Linear amplifiers designed using this technique exhibit both high gain and unconditional stability at all frequencies without needing low frequency bypass capacitances greater than 100 picofarads. These amplifiers are independently stable and do not need the presence of a low impedance voltage source to achieve low frequency stability.

7.3   Matching network complexity

The entire principle of impedance matching is based on the maximum power transfer theorem for passive networks, which says that given any source impedance (Z_S) with resistive (R_S) and reactive (X_S) characteristics

$$Z_S = R_S + jX_S \tag{7.4}$$

the sink impedance (Z_K) that provides the greatest transfer of power into the sink is equal to the complex conjugate of Z_S:

$$Z_K = R_S - jX_S. \tag{7.5}$$

Connecting these impedances in series cancels the reactances, leaving only the resistances. This is nothing more than resonating any source capacitance (inductance) with a sink inductance (capacitance) of equal impedance magnitude at the frequency of interest. This works well when the reactance has a discrete value, be it a discrete capacitor or inductor. When matching in to or out of a transistor, fixed reactive component values are rare. One example is seen in Figure 7-13. Both the input capacitance (C_{GS}) and the

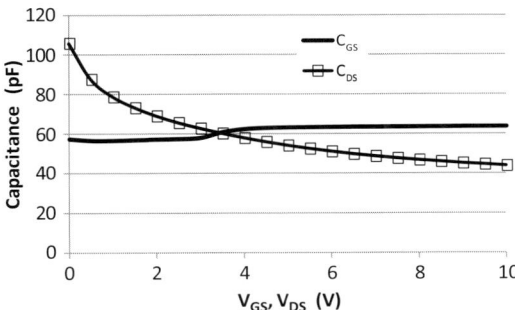

Figure 7-13   Example measured input and output capacitances of a silicon LDMOS transistor.

output capacitance (C_{DS}) are nonlinear functions of the voltage applied across them. When using a fixed supply voltage, the values of these transistor reactances are fairly well controlled. But when we use this device in an ET transmitter, the components in the impedance matching networks must vary to maintain resonance with the varying transistor reactances. How do we design a matching network to work with these voltage variable values?

7.3.1 $C_{in}(V)$: a problem for input and interstage matching networks

If the transistor input capacitance varies with the input voltage, then there is no single value of input impedance (7.4) that an input matching circuit (7.5) can be designed to

$$Z_S(V) = R_S + jX_S(V). \qquad (7.6)$$

This impedance variation directly follows the input signal because of the wide bandwidth of the RF transistor. Some examples of the input capacitance variations that exist are shown in Figure 7-14, one for a depletion mode GaN HEMT transistor and the other for an enhancement mode GaAs pHEMT device. Both of these particular transistors have input capacitance variations that exceed 2 octaves (>4:1) through the transistor active region. There is no steady-state (fixed value) input impedance, particularly as the input signal gets large for high power output. A usual strategy is to design to some average value of this varying capacitance and hope that the actual real-time capacitance variation provides equal and opposite perturbations. Another possibility is to calculate a capacitance "waveform" by applying the voltage waveform to the capacitance voltage-variation profile, take the Fourier transform of this capacitance "waveform," and design to the DC term value.

Another effect of the transistor input capacitance is particularly troublesome for bipolar transistors. With the circuit model of Figure 7-15(a), any input RF signal will have its current divided between two paths: through the "beta generator" resistance r_π, and a shunt path through the input capacitance C_{BE}. Only the current flowing through r_π influences the output current in the collector. This means that any bipolar transistor has a

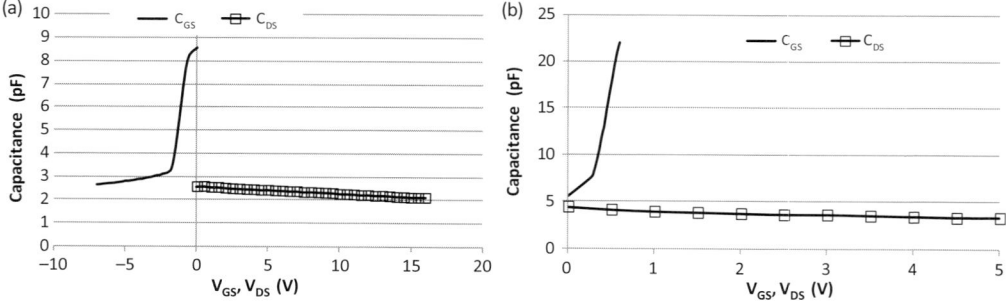

Figure 7-14   Input capacitances can vary by several octaves: (a) GaN can have large variation of the input capacitance; (b) the same is true for GaAs EpHEMT.

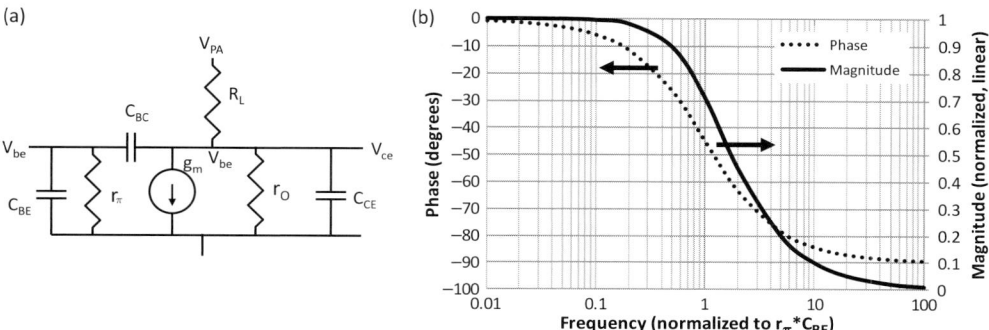

Figure 7-15   Input capacitance shunts a frequency dependent amount of the input current away from transistor action, reducing input impedance along with the high frequency gain of the transistor. Both effects are very bad for overall energy efficiency.

frequency dependent β based on this single pole RC filter. The drop in β, normalized to this pole frequency, is shown in Figure 7-15(b).

The important number to understand now is how this RC pole frequency relates to the typical input frequency applied to the transistor. Using GaAs HBT for this example, the DC V_{BE} is typically 1.3 V and the DC beta (β) is near 100. For a collector current of 500 mA, the value of input resistance r_π at DC is 260 ohms. For an input capacitance of nominally 10 pF, the RC pole frequency is near 60 MHz. Running this as an amplifier at 1800 MHz, the ratio of operating to pole frequency is 30, and this RC model predicts that the RF β is actually closer to 3 than to 100. Measurements of RF β on power GaAs HBT transistors at 1800 MHz typically fall within 3 to 4, which is consistent with this interpretation.

This RF gain drop is a problem for any interstage matching network, as well as requiring the driver stage to supply nearly 30 times more RF current than the DC β would have us expect. Experience shows that amplifier bandwidth is often limited not by the output circuit in HBT-based PAs, but by the interstage network as it tries to impedance match into the very low RF input impedance of the output HBT. The larger driver stage also lowers overall PA energy efficiency, particularly at RF output powers below P_{SAT}.

7.3.2 $C_{out}(V)$: a problem for output matching networks (OMN)

At the transistor output, there is another matching network for any amplifier that cannot work directly at 50 ohms while generating the required output power. As for the input and interstage matching networks, if the output capacitance varies much with the applied supply voltage, then there will not be a steady (fixed value) output impedance for the output network to "match" to. Because of the bandwidth of the RF power transistor, any variation in the output capacitance directly follows the RF output waveform, and does not exhibit any averaging.

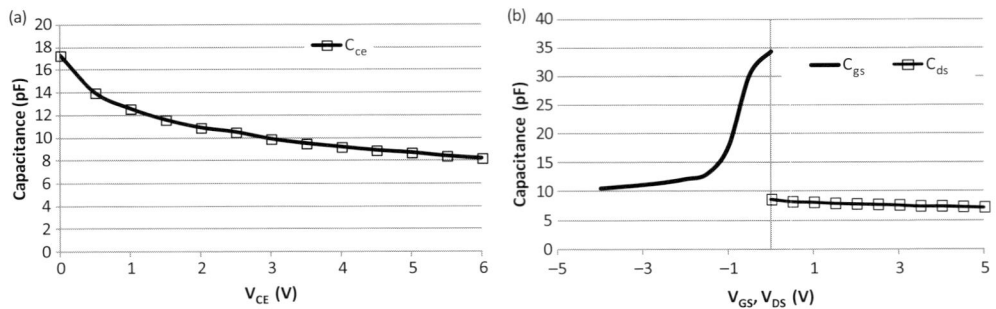

Figure 7-16   Output capacitance examples: (a) silicon RF power transistor can vary significantly; (b) a GaAs pHEMT is much more consistent in value.

Two examples of output capacitance variation are shown in Figure 7-16. The silicon bipolar transistor data in Figure 7-16(a) vary by 2:1 over this range of collector voltage. This is bad for a matching circuit, though a significantly smaller variation was seen for input capacitance variation (of other transistor types) in Figure 7-14. The LDMOS transistor capacitance data in Figure 7-13 also show a 2:1 variation in the output capacitance, though its input capacitance is much more stable.

The GaAs pHEMT data in Figure 7-16(b) show a nearly constant value of output capacitance with variations in the drain voltage. Unlike for the other transistor types so far, this transistor actually can have a well-matched output circuit in a DPS transmitter design.

Output capacitance variation with applied voltage has another problem for amplifiers that are designed with output networks that intentionally craft operating impedances to achieve specific effects. These amplifiers are all among the group of designs where the operating class is defined by the characteristics of this output network, and particularly include class E, all versions of class F, and class J. Voltage variable device capacitances make it much more difficult to hold the resonances on-frequency for the waveform manipulation that these OMN designs must adhere to. *A note of caution*: class E is not really to be considered as a linear amplifier because it is only defined for when the transistor operates as a switch, so it is actually a topic for Chapter 8.

7.4   Gain flatness and control

The very definition of amplifier linearity is that the output signal is directly proportional to the input signal. To have a linear amplifier, it is essential that the gain be consistent and constant for all values of the input signal. And for envelope tracking DPST use, it is strongly desired that the amplifier gain stays consistent across all possible values of the DPS. Indeed, this is what the core principle of ET defined by (7.1) and (7.2) requires.

An excellent tool for evaluating the performance of any linear amplifier as a 3-port for DPS use is the Booth chart. The Booth chart shown in Figure 7-17 is from a GaAs HBT

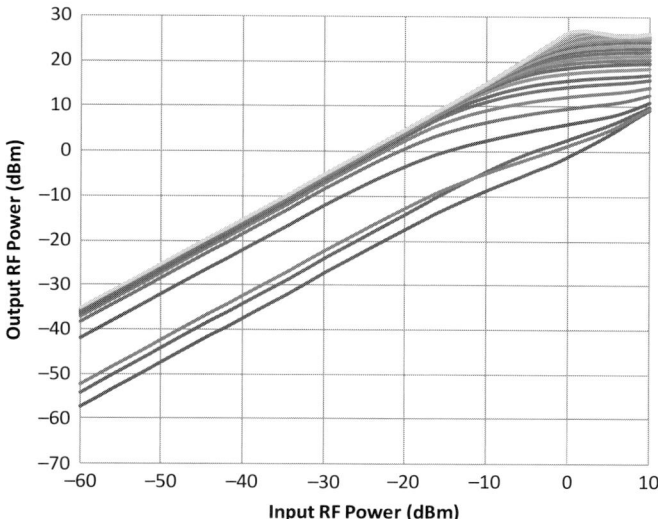

Figure 7-17   Booth chart for a typical mobile device PA based on GaAs HBT technology. Each curve is taken for a different value of supply voltage, from 0 V (bottom curve) to 3.4 V, stepping uniformly by 0.1 V.

mobile device power amplifier, and it exhibits the presence of all three PA operating modes (Section 4.1.6). From several earlier discussions, it already is well established that the ET core principles of (7.1) and (7.2) are only descriptive of L-mode, and are not compatible with either C-mode or P-mode.

This Booth chart data can be readily analyzed to provide the conventional gain chart seen in Figure 7-18, which shows the ratiometric power gain profiles (Section 4.1.5) of this PA across all power supply values. Here it is more evident that there is gain expansion at the highest gains and power supply values, along with both output compression and "gain collapse," a common colloquial name for P-mode encountered during the ET operation. We do not know yet from this chart whether the compression is a proper C-mode operation, though for ET purposes this does not matter because it is essential to avoid both P-mode and C-mode regions.

Another aspect of amplifier linearity is low distortion. Of the four different gain measures, slope gain g_S is the most indicative of the presence of waveform distortion, with ratiometric power gain being the least indicative to the onset of waveform distortion (see Figure 4-20). The gain profiles in Figure 7-18 are therefore not directly indicative of the absence of, or a measure of, the presence of waveform distortions. Additional measurements are required to make these determinations.

7.4.1   Across frequency

Gain flatness across frequency is the primary measure of amplifier bandwidth. RF power transistors are generally wideband devices, so amplifier bandwidth limitations are

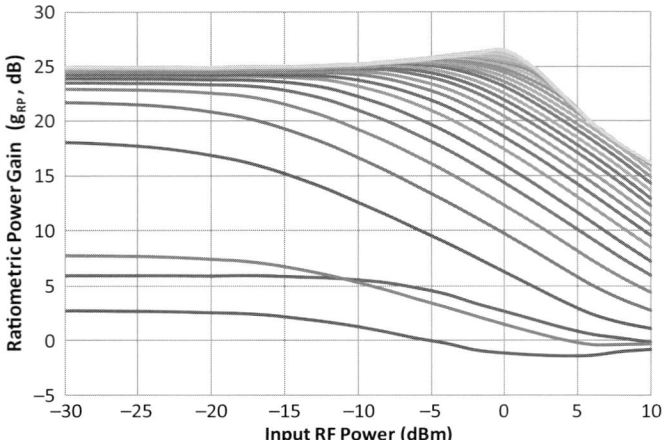

Figure 7-18   Ratiometric power gain for this amplifier derived from the Booth chart of Figure 7-17, showing some gain expansion, compression, and P-mode operation. Each curve here corresponds to the same set of supply voltage values in Figure 7-17.

usually the consequence of impedance matching networks within the amplifier. This is particularly true when sections of transmission lines are used to realize particular reactive impedances.

A particular issue for the envelope tracking DPST operation is the impact of voltage variable capacitances in the transistor on the amplifier frequency response. Section 7.3 shows that most transistors exhibit voltage variable capacitances, but not all in the same way. Some have reasonably stable input capacitances and more variable output capacitances, such as LDMOS. These will exhibit frequency variable behavior as the DPS varies primarily through the output networks. Others have large variations in the input capacitance and fairly constant output capacitance, such as GaN HEMT and GaAs EpHEMT. These transistors will have frequency variable behavior primarily driven by their input or interstage networks.

7.4.2   Across P_{in} and V_{DPS}

Gain flatness jointly across varying input power and varying power supply is largely a property of how flat the transconductance surface is for the transistor, as the output supply and the bias supply are both varied. Achieving a low output signal distortion while ET therefore prefers adoption of transistor technologies that provide a flat transconductance (or β, as appropriate) surface measured across the range of bias and supply voltages. This property is included in the transistor technology survey in Chapter 10, for example see Figure 10-42(a). Spoiler Alert: bipolar devices tend to be much better in this measure than FET devices.

Another measure of amplifier gain flatness as the power supply is varied is presented in Figure 7-19. This particular amplifier provides a much flatter response than the

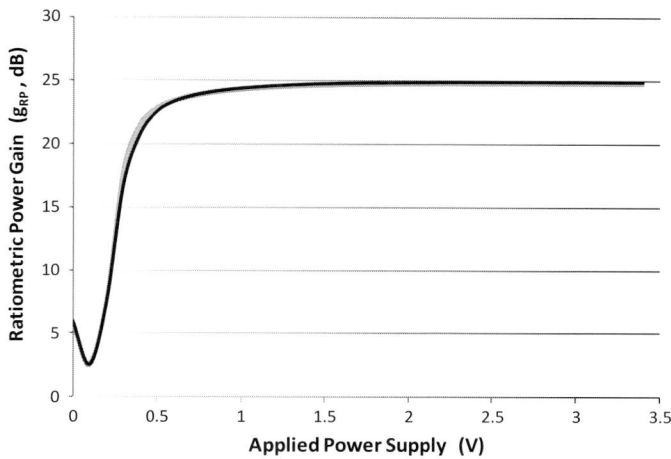

Figure 7-19   Ratiometric power gain for this amplifier as the power supply is varied, derived from the Booth chart of Figure 7-17. This transistor is very flat when the power supply exceeds 1.3 V. Below 1 V, the gain drops rapidly as P-mode operation is entered, a region that ET transmitters must avoid.

amplifier measured for Figure 7-3. This flatness is a property of L-mode operation, and is a manifestation that the transistor has very low output conductance in CCS mode (see Figure 4-25).

7.4.3   Gain control

Control of amplifier gain with the DPS value is fundamentally incompatible with (7.1), one of the ET core principles. This fact calls into question any claim that an envelope tracking DPST is able to use variation of the power supply to effect gain control. In this section, this particular mode of operation is carefully examined to determine exactly what physically is happening at the transistor level under these operating characteristic claims.

The starting point is the family of ratiometric power gain curves like those in Figure 7-18. These curves are repeated in Figure 7-20, with the addition of a line representing constant gain that crosses all of the profile curves except those exhibiting very low gain. It would appear desirable to operate the envelope tracking DPST along this line so that the output signal will experience constant gain and not be exposed to the amplifier gain distortions this amplifier inherently presents.

The situation in Figure 7-20 can be directly translated back to the Booth chart, giving the graphic in Figure 7-21(a). With this Booth chart, there are many profile curves, making a close-up view of the upper portion of this Booth chart useful for better identification of what actually is going on. One such close-up view is provided in Figure 7-21(b), where it is clear that at input power levels exceeding around 0 dBm along this "constant gain" line this amplifier is operating on the edge of C-mode. There is

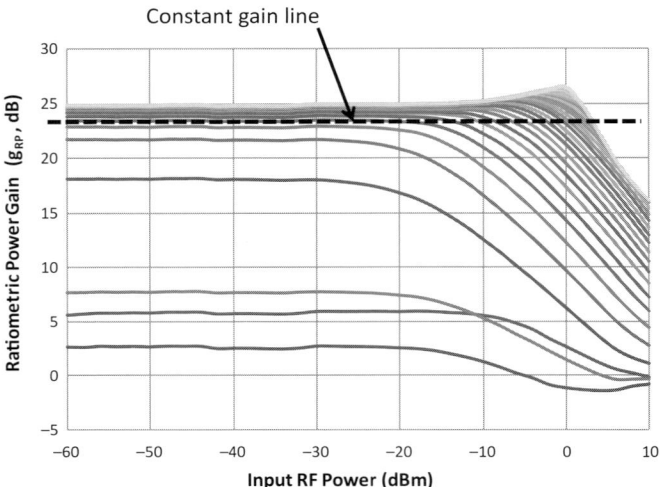

Figure 7-20   Power ratiometric gain curve family for this amplifier from Figure 7-18, here with a line drawn at a fixed gain value within the capability of this amplifier. Each curve corresponds to the supply voltage set for the Booth chart in Figure 7-17.

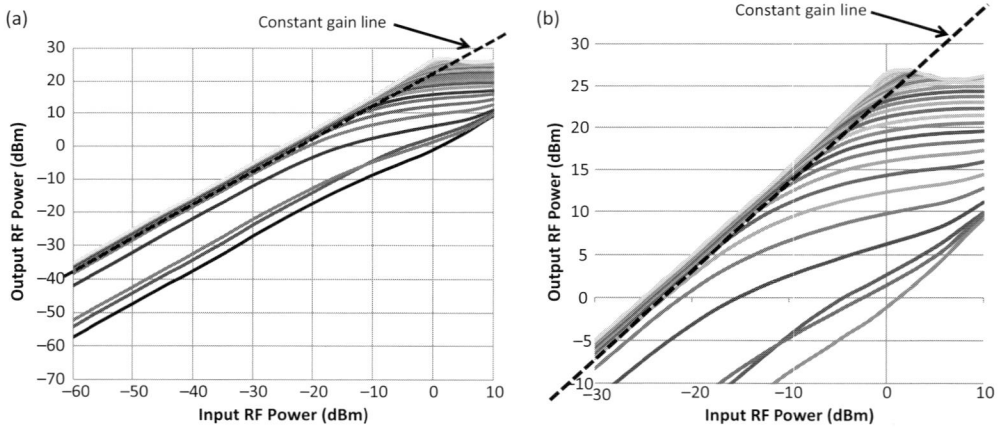

Figure 7-21   The constant gain line in Figure 7-20 becomes a line with fixed slope when translated back on to the Booth chart: (a) applied to the entire Booth chart; (b) close-up view to more easily identify the operating modes corresponding to this line of constant gain.

a strong influence of the power supply on the output signal magnitude, in violation of (7.1), and a weak influence on the input signal in partial violation of (7.2).

At lower input powers, it also appears that this amplifier is operating in, or at least along the edge of, P-mode. Discerning whether this is true or not requires another view on the Booth chart, in this case a vertical slice taken at the input power of −20 dBm. This slice provides the gain vs. power supply profile shown in Figure 7-22. Adding the

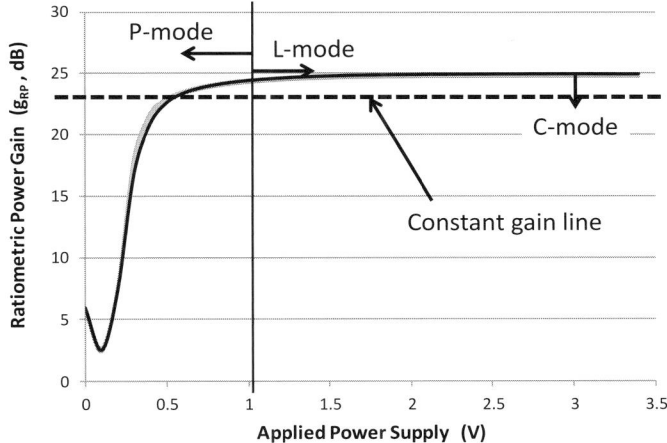

Figure 7-22 Placing the constant gain line from Figure 7-20 on to a Booth chart vertical slice at −20 dBm input power shows that this method never operates in L-mode as required by envelope tracking, but is in P-mode operation at low input powers and C-mode at higher input powers.

constant gain line at the same gain value confirms that at lower input powers this configuration is already in P-mode operation.

This "constant gain" method of operation does not exhibit any of the properties of ET from Chapter 5 and summarized in Section 5.16. Instead, it exhibits the properties of the polar operation from Chapter 6, which are summarized in Section 6.17. Thus ET operation is not something we can choose to arbitrarily claim – there are very clear physical definitions for this operating mode (along with many others) that have corresponding tests which will confirm whether or not this mode is actually realized in the hardware. It is the result of these tests that informs us what operating mode is present. This being said, it is consistent with (7.1) and (7.2) to also include variable gain amplifier (VGA) controls that operate in L-mode that are used together with the DPS that will keep the entire ET system in the required L-mode to meet the ET core principles.

7.5   Bias network dynamics – or not

The ET principles discussion in Section 5.8 shows that if the bias to the RF power transistor is held constant, then the operating class of the amplifier when it is used in ET will vary as the supply changes across its range. This fixed bias must be set for class A operation at the minimum power supply value. As the power supply is increased, this same bias operates the ET PA in class AB_1, then class AB_2. If the class A bias is low enough at the minimum supply value, when the amplifier operates at maximum supply it could almost be operating in class B (see Figure 5-34). One consequence of this bias shift is shown in Figure 7-23, where a test amplifier using a GaAs pHEMT device is biased in class AB_2 with respect to full power.

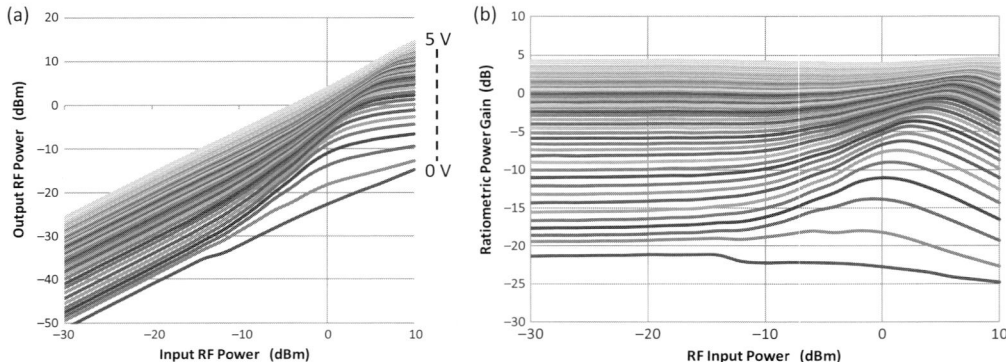

Figure 7-23   GaAs pHEMT amplifier with deep class AB_2 bias when operated at high power: (a) Booth chart; (b) associated g_{RP} results. Supply voltage for each curve is stepped in 0.1 V increments from 0 V to 5 V.

This Booth chart raises two red flags. One is that even at low power and high supply, this amplifier is operating in P-mode since the output magnitude is sensitive to both amplifier inputs. Such a strong influence from the power supply value makes this particular amplifier useless to the ET operation. This amplifier though is fine for a fixed supply operation as long as the supply is relatively high.

An additional gain expansion is seen in Figure 7-23(b) for low power supply values and larger input power levels. This is signal dependent gain, which is yet another distortion mechanism. Either one of these effects is very bad for the ET operation. Having them both present is completely unacceptable.

In order to maintain linear class A operation across all power supply values, it is necessary to also dynamically vary the bias applied to the RF transistor, in a manner similar to that shown in Figure 5-35. Not only must the output circuitry be designed for dynamic supply operation, these dynamic operating requirements also apply to the input circuitry. The usual practice of designing bias circuitry for rigid stability and low bandwidth does not apply to a proper envelope tracking PA.

7.5.1   Bias impedance

Circuit design procedures generally assume that bias networks behave as ideal low impedance voltage sources or high impedance current sources. As the bandwidth of the signals used in wireless communications applications gets wider, the need for power amplifiers to dynamically draw current from these bias networks also increases. If that current cannot be drawn in a timely manner, then the bias network injects a memory process into the overall transmitter [7-2]. Any memory effect is a distortion mechanism that is best avoided. If at all possible, eliminating memory effects is preferable to having to use additional signal processing in an attempt to adapt to them.

Envelope tracking PA design requires that not only the RF path be tested for bandwidth and flat group delay, but also the associated bias networks must be evaluated across the total signal bandwidth for flat response and no delay. In the case of multicarrier amplifiers (and particularly for LTE carrier aggregation), this evaluation bandwidth must exceed one half of the bandwidth extremes: the lowest frequency end of the lowest frequency signal to the highest frequency end of the highest frequency signal.

One good way to do these bandwidth characterizations is to follow the techniques described in [7-2]. Developed by engineers with decades of experience, these techniques have many years of proven utility in the removal of bias network memory effects. Circuit memory is a consequence of reactances (remember ELI the ICE man from the first circuits course?[1]). It is particularly troublesome when resonances exist, a natural result from the simultaneous presence of both capacitance and inductance. Even a fixed capacitor component has some parasitic inductance, and it will resonate at some frequencies. It is extremely important to check and be sure that all components used in a particular amplifier do not exhibit resonances within or near the bandwidth of their associated circuits.

Not to be left out, there can be "parasitic resistance" in a bias circuit that can play havoc with amplifier performance if not paid attention to. Power amplifiers are particularly susceptible because they operate with large signals. If the input drive signal becomes sufficiently large that an FET transistor input begins to conduct, that current must come from somewhere. When the RF input is AC coupled, this current will come from that coupling capacitor. If the bias network has an impedance that is too high to replace the charge on the coupling capacitor from that brief current flow, then the voltage on the capacitor will change. In effect, the input signal (if large enough) can change the amplifier bias unintentionally due to unintended current flow into the transistor.

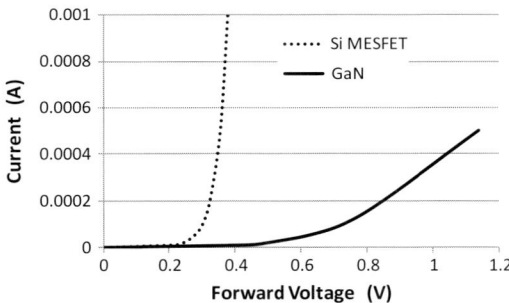

Figure 7-24   "Parasitic resistance" is the conduction within a circuit that should be capacitive and therefore very high impedance to DC (the bias). A common culprit is the Schottky diode present in the gate of many GaN HEMTs. The new silicon MESFETs also have Schottky diodes at their gate terminals.

---

[1] This is a very old memory aid where ELI refers to voltage leads current in an inductance, and ICE refers to current leads voltage in a capacitance.

7.5.2　Impact on PA stability

Maintaining circuit stability is a challenge even when the power supply is fixed. When the power supply is allowed to vary, the task of circuit stability can become a nightmare. The root cause of this difficulty is that transistor port impedances are all dependent on what the voltages on the device actually are. Remembering the s-parameter circuit stability factor (4.20), modified here to emphasize the parameter voltage sensitivities

$$k(V) = \frac{1 - |s_{11}(V)|^2 - |s_{22}(V)|^2 + (s_{11}(V)s_{22}(V) - s_{12}(V)s_{21}(V))^2}{2|s_{21}(V)||s_{12}(V)|} \tag{7.7}$$

we note that all four two-port s-parameters are present. PA designers are encouraged to select transistors that exhibit port parameters that are weak functions of voltage. It is not at all clear that such transistors exist.

The results of circuit instabilities vary across the range of small perturbations from expected performance, which is annoying, to dramatic and spectacular failures when an oscillation mode draws sufficient current to destroy the transistor and possibly other components nearby. Two examples of annoying small perturbations are shown in Figure 7-25. The characteristic curves of Figure 7-25(a) generally look fine, except for small sudden curvature changes between 0.5 and 2 V, depending on the current in the transistor. Transistor theory does not predict these "kinks," and rightly so. They are the result of small bias shifts due to low-level circuit instability. Curing the instability removes the kinks and restores the characteristic curves to their expected shapes.

Other sudden changes in measured behavior are nearly always the result of oscillations. The power supply sensitivity surface in Figure 7-25(b) is another example of a localized oscillation. Over most of the operating conditions, this FET behaves well. Yet for only a particular combination of bias and supply voltages, circuit instability appears.

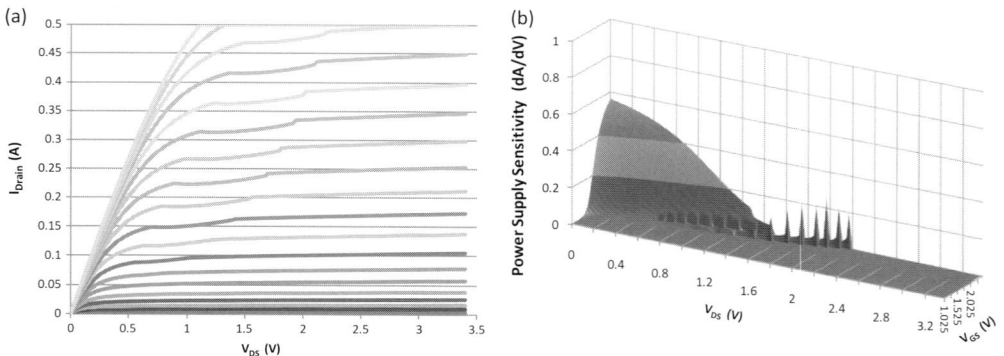

Figure 7-25　Characteristic curve data can show the presence of circuit instability: (a) characteristic curves show sudden slope changes (V_{GS} is stepped by 0.05 V from 1.0 to 2.5 V); (b) performance surfaces with higher order derivatives and anomalous responses. The data here are taken from silicon LDMOS transistors.

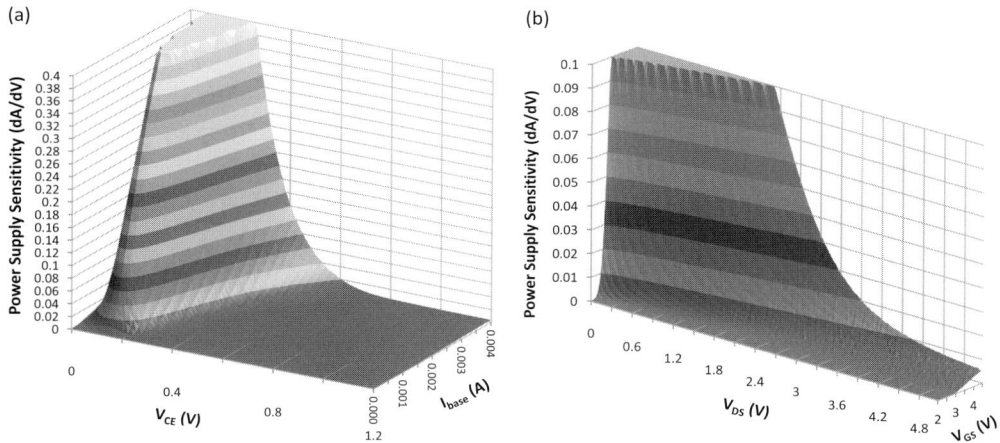

Figure 7-26 Examples of PSS surfaces for two different transistor types: (a) silicon bipolar RF power transistor type MS1649; (b) silicon LDMOS RF power FET type PD57006.

For any DPST design, it is essential to carefully test across all possible combinations of operating conditions. Unpleasant surprises are not welcome.

7.6   Optimum DPS profiles for ET

Once the device IV characteristic curves are generally known, it is possible to directly determine the optimum DPS voltage profile for the ET operation. There is no need for guesswork. The available independence of the output signal magnitude from the power supply value, along with the corresponding suppression of any noise on the power supply, becomes well known at design time with the device IV characteristic curve data in hand.

The starting point is with the transistor pss surfaces (Section 5.2), preferably measured results like those provided in Figure 7-26. Since the core principle of ET is to operate with low sensitivity to the power supply value, the DPS trajectory must be derived from regions of these surfaces that have small values.

Two examples of acceptable ET trajectories on these PSS surfaces are shown in Figure 7-27, here with the PSS surfaces rescaled to show decibels of power supply noise suppression. The trajectory in Figure 7-27(a) runs along the base of this surface and is 40 dB below the peak PSS value. The trajectory in Figure 7-27(b) runs along the lowest part of this surface and is 30 dB below the peak PSS value. This surface never goes completely to zero.

These PSS surfaces are based on device power supply and controlling port bias values. Determining the corresponding device currents so that the process presented in Section 5.4.2 can be used requires that the trajectories in Figure 7-27 be identified on the device characteristic curves. The IV characteristic curves for these example devices are shown in Figure 7-28.

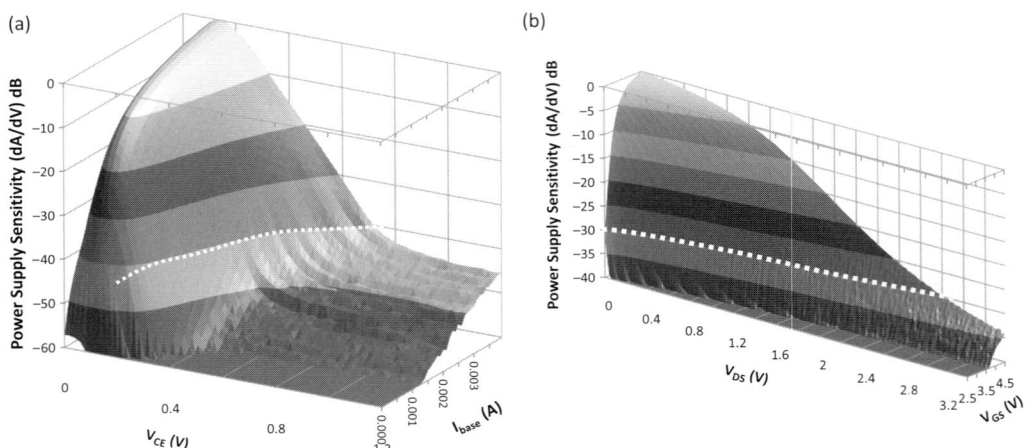

Figure 7-27   Optimum ET trajectories (dotted curves) on these decibel scaled PSS surfaces from Figure 7-26: (a) silicon bipolar RF power transistor type MS1649; (b) silicon LDMOS RF power FET type PD57006.

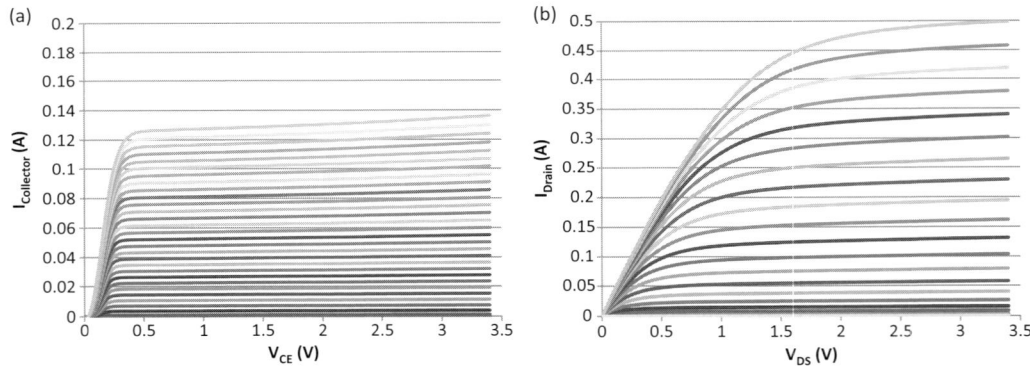

Figure 7-28   Corresponding IV curves for the transistors of Figure 7-26: (a) MS1649 (base current is stepped by 100 μA from 0 to 2.8 mA); (b) PD57006 (V_{GS} is stepped by 0.1 V from 2.0 to 6.0 V).

The optimum ET trajectory is now derived from these points on the IV curves along with the corresponding load lines for the designed R_L values at the transistor. Optimum is a "soft" word, which means that on its own we have no idea what is being optimized until it is clearly stated. Here the optimization is with respect to the ET core principle: minimum power supply sensitivity (PSS). The resulting ET profiles and the corresponding CCS operating region limits needed to achieve the optimum power supply tolerance are presented in Figure 7-29. The offset of the CCS limit trajectory on the IV characteristic curves from the I-axis is the actual operating knee voltage for each possible power supply value, in correspondence with the load line used in the PA design. The value of the knee voltage profile

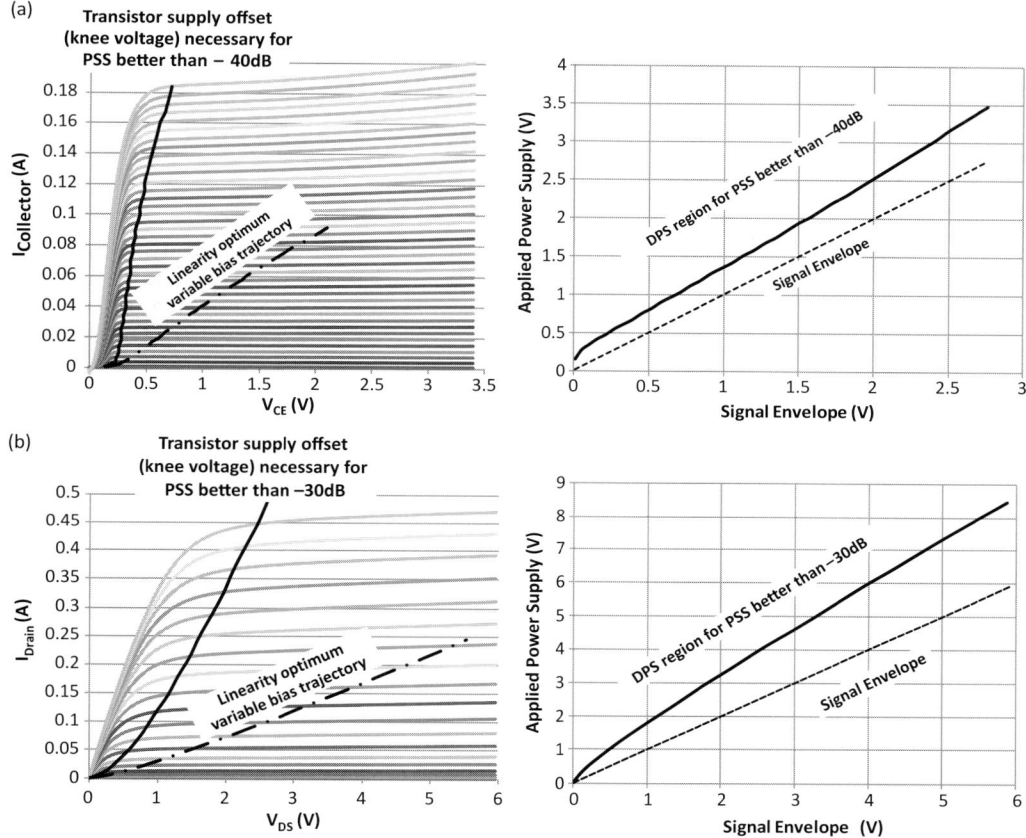

Figure 7-29 Noise-suppression optimum ET DPS trajectories for the transistors of Figure 7-28: (a) MS1649 (base current steps are continued up to 3.8 mA from those in Figure 7-28(a)); (b) PD57006.

does not change with different load lines because the knee voltage trajectory is a property of the transistor itself. What does change with the load line is the required power supply voltage necessary to get the needed device current. For the results shown in Figure 7-29, 15 ohms is used for the MS1649 and 12 ohms is used for the PD57006. The results here should be compared with the theoretical expectations in Chapter 5; for the bipolar transistor in Figure 7-29(a), the comparable theory plots in Figure 5-25 and Figure 5-26, and the FET comparable plots in Figure 5-22 and Figure 5-24.

The discussion in Section 5.4.2 stated that power supply noise suppression is reduced when the DPS voltage profile is brought closer to the signal envelope in an effort to improve energy efficiency. Here the measured data can be evaluated to determine how quickly the PSS changes for these two example transistors. One set of results is presented in Figure 7-30. Two forms of the same information are provided, building an intuitive understanding of the physical process involved in PSS degradation.

These results provide a means to unambiguously determine exactly what the knee voltage is for any transistor. The proposal here is to define the transistor knee voltage as the profile on the characteristic curves where PSS (5.1b) evaluates to −40 dB, corresponding to (7.1) evaluating to less than 0.01. This works for the MS1649 transistor in Figure 7-29(a). The PD57006 never gets to a PSS value of −40 dB, so in Figure 7-29(b) the knee voltage profile is evaluated at the degraded value of −30 dB.

Minimum DPS value (Section 5.4.3) is evaluated in Figures 7-29 and 7-30 using the variable bias technique of Section 5.8.2 for optimizing amplifier linearity. If this variable bias technique is not used, then the minimum available power supply value for the ET transmitter will increase following the mechanisms presented in Section 5.4.3. This is particularly true when a fixed-bias technique is used in the amplifier for design simplicity.

When higher energy efficiency is required than the ET technique can provide, then the transmitter operation must transition toward polar modulation, where PSS = 0 dB, as described in Chapter 6. The rapidity of this transition is evaluated using the technique presented in Figure 7-30. A succession of knee voltage profiles are evaluated from the transistor IV characteristic curves for progressively lower values of PSS. The preference is to start at PSS = −40 dB and go down. The evaluation in Figure 7-30(a) follows this preferred procedure and evaluates three knee voltage profiles at PSS values separated by 10 dB each. Applying the selected load lines to these knee voltage profiles provides the corresponding power supply profiles shown in Figure 7-30. For the MS1649 silicon bipolar transistor, the transition is rapid from ET toward polar modulation, because the three PSS-value curves are very close together. For the LDMOS transistor, "true" ET is never reached because this device never quite achieves PSS = −40 dB. Still, the transition away from its best PSS performance toward polar modulation is much more gradual than that seen for the MS1649.

One important result seen in Figure 7-30 is that for a true ET operation, the knee voltage can get very large for some transistors. This has a very significant impact on the available energy efficiency gains that the ET design can achieve – and not a good one. The principal mechanisms involved are described in Section 5.11. The energy efficiency impacts for the example transistors in this section are shown in Figure 7-33.

7.7   PA operating voltage and energy efficiency

Because power is the product of (in-phase) voltage and current, it is possible to generate the same amount of power from high voltage and low current, high current and low voltage, or any combination in between that provides the same product of voltage and current. With this continuity available, the natural question to ask is if there is one particular combination that achieves the necessary output power at the best possible energy efficiency. To address this question, the natural place to begin is with transistor characteristic curves and load lines.

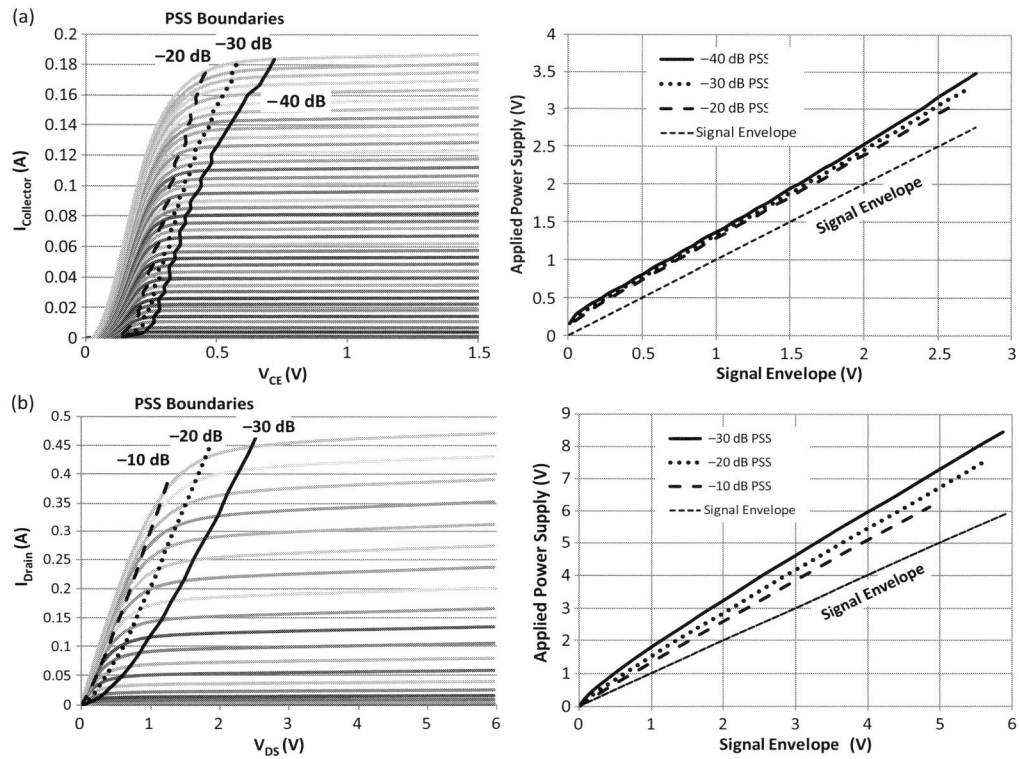

Figure 7-30   Designing for higher energy efficiency forces the ET transmitter to lose its tolerance of power supply variations: (a) MS1649; (b) PD57006.

Two different load lines, each providing the same output power, are drawn on the example characteristic curves in Figure 7-31. One load line represents the low voltage high current case, and the other is a high voltage/low current design.

Operating waveforms for both of these amplifiers are shown in Figure 7-32. The low voltage/high current case is shown in Figure 7-32(a). The knee voltage is seen at the minimum of the output signal voltage waveform, and in this case it is just above 0.5 V. The instantaneous power dissipation of the transistor operated in this way is shown, along with the instantaneous amplifier supply power and the output signal power waveform. The high voltage case is shown in Figure 7-32(b). Here the output voltage waveform is so large that it is truncated in the figure to keep the power waveform scaling identical, allowing easy comparisons. Compared to Figure 7-32(a), both the amplifier input power and the power dissipation are significantly lower. This is a direct consequence of the lower transistor power dissipation because of the lower knee voltage at this lower operating current. These graphics demonstrate clearly that the best energy efficiency for an RF power amplifier is the high voltage/low current case.

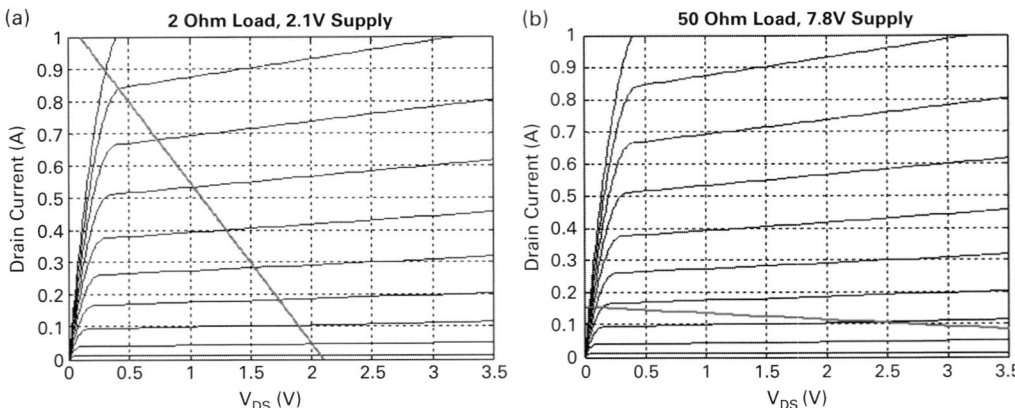

Figure 7-31   Two load lines that provide the same output power from the same FET transistor: (a) 2 ohm load with a 2.1 V supply; (b) 50 ohm load with a 7.8 V supply.

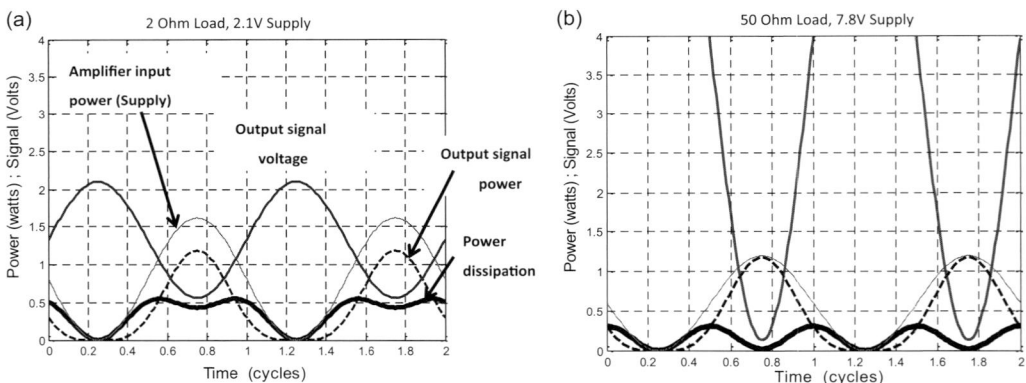

Figure 7-32   Amplifier performance for each of the two load lines showing the same output power: (a) low voltage high current; (b) high voltage low current.

This is the exact opposite of the high efficiency operating condition for CMOS digital circuitry, where low voltage operation is greatly preferred. The physics operates in the same way, but the major difference is that RF amplifiers must provide a specified amount of power while any CMOS logic gate is charging and discharging capacitors to represent the digital bits. Digital circuitry uses energy-based signal processing, and not the power-based signal processing of RF transmitters. Rather, the physical processes of transmitters are much more closely aligned with the physics of utility power distribution, which is done at very high voltages. This is no accident.

The impact of increasing knee voltages from Figure 7-30 is evident in Figure 7-33, where the energy efficiencies of the amplifiers from Section 7.6 are presented. These curves are noticeably different from the simulated ET efficiency curve in Figure 5-37. This difference is completely due to the actual behavior of the knee voltage profile

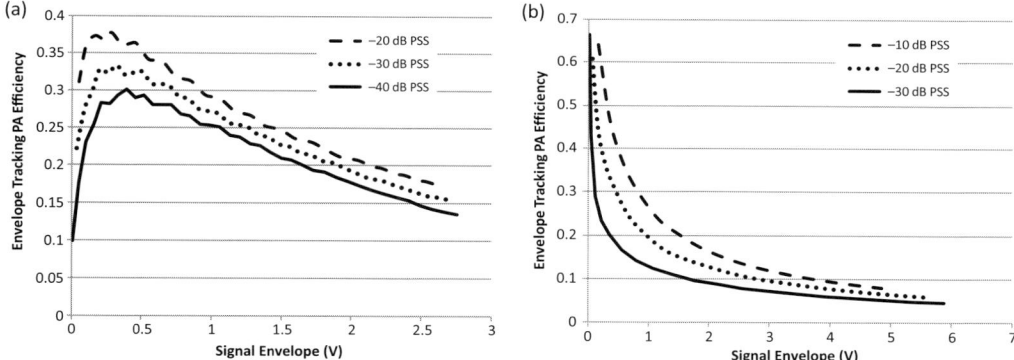

Figure 7-33 Measured impact on ET PA energy efficiency from high knee voltage, using the example transistors from Section 7.6: (a) MS1649 silicon BJT; (b) PD57006 LDMOS FET.

in the example transistor. Particularly noticeable is the rapid increase in the knee voltage seen in Figure 7-30(b), which results in the rapid energy efficiency decrease in Figure 7-33(b). Equally noticeable is the rapid increase in energy efficiency at the lowest output powers, where the knee voltage of this transistor goes to zero. Also, as expected these transistors do exhibit higher energy efficiency when the design PSS value limit is reduced, which lowers the operating knee voltage (and decreases tolerance of variation in the power supply value).

7.8   Noise figure and wideband PA noise

All linear amplifiers have a noise figure, whether the linear amplifier is designed to be a low-noise amplifier (LNA) at a receiver input or to be a linear power amplifier. This noise figure is present at all times, whether an intentional signal is present at the amplifier or not, producing an ever-present wideband noise "floor" at the output of the amplifier. In a receiver, this wideband noise is not a problem because receiver signal channel filtering prevents it from reaching the demodulator. This is not true for a transmitter power amplifier, where wideband noise is a big problem. If there is no filter at the output from the PA, a usual case, then this wideband noise is radiated by the antenna and can cause interference to any other radio nearby. Interference can happen on any frequency where this noise exists, irrespective of the intended PA operating frequency.

The value of the wideband noise density follows the conventional linear amplifier performance:

$$\text{output noise} = \text{noise figure} \times \text{amplifier small signal gain.} \tag{7.8}$$

One measurement of wideband PA output noise with no PA input signal is shown in Figure 7-34(a). This noise measurement shows significant variations across frequencies. According to (7.8), the relationship between wideband noise and amplifier noise figure is

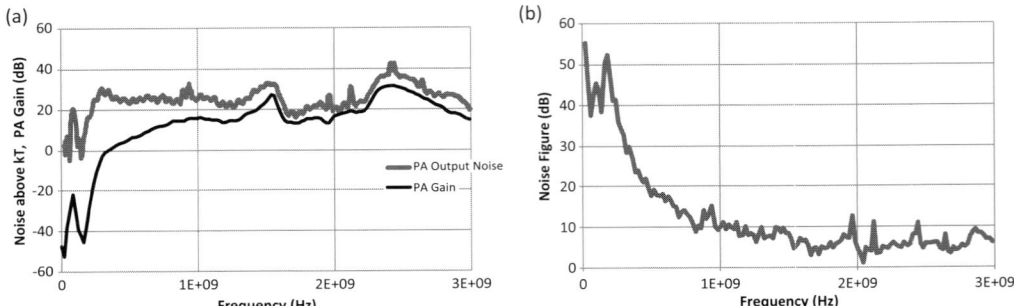

Figure 7-34   Wideband output noise from a power amplifier designed to operate at 2.5 GHz: (a) hardware measurements of PA gain and output noise floor vs. frequency; (b) extracted PA noise figure with its frequency-varying characteristic.

the wideband small signal gain. The result of this gain measurement is also presented in Figure 7-34(a). There is correlation apparent between these two measurements.

The noise figure of this PA as a function of frequency is the difference between these two curves. Extracting this difference yields the frequency dependent noise figure of this PA, which is presented in Figure 7-34(b). The noise figure is very high at low frequencies, and falls as the intended operating band is reached, eventually settling near 6 dB. Fortunately, the PA gain is very low at the frequencies where the noise figure is high, so the total output noise is not too bad. Such a result could be different for a PA designed for wideband use. Taking these wideband measurements is very important for transmitter PA characterization.

7.9   Alternative PA circuit architectures

Maximizing efficiency in any linear amplifier is only possible when the amplifier can operate at low current and high voltage, as was shown in Section 7.7. Many transistor technologies cannot support high voltage operation, particularly CMOS and SiGe HBT. Yet these low voltage technologies have plenty of speed available to operate at RF frequencies. Several circuit topologies have been developed to get inherently low voltage technologies able to produce useful amounts of RF power. The more common approaches, both series and parallel combinations, are presented in the following three sections.

7.9.1   Cascode

The cascode circuit is very old, and was originally developed in the early days of vacuum tube technology to increase the output impedance of an amplifier. Since then many other uses have been realized for this versatile circuit, shown in FET form in Figure 7-35. Selection of the bias voltage on the upper transistor provides an additional degree of freedom in this amplifier design.

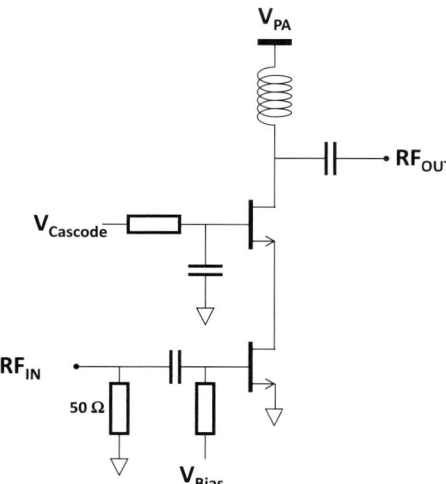

Figure 7-35   The cascode circuit is a series combination of two transistors of nominally the same type, with the upper transistor operating as a common gate (base) amplifier sharing the same current as the lower transistor operating as a common-source (emitter) amplifier.

This series connection allows the RF output voltage to swing at larger levels than any one transistor can handle alone. How the cascode stage behaves at maximum load current depends strongly on the transistor types used. For the FETs shown in Figure 7-35, as long as the source of the upper transistor can be pulled all the way to ground by the lower transistor without exceeding its voltage ratings with the presence of the constant bias voltage $V_{Cascode}$, then the upper FET simply gets considerably enhanced and its drain can also go nearly all the way to ground.

This large voltage swing is not possible when the transistors are of bipolar type. In this case, the collector of the upper transistor can only go low enough to forward bias the base–collector junction based on the value of the constant bias voltage $V_{Cascode}$. To the extent that $V_{Cascode}$ is set high to increase the voltage capability of the entire stack, this choice of transistor technology is more limiting than is available using FETs.

In all cases, the very high output impedance of a cascode circuit is very good for meeting the ET core principle of (7.1). Output conductance is very low, and the characteristic curves of the entire circuit are almost ideally flat in the CCS region. These are all very good characteristics for ET operation. Traditional definitions of the cascode circuit do not allow for varying the bias on the upper transistor. It would be interesting to investigate the effects of slow variations of $V_{Cascode}$ during ET operation.

7.9.2   Stacked transistors

It was frustrating to attend meetings in the later 1990s when participants believed that what was good for digital circuits had also to be good for RF designs – aren't the same

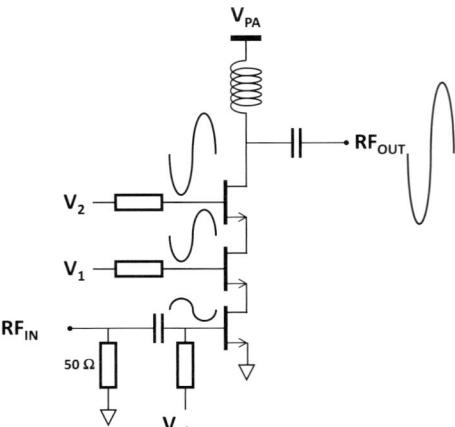

Figure 7-36   Series stacked transistors designed to have identical voltage drops across all ports. Gate voltages of the upper transistors must have increasingly larger replicas of the output signal on them so that the voltage division remains equal across each transistor.

transistors being used anyway? This led to efforts to build low voltage power amplifiers, which all had very poor energy efficiency (not a surprise at all for any engineer familiar with Ohm's Law and the discussion in Section 7.7). Realizing that high voltage operation is actually a better approach is being heralded as a new development in recent years. It is not new, but is actually a century old.

What is new are the technologies available to implement this very old idea. The present approach is to stack many transistors in series in the RF circuit, and to try to make them work together as if the entire stack was a single high voltage transistor. This is not an extended cascode approach, but a true series combination transistor. The basic approach is shown in Figure 7-36.

A common query when first encountering this is something like: "Hey, wait a minute. Don't you lose power in this series stack because the net resistance has to be n times the resistance of a single transistor?" Yes, but that is not a real problem here because the real win is achieving high voltage operation. How this is so is seen in the following brief analysis.

Let V_1 be the rms RF voltage allowed across any one transistor, and R_{ch} be the absolute resistance of that single transistor. Assuming that the voltage division along the stack is exactly equal (this is an *essential* design requirement for this amplifier approach), the total stack of the net voltage and resistance characteristics is

$$V_{NET} = nV_1 \quad \text{and} \quad R_{NET} = nR_{ch} . \tag{7.9}$$

Power dissipation in the stack is, for a sinusoidal waveform (as the RF must be)

$$P_D = \frac{V_{NET}^2}{2R_{NET}} = n\frac{V_1^2}{2R_{ch}} , \tag{7.10}$$

which is a linear scaling of the dissipation in any single transistor. So the early hunch is correct. But we are actually interested in the output power available into the load resistance R_L, which is

$$P_{OUT} = \frac{V_{NET}^2}{2R_L}.$$ (7.11)

The expected energy efficiency for the stacked transistor is then

$$\eta_{STACK} \approx \frac{P_{OUT}}{P_{OUT} + P_D}$$

$$= \frac{1}{1 + \dfrac{R_{NET}}{R_L}}.$$ (7.12)

To the extent that the high voltage operation allows use of a significantly higher value for R_L, this stacked approach is beneficial. Now we need to answer this question of how much higher R_L can be.

For identical powers, we need to now redefine R_L as R_{LH} for the load of the high voltage stack, and as R_{LL} as the load for a single transistor. Equating powers gives

$$\frac{V_{NET}^2}{2R_{LH}} = \frac{V_1^2}{2R_{LL}}$$ (7.13)

from which we get the conclusion that

$$R_{LH} = n^2 R_{LL}.$$ (7.14)

This is our answer: the load resistance value for the series stack is larger than the original load resistance by n^2, which is significantly larger for any value $n > 1$ and increasingly so for larger stack heights n. Combining (7.12) and (7.14) we get the net energy efficiency benefit of the stacked approach as

$$\eta_{STACK} \approx \frac{1}{1 + \dfrac{R_{ch}}{nR_{LL}}}.$$ (7.15)

The overall loss factor is reduced from its original value by the stack height n. As long as the voltage division along the stack can be kept equal under all conditions, adding more transistors to the stack will always be a win in both output power and energy efficiency. Again, quite the opposite of CMOS digital circuitry, the most efficient RF power amplifiers must operate from high voltage.

7.9.3 Transformer combining

If circuit structures are not available to allow access to high voltage operation, a lower efficiency option is to generate multiple instances of smaller RF powers and to combine

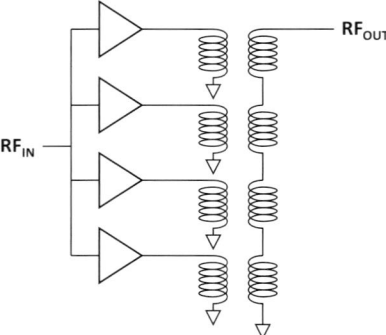

Figure 7-37   Parallel small power amplifiers with outputs combined in a set of transformers having series connected secondary windings.

them in-phase into one higher power output signal. One common architecture is shown in Figure 7-37. Each of the smaller amplifiers is usually called a "cell," and each of these cells must be driven in-phase from the amplifier input signal. The in-phase output signals from each cell drive a small transformer, where each of the transformer primary windings are in parallel. The secondary windings from each transformer are all connected in series, causing the individual transformer output voltages to add into the larger output signal. The in-phase requirement among all the constituent signals is critical to this last step.

This is a lower performing option for two major reasons. First, generation of the lower powers must be done in a higher current/lower voltage environment, which itself is less efficient. Operating the cells at low voltage and low current for improved individual efficiency results in correspondingly low RF power available from each cell, making it necessary to have more cells to reach a specified output power. This exacerbates the second reason: flux linkage among the windings in any transformer is never perfect, so there is always loss in magnetic circuitry, such as this signal combiner. When more cells are used, there are correspondingly more points of loss in this transformer-based combiner. The optimum number of cells for a particular output power depends strongly on the actual implementation of the transformers, and whether their coupling is 60%, 70%, or some other value. Coupling at 90% or higher is very rare, particularly using integrated circuit construction.

7.10   Output mismatch consequences

Section 5.15 presents the issue of output impedance mismatch for any envelope tracking DPST. Any transmission line will have a position-varying RF magnitude where the maxima and minima are separated by $\lambda/4$, one quarter of a wavelength on the line. This ratio is called the voltage standing wave ratio (VSWR) and is defined by

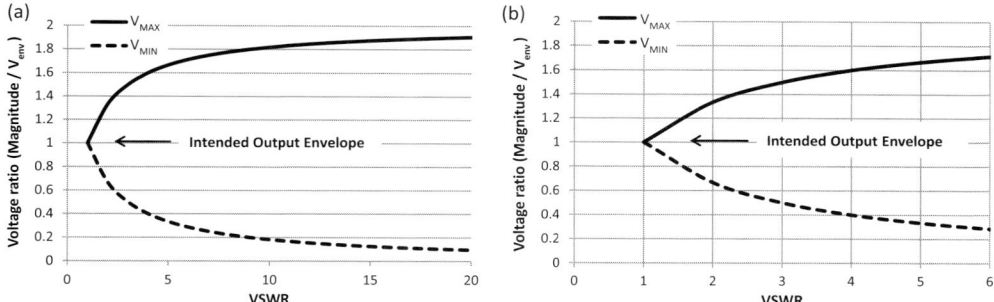

Figure 7-38   Signal magnitude variation range along a transmission line in the presence of an output impedance mismatch: (a) large range view; (b) closer detail for low and moderate VSWR values.

$$VSWR = \frac{V_{MAX}}{V_{MIN}}. \tag{7.16}$$

The nominal RF magnitude if there is no impedance mismatch is the arithmetic average of these extrema

$$V_{NOM} = \frac{V_{MAX} + V_{MIN}}{2}. \tag{7.17}$$

The circuit design problem is that we never really know where along the transmission line the point of interest actually is, making it impossible to predict the exact output RF magnitude that is actually present. The possible variations are seen in Figure 7-38, normalized to the output under perfectly matched conditions. From the DPST design perspective, one particular point of interest is the transmitter PA output. Taking for example a VSWR = 2, Figure 7-38(b) shows that the actual output RF signal magnitude at the PA can be exactly the expected value, 133% of the intended value, or as small as 67% of the value when there is no impedance mismatch. When the actual output RF magnitude happens to be smaller than expected, the envelope tracking DPST continues to work fine, though with reduced efficiency. The real problem arises when the magnitude at the PA becomes too large.

ET operation depends on having sufficient voltage headroom to ensure that the PA transistor operates in L-mode. The power supply voltages necessary to achieve this are determined using the methods in Section 5.4.2. Adding to that method now is the need to continue operating properly in the presence of output impedance mismatch. To ensure that operation into a mismatched output will keep an envelope tracking DPST working properly at all times, the power supply voltage into the PA must be scaled by the V_{MAX} value of the maximum expected VSWR from Figure 7-38. For a VSWR of 2, this scaling factor is 1.33. Similarly, we find that to handle a VSWR of 3 the scaling factor is 1.5, increasing to 1.67 at a VSWR of 5, and so on up to the asymptotic limit of 2 for an infinite VSWR. The impact of this scaling requirement on the PA voltage profile optimized for minimum power supply sensitivity from Figure 7-29(a) is shown in Figure 7-39. If one is to manage the possible presence of PA output VSWR by providing an always sufficient PA voltage headroom, the overhead voltage V_{OS} must be much higher than the minimum

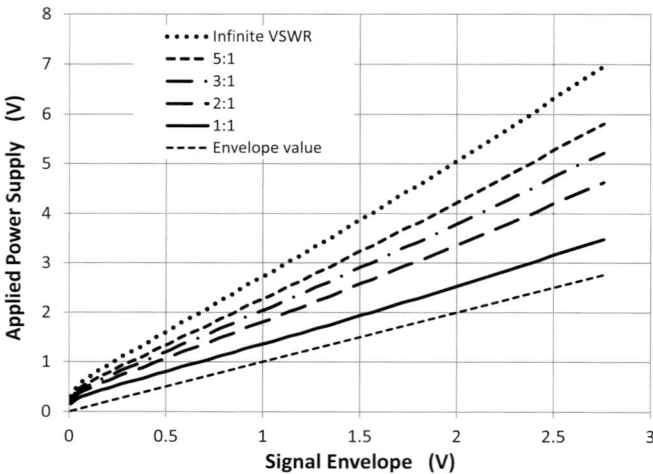

Figure 7-39   Scaling required of the envelope tracking PSS-minimizing optimum profile from Figure 7-29(a) to innately handle peak voltages due to PA output VSWR.

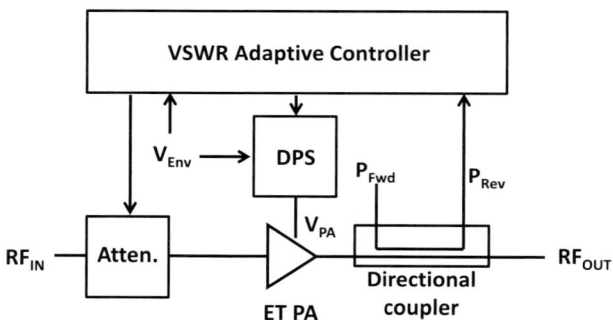

Figure 7-40   Adapting an ET transmitter to the presence of output VSWR requires measuring the reflected power returning to the PA output and using that result to adjust the DPS output, and possibly also the PA input signal power, to maintain in-specification performance.

necessary for nominal operation. The consequence is lower energy efficiency for all operating conditions that are not at the worst case VSWR.

Buying innate VSWR tolerance such that all DPST specifications are met at the price of lower energy efficiency for all other cases is almost always unacceptable. The alternatives are either to let the transmitter perform out of spec under most VSWR conditions (also generally unacceptable) or to have the DPS adapt to the then present VSWR and only adjust the voltage headroom as immediately needed.

Adapting an ET transmitter to jointly maintain specification compliant performance and high energy efficiency in the presence of unknown and variable VSWR conditions involves steps that are highlighted in Figure 7-40. This significantly increases the complexity of the envelope tracking PA stage – but again, nothing comes for free.

To adapt to a possible VSWR, the directional coupler must be arranged to provide the reverse power magnitude (P_{Rev}) to the adaptive controller. Forward power is not necessarily important for this purpose. The adaptive controller rescales the DPS output, either up or down, to maintain the design headroom and keep efficiency up. If there is not enough voltage available to maintain specification, the only recourse is to attenuate the input signal magnitude just enough to ensure that signal peaks and the necessary ET voltage headroom all still fit below the available peak supply voltage.

7.11 References

Note: Patent references are provided for bibliographic use only. Citation of specific patents here is not indicating any view on priority issues.

[7-1] R. Meck, "Power Distribution and Biasing in RF Switch-Mode Power Amplifiers," US Patent 6995613, issued Feb. 7, 2006, US Patent 7227419, issued June 5, 2007, US Patent 7250820, issued July 31, 2007.

[7-2] B. Noori and S. Rumery, "A New Technique for Measuring the Resonant Behavior of Power Amplifier Bias Circuits," *Proceedings of the ARFTG Conference*, June 2007, Honolulu HI.

8 Intentional circuit compression

Designing circuits for intentional compressed operation is a much rarer task than designing for conventional linear amplifiers. It is no surprise that many of the tools and techniques developed for linear circuit design do not apply to these compressed designs. This chapter goes into the important design details for operation well into the transistor compressed operating regimes. More importantly, details on what is required to actually get to a switching RF power stage are worked out. The design procedures also point out if such a design with the available transistor technology is physically impossible.

Since linear (L-mode) operation, defined in Chapter 4, requires operating within the CCS region of transistor operation, here we define nonlinear operation as any other mode of operation. This means that both C-mode and P-mode are nonlinear operating regimes. From the operating mode discussion of a 3-port amplifier in Section 4.1.6, C-mode operation is more apparently nonlinear than P-mode. Chapter 6 demonstrates how both of these modes are useful only to intentionally compressed transmitters, just as Chapter 5 showed that both of these nonlinear modes force ET transmitters to violate the fundamental tenets of ET.

Device characteristic curves for both C-mode and P-mode are in the transistor resistive region, away from its CCS region. All locations in the resistive region of the characteristic curves are closer to the I-axis than any part of the CCS region, meaning that transistor power dissipation is always lower in the resistive region than anywhere in the CCS region. This also holds for transistor cut-off (no current) operation, which is a nonlinear operation at the voltage axis, where instantaneous power dissipation is zero. Lower power dissipation corresponds directly to higher energy efficiency. Therefore, nonlinear transistor operation provides better energy efficiency than any linear (meaning continuous current) transistor operation. This is the fundamental cause of the trade-off between linearity and energy efficiency in any amplifier – the *only* way to improve circuit energy efficiency at a particular output power is to increase transistor nonlinear operation in C-mode, P-mode, or in the transition regions between these modes and L-mode.

8.1 C-mode operating requirements

The fundamental operating requirements for C-mode are defined in Section 4.1.6. Stated mathematically, they are: first that the sensitivity of the output power to changes in the input power is negligible

$$\frac{\partial P_{OUT}}{\partial P_{IN}} = 0 \tag{8.1}$$

and that the sensitivity of the output power to changes in the power supply input is the DPS to AM transfer gain

$$\frac{\partial P_{OUT}}{\partial V_{SUPPLY}} = g_{DPS\_AM}. \tag{8.2}$$

These are exactly opposite to the sensitivities for the linear operation (7.1) and (7.2), emphasizing the very different operating regime of C-mode. It is operationally not a simple extension of linear operation, but is an inversion. There is no reason at all to expect that design techniques used for linear design should work equivalently for C-mode design.

Careful consideration of C-mode design leads to the principles presented in the following sections of this chapter.

8.1.1   Bias vs. power supply variation

For nearly all of the history of DPS transmitter design – nearly one century at the time of this writing – the RF power switch was biased in class C (Section 4.1.1). This ensures that if there is no input signal, then there will be no quiescent current. Another consequence of this bias selection is presented in Section 6.8.1, which shows that for a given input signal magnitude the output power from a class C biased amplifier is very low, the lowest of any bias class. The high energy efficiency that class C is famous for comes at a high price: Efficiency improves only as the output power falls. Energy efficiency for class C approaches 100% only when the output power approaches zero, which is a completely useless case.

As a result, the output power from the class C biased amplifier is recovered by increasing the input signal drive. And for the power supply to actually control the output envelope, the drive must be increased high enough to drive the power stage all the way into C-mode, a fact well understood at the time [1-1] [6-2]. Such a large drive signal restricts the achievable envelope modulation dynamic range through the mechanisms identified in Section 6.11.

In Section 6.8.1, Figure 6-19 shows that it is much easier to achieve C-mode operation by starting from class A bias. By utilizing all of the gain available in the RF power device, getting into the device resistive region of operation is therefore much easier to do with smaller drive signals. A little thought connects this need to the ET bias design of Section 5.8.2, where amplifier operating class is maintained as the DPS varies during ET operation.

What about the resulting quiescent current from class A biasing? This is used simply because the high gain available in class A corresponds to the smallest possible input signal needed to drive the transistor into compression. Even though class A bias is used, in any C-mode design there cannot be quiescent current for two reasons:

1. when a zero output magnitude is desired, the DPS will be zero, and
2. at all times, there must be sufficient input drive level to ensure that the transistor operates in C-mode.

Therefore, the normal class A quiescent conditions, having full supply voltage and bias present and the absence of an input signal, is never legitimate in any C-mode design.

8.1.2   Drive into resistive operation

Section 4.1.6 shows that in order to get a transistor into its resistive operating region, the drive signal must be larger than any drive signal magnitude supporting L-mode operation at the given supply voltage. Output power does not increase proportionally. Therefore, the ratiometric power gain g_{RP} (the most common measure for RF gain; see Section 4.1.5) must drop from its value in L-mode. Transistor specifications for maximum available gain may never be achieved, since this design is intended to stay far away from any linear operation. Circuit linearity is actually intentionally suppressed.

We do need to be careful here. The primary objective in driving a C-mode stage is to get it into C-mode. The usual design technique of designing the input circuit for maximum power transfer does not necessarily apply. The circuit designer only needs to get the transistor to switch, a topic that is discussed in more detail in Section 8.9.2. For an FET, this means having sufficient input voltage, and a bipolar transistor must have sufficient input current to saturate. Neither of these criteria necessarily calls for input *power*. Just input voltage or current, as appropriate. Therefore, the input impedance that is best to design for is not the RF signal impedance match (50 ohms) but rather the two extremes: infinity for the FET voltage drive, and zero for the bipolar current drive [8-1].

Transfer curve characteristics

To comply with (8.1), it is necessary to apply sufficient drive to the power stage so that any further variation of input drive levels has a negligible effect on the amount of circuit current flowing through the load. Looking at Figure 4-26, for only one of the two load lines presented there we note that the drain current changes imperceptibly for the labeled input drive variation ΔV_{GS}. This is where we need this circuit to operate: at its resistive region.

One measure of how well the transistor is driven into its resistive region is to measure its compressed transfer curve. One example is presented in Figure 8-1, here for a GaAs pHEMT transistor. For any particular V_{DS} value, as the control input V_{GS} is increased the transistor enters its resistive region and asymptotically approaches the limiting channel ON resistance R_{ON}. The best transistors for C-mode operation approach this asymptote faster than any others.

Taking the first derivative of the transfer curves in Figure 8-1(a) shows how flat these curves become, which is a measure of the C-mode operating quality. The set of these first derivatives matching the curves in Figure 8-1(a) is provided in Figure 8-2. These are g_m curves for this transistor at the V_{DS} values selected as shown in Figure 8-1(b). The drop in transconductance predicted from Figure 4-26 in C-mode operation is clearly evident.

Current waveform vs. sinusoidal overdrive

Changing the power supply applied to a transistor between cutoff and resistive ON regions inherently results in a current flow duty cycle that depends on the power supply

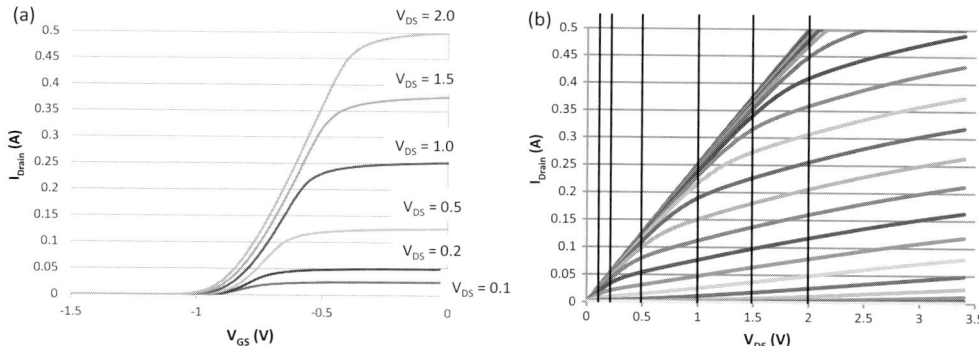

Figure 8-1 Device transfer curves (here for GaAs pHEMT) for C-mode design: (a) the transfer curves for six values of V_{DS}; (b) the vertical lines on the characteristic curves along which these device transfer curves are evaluated, where V_{GS} is stepped by 0.05 V from -1.5 V up to 0 V.

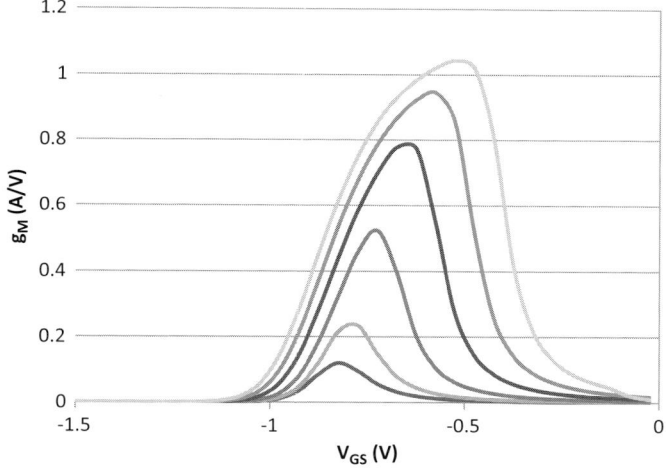

Figure 8-2 First derivative of the curves in Figure 8-1(a) provide the transconductance of the transistor, as well as show how quickly the transfer curves enter the resistive operating region. Each curve corresponds to the same V_{DS} value shown in Figure 8-1(a).

value. One example was shown in Figure 6-21(a). The reason for this follows from using a sinusoidal drive waveform interacting with the various slopes of the transistor transfer curves such as those in Figure 8-1(a). The finite slopes of the input sinewave take longer to achieve high current ON and OFF conditions than the time needed to transition between the lower current ON and OFF conditions. Current therefore flows at high values for less time than it does at low values. This supply-dependent load current duty cycle appears as a distortion in the DPS to AM transfer function (see Figure 6-21(c)).

Solving this duty cycle envelope distortion mechanism requires minimizing the duty cycle variation of the load current. Fortunately, there are two solutions to eliminating this DPS-AM distortion mechanism due to RF duty cycle variations that can be implemented

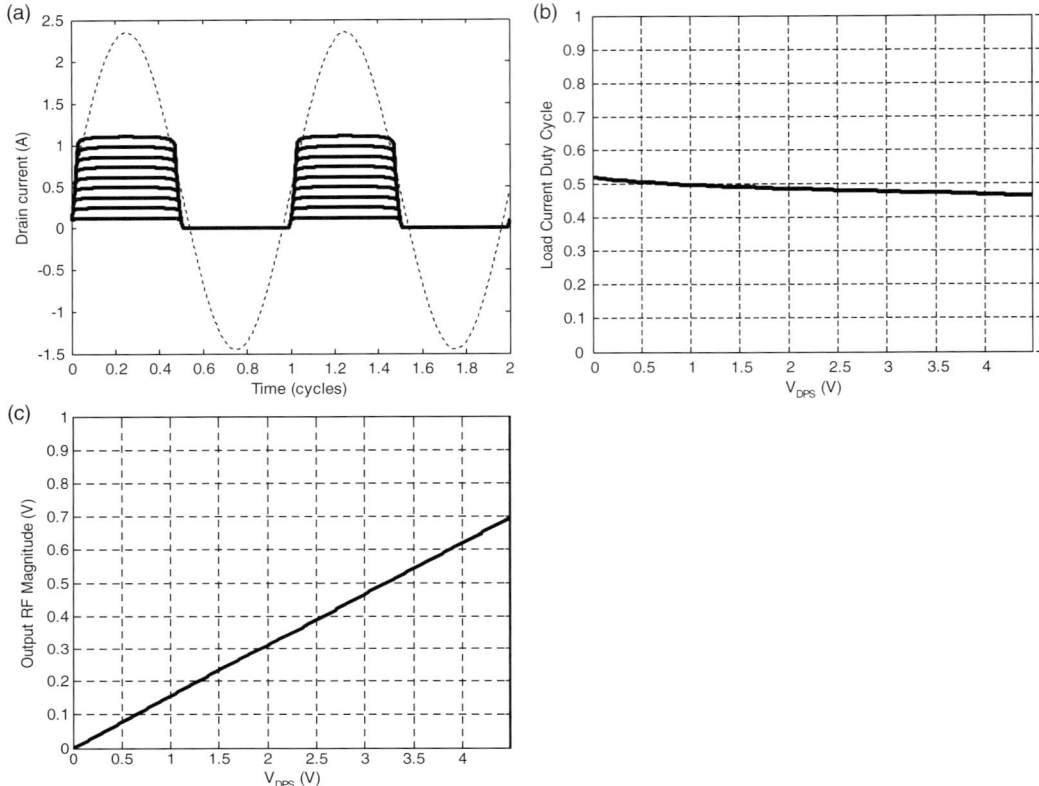

Figure 8-3   Output power changes with variations of transistor current are reduced when the drive into the transistor supports class PFS operation: (a) transistor current (solid lines) compared to the input signal (dashed line) as V_{DS} is stepped by 0.5 V; (b) duty cycle of the transistor current as the power supply is varied; (c) RF output magnitude as the power supply varies (solid line) compared to perfect linearity (dashed line).

at design time in the C-mode circuit. The first and most obvious solution is to operate the transistor in class PFS (see Section 6.8.2). Results of simply increasing the drive to the stage from Figure 6-21 are shown in Figure 8-3. Variation of the transistor current duty cycle is reduced to within the 0.1 dB variation range from Figure 6-22(b). The resulting significant improvement in output signal "linearity" (magnitude accuracy) is seen in Figure 8-3(c).

The large drive signal of class PFS has circuit operation consequences beyond the desired effect of switching the transistor with a consistent duty cycle near 50%. In particular, the presence of parasitic capacitance C_μ (shown in Figure 6-44) couples some of the drive signal directly across the transistor to the output node. The larger the input signal is, the larger the undesired coupled signal at the output is. This is the main problem with class PFS operation, because the drive signal needs to become very large. Still, this class PFS operation is better than using the traditional class C bias for the input of a compressed power amplifier [6-2]. With class PFS operation, the entire

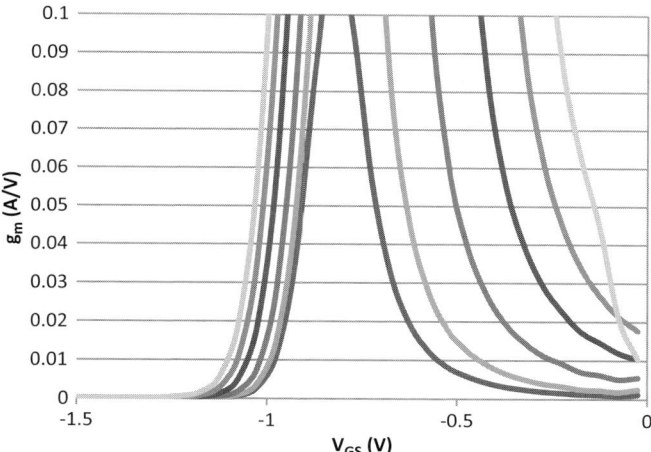

Figure 8-4 Close-in view of the transconductance curves of Figure 8-2 shows both the "class B distortion" as the transistor turns on due to curvature of the transfer curve at threshold (at $V_{GS} < -1$ V), and the slower drop in g_m as the transistor gets deeper into compression. (Distortion of the highest curve as V_{GS} approaches 0 V is due to current limiting in the measurement system at 0.5 A.)

waveform is contributing to either ON transitions or OFF states. Class C keeps the transistor conduction angle below 50%, meaning that the entire lower half of the input waveform has no contribution other than keeping the transistor "more OFF." Class C also requires more drive signal magnitude to get the transistor to its ON state, making the input signal leakage problem worse.

It is important to note that in both the cutoff region and the ON region the device transconductance is very small or zero. In Figure 8-4, this close-up view shows that the ON-state transconductance falls to near 0.01, about 1% of its peak value. These transconductance curves do not have constant values as the input signal varies, showing that this particular transistor will not be a good linear amplifier, at least at these low V_{DS} values. Fortunately, in C-mode all circuit linearity is intended to be suppressed, so this nonlinearity from varying gain is of no consequence.

Even though waveform distortion is not worried about in C-mode designs because circuit linearity is suppressed, there remains some energy transfer during nonideal behavior that can become an issue. With DPS operation of a C-mode stage, the ideal switching waveform is square, and much "analysis" of modulation distortion assumes a perfectly square waveform. Such a waveform can never happen, and the curves in Figure 8-4 show two of these limits. One is the slow start of the transistor conduction near its threshold, along with the transition timing shift that happens as the power supply varies at any particular input voltage value. Another nonideal waveform rounding follows from the slower drop in gain as the current reaches its ON value. These do manifest as nonlinearities in the envelope setting DPS-AM transfer function. Whether these nonlinearities are material to the performance or not depends completely on the application using this DPST and its selected signal modulation type.

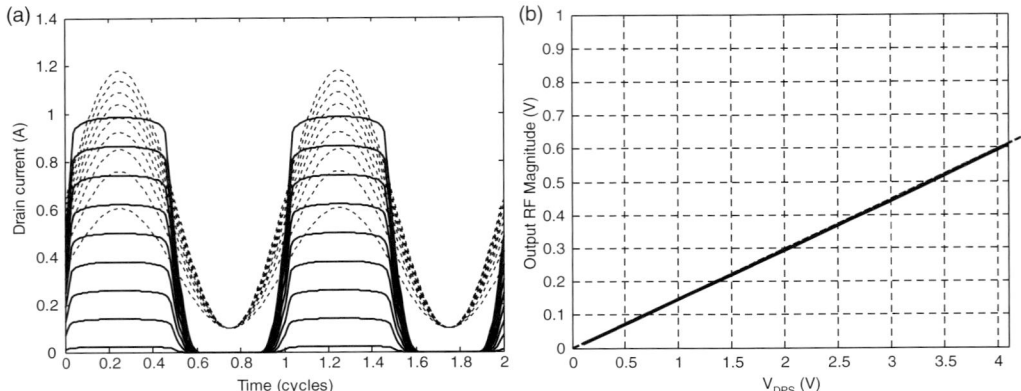

Figure 8-5   To eliminate the C-mode DPS-AM distortion mechanism described in Figures 6-21 and 6-22, the bias and magnitude of the drive here are modified in relation to the changes in the power supply as it sets the varying signal envelope or varying-envelope signal: (a) drain current and input drive waveforms; (b) control linearity of the output envelope with varying power supply.

Variable bias with sinusoidal drive

The second solution to this distortion problem is more elegant, and is important to the discussion in Section 6.11.2. Two steps are involved in this second solution: (1) vary the input signal bias in relation to the varying current the transistor must provide, and (2) preferably also vary the magnitude of the input drive so that enough signal is applied to the transistor, but not too much [6-12]. One design using this approach is shown in Figure 8-5. Compared to the class PFS design in Figure 8-3, this also has excellent envelope modulation linearity, with greatly reduced input drive magnitude requirements. This reduced drive provides slightly reduced output power from the strongly overdriven design. It is interesting to note that this approach is functionally similar to the variable bias needed to maintain linearity across a wide envelope dynamic range in ET discussed in Section 5.8.1.

Details of the variable bias and drive used in the design of Figure 8-5 are provided in Figure 8-6. The optimal relationship between the power supply and the varying bias is not a linear one. Rather, this curve follows the forward transfer relation $I_D(V_{GS})$. For this particular case, the drive magnitude increases at a fixed offset to the varying bias, but this will change depending on the particular transistor used. The optimum drive holds the input signal minimum at a fixed value, which keeps the switch OFF time nearly constant.

Non-sinusoidal drive

Nobody says that the input waveform to this C-mode circuit has to be sinusoidal. In principle, all that is needed to operate a C-mode circuit is a square wave with values between the transistor threshold voltage (or preferably just slightly below it) up to the input voltage needed to get into C-mode for the required current, given the load impedance [8-2][8-3][8-4]. Results of using such reduced-magnitude square drive waveforms are shown in Figure 8-7.

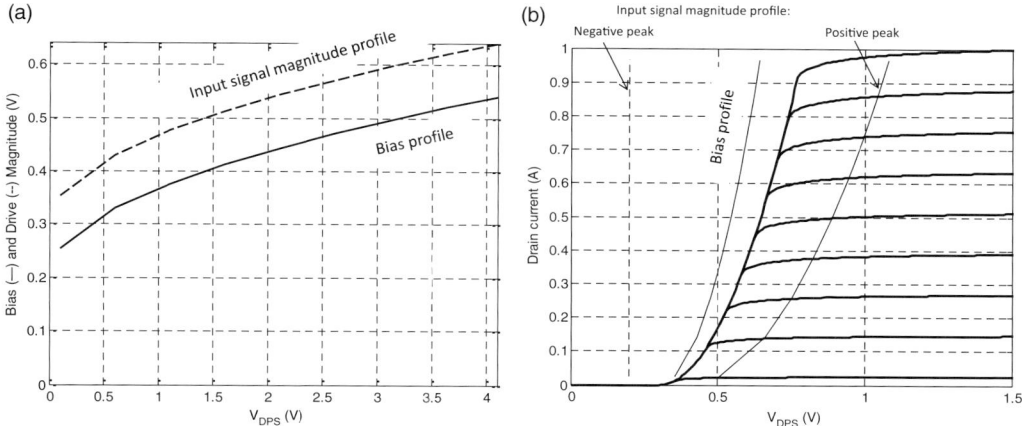

Figure 8-6 Bias and drive magnitude variations used to eliminate the C-mode DPS_AM distortion as shown in Figure 8-5: (a) how the bias (solid line) and magnitude of the drive (dashed line) are modified in relation to the changes in the power supply for the performance shown in Figure 8-5; (b) overlay of the drive conditions on to the drain current relationships for the stepped power supply values used in Figure 8-5 showing that the minimum voltage of the drive stays constant, the optimum bias starts at class AB_2 and slides up to class A, and how much overdrive is used to guarantee C-mode operation.

Comparing Figure 8-7(a) to Figure 8-6(b) shows the reduced input drive level using square waveforms. Drain currents are the same in Figure 8-7(b) and Figure 8-5(a). It is particularly interesting to compare the output signal magnitude (at fundamental frequency) in Figure 8-7(c) with that of Figure 8-5(b). Both show excellent control linearity. But the output RF signals using the square drive waveforms is slightly higher. This is an artifact of removing the drain current rounding seen with the sinusoidal drive waveforms.

For C-mode operation, this smaller input drive magnitude is advantageous for multiple reasons. First, the smaller variation in V_{GS} requires less charge transfer from the drive circuitry into the FET parasitic capacitances. Second, leakage from the input drive to the RF output is correspondingly lower due to the smaller fundamental component of the input square wave. This second point is very important to the discussion in Section 6.11.

The desired physical principle here is actually that the transistor must be operated to switch its resistance in as square a manner as possible. Transitioning between the high resistance of cutoff and the low resistance ON condition in minimum time is actually how switching operation must be defined. Once this resistive change occurs, how the surrounding circuitry reacts to this (sudden) change is the real difference between the multiple switching amplifier classes (D, E, F, S). This reaction is strictly governed by Ohm's Law.

8.1.3   FET or bipolar: behavior differences

The major difference between FET and bipolar transistors is in their operating physics: any FET is controlled by the electric field from the voltage applied across (from the

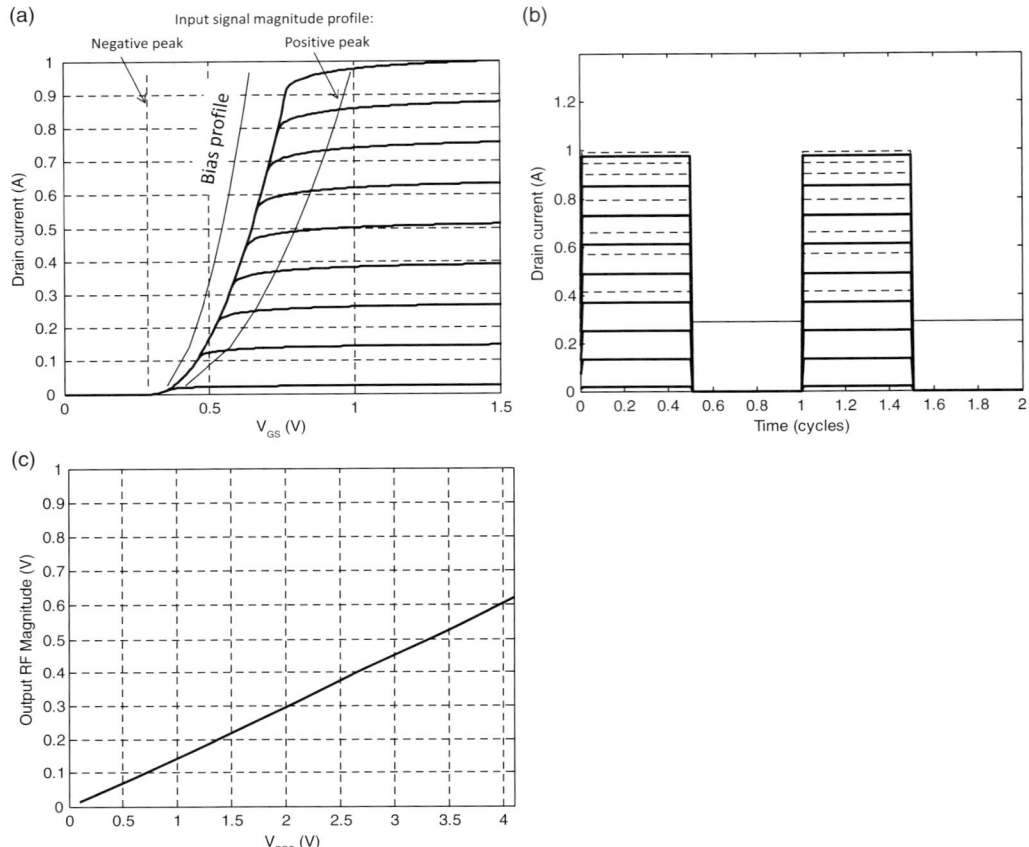

Figure 8-7   Rectangular input signals reduce leakage by having smaller drive signal magnitudes and increased efficiency by keeping transition times short: (a) smaller drive magnitude overlaid on the FET transfer characteristics (compare to Figure 8-6(b)); (b) square drive voltage provides square device currents; (c) control linearity is excellent, and slightly larger than in Figure 8-5(b).

charge stored within) the gate and source terminals (capacitance), and bipolar transistors are controlled by a current that flows through the base-emitter junction. For any FET, this voltage must be maintained to keep the transistor ON, and for any bipolar this current must be maintained to keep it ON [6-10]. In order to get the bipolar transistor away from CCS region L-mode operation and into resistive region C-mode operation, it is necessary to drive additional current through the base-emitter junction. This is shown in Figure 8-8, though it is not required to double the drive current (as shown here) to ensure and maintain bipolar saturation.

When operated in saturation, the bipolar transistor has both of its junctions forward biased, now additionally the collector-base (CB) junction. In conventional silicon BJTs, this forward CB bias injects charges into this junction that take time to be removed to turn the transistor OFF, as seen in Figure 8-9. This is called bipolar saturation delay, and is a problem for bipolar transistor use in polar (C-mode) operation.

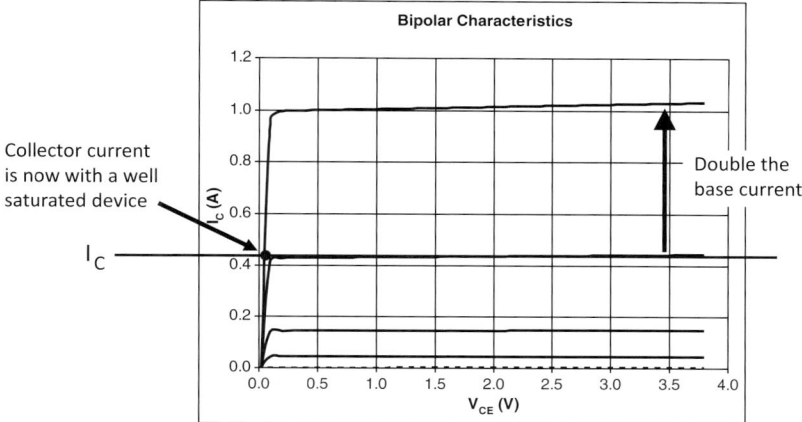

Figure 8-8   Bipolar transistors are controlled by current flow through a forward biased base-emitter junction. In order to ensure that the bipolar transistor operates in saturation, it is necessary that the base current is increased to force this condition.

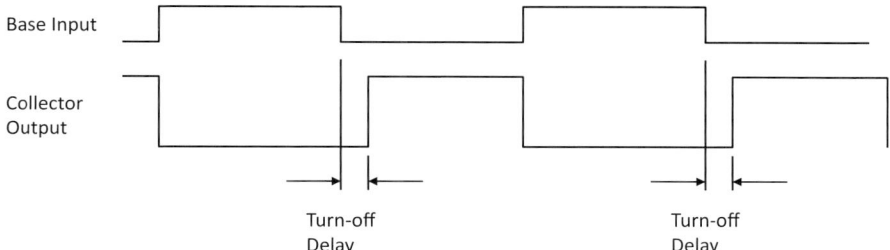

Figure 8-9   Bipolar transistors are controlled by current flow through a forward biased base-emitter junction. In order to ensure that the bipolar transistor operates in saturation, it is necessary that the base current is increased to force this condition.

Transistor ballast

Any bipolar power transistor, like any FET power transistor, is manufactured as a large collection of smaller transistors (called cells) that operate in parallel. For all of these smaller cells to work together as one large transistor, it is necessary for the total current flow to be uniformly distributed among all of the cells. For bipolar transistors, this uniformity of current flow is achieved through a technique called ballast.

The temperature coefficient of a bipolar transistor is negative, meaning that as the transistor heats up the V_{BE} value drops and it tends to draw more current. This additional current leads to increased power dissipation (because the voltage usually does not change), which means the transistor heats up some more, causing it to draw yet more current. If left unchecked, this produces the condition called thermal runaway, where the transistor eventually draws so much current that it is destroyed.

Thermal runaway applies to any individual cell, as well as to the entire transistor. Locally applied negative feedback helps stop thermal runaway by inverting the natural

(a)   (b)

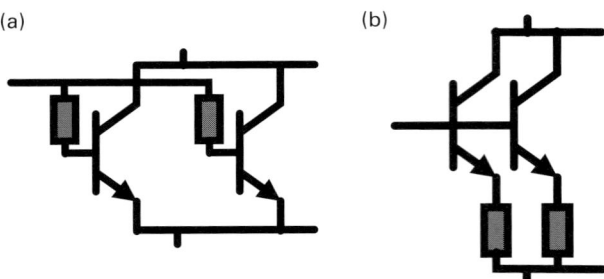

Figure 8-10   Large bipolar transistors are comprised of large arrays of smaller bipolar transistors. To keep current flow evenly divided among all of the constituent parts, there needs to be ballast applied to each sub-transistor. Ballast provides negative feedback, reducing the current drawn by particularly "hot" units to stop thermal runaway: (a) applied at the base terminals; (b) applied at the internal emitters.

tendency. Two common techniques are illustrated in Figure 8-10. In the example in Figure 8-10(a), placing small resistors in each cell's base terminal reduces the base voltage if the collector current (and therefore the base current) increases. Reducing V_{BE} reduces the collector current of the cell, imposing a negative feedback around the cell. Similarly, placing a small resistor in the emitter of each cell as shown in Figure 8-10(b) raises the emitter voltage, reducing V_{BE}, as collector current increases. This is also a negative feedback process. Both of these techniques work well to balance the current flow among the cells in a bipolar transistor. And, of course, each ballast resistor is a location of simultaneous in-phase voltage and current, so each dissipates a little bit of power.

Ballast for most FET designs is not needed separately as long as the FET cell's conducting channels have positive temperature coefficients. The ON resistance of an FET cell tends to increase as it gets hotter, which will reduce the current drawn when the voltage across the channel stays constant. This tends to naturally ballast the FET to keep current flow uniformly distributed across the FET cells. This also spreads the power dissipation across the transistor's constituent cells.

At large scales, the characteristics of bipolar and FET transistors are very similar: both have CCS and resistive operation regions. The IV characteristics of bipolar transistors are most different as the device turns off in bipolar saturation. This region of operation is near to the IV plane origin and has two characteristics that are fundamentally different from FET behavior. Figure 8-11 illustrates this difference using two GaAs transistors for illustrative yet representative examples.

The most apparent difference is that there is an offset along the voltage axis in the bipolar transistor characteristics. Consequences of this are presented throughout the remainder of this chapter in appropriate places. The second major difference is the curvature in the bipolar characteristics as the device goes toward the zero collector current, which is not usually seen in FET device characteristics. Both of these are regions of P-mode amplifier operation. ET avoids P-mode operation, so this region is never encountered in that form of DPST. Polar DPST designs can successfully use P-mode

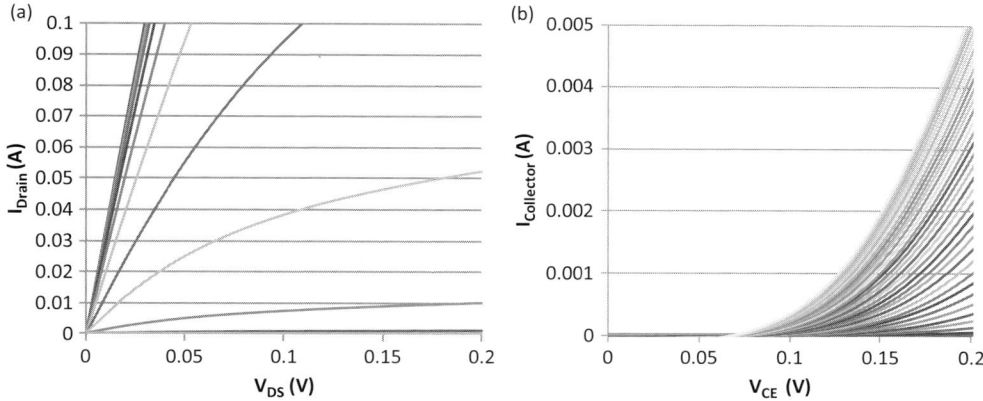

Figure 8-11   Bipolar transistors behave very differently from FETs near the IV plane origin: (a) FET (here a GaAs pHEMT with coarsely stepped V_{GS}); (b) bipolar (here GaAs HBT with finely stepped I_B).

operation, so Figure 8-11(b) provides notice that there will be an additional distortion mechanism when bipolar transistors are used this way.

8.2   Gain behavior in compression

Section 4.1.5 shows that there are actually four separate measures of gain, two that are based on the amplifier voltage transfer function and two that are based on the amplifier power transfer function. When the amplifier is operating linearly, the two voltage gain measures give the same answer for voltage gain, and the same is true for the two power gain measures. This alignment of results has allowed the amplifier engineering community to effectively get lazy, and over the past century to use these four gain metrics interchangeably as if they are identical. Now that compressed operation is desired for its associated energy efficiency, these gain measures diverge. It is vital now that they be treated as the individual metrics that they really are.

When operating any transistor in C-mode, the gain must be reduced no matter what the measure is. The measure used, and reported in the literature, is nearly always g_{RP}. When driven into compression, by definition the output signal magnitude is less than it would be if the amplifier still behaved linearly. The physical effect is the drop in effective device transconductance resulting from the instantaneous operating point moving outside of the device CCS region and toward the resistive operating region seen in Figure 8-2. All of the four gain measures reflect this fact by dropping in value, but not all by the same amount.

Figure 4-20 showed that the value of g_S falls the fastest as compression is entered. Being a derivative of the voltage transfer function, this measure is particularly sensitive to any waveform perturbations. As a result, this gain measure reflects waveform distortion most acutely. The ratiometric voltage gain measure g_R changes the slowest, making the gain measure g_R the least indicative measure of waveform distortion

As long as the amplifier operating mode does not change, this lower gain is not a problem. The discussion in Chapter 11 on hybrid DPST systems looks at the problem of gain change across operating modes in detail.

8.3   IV characteristic impacts

Intuitive understanding of what the circuit is doing, and particularly what the transistor is doing, requires mapping the device port currents and voltages on to its IV characteristic curves. This is the best way I am aware of to stay physically honest and not allow myself to "run off into the math." It always helps to think like an electron.

The fundamental description of C-mode operation on a transistor's IV characteristic curves is shown in Figure 4-26. The key requirement is to get out of the CCS operating region and well into the resistive operating region, past the curvature of the transition between these two operating regions. From the perspective of the gain measures, $g_S = 0$ and g_{SP} is very nearly zero. In this section, the focus is on additional circuit design objectives that follow from C-mode operation.

8.3.1 R_{ON} vs. R_L ratio

Any energy-efficient power amplifier converts the power drawn from the power supply into RF power into the load. The simplified C-mode view of this process is shown in Figure 8-12, which is closely related to Figure 4-27. Power supply current flows through both the load resistance and the transistor, so to be energy efficient the power dissipated in the load must be much greater than that dissipated in the transistor. This can only happen when the value of the load resistance is proportionally that much larger than the resistance of the transistor. This is yet another view of (6.11), showing additionally that high amplifier energy efficiency requires large value load resistance.

8.3.2   Operation at high voltage is preferred

Results from Section 8.3.1 show that the efficiency of an amplifier is improved when the load resistance is increased, all else being equal. But because amplifiers must usually

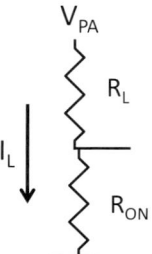

Figure 8-12   When viewed as a series circuit, efficiency of an amplifier is always enhanced by increasing its R_L to R_{ON} ratio.

provide a specified amount of power, if the load resistance increases, then the power supply voltage must also increase. Ohm's Law is absolute.

There is no point in increasing the value of R_L above the ultimate value of the load R_O that the amplifier must supply its power into. This is because any difference between R_L and R_O must be transformed across an output matching network, the topic of Section 8.9.1. Any impedance transformation, whether up from a lower R_L value or down from a higher R_L value, incurs a bandwidth limiting process that depends inversely on the size of the resistance value step. No matching network is needed if $R_L = R_O$, making this the condition of maximum operating bandwidth.

It is fortuitous that the discussion on linear amplifier efficiency in Section 7.7 also directly applies to the compressed amplifier, the interest of this chapter. One immediate observation is that if a power supply voltage limitation forces a reduction in R_L at the transistor, then according to Section 8.3.1 there must be some reduction in the output efficiency of the amplifier from the decrease in the R_L/R_{ON} ratio. Of course, if the ratio is already very large, this decrease may be only slight and immaterial to overall amplifier performance. But it *is* there. And it will be made worse by the additional need for the impedance matching network which comes with its own losses.

8.3.3   Special dissipations in bipolar transistors

Current and voltage definitions for a bipolar transistor are shown in Figure 8-13, where an NPN transistor is the example. Currents and voltages together, when in phase, lead to power dissipations which are loss mechanisms that we need to identify and minimize.

When current flows through the forward biased base-emitter junction, there is necessarily an associated voltage across this junction. According to Ohm's Law, the in-phase voltage and current waveforms cause dissipated power. This power dissipation source is present only in bipolar transistors. This base-emitter junction power dissipation is

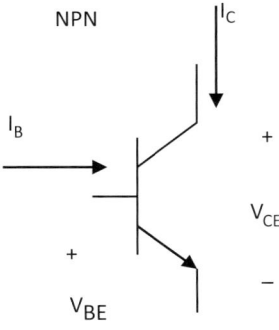

Figure 8-13   The forward voltage drop of the base-emitter junction (V_{BE}) in the presence of the base current (I_B) means that the product of the base drive current with this voltage is an additional power dissipation (loss) mechanism.

minimized when the transistor β is high (reducing the base current magnitude), and by selection of a bipolar technology with a lower forward voltage V_{BE},

$$P_{D,Base} = \left(\frac{I_Q}{\beta_{DC}} + I_{RF} \right) V_{BE}. \tag{8.3}$$

Another loss mechanism that is unique to bipolar transistors is a result of $V_{CE,SAT}$ when the transistor is ON. All of the collector current must flow in the presence of this voltage, and again Ohm's Law states that the product of voltage and current leads to power dissipation,

$$P_{D,SAT} = I_C V_{CE,SAT}. \tag{8.4}$$

In C-mode operation, there is no interest in the knee voltage V_k because transistor operation is only in the resistive region or in cut-off.

Operating in the resistive region, the power dissipation due to the conducting path (channel) resistance is

$$P_{D,Rch} = I_C{}^2 R_{ch} \tag{8.5}$$

where R_{ch} is the dynamic resistance, not the absolute resistance. This selection is made for two reasons. First, the voltage offset has already been accounted for separately in (8.4). Second, the dynamic resistance is the slope seen in the characteristic curve which provides more intuitive insight into the individual physical processes.

At quiescent (no signal) operation, these three power dissipations are active in any bipolar transistor. Their relative magnitudes are seen in Figure 8-14 evaluated for a GaAs HBT device. At low bias, the offset loss (8.4) dominates, followed by the base dissipation (8.3) and then the resistive loss (8.5). As bias increases, the resistive loss increases quadratically until it dominates at the higher powers.

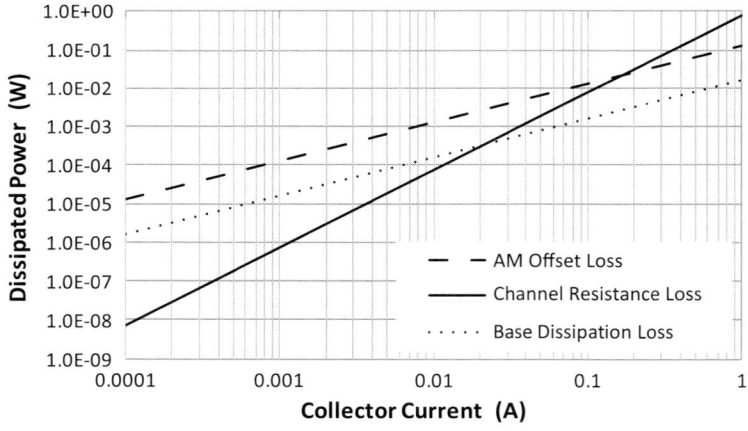

Figure 8-14   DC power dissipations in a bipolar transistor as collector current varies, here a GaAs HBT with DC $\beta = 80$, $V_{CE,SAT} = 0.12$, $V_{BE} = 1.3$, and ON dynamic channel resistance = 0.75 ohms.

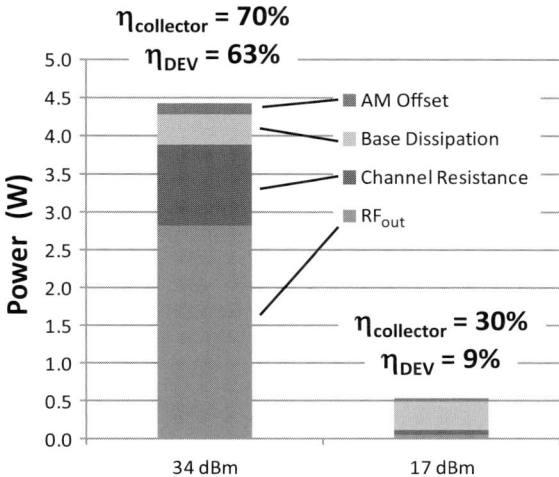

Figure 8-15 DC power dissipations in a bipolar transistor as collector current varies, here a GaAs HBT with DC $\beta = 80$, $V_{CE,SAT} = 0.12$, $V_{BE} = 1.3$, and ON dynamic channel resistance = 0.75 ohms.

When an RF signal is applied, then the energy efficiency can be evaluated. For the transistor itself, there are two efficiencies of interest, the collector efficiency

$$\eta_{Collector} = \frac{P_{OUT}}{P_{OUT} + P_{D,Rch} + P_{D,SAT}}, \tag{8.6}$$

and the device efficiency

$$\eta_{DEV} = \frac{P_{OUT}}{P_{OUT} + P_{D,Rch} + P_{D,SAT} + P_{D,Base}}. \tag{8.7}$$

For the GaAs HBT used in Figure 8-14, two signal cases are used to evaluate these power dissipations and efficiencies. The results are presented in Figure 8-15. Similar to the bias-only chart in Figure 8-14, at high RF output power the largest transistor power dissipation is from the resistance of the channel itself. It is interesting to note that at RF the base dissipation is greater than the AM offset loss, which is opposite to the quiescent case. This results from the additional RF signal input current due to the HBT input capacitance presented in Section 7.3.1. This signal current can get very large as the RF current gain gets low through the process described there. From (8.6), the predicted collector efficiency is 70%, which matches the measurement. This evaluation identifies where the loss mechanisms are and their relative importance. Including the base dissipation, the device efficiency falls to 63%.

At lower output power, the base dissipation becomes dominant. This follows from C_{be} not changing its value, so the RF β reduction remains at all input powers. The other currents roughly scale with the smaller output. The net result is that the device efficiency falls much faster than the collector efficiency at backed-off output power. Both of the power dissipation mechanisms (8.3) and (8.4) are not present in FET amplifiers.

8.4   Device speed considerations

Section 6.10 discusses power dissipation in a switching circuit, pointing out that power dissipation in the fully compressed RF power transistor is lower than either ET or linear amplifiers can ever provide because:

1. the voltage across the transistor is lower when it is ON, and
2. the amount of time spent in the power dissipating region is smaller.

Power dissipation is globally minimized when the transistor achieves this lowest power dissipation condition in the shortest possible time. Here the conditions to achieve this speed are investigated.

Control of this switching action happens with the normal transistor input. Changing the input conditions moves the circuit instantaneous operating point along the load line. The transition between ON and OFF states for switching action must therefore also proceed along the load line, illustrated in Figure 8-16. In essence, what we strive to do is to have the transistor *channel resistance* instantaneously transition between the very high impedance OFF state and the (very) low resistance ON state. What the corresponding node voltages and branch currents are in reaction to this fast resistance change is governed by Ohm's Law and the initial condition energy stored in reactive circuit elements around the transistor.

To start this analysis, we posit that the driver into the transistor is capable enough of "instantaneously" changing the voltage V_{GS}. Assuming that the transition of the input signal takes zero time, the transition of the output node from one state to the other takes a time that is set by the transistor itself along with all the external components and parasitic reactances. Circuit models of the two endpoints are shown in Figure 8-17.

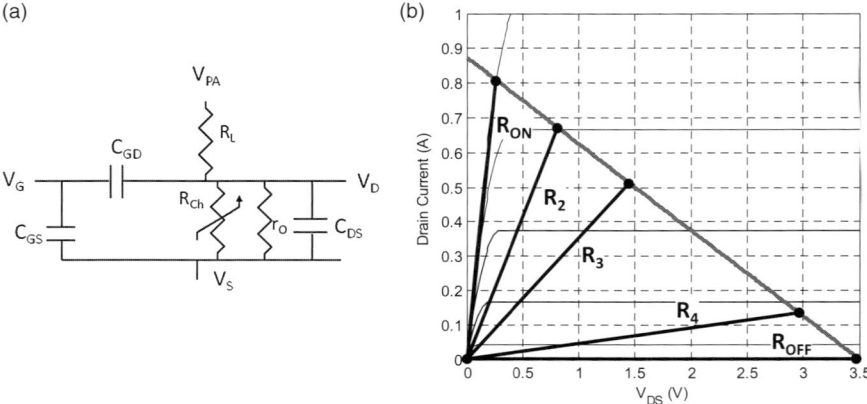

Figure 8-16   Transistor switching action is governed not by voltages and currents in the output, but by behavior of the channel resistance (R_{ch}) driving responses in the associated circuitry: (a) transistor circuit model with load R_L; (b) channel resistance variations along the load line.

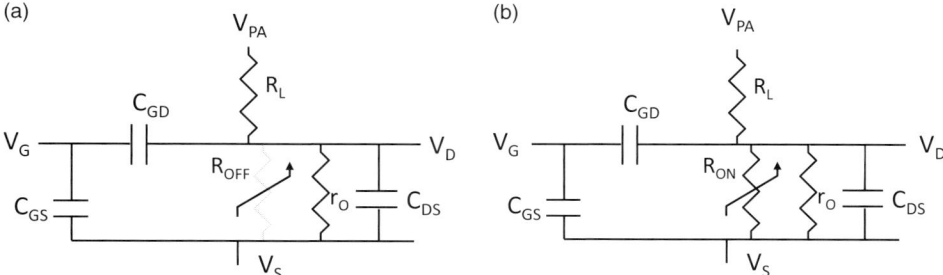

Figure 8-17 Transistor circuit models for the two switch stable states: (a) OFF; (b) ON.

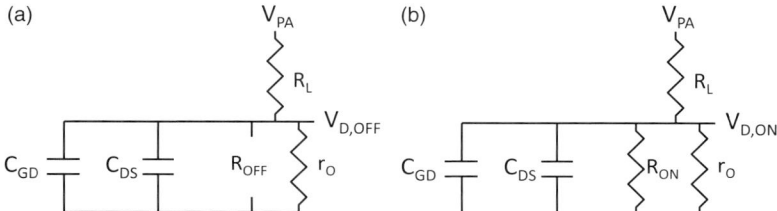

Figure 8-18 Transistor equivalent circuit models for the two switch stable states: (a) OFF; (b) ON.

Beginning with the driver assumption as a voltage source, it controls the voltage across C_{GS} and therefore removes that component from these models. The equivalent circuit models for these conditions are shown in Figure 8-18.

For the OFF condition in Figure 8-18(a), the steady state output voltage is

$$V_{D,OFF}(\infty) = V_{PA}\frac{r_o}{R_L + r_o} = V_{PA}\frac{1}{1 + \dfrac{R_L}{r_o}} \tag{8.8}$$

which is very nearly V_{PA} when $r_o \gg R_L$. Similarly, for the ON condition the circuit model in Figure 8-18(b) provides

$$V_{D,ON}(\infty) = V_{PA}\frac{R_{ON}}{R_L + R_{ON}} = V_{PA}\frac{1}{1 + \dfrac{R_L}{R_{ON}}}, \tag{8.9}$$

which is simplified for the common condition that $R_{ON} \ll r_o$. By inspection, (8.9) shows that $V_{D,ON}$ tends toward zero as R_{ON} gets much smaller than the load R_L, as expected.

The circuit models from Figure 8-18 now provide the time constants for the transitions of interest in this analysis. For the transition when the transistor suddenly turns OFF, the circuit of Figure 8-18(a) starts from the initial condition with the capacitor voltages at $V_{D,ON}$. This is a simple RC circuit which charges up to $V_{D,OFF}$ with time constant

$$\tau_{OFF} = (R_L || r_o)(C_{GD} + C_{DS}). \tag{8.10}$$

Defining $C_T = C_{GD} + C_{DS}$ and taking the usual condition that $r_o \gg R_L$ (an approximation which is not valid for nanometer CMOS), this simplifies to

$$\tau_{OFF} \approx R_L C_T. \tag{8.11}$$

For the other transition, we have the circuit model of Figure 8-18(b) with the initial condition that the capacitor voltages are at $V_{D,OFF}$. As before, this is a simple RC circuit that transitions to $V_{D,ON}$ with time constant

$$\tau_{ON} = (R_L || R_{ON})(C_{GD} + C_{DS}). \tag{8.12}$$

Taking the usual condition that $R_L \gg R_{ON}$, this simplifies to

$$\tau_{ON} \approx R_{ON} C_T. \tag{8.13}$$

Time constant (8.13) is based only on transistor characteristics, so it may be a useful additional specification for RF power transistors that are intended to be used in C-mode transmitters. Time constant (8.11) depends on external component R_L and therefore is not a useful transistor metric.

Examining these two time constants, it is apparent that the transition times are not symmetrical. Transitioning from OFF to ON happens much faster than the transition from ON to OFF. This asymmetry in switching time as the output circuit reacts to (sudden) changes in transistor resistance between R_{OFF} and R_{ON} is illustrated in Figure 8-19. The ratio of these transition times is related to the time constants by the ratio of (8.11) to (8.13) giving

$$\frac{t_{OFF}}{t_{ON}} \approx \frac{R_L}{R_{ON}}. \tag{8.14}$$

The C-mode design rule (6.11) in Section 6.7.5 call for this ratio to be maximized, at least to the point where R_{ON} is negligible when compared to the value of R_L. Note that

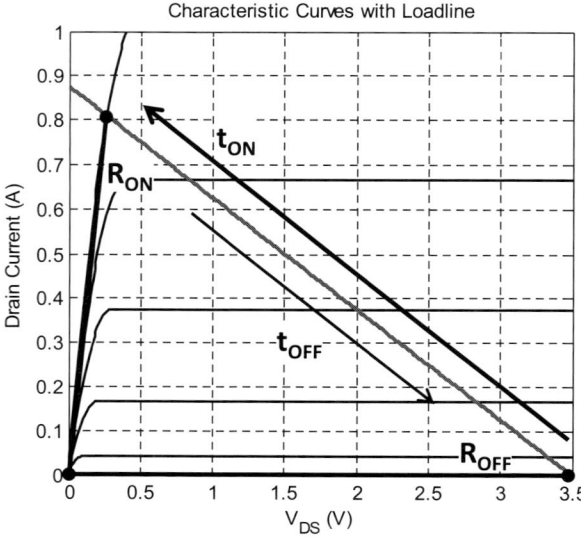

Figure 8-19   Transition times along the load line for switching action are not the same.

this C-mode design rule inherently spreads the transition times of the desired switching operation.

The natural question here is to ask how these results relate to the common metrics of transistor speed, f_T and f_{MAX}. The answer is that there is no direct relationship. Not any direct one anyway. Here is why.

The transition frequency f_T is defined as the unity current gain frequency with the output shorted out at AC. The well-known equation of f_T for a MOSFET is

$$f_T = \frac{g_m}{2\pi(C_{GD} + C_{GS})}. \tag{8.15}$$

Only *input* capacitances are involved in f_T. The critical output capacitance from (8.10) does not appear, because by definition it shorted out. The other parameter, f_{MAX}, is the frequency where the maximum power gain equals 1, having both input and output ports conjugate matched (which resonates out the input and output capacitances). The equation for a MOSFET is

$$f_{MAX} = \frac{1}{2}f_T\sqrt{\frac{R_L}{R_G}}, \tag{8.16}$$

which is related to f_T, so the influence of the input capacitances is not entirely eliminated. Still, f_{MAX} contains no consideration of the output capacitance either, and does contain device transconductance. To the extent that C_{DS} is related to C_{GS}, and the general trend that transistors with higher f_T and f_{MAX} values are faster, there is some relationship. But the only predictive relationship that is directly useful for C-mode PA design is (8.13).

The time required to transition along the load line is inversely related to the transistor parameter f_T. Specifically, the higher that f_T is, the shorter this transition time can be. Comparison of the achievable energy efficiencies for various ratios of f_T to the operating frequency f_0 are shown in Figure 8-20. At lower values of this ratio, this

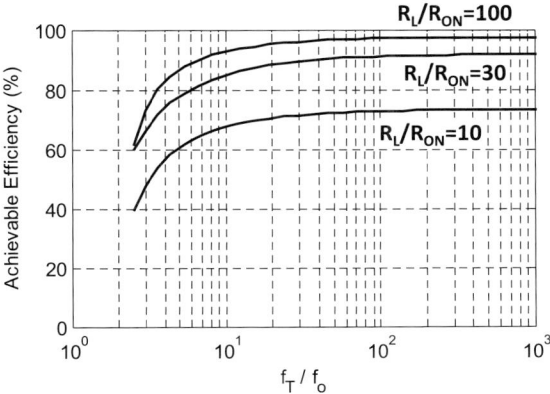

Figure 8-20 When the switching speed of the RF power transistor is greater than the operating frequency (related to the f_T/f_o ratio), the achievable energy efficiency increases until the limit set by R_L/R_{ON}.

transition time is a significant fraction of the signal period, and the time spent by the transistor in the high power dissipation region along the load line significantly degrades operating efficiency. Conversely, with a high f_T/f_0 ratio the transition between OFF and ON states is very rapid and the time spent dissipating power along the load line is very short. This results in an efficiency performance that increases with higher values of R_L/R_{ON}. For any particular output power, again we see that the energy efficiency improves with higher operating voltage.

8.5   Envelope modulation accuracy

Just as circuit linearity is very important for ET in order to provide precise output signal characteristics, the envelope modulation aspect of C-mode is critical to achieving output envelope precision in polar operation. The inversion in circuit operation between the ET core principles (7.1) and (7.2) compared to the polar core principles (8.1) and (8.2) hints that the circuit operating processes are very different. This is true, and the analysis methods for guaranteeing accuracy are likewise different. In this section, the analysis and operation requirements for accurate polar envelope modulation are discussed.

8.5.1   Case when $V_{AMO} = 0$

Implementation of (2.9) can be viewed geometrically as a triangle overlaid on the transistor characteristic curves as shown in Figure 8-21(a). This triangle is the complete extension to the IV origin of the trapezoid in Figure 6-7. As the DPS varies to effect envelope variations on the output signal, the side along the load line shifts left and right while the side aligned with the voltage axis and the side corresponding to the ON resistance both stay fixed. Any deviation of the actual transistor characteristic from the straight side of the triangle represents a distortion of the polar modulated envelope. A slight variation is noted from characteristic

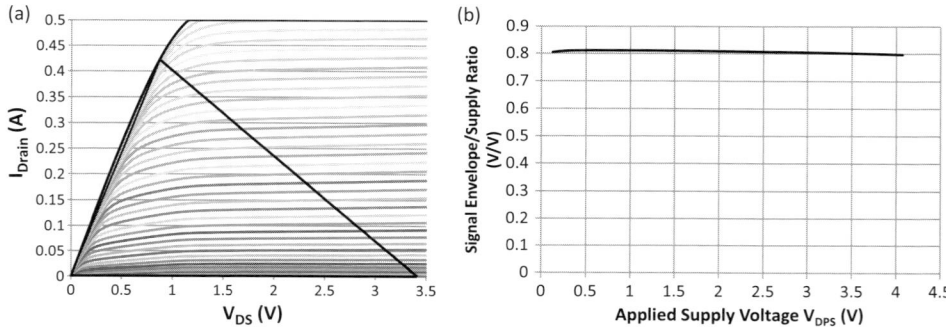

Figure 8-21   Polar modulation triangle and modulation accuracy for an NMOS transistor having $V_{AMO} = 0$, after calibration at peak output power: (a) modulation triangle to implement (2.9); (b) measured modulation accuracy for this transistor using polar modulation.

curve curvature in Figure 8-21(a), which does manifest as a distortion in the modulation multiplication of Figure 8-21(b).

With no voltage offset but finite resistance R_{DEV} for the transistor, the ON current through the load is expanded from (6.14) to yield

$$I_L(V_S) = \frac{V_S}{R_L + R_{DEV}}.$$ (8.17)

The signal voltage from this current flowing through the load resistance is

$$V_{SIG}(V_S) = R_L I_L = \frac{V_S}{1 + R_{DEV}/R_L}$$ (8.18)

which is linear with V_S only if both the load resistance and the device resistance characteristic are constants over the operating range of V_S. Additionally, if R_{DEV} is not negligibly smaller than R_L, there will be attenuation of the output signal from the ideal predicted by (6.17). Small instances of both of these effects are seen in the silicon NMOS PA transistor data in Figure 8-21.

8.5.2   Differences when $V_{AMO} > 0$

When the voltage offset term V_{AMO} is not zero, another set of distortions appears in the amplifier characterization for C-mode operation. Starting again from (6.14), but here including all terms (no approximations), we get the polar load current of

$$I_L(V_S) = \frac{V_S - V_{AMO}}{R_L + R_{DEV}}.$$ (8.19)

The output signal (V_{SIG}) is this current flowing through the load resistance (6.10)

$$V_{SIG}(V_S) = R_L I_L = \frac{V_S - V_{AMO}}{1 + R_{DEV}/R_L} = \frac{V_S}{1 + R_{DEV}/R_L} - \frac{V_{AMO}}{1 + R_{DEV}/R_L}$$ (8.20)

where the generic device absolute resistance R_{DEV} is used instead of the completely ON resistance R_{ON}. If $R_{DEV} \ll R_L$, then (8.20) simplifies greatly to

$$V_{SIG} = V_S - V_{AMO}$$ (8.21)

showing that even in the best case the presence of V_{AMO} restricts the possible output signal value from any amplifier using that transistor to be less than the available supply voltage.

There is signal distortion from this offset if it is not accounted for properly. Why this happens can be seen from the two triangles in Figure 8-22. These triangles are the complete extension of the trapezoid in Figure 6-7. As the DPS varies to effect envelope variations on the output signal, the side along the load line shifts left and right while the side aligned with the voltage axis and the side corresponding to the ON resistance stay fixed.

The C-mode envelope accuracy from DPS variations for a GaAs HBT that has a nonzero value for V_{AMO} is shown in Figure 8-23. If there is no accounting for any nonzero

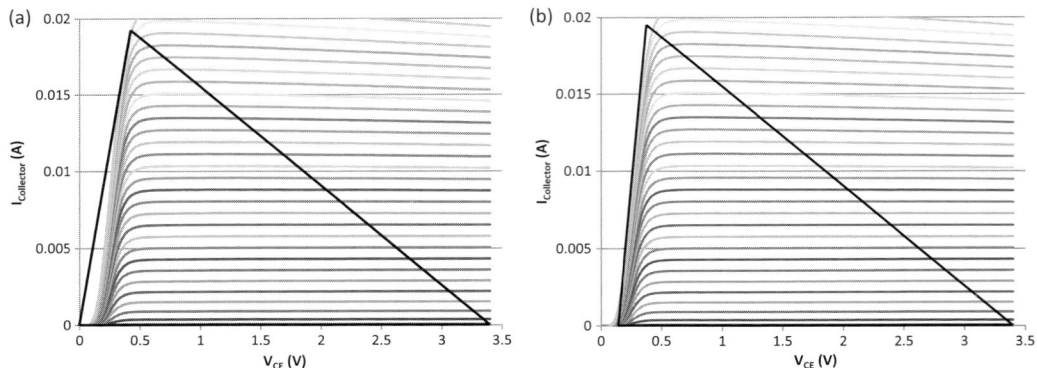

Figure 8-22 Polar modulation triangles for a transistor having nonzero V_{AMO} value after calibration at peak output power: (a) without correction for V_{AMO}; (b) with a fixed offset to correct for V_{AMO}.

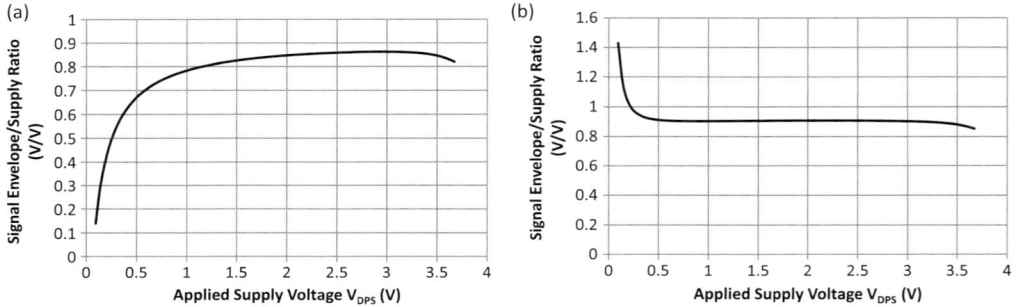

Figure 8-23 Polar distortion from using a GaAs HBT RF switching transistor having nonzero V_{AMO} value: (a) without correction for V_{AMO}; (b) with a fixed offset of 120 mV to correct for V_{AMO}.

value of V_{AMO}, then there is significant signal distortion as shown in Figure 8-23(a). This distortion is particularly severe whenever the power supply approaches and passes below the value of V_{AMO}. When the modulating envelope signal is offset by V_{AMO} in accordance with (8.19), there is a significant improvement in output envelope accuracy which is seen in Figure 8-23(b).

This accuracy does not extend to both extremes of the power supply range. The upturn seen at low supply values results from the curvature of the IV characteristic curves to higher values of channel resistance as this transistor turns OFF.

Additional power dissipation in the transistor also arises from the presence of $V_{AMO} > 0$. This follows from the fact that the transistor current flows through this voltage offset. The product of this current and voltage gives additional power dissipation of

$$P_{D,AMO} = I_L V_{AMO} \qquad (8.22)$$

which reduces the amplifier energy efficiency. Whether this reduced efficiency is material or not largely depends on the value of V_{AMO}. For example, if V_{AMO} is less than 10 mV, then with 1 amp of current this additional power dissipation is below 10 mW. This is likely to be immaterial to the application. But if $V_{AMO} = 120$ mV, then that same current causes an additional power dissipation of 120 mW. This might well be material to overall PA energy efficiency performance.

8.6   Mode identification and model validation

The circuit model of Figure 4-27 describes unconventional operation for any transistor designed for linear amplifier applications. Verifying if that operation is actually in C-mode, so that the model in Figure 4-27 is valid becomes an important aspect of intentionally compressed amplifier design. What the necessary information is to make this validation – and learning what may look like C-mode operation but really is something else – is the focus of this section.

8.6.1   Test strategy to validate C-mode operation

The principles of C-mode power stages are presented in Chapter 6. Among all of the properties that are identified there, four are particularly easy to test for. This subset of four properties of C-mode operation contains:

- no sensitivity to input RF power variation (power saturation),
- square-law transfer function from V_{DPS} to RF output power,
- constant stage series resistance value,
- strong sensitivity to power supply variation and supply noise.

All of these property tests must be successful for the proper identification of C-mode operation. All of them are necessary, but each one on its own is not sufficient. Taken together these do provide sufficient confidence for this determination. These tests are discussed below following their order in this list.

8.6.2   Power saturation

One of the easiest tests is to check for amplifier power saturation, which is a test of (8.1). This can be readily done through the amplifier Booth chart measurements. In particular, it is valuable to identify the region of supply voltage variation and input power ranges that define the region satisfying (8.1).

With power saturated operation, the power supply sensitivity (pss) is highest. In all cases, the pss evaluates according to

$$\text{pss}_{FET} = \frac{\partial I_D}{\partial V_{DS}} \qquad (8.23)$$

for an FET and

$$\text{pss}_{Bipolar} = \frac{\partial I_C}{\partial V_{CE}} \qquad (8.24)$$

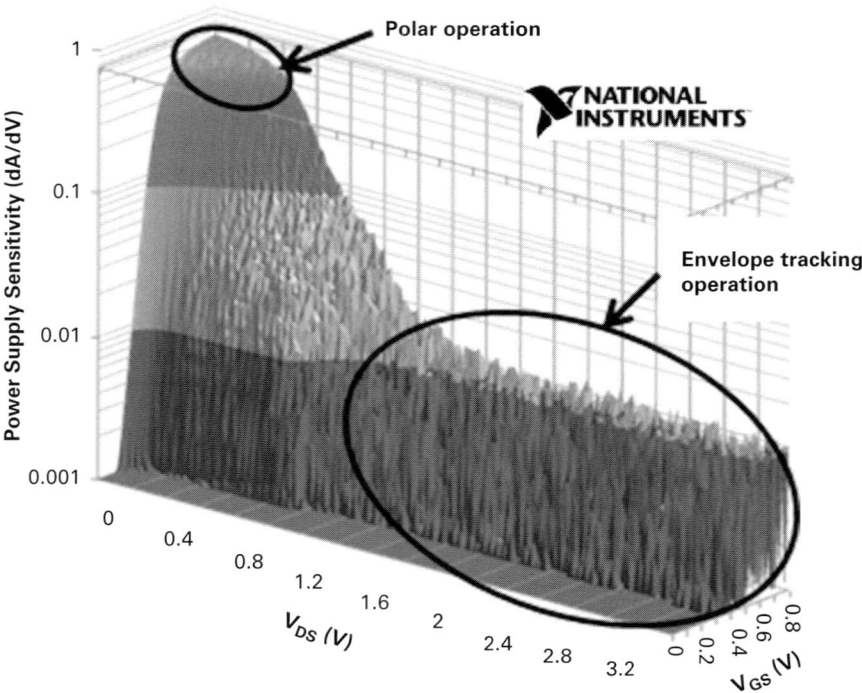

Figure 8-24   Power supply sensitivity for an RF amplifier showing the high sensitivity region that is used for polar modulation.

for a bipolar transistor. One example pss surface for a power transistor (here a CMOS PA) is shown in Figure 8-24. Polar operation is at lower V_{DS} values, so for the same load current (same output signal magnitude) the transistor power dissipation is lower and the PA is more efficient. This figure also shows the region useful for envelope tracking, where the power supply sensitivity is about 1% of the polar control, which provides a nearly 40 dB suppression of power supply noise.

8.6.3   Square-law power relationship

Section 6.2.2 shows that when the transistor is not regulating current, then Ohm's Law (6.1) is an appropriate model for the relationship between the value of the power supply and the measured output power. For purposes of this test, there is no need to have unique values for the two constants α and R_{PA} in (6.1). It is fine to treat them as a single constant of proportionality for this validation test. The important check here is that this is indeed a constant. Once one value is determined to match the model (6.1) with one measured power, this same constant must be valid for all other variations of input power and power supply value to validate this test.

One is tempted to specify the input power needed for this test. The point though is that the input power should not matter, as long as (8.1) is satisfied and the combination of input power and power supply is within the region measured according to Section 8.6.2.

Experience shows that several transistors have gain variations with varying power supply that can closely mimic this square-law power relationship. This test therefore is, on its own, necessary but not sufficient. But when the other tests in this section do successfully validate C-mode operation, the profile from this square-law test is needed to determine the exact value of V_{AMO}. This last test is discussed in Section 8.6.6.

8.6.4   Stage series resistance

A third test is to check for the validity of (6.17) using the stage series resistance test. For all points inside the C-mode region identified from the tests in Section 8.6.2, the stage series resistance (SSR)

$$SSR = \frac{V_{Stage}}{I_{Stage}} \tag{8.25}$$

should evaluate to a constant. This is a measurement of R_{PA}, which now allows the two components of the calibrated proportionality constant from Section 8.6.3 to be separately identified.

The most important result here is to show that the measured SSR value is a constant independent of both RF input power and the value of the power supply. These results will be true as long as the transistor stays within C-mode operation. The quality of this independence, usually identified by having a small sample variance among all the measurements, is usually a very sensitive test for the actual extent of the C-mode operating region.

As for any other circuit performance metric, the actual boundary for this test is up to the application. Whether a variation from the mean for any measurement is "good enough" or not is not an absolute decision. Experience from production on what an acceptable definition of "good enough" should be is still being built up. Hopefully, some papers and research results will help clarify a more universal boundary definition. Then again, research may otherwise show that the transition is slow, leaving significant opportunity to move the boundary in accordance with various applications. At this moment, there are not enough data yet to provide a more universal answer.

8.6.5   Power supply noise susceptibility

A fourth test for identifying C-mode operation is to measure the sensitivity of output RF power to power supply variations (8.2). This test is defined in Figure 6-8, and consists of a sinusoid waveform superimposed on the power supply input V_{PA}. For any particular amount of RF input power, as the value of V_{PA} is swept, the transition between L-mode and C-mode will appear in the manner shown in Figure 6-10.

This test is somewhat more complicated because a spectrum analyzer is needed to measure the AM sidebands which result from having the superimposed sinusoid on V_{PA}. As Figure 6-10 also shows, measuring the total signal power, while V_{PA} is swept, also helps identify the C-mode region. In C-mode, the magnitude of the AM sidebands will be unchanged as V_{PA} is varied, as required by (5.3) and seen in Figure 6-10. The region where AM sideband magnitude is constant while the total signal power varies according to the square law (6.1) is another identification of C-mode operation.

During the design of a C-mode stage, this test is particularly useful to examine operating mode boundary behavior as internal bias investigations are performed. Looking at Figure 6-8, it is evident that changing the transistor bias will also change the results of this test. Optimizing bias profiles is made easier by having this information available for analysis.

8.6.6   Determining V_{AMO}

Once C-mode operation is verified. then the square-law curves from the tests in Section 8.6.3 may be used to determine what, if any, offset is exhibited by the C-mode stage in its envelope control profile. This offset is the actual definition of V_{AMO}: the offset shown by a C-mode power stage to its envelope control as described by (6.16).

One method to do this is a traditional curve-fit, such as that shown in Figure 6-27. The power law model (6.16) is systematically evaluated across possible values of V_{AMO} until a best fit is identified. Any of the many best-fit algorithms and measures can certainly be used.

Another method is to mathematically manipulate (6.16) into a linear model which has an axis intercept that is equal to the desired value of V_{AMO}. One possibility is to start with (6.16) and solve for V_{AMO}, giving

$$V_{AMO} = V_{DPS} - \sqrt{\frac{R_{PA}P(V_{DPS})}{\alpha}}.$$

(8.26)

This is the equation of a line with slope

$$m = \sqrt{\frac{R_{PA}}{\alpha}},$$

(8.27)

and y-intercept of

$$y_0 = -\frac{V_{AMO}}{m}.$$

(8.28)

Of particular interest is the x-intercept

$$x_0 = V_{AMO},$$

(8.29)

which must be a projection because the output power from a C-mode stage will not actually reach zero. This process is shown graphically in Figure 8-25. In actual application, this line is best generated using linear regression on the measured data set of $\left(V_{DPS}, \sqrt{P(V_{DPS})}\right)$ data pairs.

There is temptation to measure V_{AMO} from DC data and to skip this RF test and its complexity. This may eventually be validated as a measurement for accurate design, but at this moment any relation between any offset seen in the IV characteristic curves and the operating envelope control offset of (6.16) is a tantalizing coincidence. A good service will be provided to the DPST design community by those researchers who can

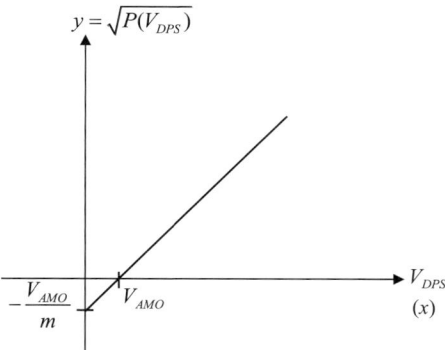

Figure 8-25   Using linear projection to estimate V_{AMO}.

definitively establish a relationship between DC and RF results that holds well for accurate DPST design.

8.6.7   Test strategy to validate P-mode operation

We must also include the other nonlinear amplifier operating mode, P-mode, in this set of tests. This is particularly true for a DPST polar transmitter. Although P-mode is shown to be a serious problem for ET and for that DPST application this region must be avoided, P-mode is uniquely useful in a polar DPST as presented in Section 6.12.3.

Three easily testable properties of P-mode operation are:

- dB for dB sensitivity to input RF power variation,
- strong output power sensitivity to power supply variation, and
- strong variation of ratiometric power gain with power supply.

The first testable property is not unique to P-mode, but it is very different from the requirement (8.1) for C-mode. The second property is in common with C-mode but in opposition to the key property (7.1) of L-mode. The third property on this list is closely related to the second property, and describes the "gain collapse" feature that must be avoided in ET.

Again, as for the C-mode tests in Section 8.6.1 there is a gradual variation in measured results from clearly P-mode operation into the transition zones toward L-mode and C-mode. Defining a boundary for P-mode operation therefore is somewhat subjective. The actual boundaries again are set by the requirements of the application. For example, the EVM specifications of cellular wireless operation require modulation accuracies that do not allow much extension away from the central P-mode region.

8.7   Impedance matching considerations

All of the established theory and practice for impedance matching follows from linear circuit experience. With C-mode operation, and particularly the core principle (8.1), linearity is suppressed. There is no reason to expect that any circuit design practice based on linear theory still applies for C-mode designs. What does "impedance" mean for a switch, anyway?

There is no one steady-state answer to this question. Of course, at any instant in time there is a state defined by the then-present voltage and the then-present current, so there is a then-present impedance. But this impedance is time-varying. Therefore, the answer to this question is: it depends on when we look.

Passive components do obey linear circuit theory. The transistor they are attached to acts with extreme nonlinearity and time variance. This makes the design of impedance transformation networks more interesting for C-mode than in traditional L-mode practice.

8.7.1   Output matching networks (OMN)

The purpose of the traditional output matching network is to transform the specified amplifier output impedance, usually 50 ohms resistive, and the load impedance required at the transistor to generate the required RF power from the available power supply voltage. If there is any flexibility remaining, the load impedance can be shifted toward a higher efficiency point from the impedance point of maximum power.

This impedance transformation is a filtering operation, and can be lowpass, bandpass, or highpass. In any case, there is a bandwidth restriction over which this impedance transformation is valid, keeping all amplifiers using OMNs from being truly wideband in their operation. Multiple separate bands may be possible with judicious design of the OMN and having increased complexity [8-1]. Otherwise the only possibility for truly wideband (allpass) amplifier operation is to develop all of the required power across a transistor load impedance of equal value to the required amplifier output impedance. Ohm's Law stipulates what the associated voltages and currents must be of such a design.

Capacitors are charged, not resonated

In a C-mode stage, the intent is to have the transistor switch, meaning that both the voltage across the transistor and the current through it should reach a steady state in a short fraction of half the carrier cycle time. One example is seen in Figure 6-39. At steady state voltage and current, any capacitance is fully charged (or discharged) and any inductance has its current limited by circuit resistances. Both of these conditions leave reactance effects which should only occur during the short transition. Once the switching steady state is reached all capacitances are opens and inductances are shorts, though both have stored energy in their respective fields. In a truly switching circuit, the current flows outside of the transient times are purely resistive. Reactive current flow is transient only.

This means that the idea of conjugate impedance matching is not applicable. Indeed, the resonance that results from conjugate matching requires that the waveform be nominally sinusoidal. For any transmitter, DPST or otherwise, sinusoidal waveform is required at the final amplifier output port, but not at the transistor.

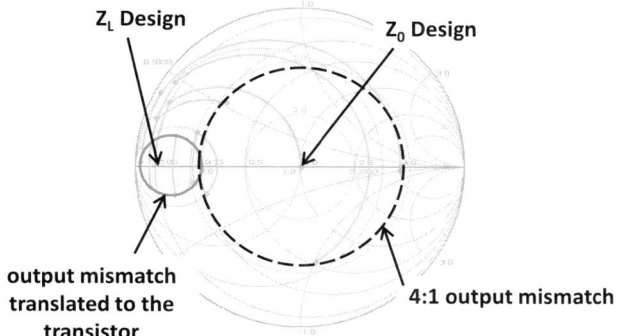

Figure 8-26 Smith chart showing that with a load impedance mismatch at the PA output, the output impedance transformation network tracks the mismatch by providing a correspondingly varying impedance at the transistor output. Here the nominal 50 ohm load impedance (Z_0) is transformed to 3 ohms resistive (Z_L). When a 4:1 mismatch is present at the output, the impedance at the transistor also varies along a circle around (but not centered on!) the nominal load resistance.

Impedance changes with output mismatch

In the presence of a mismatched output impedance, the effective impedance at the input of a passive network (like an OMN) changes from its nominal design value as a reaction to the nondesign value of the impedance at its output.

Output power still follows (6.1), here with the added variation that R_{PA} is not constant and becomes Z_{PA}

$$P(V_{DPS}, Z_{PA}) = \alpha \frac{V_{DPS}^2}{Z_{PA}(Z_0, \phi_0)}. \tag{8.30}$$

As the output impedance varies, the output power will also vary, increasing when the transistor load resistance decreases, and decreasing when the transistor load resistance increases. Output signal quality will remain as designed, as long as the design constraints from Section 6.7.5 are maintained with the value of R_L now varying.

Examples of how this happens are shown in Figure 8-26. Here the nominal transformation from 50 ohms resistive (Z_0 Design) to 3 ohms resistive (Z_L Design) is shown, along with four mismatch output impedances and their transformations through the OMN. As the output impedance increases, the load at the transistor decreases. Any reactive parts to the output mismatch stay of the same type: capacitances stay capacitive, and inductances stay inductive. On the Smith chart, the circle of possible impedances from the 4:1 mismatch (dashed) transforms to another circle around the designed transistor load impedance. Note though that this transformed impedance circle is not centered on the original intended value Z_L.

Operational impacts of this varying impedance at the transistor are presented in Section 8.7.3.

Inherent upper frequency limit

Because the resonance effects of conjugate matching are not present in C-mode designs, it is tempting to consider these power circuits for very wideband operation. This is viable to a point, but as always happens with physical reality there are effects that limit operating bandwidth anyway. Here, wideband operation of a C-mode stage is primarily limited by the output circuit, and how well the optimum switching duty cycle can be held.

In Section 8.4, it is pointed out that the physical design objective of a C-mode stage is to have the transistor resistance transition from OFF to ON values in the shortest possible time. Succeeding in this objective is largely a matter of driver design and interaction of the driver with the power switching transistor. For this discussion, we assume success of that design, and consider what happens as the output circuit reacts to these "instant" resistance changes.

Section 8.4 continues to largely answer the present question. The charging and discharging of the transistor output capacitance is described by the time constants (8.11) and (8.13). Of these two, (8.11) is by far the longer time constant, so we focus on it here. The capacitance charging time needs to be well under one-half of a carrier cycle. Taking five time constants as the minimum acceptable charging time, we get the design criterion

$$5\tau_{OFF} < \frac{1}{2f_0}. \tag{8.31}$$

Given the transistor capacitances defined in Section 8.4, this provides an upper bound to the C-mode operating frequency of

$$f_{UPPER} < \frac{1}{10R_L C_T} \tag{8.32}$$

using the substitution of (8.11). The only design variable in (8.32) is the load resistance R_L. Usually the upper frequency of operation is specified, so the load resistance must adapt using the design relation

$$R_L < \frac{1}{10f_{UPPER}C_T}. \tag{8.33}$$

If the result from (8.33) is 50 ohms or greater, then the C-mode power stage will provide wideband RF power up to the specified frequency f_{UPPER}. If not, then the maximum possible value for R_L is that given by (8.33). An output matching network is necessary to get the actual output impedance from R_L to 50 ohms. This OMN will have an additional bandwidth limit around the design frequency f_0 that follows from the quality factor Q needed by the OMN to achieve the necessary impedance ratio. Using the OMN, $Q = f_0/\text{BW}$, this bandwidth restriction is estimated to be

$$\text{BW} \approx \frac{f_0}{\sqrt{\dfrac{50}{R_L} - 1}}. \tag{8.34}$$

A plot of (8.34) is provided in Figure 8-27. If the transistor output capacitance is large enough to require a reduction in its load resistance to meet bandwidth objectives, the resulting bandwidth limitation may strongly encourage the adoption of a different transistor with sufficiently small output capacitance.

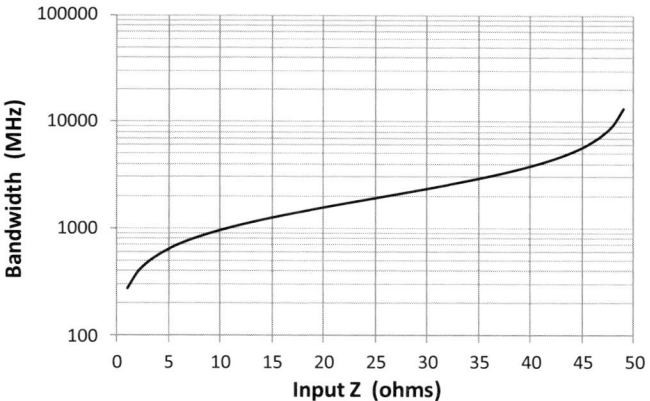

Figure 8-27   Bandwidth limitation from passive OMN depends on the impedance ratio that the network must provide.

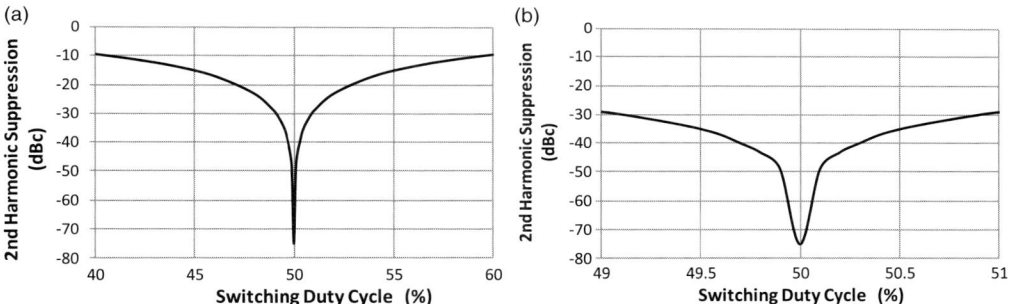

Figure 8-28   Suppression of the second harmonic by holding the switching duty cycle close to 50%: (a) within 10% of the optimum; (b) within 1% of the null.

Fourier considerations regarding bandwidth

Transmitter specifications cover much more than just output power and frequency range. They also include signal quality requirements. One of the more fundamental signal quality requirements is to not interfere with other signals at other frequencies. In particular, this means suppression of output harmonics.

The Fourier transform states that *any* signal that is not a sinusoid will have energy at harmonic frequencies from the carrier. This certainly includes the C-mode operating waveforms, where careful attention is paid to make them much more square in shape than sinusoidal. One of the best-known results of the Fourier series is that a perfectly square waveform with 50% duty cycle has no energy at even-order harmonic frequencies, and only the odd-order harmonics are present. From the point of view of transmitter design, we are primarily interested here in suppressing the second harmonic energy without need of filters by using Fourier series principles. This Fourier effect is shown in Figure 8-28.

While the idea of not needing filtering for the second harmonic is very attractive, the curves in Figure 8-28 demonstrate that this suppression null is quite narrow. For 30 dB of

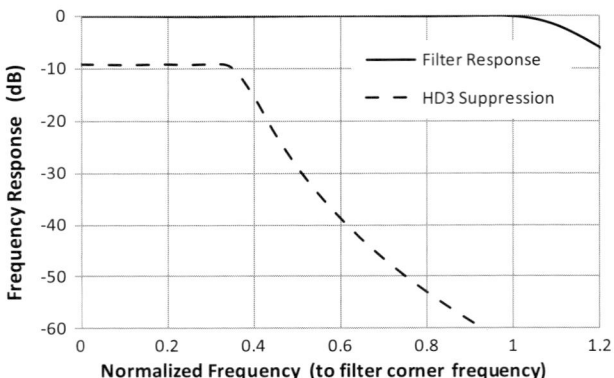

Figure 8-29   Suppression of the third harmonic (dashed curve) by using a fifth-order 0.1 dB-Chebychev lowpass filter.

suppression, the duty cycle must be held within 1% of the null, which if held perfectly greatly suppresses the second harmonic. In applications where very wideband operation must be supported, a fundamental design objective is to hold the RF switch duty cycle very tightly to 50%.

When this is successfully done, attention moves to the third harmonic, which is at –9.2 dBc. This must be there, so it can only be removed with filtering. This filter can be a separate circuit block, or it can be the output matching network if one is used. For an initial view of the available bandwidth that can be realized by simply using a lowpass filter to remove the harmonic, we consider a fifth-order Chebychev with 0.1 dB ripple in its passband. The passband of this filter, and the corresponding rejection it provides to the third harmonic (HD3) are shown together in Figure 8-29.

When the operating frequency is less than 33% of the filter corner frequency, then the third harmonic is within the passband and has no additional attenuation. As the operating frequency increases within the filter passband, the harmonic moves outside the passband and experiences additional attenuation. Near the normalized frequency of 0.6, the third harmonic is suppressed to –40 dBc. Suppression of the third harmonic to –60 dBc occurs above a normalized operating frequency of 0.92.

Operating bandwidth is therefore restricted by the harmonic suppression specification. If the suppression specification is –40 dBc, then the available operating bandwidth is between 0.6 and 1, just under an octave. Operating bandwidth is much more restrictive with a –60 dBc harmonic rejection specification, leaving between 0.92 and 1. This can be improved only by adopting filters of higher order (more cost).

8.7.2   Input matching network (IMN)

The normal design objective for input matching network design is to provide maximum power transfer into the transistor. This simply means conjugate impedances (resonance) again, which are inherently not good for wide bandwidth. But as in the discussion on output matching in Section 8.7.1, the entire idea of using resonance might not be what is needed.

For a C-mode stage, the design requirement is to get the transistor to change its resistance value from OFF to ON and back again in the shortest possible time. This is done by the input circuitry, and depends on the transistor type. For FET devices, the operating driver is the electric field from the gate, so voltage is what must be maximized – not power [3-1]. Bipolar transistors operate from the base current, so for them the input circuit must emphasize its current response.

For narrowband operation, it certainly is possible to use resonance to emphasize either voltage or current drive. With a high Q resonator, the actual input signal can be quite small, making it appear that the stage gain is amazingly high. While this may be true looking voltage to voltage, for example, the reason this happens is that the impedances at the input and the output of the transistor are very different. For an FET that really has a capacitive input impedance, resonating this can reach high impedance and high enough voltages to make the transistor switch very quickly. This technique has been used for decades in high dynamic range mixers [8-5].

The nature of input capacitance for some transistor types makes even the idea of resonance or conjugate matching nearly impossible. Two examples are provided in Figure 7-14, where during the cycle of a large magnitude input waveform, the input capacitance varies by 3:1 or more. It is impossible to produce a high Q resonance when the capacitive element continually changes its ability to store energy by octaves.

For broadband operation, the same argument used for output circuitry applies at the input. If the input circuit is a square wave as is discussed in Section 8.1.1, then the reactances at the device input are fully charged or discharged on each transition. They have no influence on steady state currents, but only during the drive transients.

8.7.3   Transistor impact of output mismatch (VSWR)

When a C-mode power stage is working into a variable impedance, then (8.29) makes it clear that the output power will change with both the value of the power supply and the load impedance. The magnitude of the load impedance varies from the nominal design value R_{NOM} with the angle ϕ of the mismatch at the point of interest (here the PA output) according to

$$|Z_{PA}| = R_{NOM}10^{\log_{10}(VSWR)\cos(\phi)}. \tag{8.35}$$

This impedance variation goes as high as

$$|Z_{PA}|_{MAX} = R_{NOM} \cdot VSWR \tag{8.36}$$

and as low as

$$|Z_{PA}|_{MIN} = \frac{R_{NOM}}{VSWR}. \tag{8.37}$$

The power variation range is proportional to this impedance range as long as the voltage applied at the C-mode stage is held constant. The resulting power variation with the angle of the mismatch is shown in Figure 8-30 for a 3:1 VSWR at the PA output.

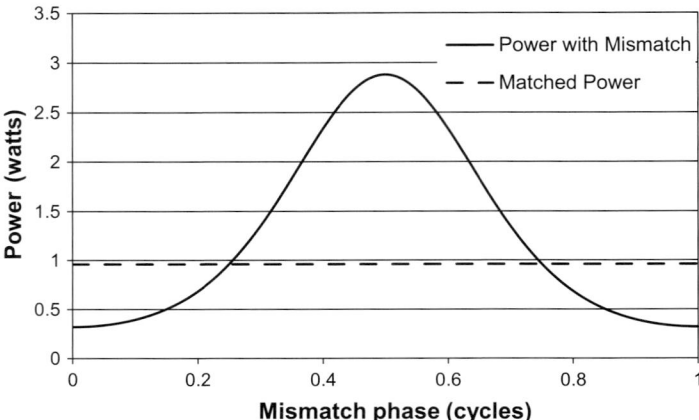

Figure 8-30 Variation of C-mode stage output power as load impedance changes with phasing of an output mismatch.

Whether this power variation is a problem or not depends on the application. Common consequences include a lack of range (reduced communication coverage) if the impedance to the PA happens to be at its high value. Similarly, if the mismatch angle happens to place the impedance at its minimum value, power goes up, but so does current consumption and power dissipation. Interference might become a problem if the radiated output power is much higher than the system expects. But usually the biggest problem at the low impedance value is rapid discharge of the battery in a mobile device. Under some realistic circumstances, the VSWR present at a transmitter output can reach 10:1, which makes all of these issues real problems. A C-mode transmitter is not immune from them.

A direct problem at the transistor happens when the mismatch angle provides its higher values to the transistor. Because the PA is designed to provide specified power into a specified load, and if it is designed to be efficient in order to also operate at higher output voltages, then when the load impedance gets higher so does the output voltage. The voltage on the PA transistor may get high enough to destroy the transistor. Avoiding this result is called ruggedizing the transistor (or the PA), and is a very important part of practical productizing for any deployed transmitter.

Just as there are design modifications necessary for an envelope tracking DPST to manage the presence of an output impedance mismatch (Section 7.10), here we encounter the design modifications needed for a polar DPST. For the C-mode to work as intended, Section 6.7.5 shows that there is a new design criterion that the transistor ON resistance must be much smaller than the load resistance it is working into. To manage the presence of VSWR, this criterion (6.11) changes to:

- The transistor load resistance must be negligibly smaller than the *minimum* load resistance the transistor could work into.

If the design requirements call for tolerance of a 4:1 VSWR, then the ON resistance of the transistor needs to be designed to still meet the C-mode resistance design criterion,

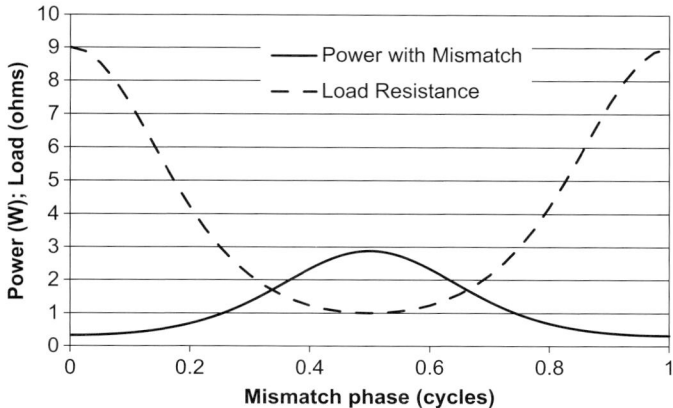

Figure 8-31 Load resistance variation at the transistor varies by $VSWR^2$:1 as the phase of the mismatch rotates through all possibilities. Here the designed load resistance value is 3 ohms, and the VSWR = 3. The maximum value now at the transistor could be 3*3 = 9 ohms, and the minimum value would be 3/3 = 1 ohm, a range of 9:1.

should the load resistance actually become $R_L/4$. This usually means the RF power transistor must be larger, decreasing its resistance but also increasing its capacitances.

The current and voltage variations at the PA due to mismatch conditions are not random. It is possible to sense them as they vary from nominal values and use this knowledge to automatically adapt the transmitter to hold output power constant. One approach to this automatic VSWR adaptation is presented in Section 9.13.

8.7.4 Multiband output capability

The ability of a transmitter to operate at multiple frequencies is of growing interest, largely driven by the fragmented frequency plans that standards committees are being allowed to adopt by their industrial clients. It is hard enough to build a transmitter that can operate at any single frequency across a wide bandwidth. It is another problem entirely to build a transmitter that can operate at more than one frequency simultaneously in carrier aggregation. Here we only consider the single operating frequency across a wide range case.

One obvious technique is to design a single amplifier to operate at power and accuracy across all possible frequencies of interest, along with any frequencies in between. This is direct multibanding and has its share of difficulties, including potential instability and high output noise from the wide-gain bandwidth product that is necessary in such a design. In this section, the discussion presents two other approaches that address the multibanding problem in different ways.

Sliced cascode

One of the operating requirements of multiband cellular handsets is to connect a single wideband modulator RFIC to multiple duplexers at the various band frequency pairs

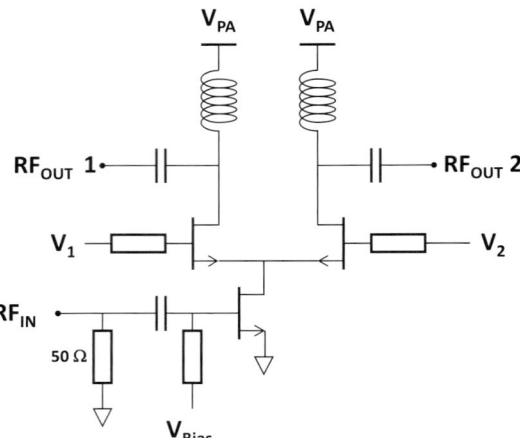

Figure 8-32   Sliced cascode architecture for isolated multiband outputs from a single wideband input.

called for in the 3GPP specifications. Because each of the duplexers has its own input for the transmitter PA output, if the PA is an inherently wideband type, there still needs to be a multipole switch to connect the PA output to the needed duplexer. Such a switch has loss and distortion from the very large 50 ohm PA output signals.

One approach developed to have multiple RF power outputs from a single wideband RF input is the so-called sliced cascode [8-6] shown in Figure 8-32. This is an extension of the linear cascode amplifier discussed in Section 7.9.1. In this approach, the cascode bias voltages (here V_1 and V_2) are used to selectively bias only one of the cascode upper stages for generating RF power. All other cascode stages are biased OFF, of which there could be many more than the two shown in Figure 8-32. This integrates the switch function within the PA itself, eliminating it as a loss mechanism and distortion source. Each RF output is isolated from all others, which makes it straightforward to include an individual output matching network (if required) on each one that is designed for the particular frequency range of its associated duplexer.

Looking at the circuit in Figure 8-32, it appears to be a tail-driven differential amplifier. Schematics can be misleading, as in this case, if circuit schematic topology is interpreted regarding how the circuit operates. We can fall into a trap very easily if the actual circuit operation is not carefully looked at.

Multibanding a single output

If frequency duplexers are not being used, then is it possible to get power at selective relatively narrow frequency bands by using an output matching network that presents low impedance to a voltage limited power transistor at the necessary frequency bands. One example is shown in Figure 8-33. Networks that have characteristics that repeat in frequency usually require including transmission lines along with reactive circuit elements. They are sometimes called re-entrant networks because of the multiple opportunities to generate RF power from one design that uses no reconfiguring switches [8-1] [8-7].

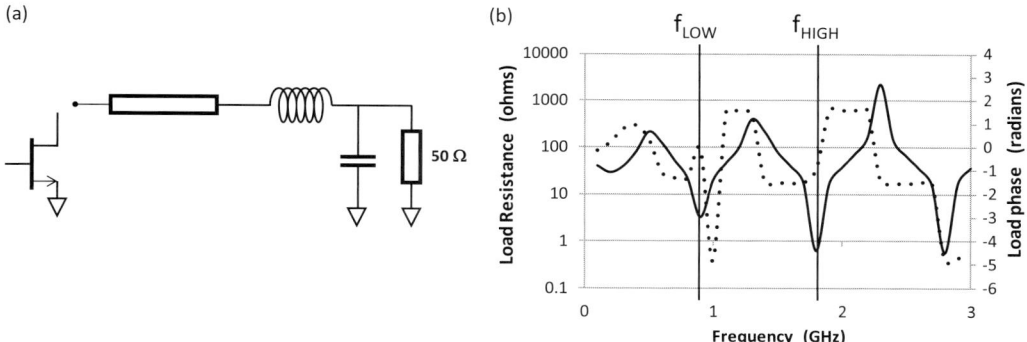

Figure 8-33   Non-continuous multiband operation from a single output using an OMN with multiple frequencies of proper load impedance: (a) RLCT network; (b) impedance magnitude and phase at the transistor across the frequency band showing the low resistive impedances for two PA bands.

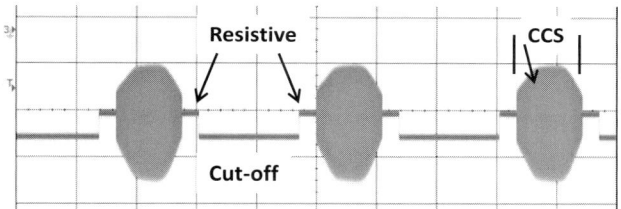

Figure 8-34   Circuit instability is only present when this transistor operated in the CCS region, and not when it operates in both the resistive and cut-off regions.

8.8   Circuit stability demonstrations

Circuit stability for C-mode operation is introduced in Section 6.15.1. There it is found that under the conditions defining C-mode, the circuit must operate with unconditional stability. More details and measurements to validate this stability are presented in this section.

8.8.1   Reduce the slope gain

The Barkhausen gain criterion is based on the voltage slope gain g_S. It is the design objective of C-mode to have a very small slope gain while simultaneously having large ratiometric gains, particularly g_{RP}. In this way, circuit stability can be readily managed while simultaneously large RF output power remains available. The price is elimination of circuit linearity.

    The validation of this stability is presented in Figure 8-34. In this demonstration, the RF power transistor (here a GaN HEMT) is operated in such a manner that the Barkhausen criteria (Section 4.3.1) are satisfied for oscillation. This means that the transistor will oscillate when the gain (slope gain g_S) gets high enough. All the possible operating modes of the transistor are encountered in the data taken for Figure 8-34, and they are labeled in the figure. No oscillation is present in the cut-off and resistive operating regions, where g_S is zero or very small. Oscillation is definitely present when g_S increases in the CCS operating region.

Even in the resistive region there remains some device transconductance, as is seen in Figure 8-2 and more clearly in Figure 8-4. Since in a practical circuit the transconductance does not actually reach zero, the slope gain

$$g_S = g_m R_L \tag{8.38}$$

also never quite gets to zero. All else being equal, circuit stability is easier to achieve with smaller values for R_L. But the RF operation needs a large value for R_L in order to enhance both energy efficiency and operating bandwidth. This establishes a design limit for the maximum allowable device transconductance of

$$g_{m,MAX} < \frac{1}{R_L}. \tag{8.39}$$

For $R_L = 50$, this means $g_{m,MAX} < 0.02$. Figure 8-4 shows that this limit is possible, with sufficient drive into resistive mode operation.

8.8.2 Reverse intermodulation

In Section 6.11.4, the presentation on reverse intermodulation performance of a polar DPST showed that the performance is excellent, particularly that seen in Figure 6-52. It is initially surprising that a strongly nonlinear circuit such as a C-mode power stage should produce so little intermodulation at its output. Thinking through the circuit operation does provide insight into the excellent performance.

Understanding why this maximally nonlinear switching circuit generates so little intermodulation requires consideration of what the signal currents are actually doing. When the switching transistor is off, there is no current flow through the OFF transistor and the circuit actually behaves linearly (though in a degenerate case). No output power is flowing, and there is nothing for the reverse flowing signal to interact with.

When the switching transistor is ON, it presents a low resistance to ground. This dumps the reverse signal directly into this ground and away from the signal load. As long as the transition time between these two *individually linear* states is short compared to the signal period, the opportunity for nonlinear interactions stays small. There is some interaction, as seen and labeled in Figure 6-39. But it is small and presently is not a problem.

8.9 Multistage issues

Section 6.11.3 presents the desirability to have some magnitude variation on the C-mode power stage input signal. Even though the RF transistor in any C-mode stage is its own RF limiter because of (8.1), this input magnitude variation is shown to reduce leakage signals across the power transistor, which also reduces output distortion through both the DPS-AM and DPS-PM processes. This benefit is seen in Figure 6-51.

One possible strategy is to simply apply the fully modulated RF signal to the C-mode input. While this is an easy approach, if the signal modulation has envelope zero-crossing events, then at those times it is guaranteed that the C-mode requirement that there always be sufficient RF drive to ensure that (8.1) remains valid will not be met. The performance stabilities of Section 6.15 are no longer guaranteed.

The better approach is to build another C-mode stage ahead of the power stage so that the input signal magnitude to the power stage can be directly controlled by a DPS on this driver stage. Since the C-mode power stage does not need much power if it is an FET, but just voltage, the capabilities of this driver stage are not severe. It need not draw much power from its DPS, which simplifies the design of both the C-mode driver stage and its associated DPS. All C-mode design rules do still apply.

The idea of amplifier pre-distortion, whether digital or analog, is to vary the input waveform in such a way that the output waveform from the transistor will be a more accurate match to a first-order scaling of the initial input waveform. This can be considered as a driver waveform modified for amplifier linearity. This optimization criterion is useful for an envelope tracking DPST, but it has no meaning for the present C-mode polar DPST. Here the appropriate optimization criterion is to distort the input waveform for efficiency [8-8] [8-4] [8-3].

This alternative optimization is only possible if there is a separate means available to ensure the accuracy of the output signal. Fortunately, this separate means is available in properly designed C-mode power stages, and it is the DPS.

Further discussion on this topic is presented in Chapter 12.

8.10   References

Note: Patent references are provided for bibliographic use only. Citation of specific patents here is not indicating any view on priority issues.

[8-1] R. Meck, "Method and Apparatus for Impedance Matching in an Amplifier Using Lumped and Distributed Inductance," US Patent 7071792, issued July 4, 2006.

[8-2] E. McCune, "Methods and Apparatus for Controlling Leakage and Power Dissipation in Radio Frequency Power Amplifiers," US Patent 8364099, issued Jan. 29, 2013.

[8-3] E. McCune, "High-Efficiency Modulating Amplifier," US Patent 6636112, issued Oct. 21, 2003.

[8-4] W. Sander, E. McCune, and R. Meck, "Driving Circuits for Switch Mode RF Power Amplifiers," US Patent 6198347, issued March 6, 2001.

[8-5] E. Oxner, "Designing FET Balanced Mixers for High Dynamic Range," Siliconix Application Note, 1985, available at http://electrooptical.net/www/mixers/EdOxnerHighDynamicRangeFET_Mixers1985.p.

[8-6] R. Meck, E. McCune, and L. Burns, "Constant Impedance for Switchable Amplifier with Power Control," US Patent 6215355, issued April 10, 2001.

[8-7] R. Meck, "Method and Apparatus for Impedance Matching in an Amplifier using Lumped and Distributed Inductance," US Patent 7206553, issued April 17, 2007.

[8-8] J. Sevic and K. Salam, "Waveform Preshaping for Efficiency Improvement in DC to RF Conversion," US Patent 6624695, issued Sept. 23, 2003.

9 Dynamic power supplies

Power supplies are traditionally designed to provide very stable and well-regulated outputs, primarily a fixed voltage that is maintained irrespective of the current that is drawn by its load. When the supply voltage is fixed in this way, there is no need to be concerned with the reactive characteristics of the load. All that matters is the value of the load resistance, which is often time-varying. Power supply regulation control loop dynamics are designed with this restriction in mind.

Here we require that the value of the output voltage from the power supply be dynamic, meaning not only that it can vary, but that it must be capable of varying at speeds matching the variation of its time-varying load. Combine this with a load having a widely varying resistance value as the voltage across it changes, as seen in Section 5.10, and the design of any dynamic power supply (DPS) becomes a formidable task.

This leads to the important concept that in any DPS transmitter there is a new interface of great importance: the interconnection between the DPS and its associated PA. To date, this interface has not received much attention, probably because of the general complacency afforded traditional design when fixed power supply voltages are used.

This chapter examines the issues involved in DPS design, including some circuit details.

9.1 Power objectives

All power into the PA flows through the DPS. The DPS must support the combination of: the PA output RF signal power, the energy inefficiencies of the PA due to its own power dissipation, and any reactive currents required by PA circuitry. Very much like the general electric utility, the DPS must source the *total* current required by the load, not just the resistive component that leads to output RF power and internal power dissipation. The utility concept of the power factor [9-1] is very important to DPS design.

SOAPBOX: It is appropriately descriptive, and therefore most valuable, to refer to the following currents as *resistive* and *reactive*. Not, and I repeat *NOT*, as real and imaginary. There is absolutely nothing imaginary about the flow of charges. The tendency of many in the engineering profession to use the mathematical terms real and imaginary (which in the mathematical use is definitely descriptive and appropriate) produces enough confusion that it should not be used at all. Similarly, in signal processing aspects the

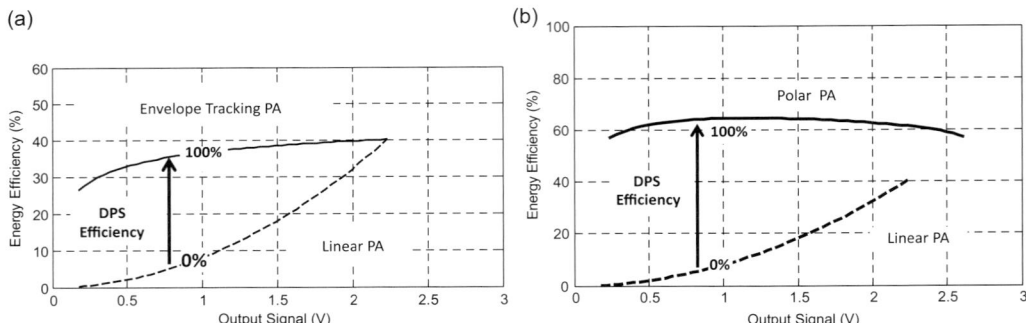

Figure 9-1   The energy efficiency gap available to DPS designers: (a) envelope tracking; (b) polar modulation.

appropriate terminology is in-phase and quadrature-phase instead of the same mathematical terms [4-2], but that is a topic for another book.

9.1.1   Resistive current

Current through a resistance leads directly to a voltage drop and power dissipation in accordance with Ohm's Law. Ohm's Law when applied to a resistance does not have a frequency term, so this voltage and power dissipation is not frequency dependent. Frequency dependence appears with the presence of reactive currents discussed in Section 9.1.2.

Absolute resistance is based on the present operating point within the IV plane. According to Ohm's Law

$$R = \frac{V}{I} = \frac{V - 0}{I - 0}, \tag{9.1}$$

which is the slope of a line between the operating point and the origin of the IV plane. One example of this is shown in Figure 9-2. Dynamic resistance is the first derivative of the absolute resistance

$$dR = \frac{dV}{dI}, \tag{9.2}$$

which is the slope of the tangent to the device characteristic curve at the present operating point. When the resistance is constant, then $R = dR$ and there is no need to distinguish between these two resistance measures. But when the device operating characteristic is nonlinear, as in the example in Figure 9-2, then the differences between R and dR are very important to system operation. This situation is completely analogous to the relationship between the gain measures of ratiometric gain (4.9) and slope gain (4.6).

The quiescent current drawn from the power supply is governed by the absolute resistance. In the presence of an AC signal, the AC current follows the small signal linearization about the operating point, which is the dynamic resistance. This is the reason we pay close attention to both of these resistance measures.

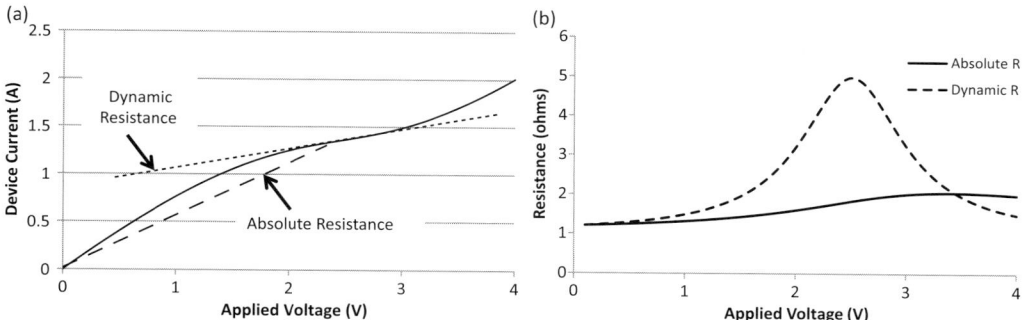

Figure 9-2   Definition and examples of absolute and dynamic resistance from any device characteristic curve: (a) characteristic curve and the slopes for each resistance measure; (b) corresponding values of absolute and dynamic resistance.

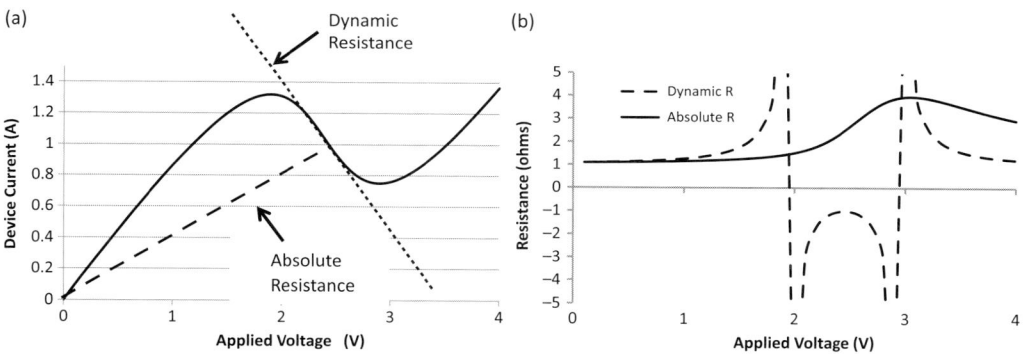

Figure 9-3   Example of absolute and dynamic resistance from a device characteristic curve with negative resistance: (a) characteristic curve and the slopes for each resistance measure; (b) corresponding values of absolute and dynamic resistance.

It is certainly possible for the characteristic curve to change slope, such as in Figure 9-3(a). When the tangential slope of the characteristic curve is horizontal, the dynamic resistance is infinite (an open circuit, or with a finite absolute resistance a constant current source or load).

9.1.2   Reactive current

When capacitances and/or inductances are present (and they *always* are), then there will also be reactive current flow. This reactive current does not dissipate power, so it cannot contribute to the generation of the output RF signal. We therefore want to minimize reactive current, because it still must be generated by the DPS and will cause power dissipation there. Reactive current in the PA does lead to an overall reduction in transmitter energy efficiency. Bypass capacitors, and capacitance added to provide low frequency stability, are the most common sources of reactive current in a PA.

For a very brief review of reactive current, we apply a sinusoidal voltage

$$V(t) = \cos \omega t \tag{9.3}$$

to reactive elements and determine the resulting currents. Beginning with capacitance, we get from basic physics

$$I_C(t) = C\frac{dV}{dt} = C\omega \sin \omega t = C\omega \cos\left(\omega t - \frac{\pi}{2}\right), \tag{9.4}$$

which shows a –90 degree phase shift from the applied voltage. Similarly, basic physics show us that for an inductance

$$I_L(t) = \frac{1}{L}\int V(\tau)d\tau = -\frac{1}{\omega L}\sin \omega t = \frac{1}{\omega L}\cos\left(\omega t + \frac{\pi}{2}\right), \tag{9.5}$$

which shows a +90 degree phase shift from the applied voltage. Relations (9.4) and (9.5) motivate the familiar vector diagram for component impedance phasors shown in Figure 9-4. Impedances of general circuitry are linear combinations of these basis phasors.

The phase shift between applied voltage and reactive current traces elliptical paths in the IV characteristic current plane. One set of ellipses for capacitors is shown in Figure 9-5, which shows that larger capacitances draw larger currents. Because current

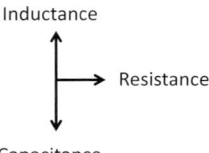

Figure 9-4   Basic phasors for resistive and reactive components.

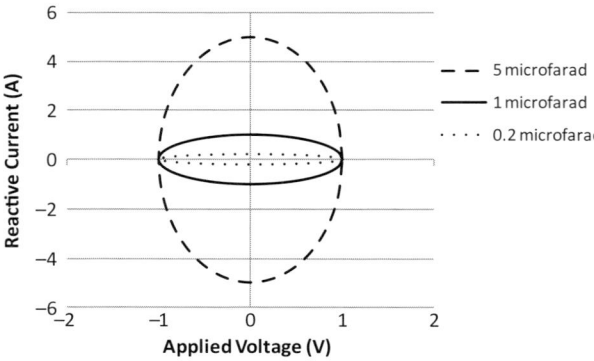

Figure 9-5   Reactive currents are necessarily out of phase with the driving voltage, appearing as ellipses on an IV diagram.

leads voltage in a capacitance, the rotation direction along these elliptical paths is clockwise.

9.1.3 Slew rate

The physical parameter in (9.4) that governs capacitor current is really the slew rate (SR) often described in the difference form

$$SR = \frac{\Delta V}{\Delta t}.$$ (9.6)

A closely related parameter of importance is the energy in the capacitor

$$E_C = \frac{1}{2}CV^2.$$ (9.7)

What really matters is the change in E_C as the voltage changes

$$\frac{dE_C}{dV} = CV,$$ (9.8)

which shows that energy change is dependent not only on the capacitance value but also on the voltage already in that capacitor. Conservation of energy (CoE) requires that any such energy change in a capacitor must be matched by a corresponding energy change in the driving circuit – here the DPS. The practical issue is that providing current in support of a slew rate by (9.4) has very different energy requirements whether that slew rate is at a low output voltage, or at a higher one. From the DPS point of view, fast slew rates at higher voltages are more difficult to support than the same slew rate at lower output voltages.

This opens up interest into the slew rate characteristics of the signals, because these impact the design of successful DPS products. The two-tone signal envelope of Figure 5-24 has its highest slew rate (both up and down) very near to 0 V. According to the present discussion, this represents the easiest case to actually build circuitry for, given a zero-crossing envelope-varying signal. Other signals are not so "friendly," such as the HSDPA signal example presented in Figure 9-6. This signal exhibits two regions of high slew rate envelope activity, one near zero magnitude and another in the upper half of the signal magnitude range.

All signals have different slew rate profiles, and a completely comprehensive survey of all possibilities is a very large task. This survey would be easier if signal envelope slew rate data, such as the histogram in Figure 9-6, were published, but unfortunately such publications are very rare. A spot-check on the various signal types that a particular DPS design is to support is shown here to be an important task for the DPS design document, because the DPS design team must completely understand the requirements their product must meet.

This slew rate requirement is in addition to the conventional bandwidth considerations for the DPS. Discussion of DPS bandwidth determination and measurement is given in Section 9.6.

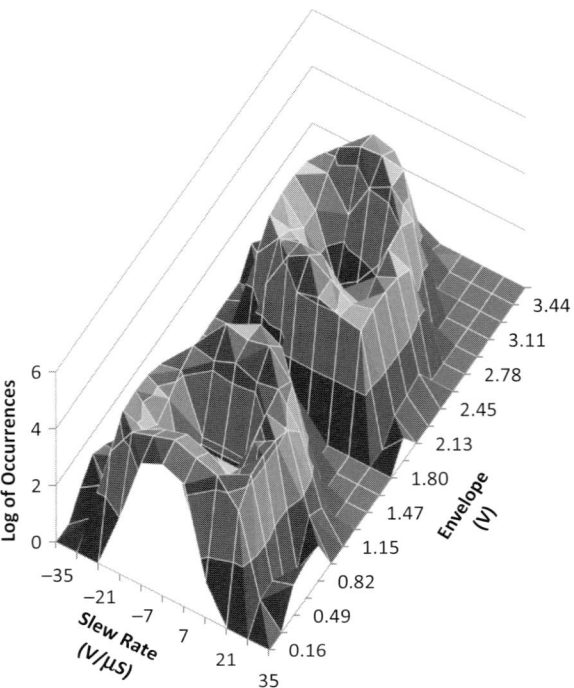

Figure 9-6   An example of envelope slew rate characteristics for an HSDPA uplink signal: DPS output slew rate capabilities must exceed these values at their corresponding output voltages for it to accurately reproduce this HSDPA signal.

9.1.4   Power conversion

Ideally, the DPS operates in a power maintaining way, which means that the ideal DPS draws the same power at its input that it supplies to the PA

$$V_{IN}I_{IN} = V_{OUT}I_{OUT}. \tag{9.9}$$

One interesting consequence of (9.9) is that it behaves as a negative resistance load to the primary power source when the output power is constant. Defining this fixed output power as P_K, we get

$$I_{IN} = \frac{P_K}{V_{IN}}, \tag{9.10}$$

which has the property that the current drawn from the input goes down when the input voltage increases

$$\frac{dI_{IN}}{dV_{IN}} = -\frac{P_K}{V_{IN}^2}, \tag{9.11}$$

even though the absolute resistance is always positive because the voltage and current into the DPS both always have the same sign. This property holds even if the power conversion has a finite efficiency (3.1).

Fortunately, a battery can manage having this negative dynamic resistance because its voltage changes very slowly. If a constant power DPS is driven by an active regulator, it is very important to validate the stability of this regulator in the presence of this negative dynamic resistance.

9.2   DPS application classes

Any DPS is an increase in circuit complexity from the traditional transmitter designs of the twentieth and early twenty-first centuries. This means that while the technology is certainly interesting, proper consideration of these designs cannot be fully separated from all the applications within which these designs are incorporated. This discussion of DPS application classes therefore begins with consideration of the technical and business issues involved in the first mobile wireless communication devices, and then considering the same issues moves on to communication infrastructure designs. An even deeper dive into the technical issues and details then follows.

9.2.1   Mobile devices

With billions of mobile wireless communication devices manufactured each year, performance certainly is important. But there are two things that are even more important: cost and risk. The cost of any product is related to the number produced – and manufacturers are strongly motivated to keep the cost for each one as low as possible. Closely related to this cost per item issue is business risk for any particular product implementation. Methods and technologies that have long histories and are well understood have the lowest risk. *Anything* different has higher risk.

Product manufacturers will trade product performance in order to achieve lower costs or lower risk. They are not rewarded for product performance above a minimum. On the contrary – as long as their customers continue to buy the product they are currently producing, the motivation to do anything different is *zero*. Any new technology, however great the performance, will not be of any advantage to the manufacturer if the costs of producing it do not lead to profits greater than currently being achieved with the existing product. This is the reason why DPS transmitters are being adopted now and were not adopted 15 years ago. This technology is not new. Rather, it is the fact that the minimum performance requirements of customers have increased that has driven manufacturers to consider DPS technologies. It is this changing demand which is raising interest in DPS technologies

In a mobile device, the available energy is strictly limited by local storage (the battery). Every electron in that battery matters [9-2]. This encourages the adoption of high performance, to a point. If the battery life is unacceptably short, the design interest in energy efficiency is extreme. Once the complaints about low battery life begin to ebb, manufacturer interest in better energy efficiency disappears in favor of lower cost. The adopted technology will have a combination of barely acceptable performance, minimum business risk, at the lowest available cost. In that order.

9.2.2   Infrastructure

Infrastructure is generally considered not to be energy constrained. The number of infrastructure installations is significantly smaller than the number of mobile units it supports. Their operating power is much higher, and the cost for each installation is significantly higher than any mobile unit. Performance of the infrastructure transmitter and receiver is paramount, because the infrastructure exists to provide communication capability to the largest possible number of customers. The infrastructure operator is rewarded for supporting the greatest number of customers. They are not directly rewarded for consuming the least amount of power while providing these communication services.

This situation is changing somewhat, because buying power is a cost and not needing so much power is one way to reduce operating cost. Lower power hardware will never happen if it means that the number of customers goes down. If the number of customers does not change, then the risk to the existing business operation of adopting a different technology becomes the primary concern. When a new technology brings improved performance and identical system capacity, but requires a change in operating procedures, this is a significant problem.

One of the newer factors in the wireless communication business environment is "being green," which means requiring less carbon-based energy to maintain operations. Improved energy efficiency in the transmitter hardware is a significant factor (6.22). DPS technology is one excellent way to achieve this goal, which partly explains the growing interest in this approach.

9.2.3   Common DPS architectures

The two top application classes of variable power supplies inserted between the main power source and the RF power amplifier are shown in Figure 9-7. These top application classes are average power tracking (APT) and dynamic power supply (DPS). Each of these top application classes has sub-classes as shown in Figure 9-7.

APT is a misnomer, because the applied power supply to the PA actually tracks the signal *peak* power. The idea is to reduce the voltage applied at the PA if the signal peaks will not approach the value of the main power source. This can be done using either linear or switching circuit techniques. At any particular RF output power setting, this power supply stays constant, forcing the PA to operate in L-mode. No attempt is made to track the signal envelope variations during modulation. This is simple to implement and a low business risk (Section 9.2.1), so it is of interest to manufacturers. However, it is not the interest in this work and will not be developed further here.

The other top application class is of interest here. Requirements developed in Chapters 5 and 6, each correspondingly refined in Chapters 7 and 8, point out that

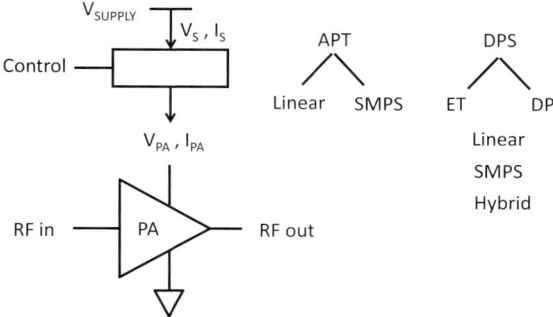

Figure 9-7   General classes of DPS designs: APT can be implemented with either linear or switching variable voltage power supplies; DPS can support envelope tracking or polar, and either can be implemented using only linear, switching, or hybrid combinations of these technologies.

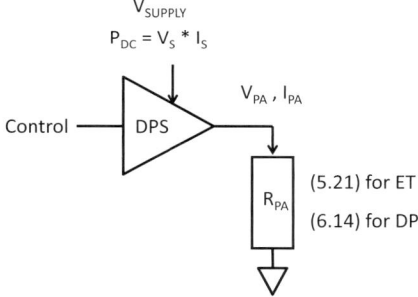

Figure 9-8   Using a linear dynamic voltage regulator as a simple DPS.

there are two sub-classes of DPS designs: one for ET, and another for DP transmitters. These are separated because of the very different DPS load experienced if an amplifier is being envelope-tracked or if it is being polar modulated. There also are large differences in the DPS output accuracy and dynamic range requirements. These are not identical design problems.

Implementation of any DPS, for either sub-class, can in principle be successfully done using any of the three techniques listed in Figure 9-7. The DPS can be implemented using only linear circuit techniques, which results in the linear dynamic voltage regulator (LDVR) of Figure 9-8.

Or a design can solely use switch-mode power supply (SMPS) techniques which results in the dynamic SMPS structure of Figure 9-9.

Often a hybrid combination of these techniques is used because the properties of linear and switch-mode power circuits are not identical and do work well together. There are two core combinations of these linear and switching circuits: the series combination of Figure 9-10, and the parallel combination of Figure 9-11.

The series connection of Figure 9-10 uses the SMPS to reduce the input voltage to the LDVR. The LDVR applies its PSR to remove the SMPS output voltage ripple, and to

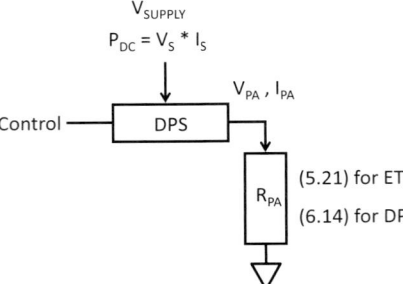

Figure 9-9   Using a dynamic switch-mode voltage regulator as a more energy-efficient DPS.

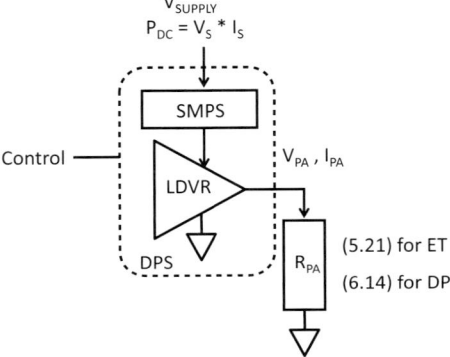

Figure 9-10   Hybrid DPS architecture using a series connection of a dynamic switch-mode voltage regulator driving a LDVR.

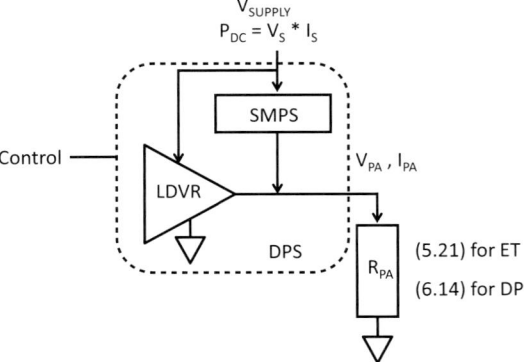

Figure 9-11   Hybrid DPS architecture using a parallel connection of a dynamic switch-mode voltage regulator and a LDVR, sharing the same output node.

provide a cleaner supply voltage to the PA. Essentially, the same current flows from the SMPS into the PA through the LDVR. Control is applied to both the SMPS and the LDVR such that the voltage drop across the LDVR is never very large while it provides a very accurate supply voltage to the PA. So many different ways exist to partition the control input information to the SMPS and the LDVR that these details are not included in Figure 9-10.

The parallel connection of Figure 9-11 operates fundamentally differently, even though the same circuit blocks used in Figure 9-10 may also be used here. The entire ripple from the SMPS, both voltage and current, is exposed to the PA. This ripple must be canceled by action from the linear dynamic voltage regulator (LDVR) so that the PA never experiences it. The LDVR output current must be the inverse of the SMPS ripple current. Any error in the SMPS output current must also be supplied (or absorbed) by the LDVR. Any error in the output voltage must also be corrected by the LDVR, making it very important that the output impedance of the SMPS be high enough that the voltage difference from the LDVR does not supply current back into the SMPS. Similarly, if the LDVR needs to lower the output voltage, it should not have to absorb much current to do so.

Each of these DPS architectures is individually discussed in the next three sections.

9.3   Linear dynamic voltage regulators (LDVR): the simplest DPS

In the spirit of the business realities presented in Section 9.2, the DPS architecture that has the lowest cost and lowest business risk is a linear regulator with sufficient dynamic capability to meet the immediate DPS requirements. All linear voltage regulators are variable resistors which adapt their pass resistance R_{LDVR}

$$R_{LDVR} = \frac{V_S - V_{PA}}{I_{PA}} \tag{9.12}$$

such that the output voltage remains fixed as the load current varies. Because this is a simple series circuit (Figure 9-12), the current through both resistances is the same

$$I_S = I_{PA}. \tag{9.13}$$

Breaking (9.12) into two separate terms shows that R_{LDVR} is the difference of two resistances

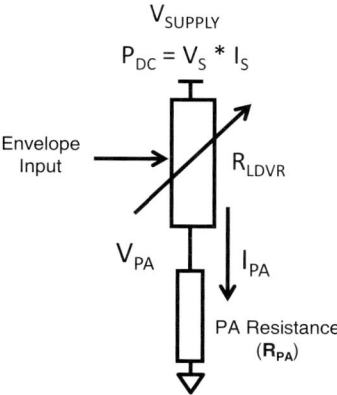

Figure 9-12   Linear regulators all perform as a variable series resistance.

$$R_{LDVR} = R_{TX} - R_{PA}. \tag{9.14}$$

Power dissipation in the LDVR pass resistance is (see Figure 3-3)

$$P_{D,DPS} = I_{PA}(V_S - V_{PA}) \tag{9.15}$$

which, when combined with the power supplied into the PA gives a total power consumption of

$$P_{TX} = P_{D,DPS} + P_{PA} = I_{PA}V_S. \tag{9.16}$$

This is the same power from the source supply that is provided to the transmitter if the same PA is directly connected to the source supply without a DPS. But this power dissipation is now split into two places, demonstrating the parallel thermal paths in Figure 3-5(b).

A power consumption reduction from the top supply is realized if the PA current does vary with its applied voltage according to the variable bias operation in Figure 5-30. In this case, the currents of (9.13) are not constant, falling as the output signal and applied PA voltage get smaller. Therefore, the power dissipation (9.16) also varies along with the output envelope, and has an average value lower than the fixed-bias linear amplifier case.

9.3.1   LDVR design principles

The basic LDVR circuit is shown in Figure 9-13. An operational amplifier senses the output voltage V_{PA} and controls the pass transistor, here an N-channel MOSFET, as required to get the channel resistance equal to the needed R_{LDVR}. The closed-loop bandwidth of this control loop needs to meet the requirements of Section 9.6. With a large enough loop gain (the difference between the open-loop gain and the closed-loop gain), this combination behaves like a single op-amp that is capable of providing high output power.

The problem with the circuit in Figure 9-13 is that when the output voltage needs to get close to the input supply voltage, the pass transistor needs to be considerably enhanced so

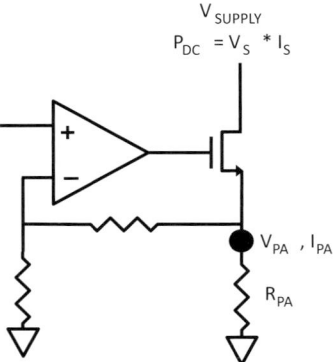

Figure 9-13   Feedback control applied to a pass transistor provides the desired voltage regulation. The output of the op-amp must be capable of providing this N-MOSFET with the necessary gate voltage to exactly set the required channel resistance.

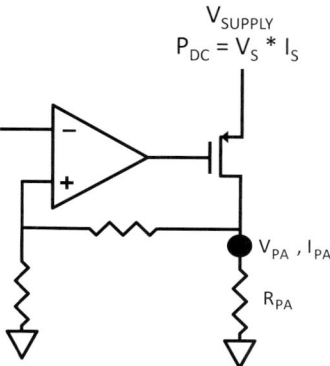

V_{SUPPLY}

$P_{DC} = V_S * I_S$

V_{PA}, I_{PA}

R_{PA}

Figure 9-14   An inverting output stage provides additional gain of g_m*R_{PA} to the control loop, which may cause significant stability issues.

that its channel resistance becomes very low. This requires the op-amp output voltage to the N-MOSFET gate to reach a voltage above the applied supply voltage V_S. To avoid this problem, the low-dropout (LDO) design of Figure 9-14 is often used.

This solution comes with a new problem – circuit stability. The series combination of the P-MOSFET and R_{PA} (the power path) is a common-source amplifier that can have significant gain when R_{PA} has a large value. This is very different from the common-drain power stage in Figure 9-13, which always has a voltage gain less than 1. Because the loop gain in the LDO design is multiplied by the gain of the power path, the control loop crossover frequency moves out and the phase margin is reduced. Any wideband LDO needs additional compensation to restore this phase margin and regain stability. Unfortunately, this compensation is generally a strong function of the value of R_{PA}. This makes it impossible to test an LDO style DPS at low output power (high R_{PA} value).

Like many of the problems in circuit design, the two LDVR designs of Figure 9-13 and Figure 9-14 have different properties. The conventional architecture of Figure 9-13 works best at lower output voltages and has a problem when V_{PA} needs to closely approach V_S. The LDO has opposite preferences, preferring to provide output voltage close to the input supply, and not working as well when the output voltage gets small. One natural approach toward optimizing this situation is to selectively use the most appropriate structure for the needed output voltage value [9-3]. More discussion on this option is in Section 9.5 and particularly in Figure 9-32.

9.3.2   Efficiency effects

The LDVR has a conversion efficiency which is given by

$$\eta_{LDVR} = \frac{P_{OUT}}{P_{IN}} = \frac{V_{PA}}{V_S}. \tag{9.17}$$

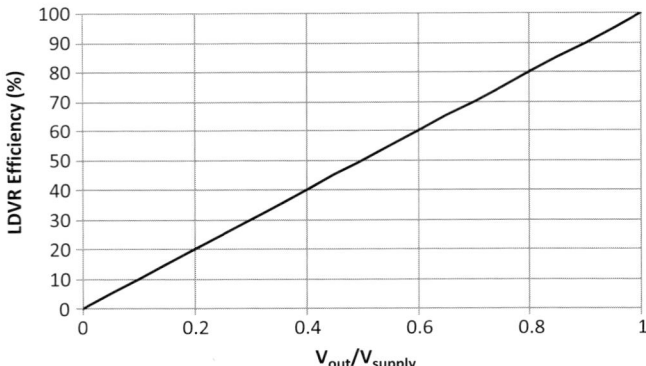

Figure 9-15   Transfer efficiency of any linear voltage regulator is equal to the ratio of the output voltage to the input voltage.

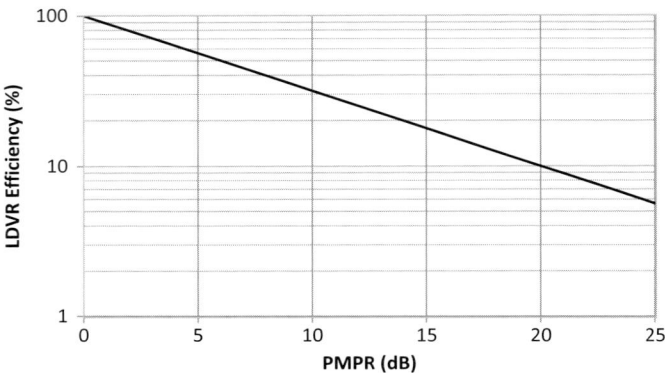

Figure 9-16   Energy efficiency of the simple LDVR DPS can be very high when the peak-to-minimum power ratio of the output signal is restricted to a narrow range.

This is plotted in Figure 9-15. The conversion efficiency of linear regulators is actually very good when its output voltage is close to the input supply voltage. Similarly, the conversion efficiency is very low with a low output voltage.

This property of high conversion efficiency at higher output voltages suggests that a signal designed to maintain high RF magnitudes [9-4] can successfully use the low-cost LDVR as its DPS. This is equivalent to restricting the peak-to-minimum power ratio (PMPR) of the signal to the smallest possible value (the opposite of what most standards committees have adopted for many years). Figure 9-16 shows how the conversion efficiency of a DPS implemented with a LDVR varies with the signal PMPR.

9.3.3   Output impedance

Ideally, the output impedance of an LDVR is zero, because the action of the feedback should sense any perturbation on the output and cancel it to maintain the output at a fixed

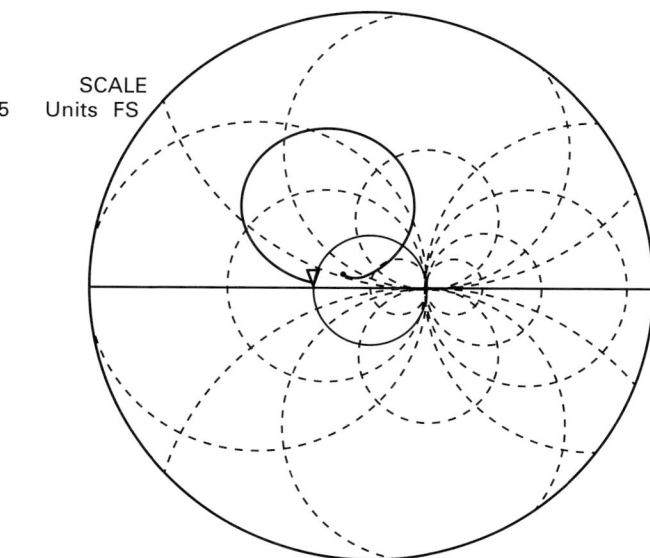

Figure 9-17   Measurement of the LDVR output impedance shows a characteristic that is all inductive, and has a significant region of negative resistance.

value. This meets the objective for low frequency stability called for in Section 7.2.2. In reality, the output impedance of an LDVR is quite a variable, as seen in the Smith chart measurement in Figure 9-17. This result shows that not only is the output impedance always inductive, for a set of frequencies the resistance behavior is negative. This is a great recipe for making an oscillator with an added capacitor.

Converting the Smith chart data to resistance and inductance values across frequency provides the chart in Figure 9-18. The resistance does start out near zero and positive, then swings to negative before rapidly returning to positive. The effective inductance value rises from its nominal value to a peak as the resistance rapidly returns positive, and then returns to the nominal value. This behavior actually does make sense.

A closer view of the data at lower frequencies is shown in Figure 9-19. This LDO style LDVR is designed to have a minimum closed-loop bandwidth of 5 MHz. At frequencies between DC and 6 MHz, the output impedance is below 1 ohm, a low value but showing that this LDO does not have enough loop gain to fully cancel the signal applied to the output for this output impedance test. As the frequency of the test signal increases, the delay of the control loop becomes increasingly significant. This delay causes the feedback current to be shifted in time (and therefore also in phase). Whenever the voltage and current are shifted in phase such that the voltage leads the current, the behavior is inductive. Resistance goes negative whenever the current goes down if the voltage increases, a very practical situation when the phase relationship of the voltage and

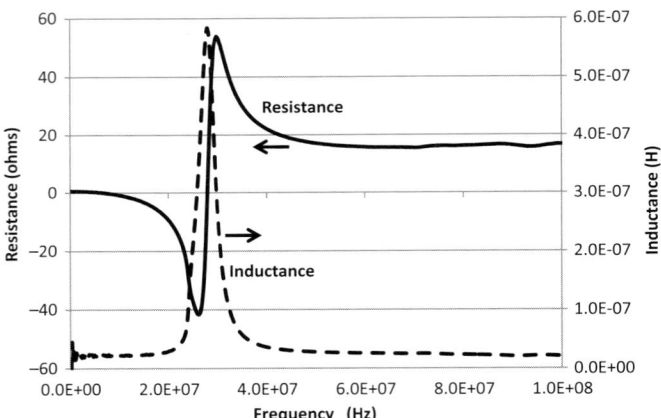

Figure 9-18   Evaluation of the resistance and inductance across frequency for the output impedance measurement of Figure 9-17.

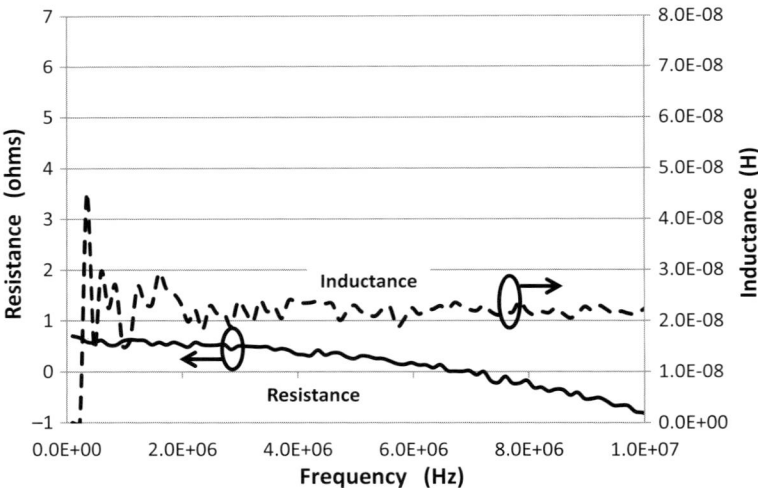

Figure 9-19   Within the designed 5 MHz closed-loop bandwidth, the output impedance is a low value resistance, in series with an inductor with a nominal value of 22 nH.

current waveforms nears quadrature. As the phase shift exceeds quadrature, the effective resistance quickly returns positive and the measurement trace returns to the inside of the Smith chart.

The measured negative resistance and inductance are actually there, proved by adding a shunt capacitor at the output that does produce a high-quality sinusoidal oscillation with frequency and power dependent on the value of that capacitance. This oscillation does not occur with all capacitor values, as is seen in Figure 9-20. Large

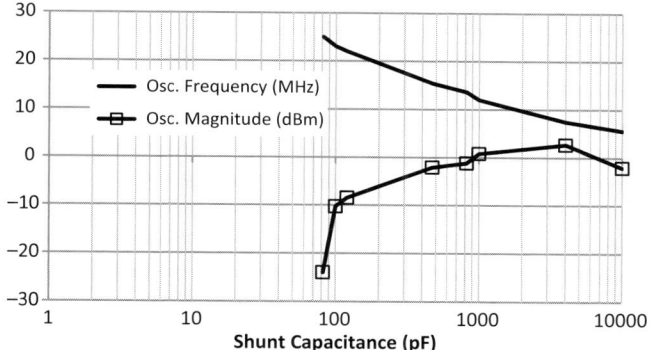

Figure 9-20   When the capacitance at the LDVR output is sufficiently large, the LDVR forms an oscillation with its output impedance. Overall system stability requires very small output capacitance values.

capacitor values provide a strong signal, while the oscillation ceases when the output capacitance is less than 82 pF. This is yet another reason to have small value bypass capacitance on the RF PA, beyond the desire to reduce capacitive dynamic current as shown in Figure 7-10.

9.4   Switching regulators

Dramatic improvements in top-level energy efficiency can only be realized when the DPS is implemented using nonresistive nonlinear (switching) techniques. Energy efficiency is improved above the resistive linear regulator by including energy storage within these circuits. In switch-mode power supplies, commonly called DC–DC converters, the energy storage is in the magnetic field around an inductor. For charge-pump switching converters, energy is stored in the electric field within capacitors. In either case, the circuit efficiency increase results from reducing transistor power dissipation by operating the device closer to the current or voltage axis at nearly all times – exactly as is done for high efficiency RF power amplifiers.

It is most illuminating to pay attention to the power dissipation of the variable supply from Figure 5-34

$$P_{D,DPS} = \left(\frac{1}{\eta(V_{PA})} - 1 \right) V_{PA} I_{PA} \qquad (9.18)$$

where $\eta(V_{PA})$ is the conversion efficiency of the switching converter, explicitly showing that this conversion efficiency is generally not constant but actually varies with the output voltage. This power dissipation goes to zero if the conversion efficiency is perfect, making the efficiency gains at the RF PA fully visible to the top-level supply. Imperfect conversion efficiency therefore "hides" some of the energy efficiency gains at the RF PA from the top-level supply.

Table 9-1 Naming of switching power supply operation possibilities

Function sense	Polarity	
	Same	Opposite
$V_{out} < V_{in}$	Buck	Inverting buck
$V_{out} > V_{in}$	Boost	Inverting boost

Unlike the LDVR analog DPS, a switching converter can also (1) operate to increase the output voltage compared to the input voltage, and (2) change the polarity at the output voltage from that of the input voltage. Any linear voltage regulator can only reduce the input voltage to the regulated output voltage keeping the same polarity. In the language of the switching power supply community, the names of power supplies that are available using switching techniques are listed in Table 9-1. Combinations of these available operation possibilities are certainly possible.

The greatest benefit here is gained when the high peak-to-average power ratio (PAPR) signals are generated with the DPST.

9.4.1   Switching transistor characterization

Designing C-mode RF power amplifiers requires abandoning the normal techniques of linear amplifier design, replacing them with procedures that were described in Chapter 8. These principles do not change when design attention shifts to developing a switching-based DPS. The C-mode operation described in Section 4.1.6 directly applies to switching power supply design.

The principle of operating the transistor for square wave resistance behavior presented in Section 8.4 is likely a more obvious objective for SMPS design since these transistors are labeled as switches instead of as amplifiers. The results of the analysis in Section 8.4 regarding limits on operating frequency of the switch therefore also directly apply here. The objective of any SMPS is achieving the highest available conversion efficiency. Low-cost implementations desire higher frequencies of operation at the switch. The methods of Section 8.4 are useful to evaluate how SMPS design has evolved to achieve high conversion efficiency.

A typical MOSFET designed to be a power switch in a SMPS design today has a transconductance near 100 A/V and an input capacitance ($C_{IN} = C_{GS} + C_{GD}$) around 6 nF. Using (8.15), this corresponds to $f_T = 2.6$ GHz for the switching transistor. Resistance of this switch is around 10 milliohms when ON, and the power supply load resistance is around 1 ohm. The conversion efficiency the SMPS can achieve when operating at 1 MHz is 92%.

This is enough information to evaluate the performance of such a switching power supply design and to compare it with the switching RF PA design theory from Section 8.4 and the corresponding predictions in Figure 8-20. The transistor speed

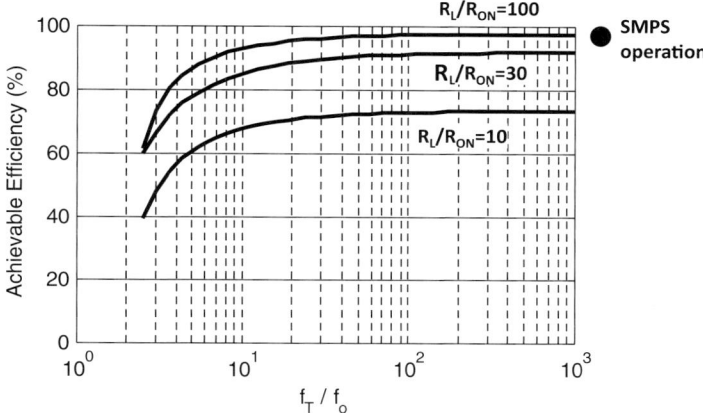

Figure 9-21   Comparison of a SMPS design with the prediction of switching RF PA design limits from Section 8.4 shows strong agreement.

metric f_T to the operating frequency ratio is 2600, and $R_L/R_{ON} = 1/0.01 = 100$. This provides a point that falls just outside the chart from Figure 8-20 which is shown in Figure 9-21. The slightly lower conversion exhibited by the SMPS compared to the prediction of 97% from Figure 9-21 demonstrates that additional loss mechanisms are present in the SMPS. These additional losses are well known, and are largely due to the limited ability to drive the switching transistor into square wave resistance behavior. Spending time in the transition between OFF and ON dissipates power.

The switching performance evaluated in Section 8.4 is focused on the output of the switching transistor by only evaluating the transistor output capacitance environment. Evaluation of the input characteristics for switching transistors across any FET technology is presented in [9-5] and summarized in the Appendix. An extension to the commonly used figure of merit for comparing switching transistors is introduced, which is designed to aid in driver design and to apply to any FET technology, including silicon and nonsilicon types.

The net result is that while silicon MOSFET switching transistors are continuously improving, this technology remains the worst performing of the FET technologies for high-speed switching. This sets up a very familiar cost–performance trade-off that is seen in many other aspects of electronic design: higher cost parts may perform better, and better performance parts almost certainly cost more. Evaluating how much better the high-speed switching performance may be is the objective of this extended figure of merit.

For the best efficiency, a key operating condition for any FET is to have its channel resistance be a "square waveform," meaning that the drive objective is to get the FET channel to change resistance in a minimum amount of time from R_{ON} to R_{OFF} and back again. Particularly when the transistor turns OFF rapidly, even as the output voltage takes its time to respond, the current through the transistor will essentially be zero, so transistor power dissipation is negligible.

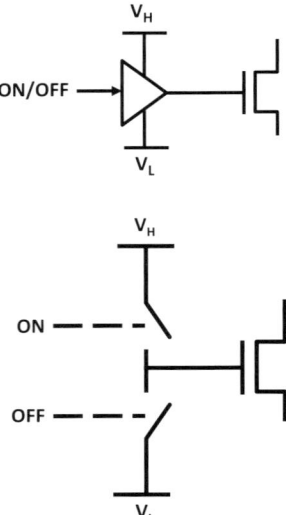

Figure 9-22   Driving a large FET switching transistor using voltage-mode switching on the FET gate.

FET "memory" for reduced-power drive

The same problem of reactive current evaluated in Section 7.2.1 for amplifier supply-bypass capacitors applies to drivers for these switching transistors. With input capacitances well within the range shown in Figure 7-10, the current required from the driver to effect a maximum slew rate at the switching transistor gate easily approaches or exceeds an ampere. The usual circuit to achieve this is a set of low resistance switches to the ON and OFF gate voltages shown in Figure 9-22.

If the driving approach of Figure 9-22 works, then the driver switches are implemented using complementary transistors, p-channel for the ON switch and n-channel for the OFF switch. If the technology being used for the driver does not have complementary transistors available, then implementations of this driver style tend to consume large amounts of power because one of the "switches" must actually be a low value resistance. One way around this problem is to use a charge-based driver, inserting or removing a fixed amount of charge to turn the big switching transistor ON or OFF respectively as shown in Figure 9-23. The actual V_{GS} value that manages the channel resistance is held by the input capacitance of the FET in between transitions. This effectively allows the driver to be OFF whenever a switching transition is not in process, significantly reducing its power consumption [9-6]. In effect, the FET input capacitance is operated here as a 1-bit dynamic memory cell, with one voltage state for ON and another one for OFF.

9.4.2   DC–DC is a current source

Using an inductance as the power conversion energy storage element makes the power converter act as a current source, something that is seen in Figure 9-24. The output

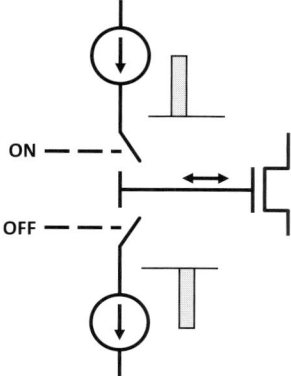

Figure 9-23 Driving a large FET switching transistor by applying current pulses on the FET gate, and using the FET input capacitance to hold the resulting V_{GS} value.

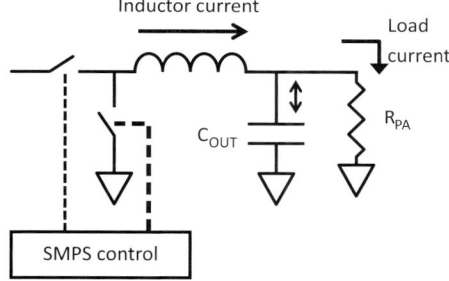

Figure 9-24 Any inductor-based switching converter is fundamentally a current source due to inductor action.

capacitor acts as a charge reservoir for some measure of voltage regulation. Excess inductor current above what the load demands flows into the capacitor (increasing its voltage and therefore the load current also). Likewise, an inductor current less than the load demands must be added to by some discharge current from the capacitor, reducing its voltage. For the output voltage to stay reasonably constant, the value of this capacitor must be very large. Yet for the output to adapt rapidly to varying signal envelope values this capacitor needs to have a small value.

Holding the output voltage fixed therefore requires a two-step process: (1) source enough charge into the output capacitor so that its voltage is at the desired output voltage, and then (2) match the inductor current to the existing load current so that the capacitor voltage is maintained, with no currents flowing into or out of it. Achieving the second step requires feedback from the load voltage as shown in Figure 9-25. Any change in the output voltage from the set value is met with duty cycle changes in the switch-mode power supply (SMPS) control, varying the ON and OFF times of the power switches to correspondingly increase or decrease the inductor current.

Depending on the dynamics of the load, or the required dynamics of the output voltage, for the supply current to change quickly the feedback loop must have a wide

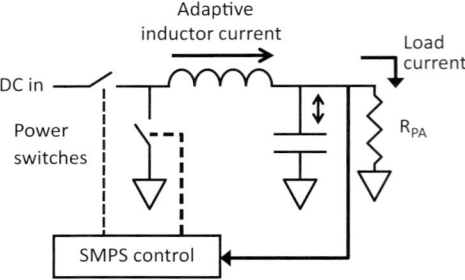

Figure 9-25   Making an inductive switching converter behave as a voltage source requires feedback, so that the power switch duty cycles can manage the inductor current to closely match the load current that a voltage source would provide to the load R_{PA}.

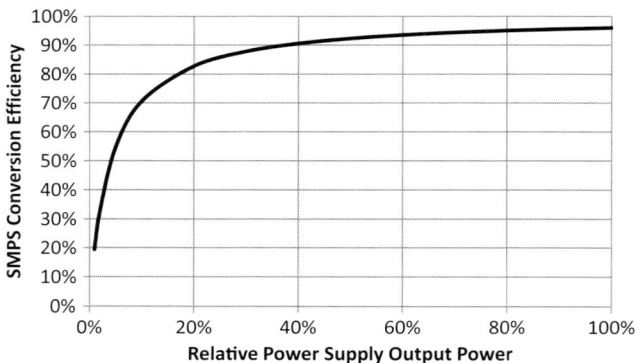

Figure 9-26   A typical conversion efficiency relationship for a SMPS as the output power varies from full down to nearly zero.

bandwidth. This also means that the switching frequency must be high and the value of the output capacitor must be small. Stable dynamics is not just nice for a DC–DC design, it is essential for it to work at all. Without an excellent feedback control system, the idea of switching DC–DC conversion does not work.

9.4.3   Efficiency varies with load and output voltage

It is tempting to model the conversion efficiency of the SMPS as a constant, as was shown in (3.1). Unfortunately, that is not how the world works. Conversion efficiency is not constant across output power levels for DC–DC converters, just as it is not constant for linear converters. Of course, the mechanism is very different here, because the conversion efficiency in Figure 9-26 is very different from that for the LDVR in Figure 9-15.

Comparing with Figure 9-15, the efficiency of the SMPS is actually lower at the higher output powers. This is because there is a lot more activity going on with any SMPS, and the dynamic operation does consume some power. The attraction of course is that as the output power requirement drops, the SMPS holds its high conversion efficiency value across a significant dynamic range. As the output power requirement decreases, the drive requirement into the large switching transistors does not change: the transistor capacitances remain the same, as they must. When the output power gets small enough, the fixed drive power begins to dominate the input power consumption, and the overall conversion efficiency begins to fall rapidly. For a signal with PAPR greater than 6 dB, this varying conversion efficiency needs to be taken into account.

9.4.4   Charge pump + SMPS

Charge pumps use capacitors to store energy, and adjust the output voltage by rearranging the capacitors. For a boost charge pump, the capacitors are usually initially charged by the input supply by being all connected in parallel, and then used at the output rearranged by being connected in series. Unlike an inductor-based SMPS, the charge pump does not need feedback to operate, particularly when the load current goes to zero. But the usual charge pump architectures do not provide for a continuously variable output voltage control.

Because the voltage-mode charge pump and the current-mode SMPS are so different, it is possible to combine their functions in such a way as to use each to its best advantage. One example of a combination approach to a boost converter is presented in Figure 9-27. This design uses a charge pump to provide a nominal doubling of the input voltage and to support most of the output current requirement. Then a smaller SMPS than would be needed to do the entire boost function on its own provides smaller adjustments to the nominal charge pump output [9-7]. Depending on the relative timing of the switch controls and their time overlap, this inductor current flow magnitude and direction can be varied, allowing an adjustment down (buck) or up (boost) from the charge pump nominal output voltage.

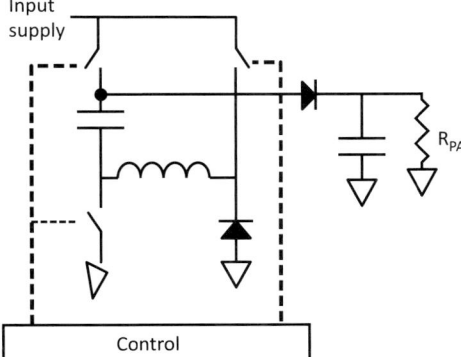

Figure 9-27   Combining a switching DC–DC converter with a charge pump voltage doubler to reduce the inductor size and value and provide variable continuous output voltage control.

9.4.5   Combined Si–non-Si

The major problem with switching DC–DC converters is achieving a small size for the inductor and capacitor. Limiting the lower value for these components is the switching frequency that the transistors can operate at with good efficiency. The result in Section 9.4.1 shows that the ratio of transistor f_T to the highest switching frequency must be nearly 3000:1 to achieve conversion efficiency above 95%. The resulting input capacitance for the low ON resistance switching transistor is generally large enough (several nanofarads) to require significant driver current (Figure 7-10) demanded by fast driver slew rates, along with a corresponding drop in conversion efficiency as both the output power drawn is reduced (Figure 9-26) or as the switching frequency increases. A typical all-silicon active component DC–DC buck converter shows a conversion efficiency vs. switching frequency profile like the solid curve in Figure 9-28(a).

These conclusions all follow from the tacit assumption that all circuitry must be implemented using silicon technologies. This is not a technical requirement, of course, but is largely an economic desire. The major loss mechanism here is driver oriented, which follows from the need to rapidly change the V_{GS} value in the presence of C_{IN}. The most expedient way to reduce this driver power requirement is to use transistors that have much smaller C_{IN} values for the same ON resistance. These transistors do exist, but are often made of nonsilicon technologies such as gallium arsenide (GaAs) or gallium nitride (GaN).

The operating benefit to a switching DC–DC converter is shown in Figure 9-28(a) when nonsilicon switches are used with no change to the controller [9-8]. To date, GaN transistors which are fabricated to have ON resistances near those of common silicon MOSFETs tend to have input capacitances reduced by 70%–85% compared to the silicon devices. Low frequency efficiency of DC–DC converters using these transistors is similar to that of traditional MOSFET designs, but now this conversion efficiency extends to higher switching frequencies before driver power begins to bring the overall efficiency down. GaAs FET devices to date are not designed for such low ON resistances

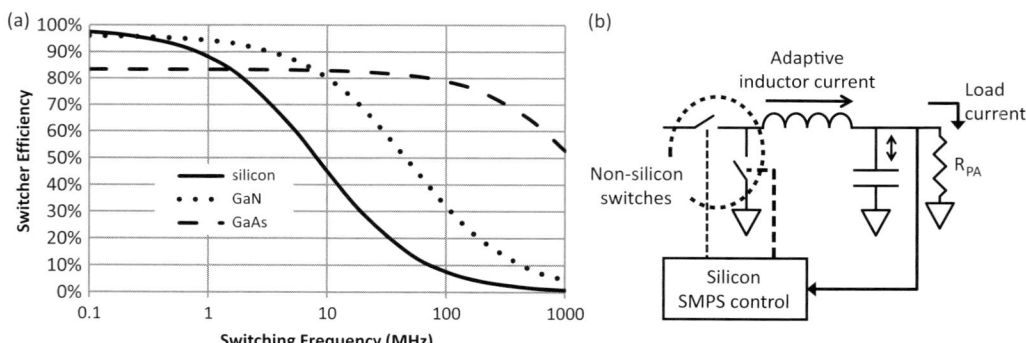

Figure 9-28   Typical DC–DC conversion efficiency profiles possible with changing the transistor technology for the chopping switches: (a) efficiency profiles for silicon, GaAs, and GaN technologies for the switches; (b) keeping silicon-based technology for the controller while changing the transistor technology for just the switching devices.

as GaN and MOSFETs, so the low frequency conversion efficiency is not as good. But the input capacitance of these GaAs FETs is much smaller than even for the GaN devices, and the operating bandwidth at high conversion efficiency gets to 10s of MHz. More details of FET technology comparisons regarding drive requirements is provided in Section 10.3.3.

9.5   Combined regulators within one DPS

Proper design of any DPS requires understanding of the signal that it must provide to the PA. Chapter 8 showed that the most stringent case is polar operation, so here we focus on successful implementation of polar modulation. The key information to base this discussion on is the spectral characteristic of the complete envelope signal. One example, here for a WCDMA signal, is shown in Figure 9-29(a). It is difficult to see the large spectral component at DC, containing most of the power in the envelope. The power integral in Figure 9-29(b) shows that this DC component contains nearly 80% of the total envelope power. Between DC and 1 MHz is 99% of the total envelope power.

The high frequency components of the envelope are certainly important for signal accuracy. The large power fraction at low frequencies suggests that a frequency-band splitting approach for the DPS may be interesting. The usual implementation provides the high power envelope components at low frequency using a high efficiency DC–DC converter, and the higher frequency envelope information, hopefully having zero mean, is provided by a lower efficiency but higher bandwidth LDVR. This idea is explored in Figure 9-30 using a WCDMA signal envelope.

The fast waveform in Figure 9-30(a) is a segment of this WCDMA envelope, including one envelope null identifying a signal zero crossing. Having identified in Figure 9-29(b) that the great majority of the envelope power is within the lowest 1 MHz of this signal spectrum, the much more slowly varying filtered envelope to 1 MHz bandwidth is also shown. Both of these waveforms are normalized to their rms magnitude. The filtered envelope is now the low frequency component of this frequency-band split. The corresponding high frequency component is the difference of these two

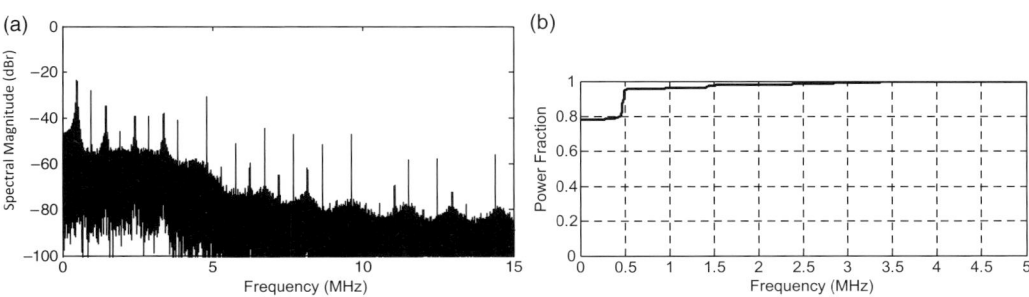

Figure 9-29   Most envelope information is at low frequencies: (a) envelope signal PSD for WCDMA has a wide bandwidth; (b) summation of total power shows that 99% is located at frequencies below 1 MHz.

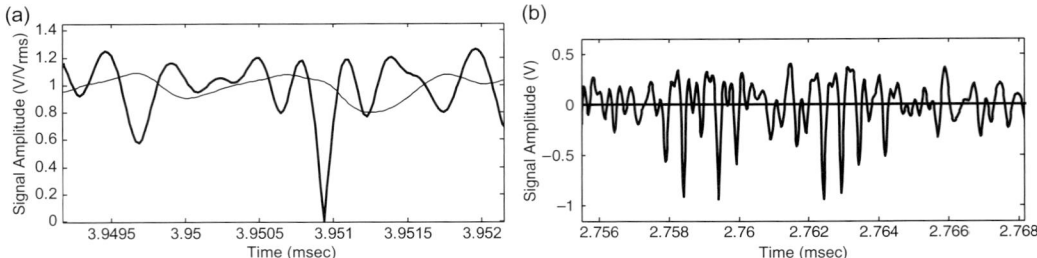

Figure 9-30   Time waveform segments of a WCDMA envelope: (a) overlay of the complete envelope waveform with the same waveform bandlimited at 1 MHz (slower curve); (b) difference signal between this bandlimited envelope and the complete envelope.

waveforms, one segment of which is presented in Figure 9-30(b). Wave shapes are not the same above or below zero, but indeed this difference waveform does have zero mean.

Whenever the difference signal in Figure 9-30(b) is greater than zero, the LDVR is both providing additional current into the PA and raising the voltage from the DC–DC converter. Additional current from the LDVR is required to raise this voltage, and the amount of current depends on the output impedance of the DC–DC converter. This additional current provides charge that eventually can be drawn upon to power the PA.

When the DC–DC converter has too high a voltage, and therefore too much current available for what the PA needs, the fact that the DC–DC converter is really a current source becomes important. According to Kirchhoff's Current Law, the sum of all currents at a circuit node must be zero. When the DC–DC converter puts in more than the PA needs to draw, the LDVR must remove the excess. Usual designs dump this current straight into ground. This is not good for overall transmitter efficiency because it represents electrons drawn from the top power supply that do not flow through the PA.

In lower voltage systems, it is advantageous to implement the LDO architecture of Figure 9-14 using a low saturation voltage PNP instead of using a P-MOSFET for the pass element. The advantage comes from the much higher transconductance of the bipolar transistor

$$g_{m,BJT} = \frac{I_C}{V_T} \tag{9.19}$$

than that of a large MOSFET. For low voltage applications, the controlling op-amp only needs to obtain one V_{BE} below the supply to operate the PNP into its resistive region. Depending on the specific structure and manufacture of the P-MOSFET, the threshold and channel enhancement required for low ON resistance can often be from several volts to 10 V. This makes the PNP particularly useful for mobile product DPS designs operating at 3 V yet needing watt-level RF power levels.

The stability problems identified in Section 9.3.1 and the preceding paragraph can be solved by selecting the most appropriate LDVR architecture for the immediate operating

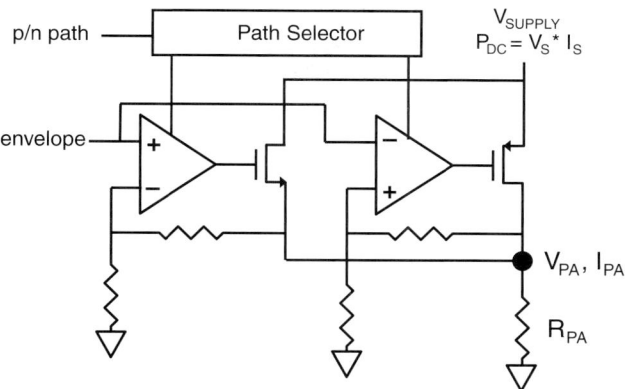

Figure 9-31   Solving LDO stability problems with an alternate path for lower output power operation LDO for PEP, emitter follower (EF), source follower (SF) for lower powers.

needs using the architecture presented in Figure 9-31, combining the circuitry in Figure 9-10 with the circuitry in Figure 9-11. This architecture applies equally well when bipolar transistors are used instead of MOSFETs [9-3]. When high output voltage is needed for the PA, the p-path (LDO) is selected. At lower powers, when there is enough headroom to operate, the n-path regulator is selected and the p-path is shut down. This avoids both the lower output stability problem with the LDO and the high power restriction of the n-path voltage regulator.

9.5.1   Series combination

The series combination of Figure 9-10 is repeated in Figure 9-32 along with more implementation details. There is one current path for the power flow to the PA from the top supply, as long as the PA impedance is low enough that it alone can pull down the DPS output node instead of requiring the LDVR to pull charge out of the PA for very low voltage events. With such a sufficiently low PA load impedance, effectively all of the current supplied by the top supply eventually flows through the RF PA.

   If the current from the DC–DC converter is too high, the interim voltage V_{DCDC} will build up, increasing the separation between the LDVR output and input voltages. And, according to (9.17), the efficiency will be reduced. This is contrasted with the power supply rejection (PSR) provided by all good voltage regulator designs. Once sufficient "headroom" (pass element voltage drop) is present, then additional voltage variations at the LDVR do not appear at the DPS output. In effect, the LDVR acts like a high rejection filter to any ripple voltage at the output of the DC–DC converter. The advantage is that this ripple rejection is independent of the bandwidth of the LDVR [9-9] [9-10] [9-11]. This characteristic is very unlike that of a high rejection analog filter, which must have a bandwidth much less than the switching frequency to achieve high ripple rejection.

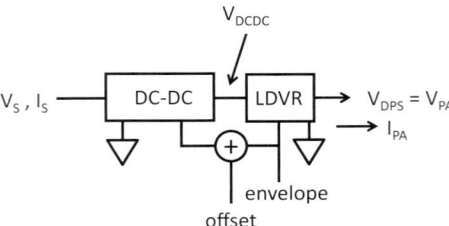

Figure 9-32   Series connection of a DC–DC converter followed by an LDVR to make up the total DPS; an offset must be added to keep the output voltage of the DC–DC converter high enough to properly operate the LDVR.

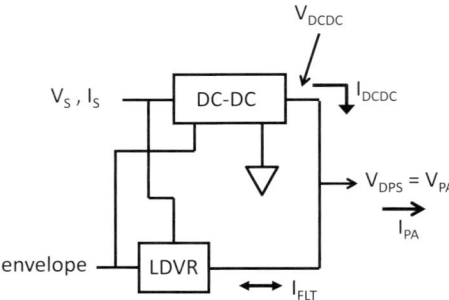

Figure 9-33   Shunt connection of switching and linear regulators, with parallel control to both circuit blocks. All ripple current (I_{FLT}) must be completely cancelled.

The obvious key to this design is to ensure that the LDVR always has sufficient headroom to operate properly. If the DC–DC converter is not providing enough current for the present load demand, then the headroom will be drawn down and this DPS will clip. This bad situation is avoided by adding an offset to the voltage setting command into the DC–DC converter, which directs the DC–DC converter to provide the necessary headroom for the LDVR. Dynamics of the DC–DC converter control loop must be fast enough to achieve the load current matching requirement described in Section 9.4.2 in order to maintain this headroom at its optimum level.

9.5.2   Shunt combination

The dual of the series circuit is the shunt circuit. This certainly is possible, and an expansion of the switching/linear hybrid DPS structure shown in Figure 9-11 is provided in Figure 9-33. The major differences from the series combination are the elimination of any need for a control offset into the DC–DC converter, and the direct summation of output currents from both the switching and linear regulators.

With the architecture of Figure 9-33, correction of any errors in the output of the DC–DC converter must be made by the LDVR. If the LDVR is to have a zero-mean output signal, then the DC–DC converter must always have an accurate output voltage [9-12]. All ripple current must be completely canceled by the LDVR: when the DC–DC current is too high, the LDVR must absorb it (and dump it into ground); and when the DC–DC current is too low, the LDVR must source the difference to provide the proper net current into the PA.

Having the LDVR absorb current is *always* bad for DPS efficiency, because the absorbed current is shunted away from the power supply and dumped into the ground return. These are electrons that do not flow through the PA for generation of more output power, but are supplied from the top power supply. More power in for no additional power out means lower transmitter efficiency.

This property suggests that having a zero-mean output current from the LDVR in this shunt configuration is exactly what we do not want, when overall energy efficiency is the goal. We ideally want the LDVR to always be contributing to PA current and not shunting it away. This implies that the DC–DC output should always be low. But we also do not want too much current from the LDVR because strictly speaking it always has a fairly large voltage difference across it, so whatever current it does supply is at low conversion efficiency according to (9.17). The dance between the switching and linear circuit blocks in the shunt combination is a very careful one.

Also, if there is an error in the output voltage of the DC–DC converter, then the LDVR is also responsible for correcting it. What the DC–DC converter control system does in this situation must be carefully designed. Normal practice is to have the control system regulate for a set voltage based on the load current. If this voltage is too low, the reverse current coming back into the DC–DC converter (in addition to the current added to the PA load by the LDVR) directly fights the inductor current. A conventional regulator will work hard to absorb the reverse current, though most DC–DC converter designs do not have any provision for absorbing current. The simple solution is to place a small resistor between the DC–DC converter output and the DPS output node. This eliminates the fighting voltage sources. It also forces an additional power dissipation from all the current flowing through this resistor that adds to overall efficiency loss.

9.5.3   Hysteretic switching

One popular architecture related to the shunt combination DPS architecture is the hysteretic switching method shown in Figure 9-34. Here the switching converter is not completely implemented, and the structure is more accurately described as an LDVR with selective inductive current assistance. During nominal operation, the LDVR provides all the current needed by the load. When this LDVR output current gets large, an inductor is switched ON to build up current that can be supplied to the PA. Kirchoff's Current Law (KCL) says that as the inductor current builds up, the amount of current provided by the LDVR correspondingly falls, even as the LDVR holds the voltage at the output node constant. To the LDVR, the load impedance appears to be increasing, since it is providing less current into a regulated voltage.

When the LDVR current falls to a design level, the inductor charging switch is opened and the magnetic field around the inductor continues to supply current from the inductor into the output node. As long as the output current from the LDVR remains below the set

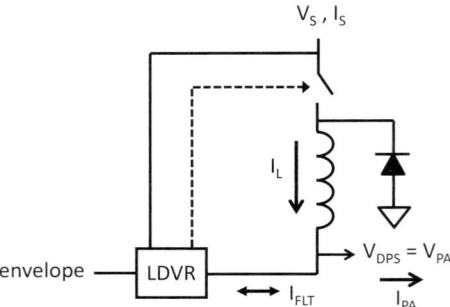

Figure 9-34   Shunt connection with waveform-dependent parallel drive using hysteretic control.

value that starts to charge the inductor, the switch remains open. Like the conventional shunt connection DPS described in Section 9.5.2, with this architecture the LDVR should always be sourcing current if high efficiency is needed. If at any time the LDVR shunts excess inductive current away from the PA load, this shunted current is provided by the top supply but not made available to the PA, which is an inherent energy efficiency loss.

Unlike a conventional DC–DC converter, here the inductor is only switched when the LDVR current gets outside the design hysteresis levels. The switching frequency is not fixed. Also the ON time or switching duty cycle is not fixed. Both the switching time and the ON time are reactive to the current load experienced by the LDVR. Any noise resulting from the switching is not periodic, but it is highly correlated to the envelope waveform.

One additional assumption is present in this design, and that is that the envelope is dynamic enough that its value never stays at a particular value for very long. One can envision an interesting situation with this circuit if the output voltage to the PA has to stay at a fixed level to meet some function of the RF protocol. If the output current stays within the hysteresis range of the LDVR, then the inductor switch is never activated and this architecture naturally reduces to the configuration of Figure 9-8. If however the required PA current is high enough to trigger the switch activation, then an oscillation begins. As the inductor current builds up, the LDVR current drops until it releases the switch. As the inductor current decays with its collapsing magnetic field, the LDVR current increases until it gets large enough to activate the switch. The cycle then repeats.

In essence, this is a switching regulator of sorts, but not exactly like that of Figure 9-9 because the LDVR is present here to manage ripple current. Efficiency of the LDVR depends on the voltage difference between V_S and V_{PA}. The frequency of this oscillation is governed by the current hysteresis of the LDVR and the charging time of the inductor. There is no external clock involved to control the switching times. It is necessary to ensure that there is no possible condition where the switch might stay ON.

9.6   DPS bandwidth

Bandwidth measures for any DPS are governed by waveform fidelity requirements, not power transfer accuracy. As mentioned in Section 3.6, this means that the conventional

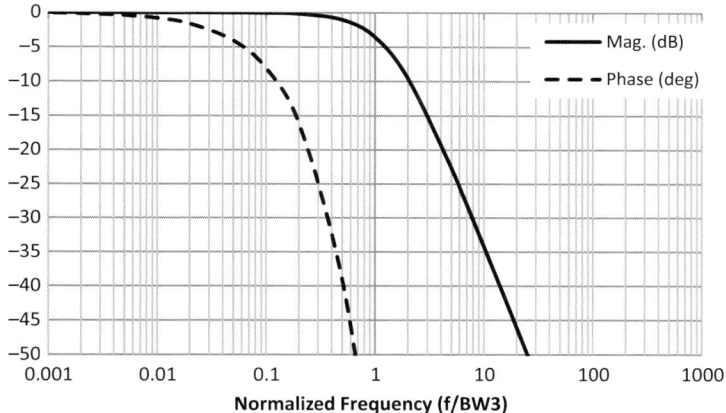

Figure 9-35   DPS frequency response modeled as a two-pole RC filter, showing that the phase response begins much sooner than the magnitude response. If 5 degrees of phase distortion is tolerable, then the actual available bandwidth for DPS operation is 0.06 of the traditional measure BW3.

bandwidth of BW3 is inappropriate for DPS use. This output waveform fidelity requirement is particularly important when the DPST energy efficiency must be maximized, because this means that polar modulation must be used. Converting that to circuit bandwidth only considers the behavior of a one-pole RC filter in Section 3.6. Most actual DPS designs are better modeled as two-pole circuits. A close-up view of the frequency response of a two-pole RC filter is presented in Figure 9-35, with frequency normalized to the traditional −3 dB magnitude bandwidth (BW3). The magnitude response is plotted in dB, and the phase response is shown in degrees of phase shift.

Properly interpreting the information in Figure 9-35 requires a system-level determination on the acceptable phase distortion on the highest frequency components of the envelope waveform. The signal bandwidth from the highest energy efficiency DPST is the result of a convolution between the power spectral densities of the magnitude signal component and the phase signal component. For this convolution to occur accurately out to the necessary occupied bandwidth, both the magnitude and phase aspects of the component power spectral densities must also be accurate. Depending on how much phase distortion is actually tolerable, the actual useful bandwidth of the DPS may be much less than 10% of BW3.

9.6.1   Relation to signal bandwidth

Much of the bandwidth expansion for the magnitude and phase modulation components is a result of the signal envelope requirements going to zero, or close to zero. This invokes the properties of the Fourier transform, that is whenever the signal itself is not continuous, or if the waveform is continuous but its derivatives are not, then there must be expansion of the bandwidth required to successfully reconstruct the waveform. There is no option. And the converse is also immutable: to accurately reproduce these waveforms, the implementing circuitry must have this corresponding bandwidth. Waveform accuracy is strictly required whenever the underlying PA operates in C-mode or P-mode. And C-mode operation is a pre-requisite for maximum PA energy efficiency.

Signal design is the only way to manage this. Recent signal designs from standards committees actually tend to maximize the presence of envelope zero-crossing events, making these implementation problems worse. Common examples are any QAM (including all PSK signals) and any OFDM-style signal. Constellations can be designed which minimize signal zero-crossing events. Modulation protocols that use constellations with restricted transition options between points are very successfully used to completely eliminate zero-crossing events; major examples in wide use include $\pi/4$-QPSK, OQPSK, and $3\pi/16$-8PSK. Also included in this strategy of restricting transitions between constellation points are the pPSK signals [9-4]. It is also possible to change the filtering strategy used to bandlimit the signal state sequence to achieve an elimination of envelope zero crossings [9-13].

Another view of DPS bandwidth requirements is taken from (2.9). This relation shows that at high efficiency the output signal is formed by a physical multiplication at the PA transistor involving the RF carrier and the voltage on that transistor. According to the Fourier transform, a multiplication in the time domain corresponds to a convolution in the frequency domain. Thus, a physically equivalent view of (2.9) is that the signal spectra for both the phase-modulated carrier and the DPS controlled envelope modulation are convolved to form the final output signal. The Fourier transform also makes clear that both the magnitude and the phase components of these component power spectral densities are important to the successful outcome of this convolution. Again, high energy efficiency in the presence of finite bandwidth circuitry necessitates a signal design that has all the waveform information contained within the allotted circuit bandwidth. Physics gives a very clear roadmap on the necessary requirements.

9.6.2   Signal slew rate: a brief survey

The discussion in Section 9.1.3 demonstrates the importance of signal envelope slew rate information in DPS design. It is regretted there that slew rate information is not generally published for any signal modulations. This leaves it up to every DPS design team to evaluate the slew rate characteristics for all the signals that their design is targeted to support. The resulting slew rate capability then becomes an important feature specification of the resulting DPS.

In this section, a very small set of signal types are evaluated for general slew rate characteristics. This brief survey is neither complete in scope nor well detailed in each signal's slew characteristics. Still the information provided here in Figures 9-36 through 9-38 is instructive in pointing out that slew rate properties do vary significantly across signal types. The figures are divided into three general groups: QAM signals, narrowband signals, and high data rate signals. In each graphic, the horizontal axis is the complete span of slew rates evaluated for that signal. These graphs are all histograms, so the details of the vertical axes are not critical. In general, these graphs are best thought of as probability density functions for their associated slew rate scales.

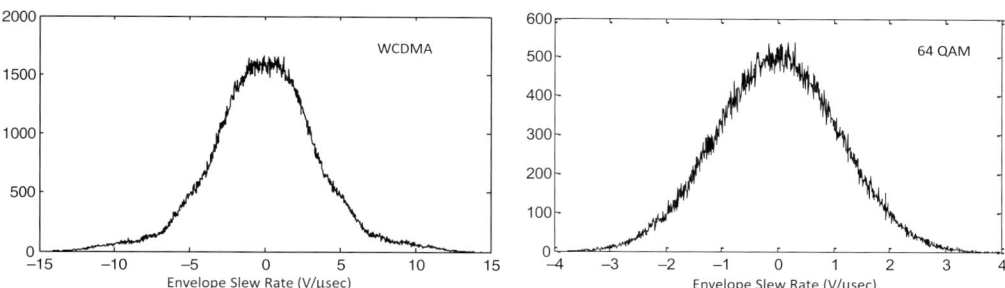

Figure 9-36   Slew rate histograms for WCDMA and 64 QAM signals of similar bandwidth and with identical Nyquist bandlimiting filters. The uniform constellation point spacing of the QAM signal keeps the slew rate span down compared to the nonuniform constellation used in this WCDMA signal.

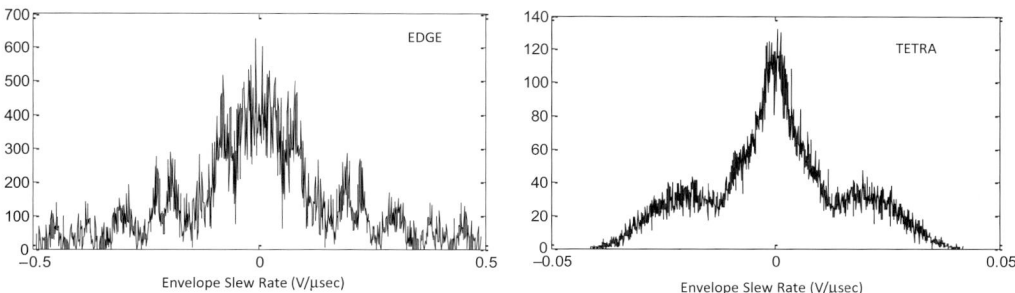

Figure 9-37   Slew rate histograms for EDGE and TETRA signals. Both of these have slow slew rates due to both their narrow occupied bandwidths and from their non zero-crossing envelopes.

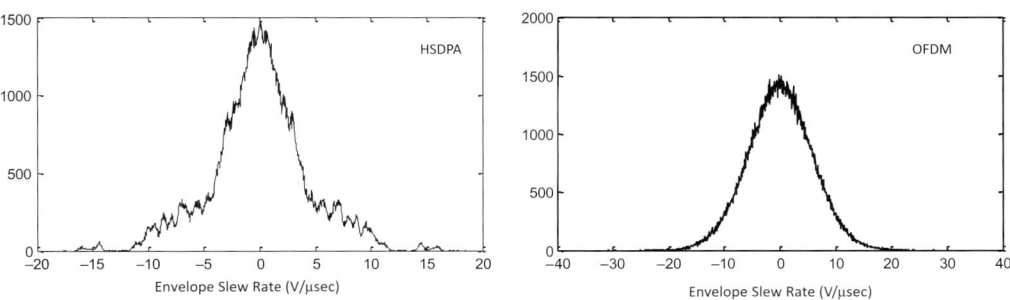

Figure 9-38   Slew rate histograms for HSDPA (uplink) and OFDM signals with similar occupied bandwidth. These signals have the widest range of slew rates, primarily due to a very nonuniform constellation for the HSDPA signal and to the large number of zero crossings in the OFDM signal.

9.7   Conversion efficiency characteristics

The best way to evaluate energy efficiency of any circuit is to "follow the electrons." This is particularly true when comparing the various DPS architecture options. Once the current flows are physically well understood, then the mathematics writing process follows. It is easy to write mathematical relationships that describe measurements but which have no direct physical analog and therefore no means to provide design insight on how performance may be improved (see Section 2.1.1). When the task is to design a maximally energy-efficient transmitter, the flow of electrons must be very clearly understood.

Whenever there is any drop in efficiency, there must be power dissipation and therefore something is "getting hot." Current and voltage relationships in ideal reactive components only store energy, and do not dissipate it. So such components cannot contribute to efficiency drops. Whenever an efficiency drop is measured, good engineering practice is to directly identify all the points where power is being dissipated – at least enough of these points to account for 90% of the measured dissipation. Only then will any ability to improve things become evident, or alternatively to know that within the limits of technology peak performance has already been achieved.

To this end, the objective of any RF transmitter is to generate the specified RF signal at required output power. To do this, current must flow through the power amplifier and eventually through the output load. Any other current flows that do not directly provide signal current flow through the load are not directly contributing to the output signal power. These other current flows reduce overall energy efficiency. This is the reason why current shunted away from the PA is paid so much attention in Sections 9.4 and 9.5.

Impacts on overall energy efficiency (top-level power)
The relative simplicity of LDVR designs compared to those of switching power converters – and the wider operating bandwidth attainable with LDVR circuits compared to switching power converters – is traded against their lower energy conversion efficiency. A tendency is appearing in the modern literature to use multiple combinations of switching converters around each LDVR, much like combining the architecture in Figure 9-10 with that in Figure 9-11. To make this work, the SMPS in series with the LDVR must have nearly the same agility as the LDVR itself. This is not easy, and such a fast SMPS generally pays a penalty by having lower conversion efficiency.

The message here is that complexity also has its costs. Not just the obvious additional cost in more parts, but the anticipated energy efficiency benefits might not appear after everything is put together. Building complexity, on its own, is not successful research. Improving overall energy efficiency is good research, and that takes attention back to earlier in this section.

9.8   DPS output noise

Output noise performance of the DPS is one of the most critical parameters in DPS design. This is hard to define specifically for general purpose DPS products because the noise tolerable by the transmitter depends strongly on how the transmitter using this DPS is operated. If envelope tracking is used, then the PA will suppress any noise on the DPS output by 40 dB as long as (10.1) applies. This makes DPS design relatively easy, particularly for switching power converter architectures.

At the other end of the operating possibility is polar modulation, where any output noise on the DPS directly modulates the output signal envelope; the polar PA noise suppression is 0 dB. There is no noise suppression at all. This makes DPS design much more difficult for use with C-mode polar transmitters. When switching power converters are used for better conversion efficiency, their output ripple must be significantly suppressed by the DPS before connection to the PA.

Basic relationships between DPS noise and polar transmitter sidebands from this noise are presented in Section 6.2.4. One of the critical PA parameters necessary for evaluating the amount of noise that can be tolerated on the DPS output is the PA value of g_{DPS_AM}, the transfer function of V_{DPS} to output envelope variation (5.5). This depends on both the external and internal impedances of the PA (5.7) and so it must be specified by the PA manufacturer. This parameter can also be directly measured in the absence of specifications from the PA manufacturer.

The other key information that is necessary is set by the application: how well do sidebands from DPS noise need to be suppressed in the communication system? A more appropriate specification is actually the maximum sideband power that can be tolerated, independent of the power in the desired signal. This follows directly from (5.3), which shows that the noise conversion to RF sidebands is independent of the value of the DPS output voltage, and therefore is also independent of the signal power. Demonstration of this property is shown in Figure 6-9 and Figure 6-10.

When the DPS is used in a transmitter that generates wideband signals, then there is no opportunity to filter the noise from the DPS. The bandwidth of the noise on the DPS output therefore transfers directly to the bandwidth of the sidebands around the transmitter output signal. This bandwidth can exceed 100 MHz as discussed in Section 9.6. With such a wide bandwidth, this modulation process can easily cause transmitter output sidebands at frequency offsets commonly used for duplex separation in two-way communication systems.

One additional thing to pay attention to – artifacts from digital construction of the envelope waveform will pass through a wideband DPS and modulate the output signal. The envelope waveform must be carefully filtered to remove all digital signal processing artifacts before it is connected to the DPS input to avoid wideband (large frequency offset) transmitter output noise surprises – bad surprises.

9.9   DPS output impedance

Section 7.2.2 presented the importance of a wideband low output impedance from the DPS to provide low frequency stability for an envelope tracking linear PA. For three of the four DPS architectures presented in Section 9.2.3, this output impedance is governed primarily by the LDVR within the DPS. Only in Figure 9-9 is there no LDVR, and the DPS output impedance needs to be evaluated with other means.

With respect to any LDVR, the output impedance measurements in Section 9.3.3 are very enlightening, as well as scary. A quick check of standard voltage regulator specifications shows that wideband output impedance is basically never measured, or, if it is, the results are not published. These measurements show that the output impedance is as expected within the bandwidth of the closed-loop gain. At higher frequencies, this is no longer true.

Because the LDVR output impedance becomes inductive along with negative resistance, the nature of the supply input impedance for the PA becomes very important for the two devices to successfully work together. When the LDVR output resistance goes negative, it can no longer ensure the low frequency stability requirements of the PA. Knowing that the LDVR output impedance behaves in this way provides a message to the PA designers that the PA itself must not depend on DPS output impedance characteristics above a certain frequency to provide the low frequency stability. Without the DPS output impedance as part of its published specification, the probability that interface problems will occur when any DPS is connected to any PA increases.

9.10   Output stability

It is imperative that the DPS remain stable no matter the load provided to it by the attached PA, which can vary widely as presented in Sections 5.10 and 6.9. This is definitely not an assured result, as aptly demonstrated in Section 9.3.3, where just the inclusion of a capacitor of sufficiently large value combines with the natural output impedance to form a resonant circuit with sufficient gain to sustain sinusoidal oscillation. There are other load conditions that can cause DPS instability, depending on the DPS design. One is if the output load impedance is too high, which is a particular problem for DPS designs that use an LDVR having the LDO architecture of Figure 9-14. Another load issue is if the dynamic load resistance becomes negative. These are all very practical situations that must be carefully designed for.

9.10.1   Load can be negative dynamic resistance

Both the absolute and dynamic resistance of the load need to be positive to have reasonable assurance that the control loops within the DPS will remain stable with good phase margin. Experience over the years shows that having a positive absolute resistance is highly probable except at very low DPS output voltages. At low PA supply values, there can be sufficient leakage signals from the RF path that the PA output

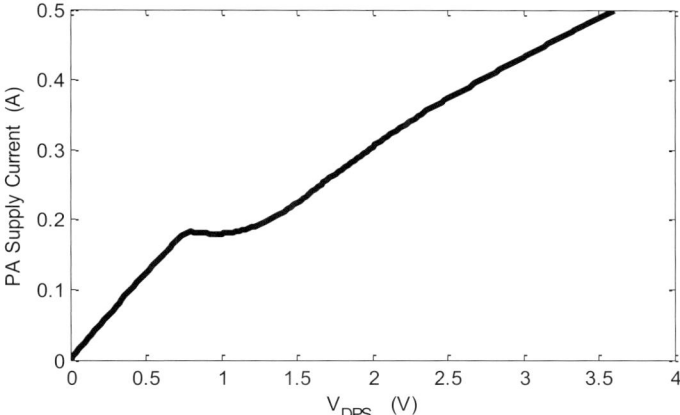

Figure 9-39   A negative dynamic resistance can appear when one DPS is used to drive two consecutive C-mode stages.

becomes a low-level power source. When combined with the DPS, also operating as a low-level power source, this combination can shift to appear to the DPS as a negative resistance load.

It is more probable that while the absolute resistance remains positive, the dynamic resistance can become negative. One example of how this occurs is shown in Figure 9-39 for two consecutive PA stages supplied by the same DPS. While the supply is low enough that the driver stage does not provide an output signal large enough to compress the final stage, the currents increase rapidly. As the final stage compresses it draws less current. It is quite practical that the compression of this final stage happens fast enough with increasing DPS output voltage that the DPS overall load current can fall briefly. This is enough to cause problems for many DPS designs. Details on how this situation practically occurs are provided in Section 12.2.3.

There is a strong economic incentive to minimize the amount of circuitry in any product. The appearance of negative dynamic resistance shows that this design is economically a bad choice. Many DPS designs cannot handle this brief excursion into negative dynamic resistance and oscillations are seen when the DPS output passes through these voltage levels.

9.10.2   Load impedance is widely varying for ET

In any ET application, the impedance provided to the DPS by the linear PA varies across a very wide range, described by (5.20) and graphed in Figure 5-33. Most circuits are designed to support a single load impedance value. Fortunately, voltage regulators are usually designed to tolerate a wider range of load resistances, but convention is for the output voltage to be held constant. The design of variable voltage supply into simultaneously varying load values is a new condition that needs expanded tools for better and faster design success.

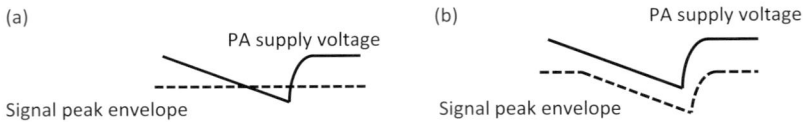

Figure 9-40   Adapting to low battery voltage events: (a) no adaptation causes signal clipping and out-of-specification performance when the battery voltage is insufficient to support the required signal envelope; (b) locally adapting the signal power while the battery voltage drops and recovers maintains performance within specifications.

9.11   Automatic low battery compensation operation

In any mobile device, the battery voltage supplying the transmitter is always varying as the charge within the battery is depleting or being restored. Operation of the wireless communication system pays no attention to this. If the communication protocol requests the wireless transmitter to provide high output power, it will attempt to do so. One particularly troublesome situation is illustrated in Figure 9-40(a), which shows the requested output peak signal envelope compared to the local battery voltage. As long as the battery voltage is higher than the peak signal envelope, then all is fine. When the battery voltage falls below the ability to support the peak signal envelope there will be severe signal distortion due to the battle between the transmitter trying to provide the high power output signal and the battery – this is a battle the battery will win. Everything returns to normal once the battery is recharged. But in the meantime the transmitter performance is unacceptably bad. That is, unless some local process intervenes. One way to solve this problem is shown in Figure 9-40(b) using a process called automatic low battery compensation (ALBC).

To completely avoid output signal distortion while ALBC operates, it is necessary to sense the pending compression of the DPS due to insufficient input voltage before compression actually happens. One circuit that meets this objective is shown in Figure 9-41. The idea is to sense the coming onset of transistor saturation, and act on that warning while the feedback control still operates and guarantees output waveform accuracy. To be effective, the ALBC function must be complete before the pass transistor gets far enough into saturation that is loses gain and causes the feedback to fail.

One design that meets this objective is a conventional simple LDO-style LDVR, shown in Figure 9-42 [9-14]. This LDVR adds an additional emitter follower as a buffer between the error amplifier and the PNP pass transistor. The emitter current of this buffer is equal to the base current of the large pass transistor. The collector current of the buffer is also equal to the pass transistor base current as long as this buffer transistor has sufficiently high β. The voltage across the resistor through which the buffer's collector current flows is proportional to the pass transistor base current.

When the pass transistor operates linearly, its base current is very small. As the pass transistor begins to approach saturation, its base current rises quickly, even as the feedback loop still has enough gain to maintain output regulation. This base current increase causes a large increase in the voltage across the buffer collector resistor,

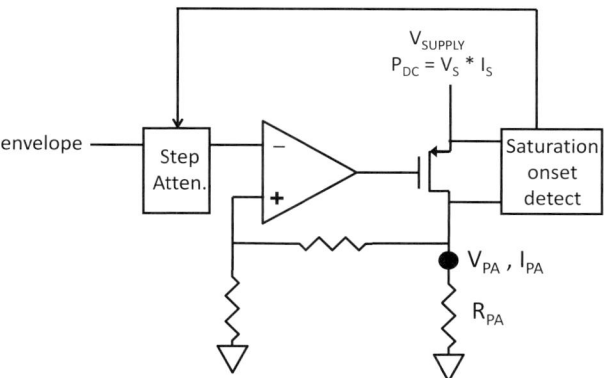

Figure 9-41   ALBC is implemented by sensing the compression status of the pass transistor, and requesting attenuation of the input signal if the output peak envelope gets too close to the available voltage.

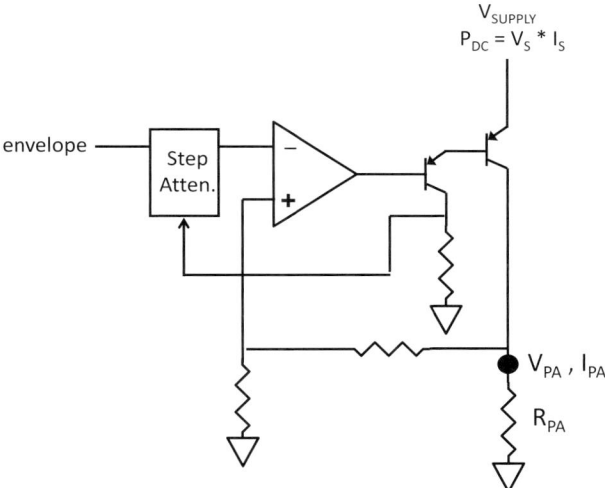

Figure 9-42   Implementing ALBC sampling in a simple LDVR by adding an emitter follower buffer between the control amplifier and the output pass transistor to sample the base current needed by the pass transistor to settle the output voltage at its required level.

providing a warning that the LDVR is running out of headroom, but not yet losing regulation. This warning signal is sent to the input circuit driving the LDVR to effect a small attenuation on the LDVR input signal.

As long as the warning signal is active, the input signal is stepped down with added attenuation until the warning signal is no longer active. This keeps the LDVR at its highest efficiency, maintaining an accurate output signal at all times. Usual practice is to later probe whether the LDVR headroom has increased by beginning to remove steps of the input attenuation. If the warning signal returns, then the attenuation is increased. If the warning signal does not recur, then the additional attenuation is removed until either

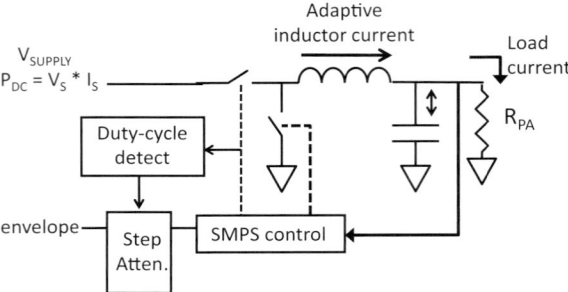

Figure 9-43   ALBC implemented in a SMPS, here monitoring the duty cycle of the switch control to sense when the output voltage is getting too close to the input voltage.

the warning signal returns or all of the attenuation is removed. This effects the desired adaptive control shown in Figure 9-40(b). And all without interrupting normal transmitter operation, and also without ever encountering output signal distortion.

The discussion so far focuses on bipolar implementation because of the ease in describing the control operation using the features of bipolar technology. Equivalent mechanisms exist for FET-based LDVR circuits. The same principles also apply to switching power converters, most easily visualized for the buck configuration [9-15] [9-16] [9-17]. The ratio of output voltage to input voltage is captured in the switching duty cycle of the series switch transistor. ALBC in this case operates by not allowing this duty cycle to get too high (e.g. below 95%), using an approach like the structure shown in Figure 9-43. When the duty cycle of this switch driver waveform gets too high, this is the warning signal used to begin the small steps of input signal attenuation.

9.12   VSWR management (DP only)

Output load mismatch is a serious problem for any DPS transmitter. The issues involved with envelope tracking are discussed in Sections 5.15 and 7.10. Maintaining the ability to operate the PA within the wireless communication system specifications requires that higher supply voltages be available to the envelope tracking DPS, up to double the nominal supply voltage, as presented in Figure 7-38. To the extent that this sufficiently high input voltage is available to the DPS, the envelope tracking DPST has no problem with the presence of an output mismatch as a result of the ET output signal independence from the actual value of the PA supply voltage. The price for this tolerance is an immediate reduction in the overall energy efficiency of the ET transmitter. This efficiency loss is greatest when an LDVR based DPS design (Figure 9-7) is used. The energy efficiency penalty is lowest when the DPS is implemented only with switching power conversion as shown in Figure 9-9. Using the combined architectures of Figures 9-10 and 9-11 achieves an energy efficiency penalty between these extremes.

Operation of the ET transmitter to be adaptive to the possible presence of an output impedance mismatch requires an architecture such as that in Figure 7-40. The output

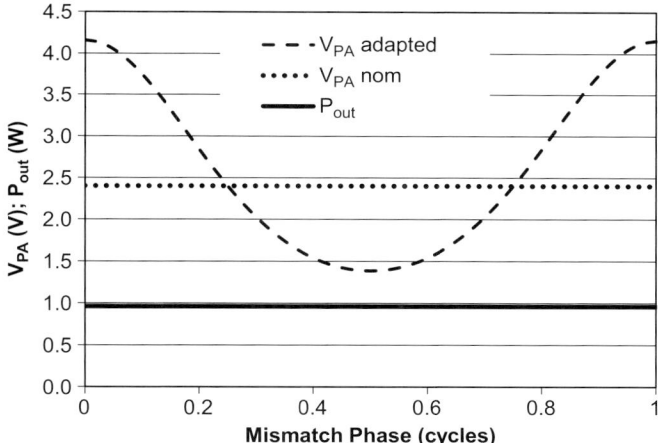

Figure 9-44 Adaptive DPS operation maintaining output power in the presence of DPST output impedance mismatch. This compares directly with Figure 8-30.

mismatch is sensed and measured, and the VSWR adaptive controller rescales the envelope command from the modulator either up or down as necessary to match the conditions at the RF PA transistor. This keeps the energy efficiency more consistent – at the price of greatly increased complexity.

Operation of polar transmitters encounters different issues as discussed in Sections 6.9.3 and 8.7.3. Presence of an output impedance mismatch changes the value of the load present at the PA output, just as it does for the linear PA ET case. But with C-mode operation, the reaction of the polar PA is to change its output power in accordance with (6.25), expanded here to emphasize dependence on final load impedance Z_L,

$$P_{OUT}(a_{rms}, Z_L) = \frac{a_{rms}^2}{R_{PA}(Z_L)}. \qquad (9.20)$$

The relation (9.20) provides instruction on how to adapt for constant output power. If the value of R_{PA} changes in reaction to an output impedance mismatch, it is possible to rescale the envelope waveform in a corresponding direction to keep the power constant. Doing this rescaling for the mismatch conditions of Figure 8-30 provides the results in Figure 9-44. Output power in the presence of this output mismatch at any phasing is now constant. The applied voltage to the C-mode PA varies according to the dashed line as the mismatch phasing changes, instead of being held constant as shown by the dotted line for the properly matched case. The adapted PA voltage here can increase up to 4.2 V. If this voltage is not available at the input to the DPS, then the DPS will not be able to control the PA to full power under that condition. This is a situation very similar to the discussion on voltage limitations in Section 9.11, so an ALBC-like approach can also be a solution here. Combining the ALBC adaptive control with this mismatch adaptive control is a very effective DPS feature for polar operation.

The VSWR adaptive control has its objective to equate the output power in the mismatch condition to the design output power that is present when the transmitter is nominally matched

$$P_{OUT}(a_{rms}, Z_L) = P_{OUT}(V_{env}, Z_0). \tag{9.21}$$

Sensing the mismatch is done by measuring the current drawn by the C-mode PA, since this current is a measure of the PA resistance in the presence of this mismatch. The control loop then causes the DPS output voltage to vary until the product of the adjusted DPS output and the current drawn by the PA at that voltage matches the power condition of (9.21) [9-18].

9.13   Power control

Whether the DPS transmitter operates in either envelope tracking or polar, as the output power required from the transmitter goes down, the voltage from the DPS also must decrease. In a polar transmitter, this decreasing power supply value is effecting the RF signal power control. But in an ET transmitter, the power control is being set by attenuating the PA input signal, and the DPS is just reacting to this change.

9.14   References

Note: Patent references are provided for bibliographic use only. Citation of specific patents here is not indicating any view on priority issues.

[9-1] R. J. Smith, *Circuits, Devices, and Systems*, 2nd edn., J. Wiley & Sons, New York, 1971.

[9-2] R. Broderson, "The Network Computer and its Future," *Proceedings of the 1997 International Solid State Circuits Conference (ISSCC 97)*, Keynote presentation, pp. 32–36.

[9-3] M. Dunsmore, E. McCune, and G. Do, "Polar Modulation Transmitter with Envelope Modulator Path Switching," US Patent 8301088, issued Oct. 30, 2012.

[9-4] E. McCune, "pPSK for Bandwidth and Energy Efficiency," *Proceedings of the 43rd European Microwave Conference (EuMC)*, Nuremberg, October 2013.

[9-5] E. McCune, "Process- and Technology-Independent Power Switching Transistor Figures of Merit," *Proceedings of the 2008 IEEE Radio and Wireless Symposium (RWS)*, Jan. 2008, pp. 195–198.

[9-6] B. Silic, "Switchmode Supply and Driving Method for Efficient RF Amplification," US Patent 6867574, issued March 15, 2005.

[9-7] W. Sander, "Boost Doubler Circuit," US Patent 6522192, issued Feb. 18, 2003.

[9-8] E. McCune, "Extremely High-speed Switchmode DC-DC Converters," US Patent 7026797, issued April 11, 2006, Patent 7227342, issued June 5, 2007, and Patent 7906944, issued March 15, 2011.

[9-9] E. McCune, "High Efficiency Modulation RF Amplifier," US Patent 6377784, issued April 23, 2002, US Patent 7099635, issued Aug. 29, 2006, US Patent 7395038, issued July 1, 2008.

[9-10] W. Sander, R. Meck, and E. McCune, "High-efficiency Modulating RF Amplifier," US Patent 6816016, issued Nov. 9, 2004.

[9-11] E. McCune and W. Sander, "High-efficiency Amplifier Output Level and Burst Control," US Patent 6864668, issued March 8, 2005.

[9-12] B. Silic and E. McCune, "Enhanced Hybrid Class S Modulator," US Patent 7558334, issued July 7, 2009.

[9-13] E. McCune, "Signaling Transition Control in a Modulated Signal Communication System," US Patent 5321799, issued June 14, 1994.

[9-14] W. Sander, "Saturation Prevention and Amplifier Distortion Reduction," US Patent 6528975, issued March 4, 2003.

[9-15] K. Cioffi, N. Tolson, and E. McCune, "Power Supply Processing for Power Amplifiers," US Patent 6781452, issued Aug. 24, 2004, US Patent 6924695, issued Aug. 2, 2005, US Patent 7038536, issued May 2, 2006, US Patent 7642847, issued Jan. 5, 2010.

[9-16] C. Pennec, "Automatic Low Battery Compensation Scaling Across Multiple Amplifier Stages," US Patent 7554395, issued June 30, 2009.

[9-17] E. McCune, "Envelope Modulator Saturation Detection Using a DC-DC Converter," US Patent 7702300, issued April 20, 2010.

[9-18] R. Booth and E. McCune, "VSWR Normalizing Envelope Modulator," US Patent 8331882, issued Dec. 11, 2012.

10 Device technologies: special issues for DPS use

Many different transistor technologies exist, each with its place and purpose. To date, there is no transistor technology specifically designed for DPST use. It is a valuable exercise then to evaluate the suitability of the transistor technologies that are available for use in a DPST RF power amplifier design. Doing just this is the purpose of this chapter.

Chapters in Part I relate the properties of various DPST architectures to features in examples of transistor characteristic curves. The extension of these to actual transistor characteristic curves is particularly important, because if the selected transistor technology does not have the needed features in its characteristic curves, then the DPST project is immediately doomed. It will never work properly simply because the transistors chosen will not support the necessary performance features. To avoid this catastrophe, it is necessary to understand what each available technology can support with respect to the necessary performance features.

This extensive chapter is very important because it draws together for the first time all of the DPST relevant characteristics of each commercially available RF transistor technology in one place. Each technology is comprehensively represented by the same 12 measurement sets to allow comparison of their relative behavior. More importantly, the applicability of each transistor technology – or not – to the needs of DPS transmitter designs are evaluated.

With this comprehensive data now available, there are two primary objectives:

- selection of the most appropriate transistor technology can be made in an informed manner to enhance success of any particular DPS transmitter design project, and
- transistor manufacturers can review the present states of the art shown here to further improve the applicability of their technologies within the growing importance of DPS transmitters.

This chapter begins with establishing a transistor characteristic wish-list from the DPS transmitter designer's point of view: in an ideal world what the transistor technologies would be like to make the DPS transmitter design projects easy to complete and put into stable production. Following this foundation is a comprehensive survey and review of ten different RF transistor technologies in three different materials: silicon (5 types), gallium arsenide (GaAs) (4 types), and gallium nitride (1 type). Comparisons among these are made following the technology survey.

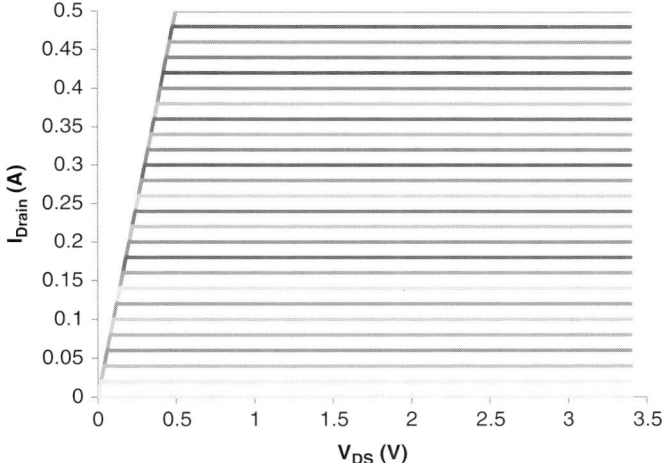

Figure 10-1   Ideal characteristic curves for a transistor (FET shown here) for application in a DPS transmitter. Different equally spaced values of V_{GS} set different equally spaced values of drain current, and the transition from current source operation to resistive operation is instantaneous.

10.1   What do we really want?

The ideal practical transistor has a fixed nonzero resistance when ON, and acts as an ideal current source when the applied voltage exceeds the voltage from Ohm's Law for the ON resistance value and the current control setting. Additionally, the transition between current source operation and resistive characteristic curves is instantaneous. One example of transistor characteristic curves that meet these objectives is shown in Figure 10-1: one curve for each step in the transistor control parameter. In this figure, the nomenclature for an FET is shown, so the control parameter is gate-source voltage. These ideal curves do not change if the transistor is bipolar, though the nomenclature does change to collector-emitter voltage, collector current, and with the control parameter becoming base current.

10.1.1   IV characteristic curves

The importance of tying everything to transistor characteristic curves is that these are a property of the transistor itself, and not of the circuit that transistor is used in. While it is possible to build circuits that improve on individual device characteristics (for example, reduced output conductance of a cascode configuration), any circuit that needs to do so adds complexity and likely increases cost somewhere. It is best to get as close as possible with the base transistor technology.

One aspect that is more important to polar operation than envelope tracking is if there is a voltage offset between where the linear projection of the transistor resistive characteristic intersects the voltage axis and the IV plane origin. There is no offset

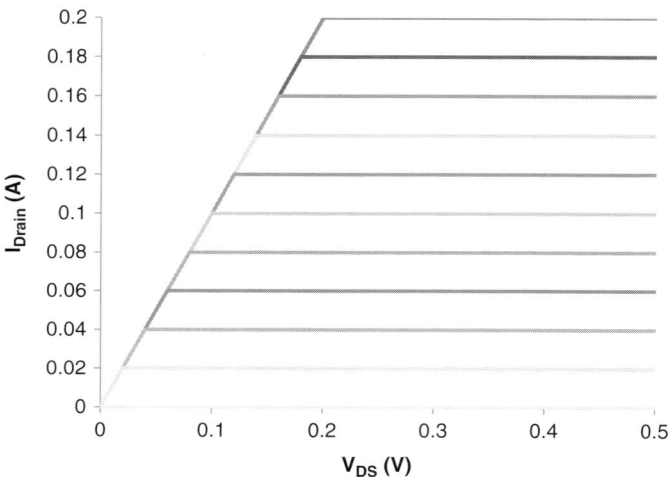

Figure 10-2   Idealized near-zero behavior of the characteristic curves from Figure 10-1.

present in the curves for this idealized transistor, evident in the close-up view near the origin in Figure 10-2.

Envelope tracking cares greatly about how horizontal the CCS region curves are. In both Figures 10-1 and 10-2, the idealized CCS region curves are perfectly horizontal. This detail is unimportant to a DPS design using polar operation.

Both techniques care about how straight the resistive characteristics are, but for different reasons. Envelope tracking must provide a power supply voltage high enough to keep the transistor in its CCS region, and a straight resistive characteristic makes this easy to predict. Polar transmitters operate along the resistive characteristic, so any deviation from straightness is a distortion mechanism.

Sensitivity of the device current to changes to the supply voltage is called power supply sensitivity (PSS). Envelope tracking ideally needs this to be zero, and in contrast polar modulation requires considerable power supply sensitivity. PSS is calculated as the first derivative of each characteristic curve, results of which can be collected together into a surface over the control parameter and the applied voltage. The power supply sensitivity for the ideal transistor from Figure 10-1 is provided in Figure 10-3(a).

Envelope tracking optimally operates very close to the transition from low to high power supply sensitivity, staying in the low PSS region. One example is shown in the close-up view in Figure 10-3(b). The offset into the PSS = 0 region must exceed the amount of power supply variation expected from the ET operation, plus the maximum difference between the DPS output and the actual output envelope. Energy efficiency is highest when this offset is small, so there is a trade-off between the tolerance of power supply inaccuracies and the overall DPST ET energy efficiency.

Absolute resistance of the transistor (the conduction (channel) resistance) is the ratio of applied voltage to the current drawn, equal to the slope of the line segment between

(a)

(b)

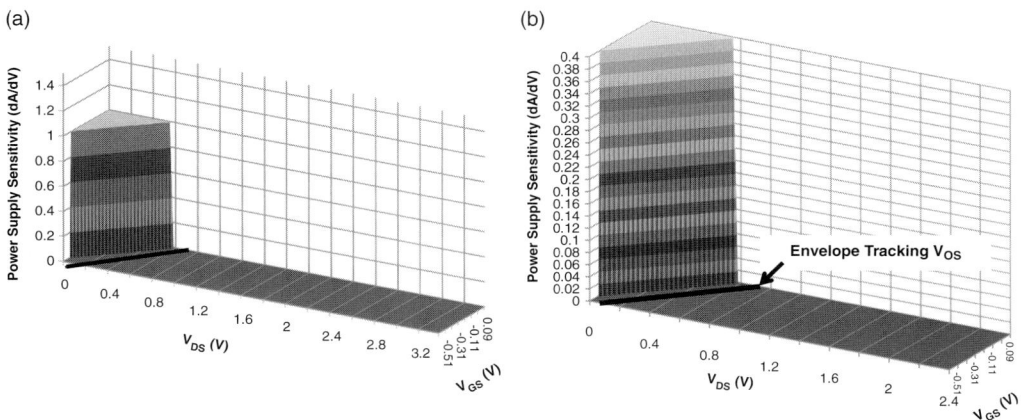

Figure 10-3   Power supply sensitivity for the idealized transistor: (a) overall characteristic; (b) close-up view of the transition region showing an optimal design for the ET voltage offset (V_{OS}). See Section 5.4.2.

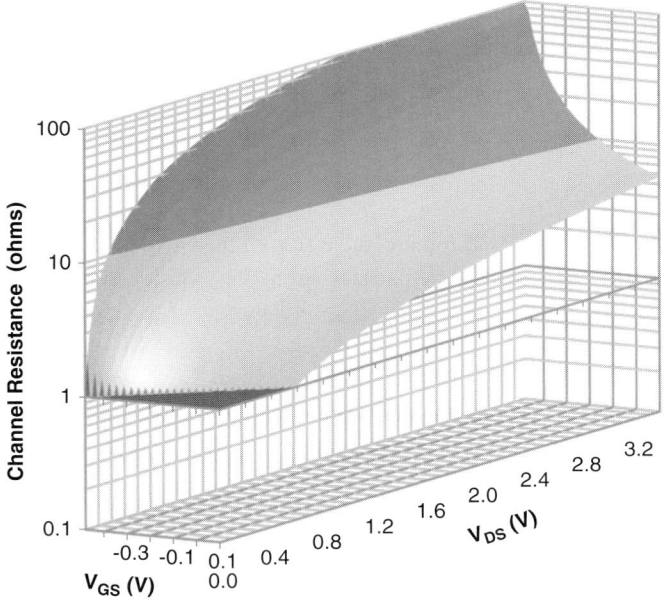

Figure 10-4   Channel resistive characteristics evaluated for the idealized transistor of Figure 10-1.

the instantaneous operating condition and the I–V plane origin. For the ideal transistor characteristics of Figure 10-1, the channel resistance evaluates to the surface shown in Figure 10-4. Channel resistance depends on both the device voltage and the control voltage (for an FET) until the transistor gets into its resistive characteristic. Then, for this transistor at least, the channel resistance remains fixed (and normalized to 1 ohm).

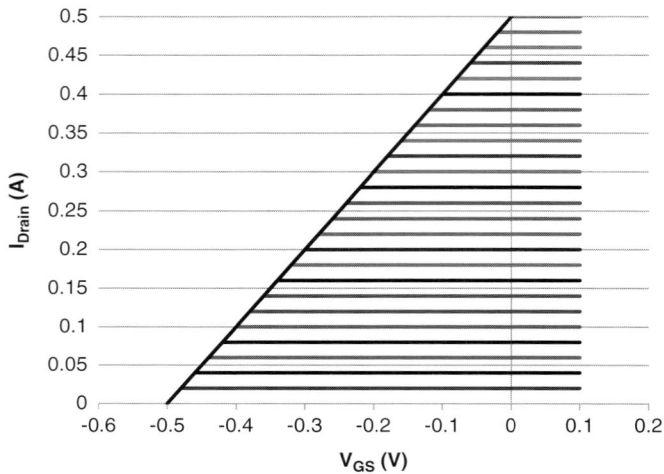

Figure 10-5 Idealized transfer curve set for this idealized transistor. One curve is shown for each different value of device supply voltage in the evaluation.

DP operation needs to have this ON resistance constant, because any variation in the ON resistance can be a signal distortion mechanism. The precise value of the ON resistance is not usually critical, as long as the conditions defined in Section 6.7.5 are satisfied.

10.1.2   Transfer characteristic and g_m

Circuit design for bias and RF drive depends on the transistor transfer characteristic, shown in Figure 10-5 for this ideal transistor. Conduction for this device begins at a negative value of V_{GS}, signifying that this particular example device is a depletion mode FET. It is important that this transfer characteristic is the same for any value of applied voltage within the CCS region when a low distortion linear amplifier for envelope tracking is needed: both at the threshold of where conduction begins, and the shape of the curve describing how conduction increases. These must be measured, because this behavior is not guaranteed for any transistor type.

To operate well as a switch for polar DPST operation, once the transistor enters the resistive region the device current must become a constant that depends solely on the device voltage (here V_{DS}). This behavior is seen in Figure 10-5, which should be compared to Figure 8-1.

The slope of the transfer characteristic is the FET transconductance (g_m). Evaluating the transconductance for this ideal transistor provides the surface in Figure 10-6. It is closely related to the PSS surface of Figure 10-3, though with opposite characteristics. Here if changes in the control signal cause no changes in the device current, then $g_m = 0$, which is true for this idealized transistor when it is operating in its resistive region.

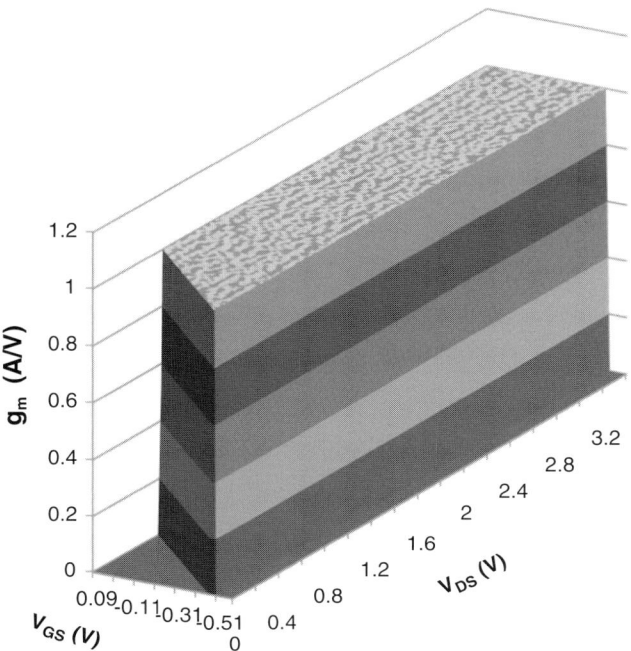

Figure 10-6 The CCS (g_m for FET, β for bipolar) characteristic for this idealized transistor.

Amplifier linearity is strongly dependent on how flat the nonzero g_m surface is. For this idealized transistor, the g_m is very consistent, evident in Figure 10-6. The voltage gain is the product of g_m and the load resistance R_L. Since R_L is fixed, the larger the value of g_m, the greater the amplifier voltage slope gain. High gain corresponds to small input signals for a given linear output signal, and therefore to small leakage signals at the output. And the driver stage design is easier.

10.1.3 Parasitic and installed capacitance

Capacitances are always present, and are a problem for broadband transmitter designs. The impacts of device capacitances are discussed in Sections 6.8.5, 6.11, 7.2, 7.3, and 8.4. Details presented in these sections are summarized here as part of the groundwork for examining the various transistor technologies.

Different instances of parasitic capacitances have differing effects. Device input capacitance C_{in} figures prominently in the device speed metrics f_T (8.15) and f_{MAX} (8.16). Resonant effects limit bandwidth due to the conjugate impedance matching used in input matching networks (Section 7.3). Additionally, input capacitance in bipolar devices shunts input signal current away from the transistor "beta generator" action, and effectively reduces transistor gain at high frequencies through this effect.

Output capacitance also incurs resonant effects that limit amplifier bandwidth due to the conjugate impedance matching used in any OMN (Section 7.3). Output capacitance is also the limiting factor for the transistor switching speed in the absence of resonant output networks (Section 8.4), which is an appropriate design situation when the transistor is switching (Section 8.7.1).

Reverse transfer capacitance C_μ contributes to stability issues in linear amplifiers because it provides a feedback path from output to input. Additionally, through the Miller multiplicative effect this capacitance effectively increases the apparent input capacitance of the device while it is operating as an amplifier. In applications when the amplifier is switching and its slope gain is zero, C_μ provides a feed-forward path that couples some of the input drive signal to the switching stage output (Section 6.11). This results in distortions in both the magnitude and phase of the switching output signal, particularly at lower output magnitudes.

Beyond these device parasitic capacitances, installed capacitors added for RF bypass and low frequency stability (Section 7.2) shunt signal currents from their attached circuit nodes. In the case of RF bypass, this is exactly the desired effect. In the case of low frequency stability, this shunts current from the DPS away from the amplifier, which reduces achievable PA energy efficiency.

10.1.4   FET-action regions

The ideal FET device model of Figure 6-23(a) has no input conductance. For most of the FET devices used in RF power amplifier applications, there are bounds on the gate-source voltages where this model applies. The input voltage span within these bounds is called the FET-action region for each FET type. This FET-action region is the V_{GS} range between when the channel conduction begins and when gate conduction begins, shown in Figure 10-7.

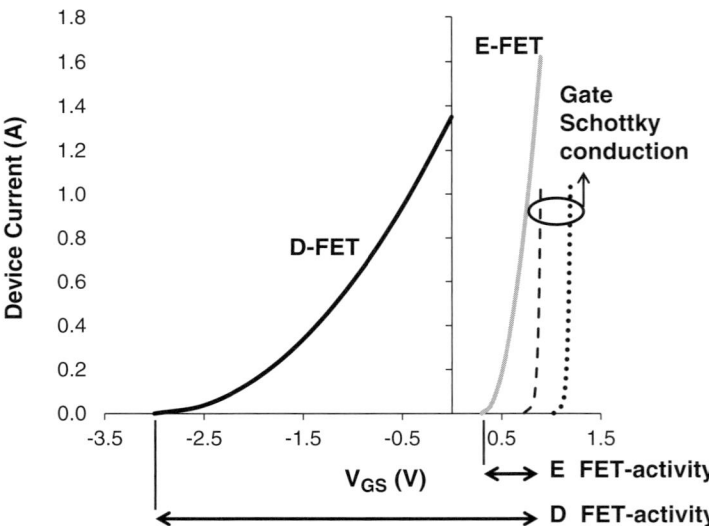

Figure 10-7   FET-action regions are the ranges of V_{GS} between the onset of drain current and the onset of gate current.

For many FET technologies, this gate conduction is the result of forward biasing a diode, either junction or Schottky. The value of the diode forward voltage is completely dependent on the technology used, and usually varies from 0.7 V for silicon junction diodes to around 1.3 V for GaN Schottky gate diodes.

Not all FET technologies have a diode conduction path, and in these technologies the upper limit of the FET-action region is often a destructive process of the gate. Silicon MOSFETs are one such technology.

10.2   Linearity and noise suppression for envelope tracking

Chapters 5 and 7 show how amplifier linearity is critical for any successful ET transmitter. It is of particular importance to satisfy (7.1) to ensure independence of the output signal from the value of the PA power supply.

Within the CCS region, amplifier linearity naturally follows from a consistent g_m value. The converse is also true: amplifier nonlinear behavior follows from a varying g_m characteristic. Within the transistor survey data, a transconductance surface (or β surface for bipolar transistor types) is calculated from each set of characteristic curves to pictorially represent the inherent linearity, or lack of it, shown by the core transistor.

10.2.1   Transconductance (or β) uniformity

From Section 10.1.2, a linear amplifier requires that its transistor transconductance (for FETs) or beta β (for a bipolar) must be constant across the values of the input controlling signal. In general, this is not true for any transistor, though some transistor types get closer than others. To the extent that the transconductance (or β) is not uniform across the applicable range of the input signal, there will be distortion of the amplifier output signal.

Cancellation or minimization of this distortion can be achieved using both analog and/ or digital means. In an in-line linearizer, the analog or digital process is designed to pre-distort the amplifier input signal in such a way that the amplifier inherent distortion cancels this pre-distortion, giving the desired (apparently linear) output. This assumes that the transistor distortion process is well known and unvarying. Neither is usually true, though variations of the distortion process are often more of a problem than understanding what the distortion process is.

A more modern tendency is to make the pre-distortion adaptive to the specific transistor behavior that it is attached to. Almost always, this adaptive pre-distortion is used with a digital process called adaptive digital pre-distortion (A-DPD). Whatever the actual means used, any adaptive process must measure the actual output from the amplifier to determine what (if any) changes to the pre-distortion are needed. Thus, a local precision receiver is always required, tasked to make these measurements within the needed accuracy.

10.2.2   Extent of the CCS region

Linear amplifier operation only occurs within the L-mode region (Section 4.1.6), where the transistor operates as a controlled current source (CCS). Ideally, the output signal is not sensitive to the value of the power supply in this region, meaning the power supply sensitivity (PSS) is zero. This is almost never true, so a more practical definition for the boundary of the L-mode (CCS) region is needed that is specific only to the transistor and not impacted by any circuit within which it is applied. One useful and measurable CCS boundary definition within the spirit of (7.1) is based on the transistor PSS surface. CCS operation is defined to be where the transistor PSS falls below 1% of its peak value for that transistor

$$\frac{PSS}{PSS_{MAX}} \leq 0.01. \tag{10.1}$$

The device voltage profile for which (10.1) holds is the ET optimum knee voltage. Operating the transistor at voltages above this knee profile guarantees that the output signal envelope is at least 99% independent from the actual value of the power supply. Equivalently, this ensures that while operating within this region, any signal envelope modulation process due to variation of the power supply is suppressed by 40 dB.

From Section 5.4.3, the minimum useful power supply voltage for an ET design using this CCS region depends on both the bias trajectory selected, and the device onset of P-mode. Staying within transistor-specific limits here, only the onset of P-mode is taken for the lower limit. The example in Figure 7-2 demonstrated how this physically operates. This lower limit is also readily derived from the transistor PSS surface because the power supply sensitivity in P-mode is very high. Therefore, any algorithm analyzing the transistor characteristics satisfying (10.1) also naturally avoids P-mode and finds this lower power supply limit.

The upper limit of the L-mode region is usually set by two physical processes. One is voltage breakdown of the transistor structures, which is usually destructive to the device. Another upper bound is excessive power dissipation as the amplifier operates far from both axes in its IV characteristics.

10.2.3   Transition between resistive and CCS regions

The knee voltage for any transistor is the boundary of the CCS region where (10.1) begins to be satisfied. For this voltage to be small, it is essential that the transistor rapidly transition from its resistive operation to CCS operation. This transition is particularly obvious on the PSS surface for the transistor.

Characteristic curves with gradual curvature between resistor and CCS behaviors have a gradual transition between these regions. Tighter curvature is needed in this transition region to allow for a faster transition. For the idealized transistor of Figure 10-1, the curvature is infinite to provide the instantaneous transition from resistive to CCS behavior.

Chapter 5 demonstrates that the fundamental tenets of ET require the transistor to be operating in the CCS region. Chapter 6 then shows that the fundamental tenets of polar operation require that the transistor operates in the resistive region. Thus, this transition region in the transistor characteristic curves between resistive and CCS regions directly corresponds to the transition between polar modulation and ET operation.

10.3   Switching characteristics for DP

Discussions in Chapters 4, 6, and 8 showed that the switching operation from C-mode, necessary for polar modulation, has no suppression of any noise on the power supply. This key result follows from (6.13). This is a direct violation of the envelope tracking core principle (7.1), pointing out that envelope tracking and polar operation are not the same thing. Indeed, the presence or absence of suppression of power supply noise is a very useful test to determine which operating mode the DPST is performing in.

10.3.1   Resistive characteristics

Resistive operation for any transistor is when the current through the device depends on the voltage applied across the conducting path (channel). This current dependence is preferably linear, but it often is not. But the transistor can be considered a switch only when it operates within this resistive region. The lowest value of resistance achieved by any transistor is called its ON resistance.

All transistors exhibit a controlled resistance region when the voltage across the device is small. Results in this technology survey show that in general the voltage controlled resistance behavior of FETs has linear resistance characteristics even though the control of that resistance is not a linear transfer function. Bipolar transistors have current controlled resistances at low device voltages, but these resistances are linear only for a segment within this region. The inherent voltage offset $V_{ce,sat}$ of bipolar transistors distorts the resistance value at very low device currents.

10.3.2   Voltage offset

There is no universally agreed definition of what the transistor voltage offset is, apparently largely because this offset has different manifestations in higher-level circuit behavior. One obvious definition is simply the voltage region in Figure 10-8, where the device current remains zero (actually nonpositive).

For evaluation of polar modulation accuracy, the projection of the linear resistive segments on to the voltage axis is a more appropriate metric. For the transistor example shown, this set of five example projections results in four different values of transistor voltage offset. Each of these needs to be individually identified if current controlled resistance is used to set the value of the transistor resistance to be used. Otherwise, just one case is selected and only its linear segment projection is used.

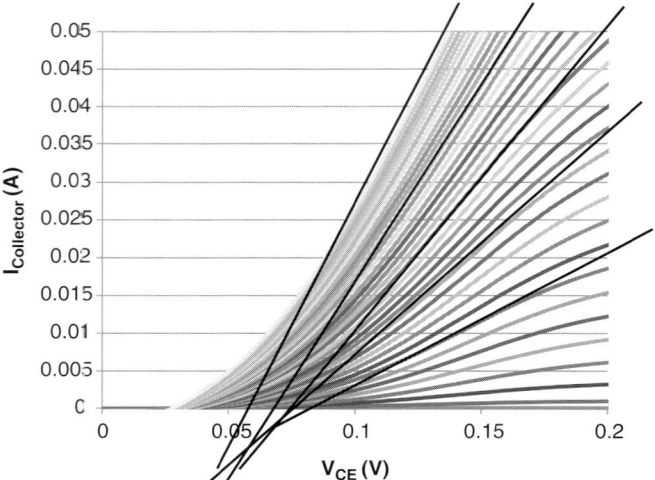

Figure 10-8 Voltage offset estimation from projection of P-mode resistance values does not produce a unique value. A finite range of voltage offset values does result from this measurement method.

10.3.3 Comparisons across switching transistor technologies

Evaluation of FET devices for use as high frequency switches (not as switches for high frequency signals, which is a different topic altogether) across various technologies is introduced in [9-5]. In this work, the focus is on driving the switch, which emphasizes the input characteristics of the FET.

There is a figure of merit used widely by the switching power supply community that relates the minimum channel resistance (R_{ON}) to the total amount of gate charge (Q_G) needed to form the channel to that resistance

$$FoM \equiv Q_G R_{ON}. \tag{10.2}$$

This FoM is useful for comparing transistors amongst themselves, as long as all other parameters are unchanged. When design of driver circuits is needed, this FoM is not much help because of the charge relationship

$$Q = C \cdot V \tag{10.3}$$

or more accurately

$$Q(V) = C(V) \cdot V, \tag{10.4}$$

which emphasizes the nonlinearity of this process. This relationship says that the required charge may be from high voltage on a small capacitance, or from a large capacitance needing only a small voltage. Driver design is a voltage-based process making it important to resolve this question.

The new FoM is a modification of (10.2), adding a factor accounting for differing threshold voltages and gate enhancement

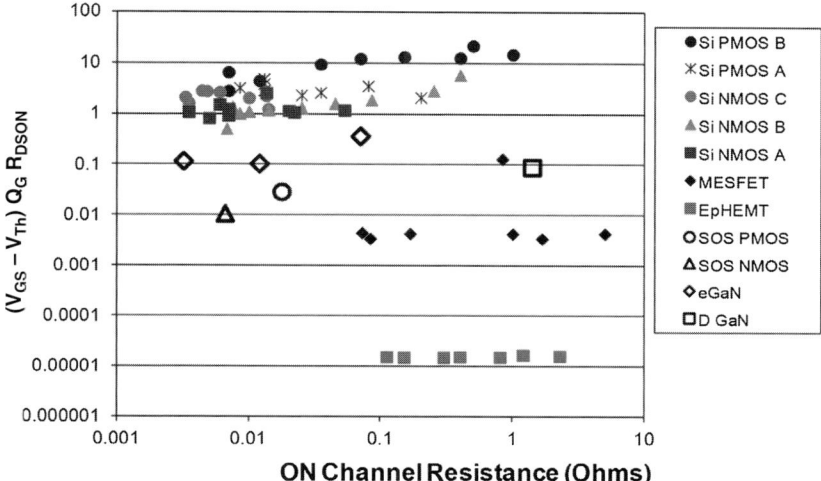

Figure 10-9 FET FoM chart comparing 11 different transistor technologies and manufacturing process variants for ease of high frequency switching use. Low values are best on both axes.

$$FoM_{FET} \equiv (V_{GS} - V_{Th}) \cdot Q_G R_{ON}. \qquad (10.5)$$

Lower values are desirable for FoM_{FET}. Achieving low values emphasizes parts with low drive voltage requirements, smaller gate charge amounts needed to form (or close) the channel, and achieving lower values of ON resistance. Units are [V^2 sec], though it is not at all clear if this has any direct physical meaning. One major benefit of (10.5) is that it is applicable across all FET technologies [9-5], allowing comparison of the various technologies for ease of use in switching applications. Comparison of 11 different FET technologies is presented in Figure 10-9. Silicon MOSFET transistors have the highest values of FoM_{FET}, and the lowest evaluated FoM_{FET} values are for GaAs EpHEMT transistors. Present GaN HEMT devices evaluate between these values.

10.4 Transistor technology survey

Projects can be immediately lost, doomed-from-the-start if you will, if the wrong technology is selected to implement the DPST. This does not imply that selecting an appropriate technology guarantees success, since this requires significant high-quality engineering in using the technology properly. With all of the transistor technologies that are available to designers, it is of vital interest to understand how each one performs with respect to the principles described in earlier chapters.

The remainder of this chapter is dedicated to just this type of comparison. All 11 of the evaluated transistors are n-type, be they n-FET or NPN bipolar. Taking the important transistor parameters identified in Chapters 4 through 8, the list used here for this mutual comparison is:

- IV characteristic curves,
- transconductance surface,
- ON resistance surface,
- transfer relation,
- PSS surface,
- device capacitances,
- knee voltage profile,
- envelope tracking bias profile,
- envelope tracking minimum DPS profile,
- DPS value changes along the transition between envelope tracking and polar modulation,
- signal envelope accuracy with polar modulation.

For the final five evaluations, the results from the device measurements are used to examine the DPST specific properties. These include evaluation of the optimum DPS voltage profile for the ET operation, where the optimum is defined as providing −40 dB power supply sensitivity (99% power supply noise suppression) – or the lowest PSS that the particular transistor will support. From this optimum DPS profile, changes to that profile that provide higher PSS are shown as the ET design is pushed to provide higher energy efficiency. This occurs as the ET operation converts to polar operation, improving efficiency as linearity is suppressed and sensitivity to power supply noise increases. A chart is also provided for each transistor type, showing how accurately it allows control of the output envelope in polar operation. Details for each of these analyses are provided in Chapter 13.

Transistor types are described generically in this technology comparison. In a few instances, the specific part numbers for the evaluated transistors are provided when this aids understanding of the differences being pointed out. In all other instances, the data are representative of the general class of transistors for that particular material and type.

10.5   Silicon technologies

Silicon transistor technologies are by far the most widely used around the world. This provides economies of scale that makes these technologies among the lowest cost options available. The efforts that have driven development of silicon technology following the establishment of "Moore's Law" in 1965 now provide integration capabilities approaching a billion transistors/mm^2 of die area. Such transistors are tiny and not individually useful for RF transmit powers. The ability to use large numbers of these to digitally assist the large RF power transistors is a unique capability of silicon technology.

Presentation of the various silicon transistor types is roughly in the order they were developed historically. The oldest type is the bipolar junction transistor. Transistor types continue to be developed, and now the new silicon MESFET is preparing to join this venerable pantheon.

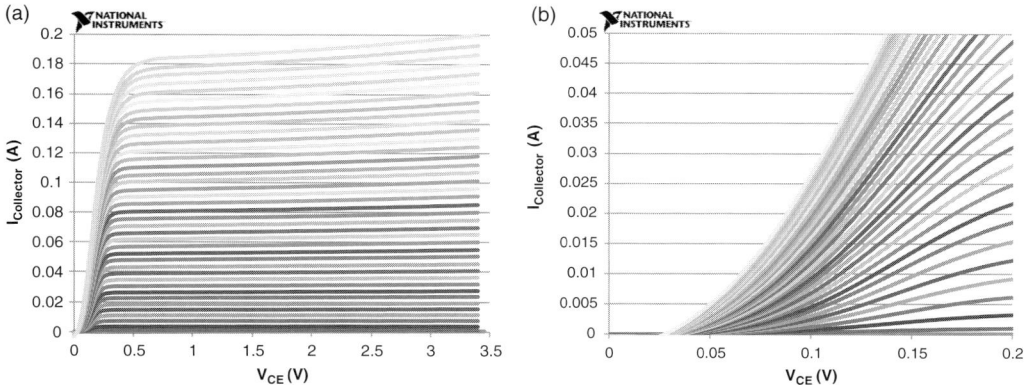

Figure 10-10 Characteristic curves for the MS1649 silicon bipolar RF power transistor: (a) measured curve family with many steps of base current; (b) close-up view of these curves near the IV plane origin.

10.5.1   Si bipolar

The original transistor type is the bipolar junction (BJT), which has evolved greatly over the past 65 years from its original point-contact form. These devices remain in wide use and are now available in a large variety of types. Characteristics of these transistors do vary widely. Two examples of the many available in production today are evaluated here, selected because of their different characteristics.

The MS1649 is an NPN RF power transistor designed for grounded-emitter class C operation to provide 3 W at 470 MHz. Its characteristic curves are provided in Figure 10-10.

The MS1649 characteristic curves in Figure 10-10(a) do look very similar to the ideal presented in Figure 10-1. Both the resistive region and controlled current source (CCS) region are readily evident. The transition between these regions is fairly rapid. Output conductance is low but not zero, as there is a slope to the curves in the CCS region which increases at higher device currents.

The close-up view near the origin has major differences with the ideal goal of Figure 10-2. The voltage offset before conduction begins, typical in bipolar transistors, is clearly evident, here at a value near 30 mV. Additional collector voltage is needed to get the transistor channel resistance to its minimum value, resulting in a curvature (nonlinearity) in the device current. Curves for each of the applied values of the base current are distinct, signifying that at these low transistor voltages the MS1649 behaves as a current-controlled resistance. The most appropriate first-order device model is that of Figure 4-29, showing that this transistor is in P-mode operation under these conditions.

From this characteristic curve, the surfaces for transistor current gain β and absolute channel resistance are calculated. These results are presented in Figure 10-11. The β surface in Figure 10-11(a) rises quickly as the transistor turns on, then becomes planar. The β plane is tilted, indicating that the characteristic curves are wider apart as the base current (and collector current) increases. Shaping on the channel resistance

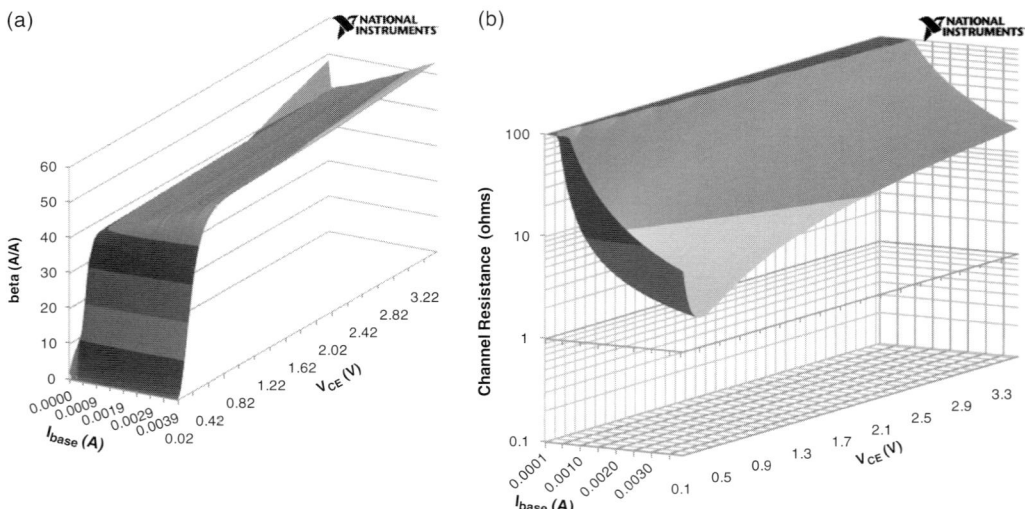

Figure 10-11   First-order analysis results from the MS1649 characteristic curves: (a) current gain β surface; (b) transistor absolute resistance surface.

surface shows that the minimum value is offset from 0 V across the device, which continues to decrease as the transistor drive increases. This shows that this transistor does not behave as a good switch.

Another view of the transfer characteristics is presented in Figure 10-12, which provides insight into the switching behavior of this transistor – or more accurately, the lack of it. Compared to Figure 10-5, here there is no horizontal aspect to any of the curves. This means that in all cases as more current is supplied into the base, more collector current flows. Within this set of conditions, this transistor never actually switches into a fixed ON state.

The PSS surface for the MS1649 is shown in Figure 10-13. Its shape looks like a sail, or a fin, which identifies the resistive region of the transistor characteristics. The peak of this surface is offset from zero, showing the resistive behavior offset seen in Figure 10-10(b). Transition from this resistive region to the CCS region at the base of this fin is rather rapid. This signifies that the transistor does not need much voltage across it to behave as a current source. We therefore expect that the analysis of its optimum knee voltage from (10.1) will return a low value.

Parasitic capacitances for the MS1649 are measured and shown in Figure 10-14. The input impedance measurements in Figure 10-14(a) show a widely varying capacitance which is actually an artifact of resonance at 25 MHz as the base current increases past 3 mA. At higher drive, the capacitance shows a negative value, which means that it is inductive under these conditions. Output capacitance C_{CE} acts like a varactor diode, changing its value by just over 2:1 across these operating voltages.

This transistor is evaluated now for its DPST performance capabilities. Using the process from Section 7.6, this transistor is found to meet the definition of knee voltage

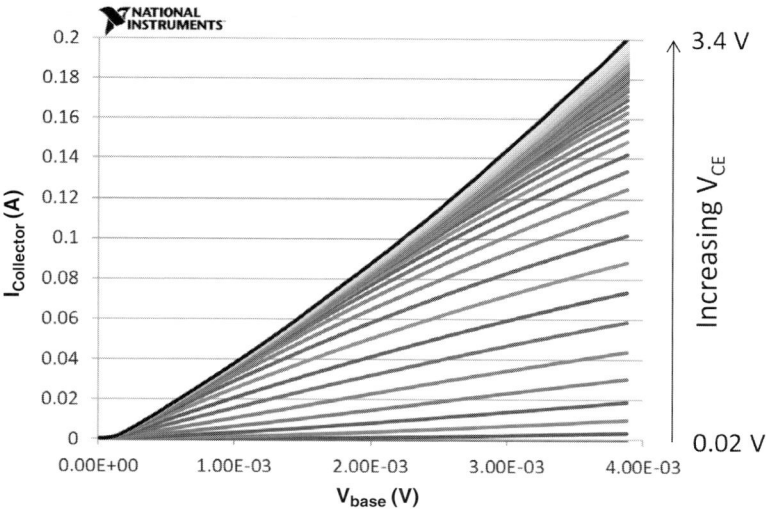

Figure 10-12   The set of current transfer functions corresponding to Figure 10-10(a). Gain is proportional to the slope of these curves, and linearity is related to their straightness. This transistor has a small offset at turn-on and shows gain expansion at the higher base drive.

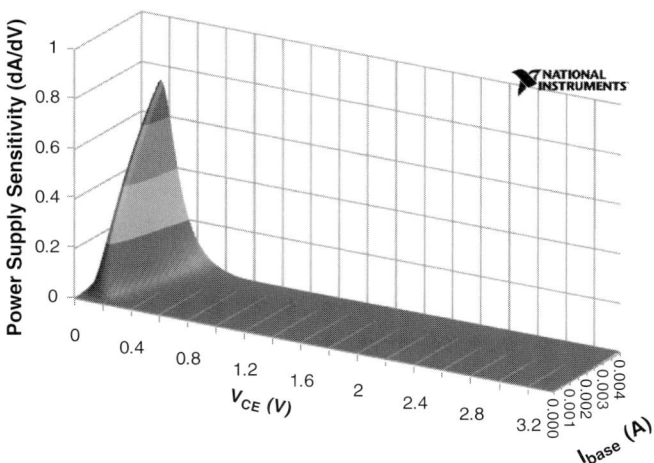

Figure 10-13   Power supply sensitivity surface for the MS1649. Both a rapid transition from resistive to CCS regions and a voltage offset of the resistive region are evident.

(10.1), so true envelope tracking DPST operation can be realized. Figure 10-15 shows the calculated boundary of the CCS region above which (7.1) is satisfied. Also shown in Figure 10-15 is the dynamic bias trajectory needed for an ET PA using this transistor to maintain both operating linearity and to maximize the ET dynamic range. Note that there is an offset in the reported minimum supply voltage.

If higher sensitivity to the actual power supply value is acceptable, the effective knee voltage can be reduced. For this transistor, the power supply sensitivity changes rapidly,

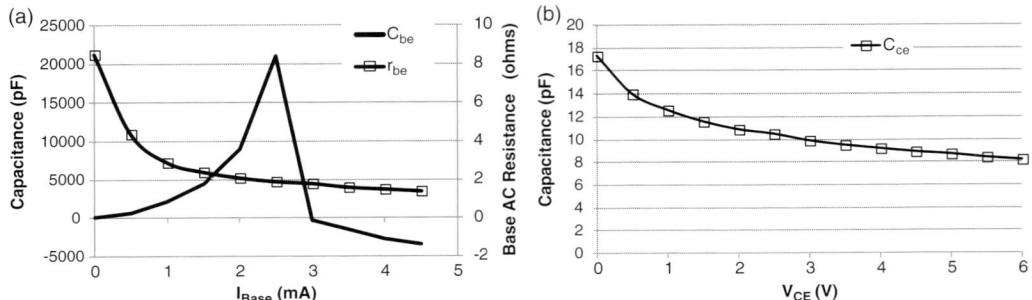

Figure 10-14   Measured parasitic capacitances for the MS1649 at 25 MHz: (a) input, which resonates at 3 mA of base current and is inductive at higher drive; (b) output capacitance varies by about 2:1 across these voltages.

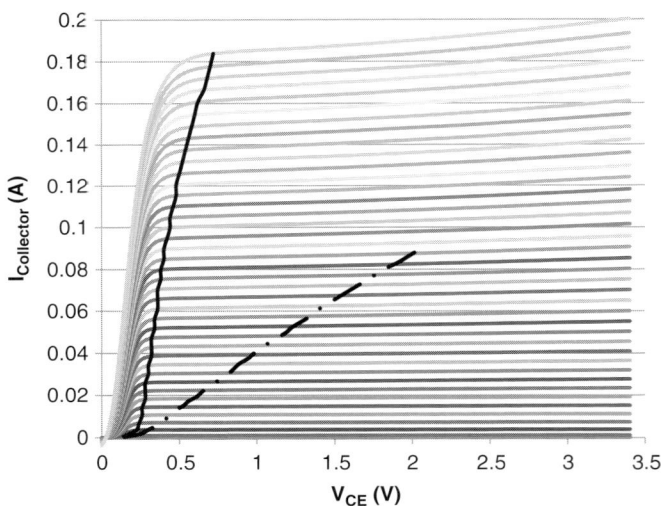

Figure 10-15   ET operation region for the MS1649 silicon bipolar transistor. The boundary for the CCS region meeting (10.1) is identified by the solid line. The dash-dot line shows the dynamic bias profile this transistor must follow to simultaneously maintain amplifier linearity and maximize the power control dynamic range.

shown in Figure 10-16(a). This begins the transition from envelope tracking to polar operation, which completes when the power supply matches the signal envelope as shown by (6.13). PA energy efficiency improves across this transition, due to the transistor operation moving closer to the current axis, as seen in Figure 10-16(b). But to be very clear – none of this extended operation is true ET as it all violates the ET core principle (7.1).

From the closeness of the three curves in Figure 10-16(a), it is apparent that the transition away from envelope tracking and toward polar operation for this transistor is very rapid. It is not a uniform intermingling, but actually here is a rapid gradient. Seeing the sharp definition of the resistive region in Figure 10-13 does support this result. Pushing all the

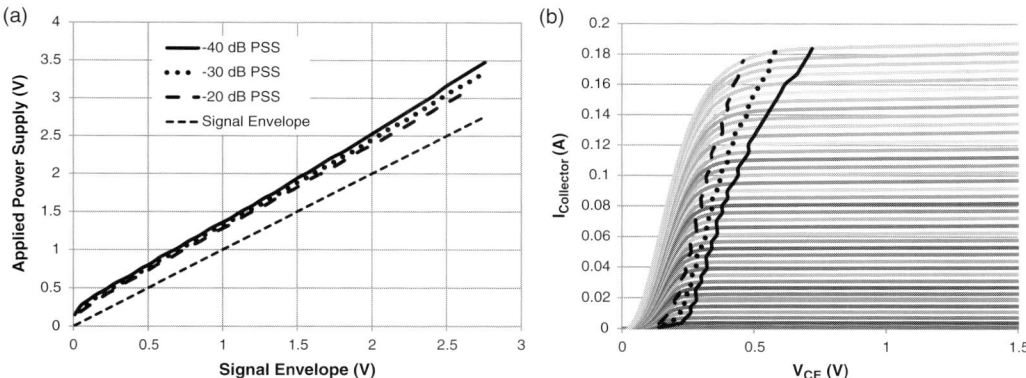

Figure 10-16 ET characteristics for the MS1649: (a) DPS voltage profiles for power supply noise suppression of –40 dB, –30 dB, and –20 dB; (b) corresponding ET knee voltage profiles for these three values of power supply noise suppression.

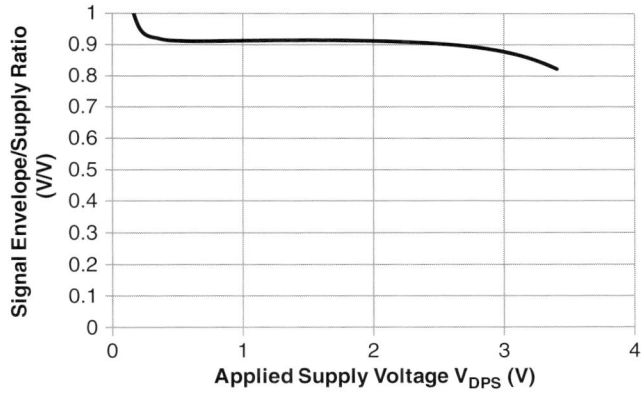

Figure 10-17 Envelope control precision and accuracy for the MS1649 when operating in polar DPST operation. Polar envelope control is accurate over a dynamic range of about 16 dB. Distortion from curvature in the characteristic curves at low voltages (Figure 10-10(b)) is evident. Best-fit is achieved with $V_{AMO} = 0.05$ V.

way to polar operation by determining the best-fit triangle as shown in Figure 8-22 for implementing (2.9) obtains the polar modulation accuracy curve in Figure 10-17. At the low end, distortion due to curvature of the transistor characteristics in Figure 10-10(b) is evident. At higher supply voltages, there is distortion due to the resistive characteristics transitioning toward CCS operation. Total polar modulation envelope dynamic range with this transistor is about 16 dB.

A different style of bipolar RF power transistor is represented by the MRF586. The MRF586 is an NPN RF power transistor designed for grounded-emitter operation providing 1 W around 300 MHz. Its characteristic curves are provided in Figure 10-18.

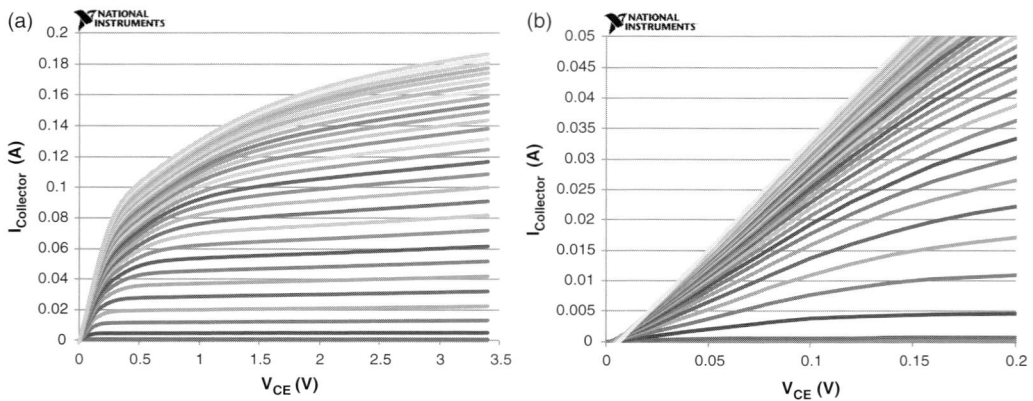

Figure 10-18   Characteristic curves for the MRF586 silicon bipolar RF power transistor: (a) measured curve family; (b) close-up view near the IV plane origin. Note the differences from Figure 10-10.

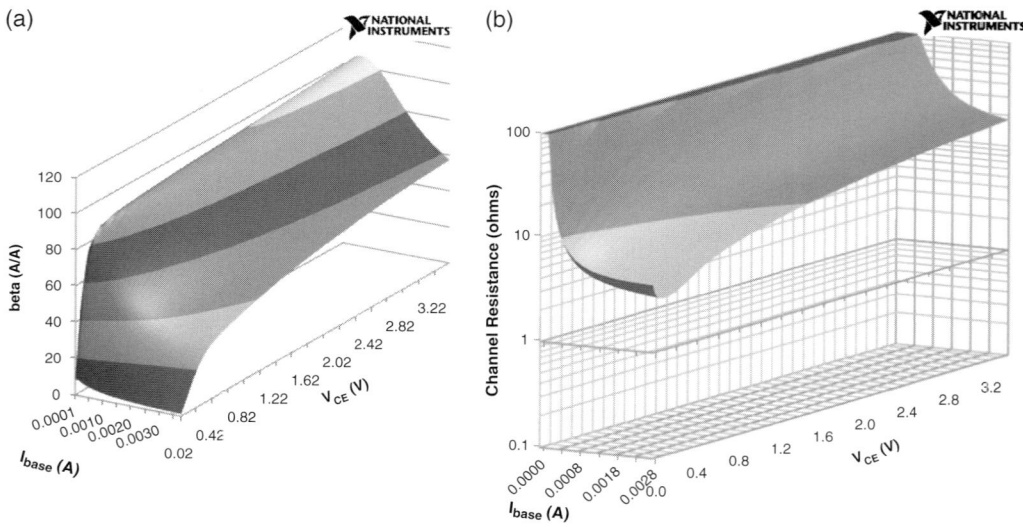

Figure 10-19   First-order analysis results from the MRF586 characteristic curves: (a) current gain β surface; (b) transistor absolute resistance surface.

The characteristic curves in Figure 10-18(a) do not closely resemble the idealized goal from Figure 10-1. Yet the close-in curves in Figure 10-18(b) are closer to the ideal of Figure 10-2. Hence, the MRF586 illustrates a different trade-off in transistor selection for DPST use.

In Figure 10-19(a), the transconductance surface is never constant. The β variation across the base current immediately demonstrates that without additional circuit design this transistor is not inherently linear as an amplifier. Conventional degeneration (negative feedback) circuit design techniques are a big help. A different conclusion is reached when examining the Figure 10-19(b) variation across device voltage V_{CE}, where the

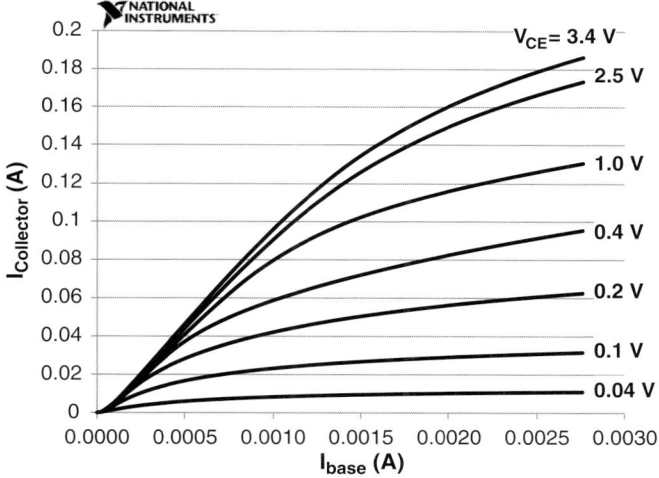

Figure 10-20   A subset of current transfer functions corresponding to Figure 10-18(a). Gain is proportional to the slope of these curves, and linearity is related to their straightness. This transistor has a very slight offset at turn-on and shows gain compression at higher base drive.

performance is more uniform. We can expect that once an acceptably linear amplifier is designed, that linearity should remain quite constant as the power supply varies. This is good for anticipated ET use.

The surface of channel resistance in Figure 10-19(b) shows the resistance minimum at an voltage offset similar to that seen in Figure 10-11(b). Here this minimum is neither as pronounced nor as deep as that for the MS1649. This easily illustrates both the lower slope of the transistor resistive region characteristics in Figure 10-18(b), and the reduced curvature of these same characteristics as channel (collector) current goes to zero.

Analyzing the transfer functions of the MRF586 with varying V_{CE} gives a selected set of curves in Figure 10-20. All of these curves have curvature toward becoming horizontal, meaning that this transistor is starting to behave as a switch across this set of conditions. This is quite different than the behavior seen in Figure 10-12. At least when the applied voltage is very small, this transistor closely approximates a switching ON characteristic.

Looking at the PSS surface for this transistor, several differences and similarities are apparent from that of Figure 10-13. The resistive region is located at the lower operating voltages, and its peak value is at an offset away from 0 V. The transition from the resistive toward the CCS region is at first distinct, but then slows to a more gradual slope. Even at the highest tested voltage, an offset above zero of this PSS surface is seen at higher drive current conditions. Quality of this CCS region is suspect, and is quantified in later analyses.

Parasitic capacitance measurements for the MRF586 are presented in Figure 10-22. These behave very similarly to the data in Figure 10-14, although here the test frequency is four times higher. An input resonance occurs near a base current of 3.5 mA. At higher drive levels, the input impedance is inductive. Output capacitance C_{CE} is less variable than that of the MS1649, here varying by about 30% instead of more than 50% in Figure 10-14.

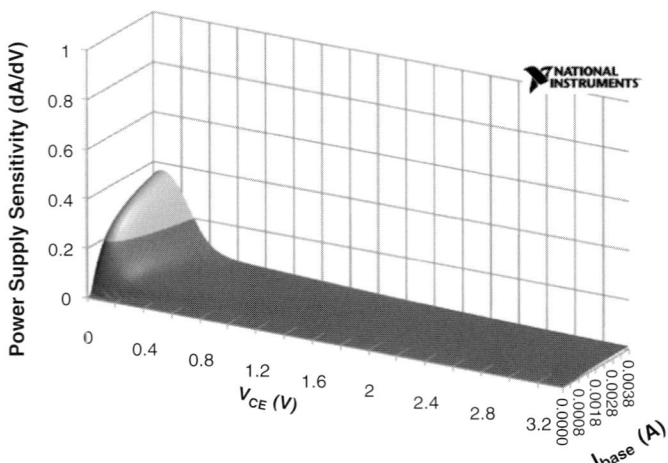

Figure 10-21   Power supply sensitivity surface for the MRF586: a smaller voltage offset on the resistive region is seen compared to Figure 10-13. A long slope is present as the CCS region is entered.

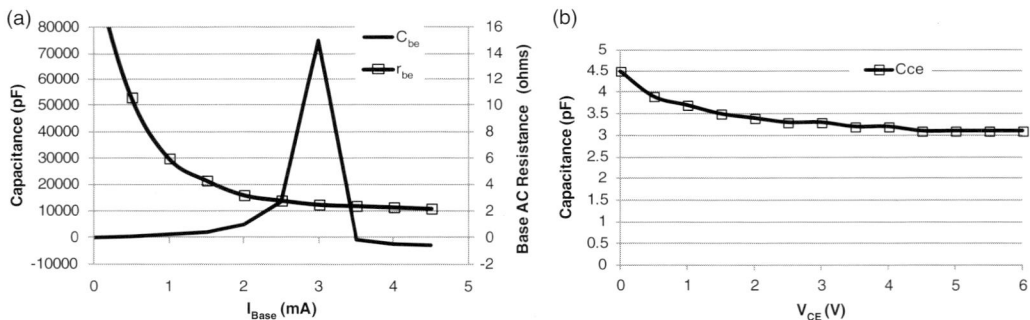

Figure 10-22   Measured parasitic capacitances for the MRF586 at 100 MHz: (a) input, which resonates at 3.5 mA of base current and is inductive at higher drive; (b) output capacitance varies by about 1.5:1 across these voltages.

Evaluation of the DPST performance capabilities of this transistor begins with the evaluation of its optimum ET voltage profile. The analysis shows that there are no conditions where (10.1) is satisfied using this transistor. As such, strictly speaking this transistor can never support ET. There is never enough independence between the output signal envelope and any power supply variations. This transistor always operates with some large measure of polar modulation when a DPS is applied.

The best power supply noise suppression provided by the MRF586 is 20 dB, and the knee voltage profile providing this reduced noise suppression is shown by the solid line in Figure 10-23. This profile is offset at zero current, showing that this algorithm does avoid P-mode. The dynamic bias profile, which maximizes both linearity and dynamic range, is included as the dash–dot curve. Bunching of the characteristic curves along the CCS boundary for 20 dB supply noise suppression shows that output signal linearity will

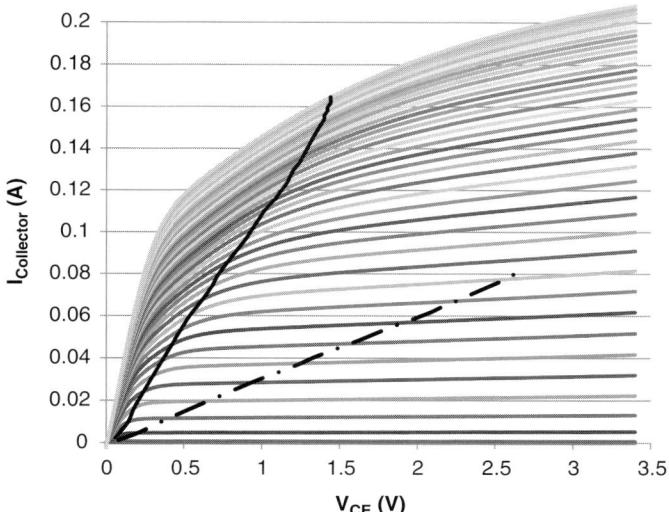

Figure 10-23   ET operation region for the MRF586 silicon bipolar transistor. This transistor does not meet (10.1) under any conditions. The boundary for the CCS region meeting at PSS = –20 dB is identified by the solid line. The dash–dot line shows the dynamic bias profile this transistor must follow to simultaneously maximize amplifier linearity and the power control dynamic range.

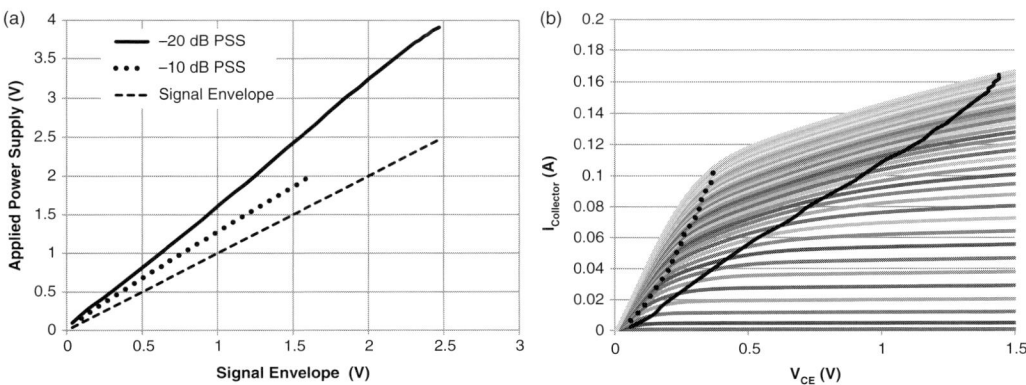

Figure 10-24   ET (barely) characteristics for the MRF586: (a) DPS voltage profiles for power supply noise suppression of –20 dB, and –10 dB; (b) corresponding ET endpoint trajectories for these two values of power supply noise suppression, with a dramatic drop in the knee voltage.

be degraded at the higher output powers. Amplifier linearity is better only at the lower output powers.

With this transistor already mostly operating with polar modulation, results from evaluating the remainder of the transition to polar are shown in Figure 10-24. Allowing the power supply suppression to fall to 10 dB (dotted curves) has a very large impact on the knee voltage profile. At this operating condition, inspection shows that this transistor

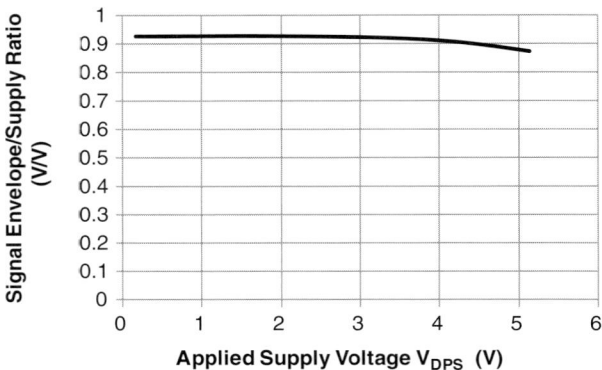

Figure 10-25 Envelope control precision and accuracy for the MRF586 when operating in polar DPST operation. Polar envelope control is accurate over a dynamic range of about 24 dB with a 35 ohm load resistance. Distortion from curvature in the characteristic curves at low voltages (Figure 10-18(b)) is not evident, confirming its relative absence. Best-fit is achieved with $V_{AMO} = 0.007$ V.

is operating completely in P-mode at this lower boundary. Distortion from this P-mode operation is potentially severe, depending on the signal being generated.

Because this transistor essentially operates in polar with large output signals, the linearity of the polar modulation characteristic in Figure 10-25 is not a surprise. This performance is much better than that seen in Figure 10-17, particularly with the absence of low magnitude distortion. The best-fit triangle to implement (2.9) needs an offset of 7 mV to achieve this good dynamic range.

10.5.2   CMOS

Interest in using CMOS technology for power amplifiers is longstanding. The major attraction is to be able to leverage the huge production capability installed to produce CMOS. There is also an added capability to include digital circuitry on the same die, to provide a "digital assist" to the RF transmitter. While the speed of digital CMOS transistors is extremely fast, their requirement to operate at low voltage is a real challenge to PA designers (Section 7.7).

Evaluation of a CMOS power amplifier provides the results in this section. A cascode structure (Section 7.9.1) is used to address the high voltage problem in small geometry CMOS. The measured characteristic curves are presented in Figure 10-26(a). These do appear to be a decent match to the idealized characteristics of Figure 10-1. The close-up view near the IV origin seen in Figure 10-26(b) is also a decent match to the corresponding idealized characteristics of Figure 10-2. The spread of characteristic curves in Figure 10-26(b) demonstrates a voltage controlled resistance behavior that is characteristic of the P-mode.

Corresponding surfaces for transistor transconductance and absolute ON resistance are provided in Figure 10-27. There is a small constant value region in the transconductance surface centered around 0.65 V V_{GS} at V_{DS} values exceeding 1 volt. In this region,

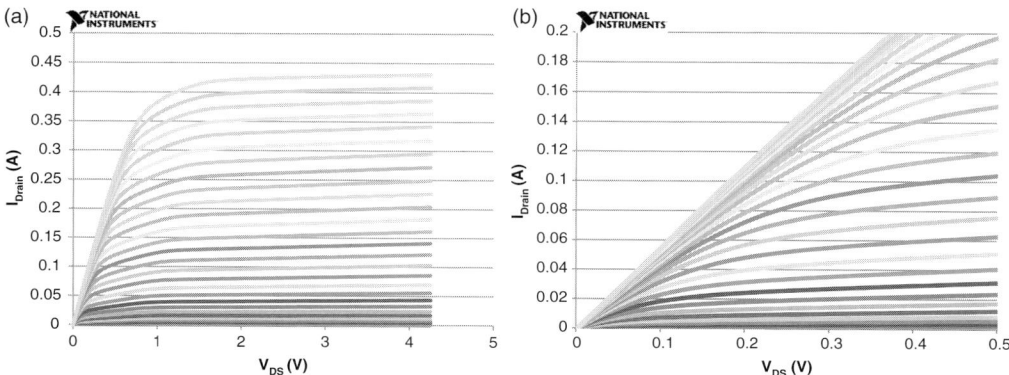

Figure 10-26 Characteristic curves for the CMOS n-MOSFET RF power transistor: (a) measured curve family with stepped V_{GS} values; (b) close-up view of these curves near the IV plane origin.

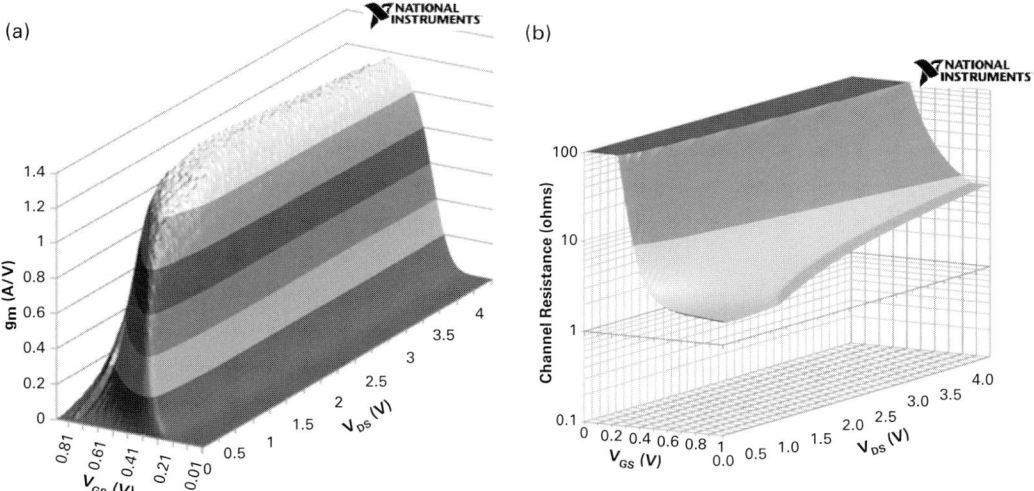

Figure 10-27 First-order analysis results from CMOS n-MOSFET characteristic curves: (a) transconductance (g_m) surface; (b) transistor absolute resistance surface.

the CMOS PA transistor provides good amplifier linearity. For larger drive signals, amplifier linearity requires additional circuit techniques beyond this transistor.

The ON resistance surface does not have a V_{DS} voltage offset to its minimum value, setting the expectation that no offset will be needed in the later evaluation of polar operation. This result is also consistent with the appearance in Figure 10-26(b) that all of the resistive characteristic curves pass through the origin. This surface does get flat where V_{DS} values are small and V_{GS} is large, signifying an independence of the ON resistance from those two parameters in that region. Such independence is a key characteristic of switch behavior.

Evaluating the transistor transfer characteristic for a subset of V_{DS} values provides the data shown in Figure 10-28. At small V_{DS} values, the transfer curves return to horizontal,

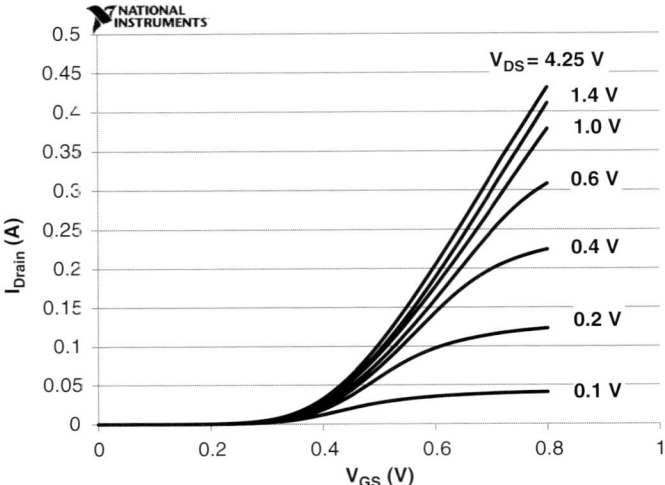

Figure 10-28   A subset of transfer functions corresponding to Figure 10-26(a). Gain is proportional to the slope of these curves, and linearity is related to their straightness. Threshold behavior is consistent across all applied V_{DS} supply values.

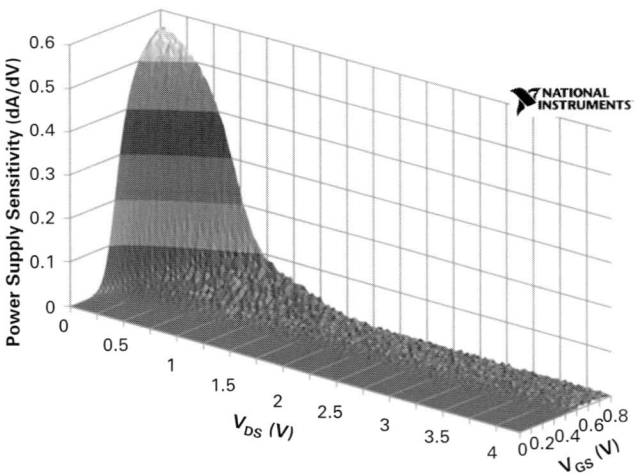

Figure 10-29   Power supply sensitivity surface for the CMOS RF power transistor.

or very nearly so at higher V_{GS} values. This confirms that this transistor is acting as a switch at these conditions. At higher applied supply voltages, these transfer curves are straight, confirming that this transistor acts as a linear amplifier. Curvature of each of these transfer relationships near turn-on demonstrates distortion to the signal as bias approaches the class B definition: at the device conduction threshold.

Figure 10-29 presents the PSS surface for this transistor. Consistent with the earlier results, the peak power supply sensitivity is at low V_{DS} and high V_{GS} operation, consistent with the minimum ON resistance. There is no offset of this peak away from

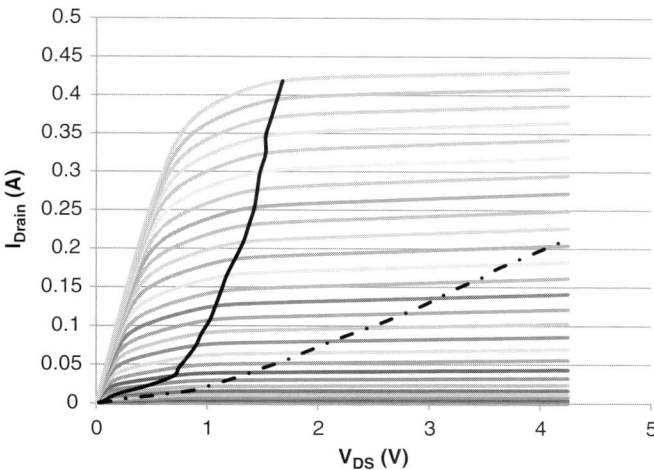

Figure 10-30   ET operation region for the CMOS n-MOSFET RF power transistor. This transistor approaches the performance of (10.1) but never quite gets there. The boundary for the CCS region meeting PSS = −30 dB is identified by the solid line. The dash-dot line shows the dynamic bias profile this transistor must follow to simultaneously maintain amplifier linearity and maximize the power management dynamic range.

zero V_{DS}. Transition from the resistive region to the CCS region is not as rapid as that seen in Figure 10-13, but the transition tail is not like that seen in Figure 10-21.

Evaluation of the envelope tracking DPST performance of this PA structure is provided in Figure 10-30. This particular transistor just barely misses the defined CCS boundary of (10.1), so instead the boundary shown (solid line) is 10 dB into the transition region toward polar operation at 30 dB of power supply noise suppression. This defines the best ET knee voltage available with this transistor. Also shown is the optimum bias trajectory (dash-dot curve) that maximizes linearity and the ET dynamic range, as the DPS varies along its chosen profile.

It is interesting to note that the optimum knee voltage has a slope shift near $V_{DS} = 0.7$ V. This indicates that this PA structure has a slope change in its characteristic curves, which increases power supply sensitivity. To maintain low sensitivity to power supply noise, the knee voltage must get around this slope change region.

Bunching of the characteristic curves below 50 mA of drain current signifies that the gain is lower in this region than it is at higher currents. This is not gain collapse from P-mode, but the result of lower transconductance from the transistor in CCS operation at these operating conditions. The envelope tracking DPST designer may choose to avoid operating in these conditions if applying gain correction is not readily available.

The knee voltage from Figure 10-30 is generally greater than 1 volt. This is larger than that of Figure 10-16, meaning that the energy efficiency benefit using ET on this amplifier will be more restricted. Should energy efficiency be more important than power supply noise suppression, this amplifier can be operated closer to polar modulation with the reductions in knee voltage seen in Figure 10-31. Unlike the rapid transition to polar operation seen in Figure 10-16(a), here the transition is more gradual.

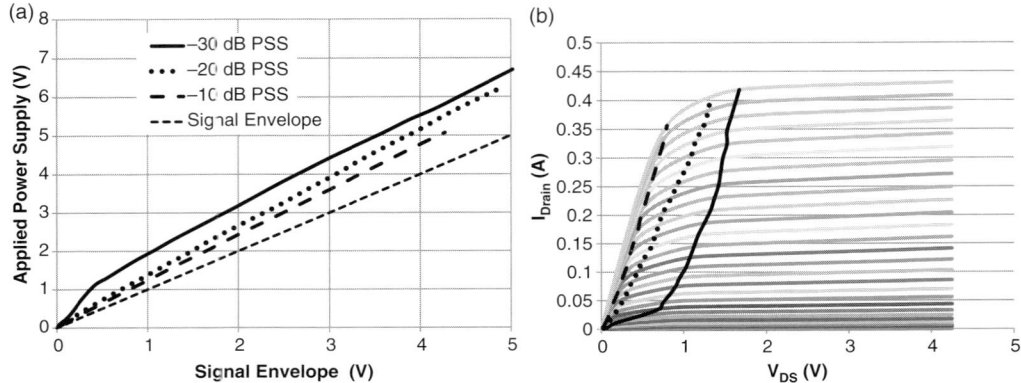

Figure 10-31 ET characteristics for the CMOS n-MOSFET RF power transistor: (a) DPS voltage profiles for power supply noise suppression of −30 dB, −20 dB, and −10 dB; (b) corresponding ET endpoint trajectories for these three values of power supply noise suppression.

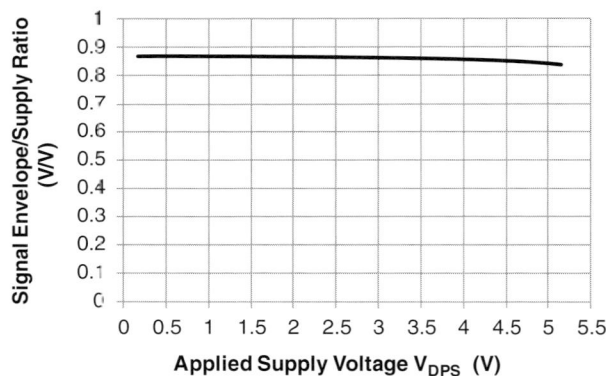

Figure 10-32 Envelope control precision and accuracy for the CMOS RF power transistor when in polar DPST operation. Polar envelope control is accurate over a dynamic range exceeding 26 dB.

When the transition to polar operation completes, the accuracy of this polar modulation is seen in Figure 10-32. This line has a very slight curvature. It is interesting to note that a small offset of 2 mV provides the best-fit polar modulation triangle for this PA.

10.5.3   LDMOS

The workhorse of RF power generation in the world today is silicon LDMOS. These FET devices are used at powers from a few watts up to kilowatts. The part evaluated here is rated for 6 W operation. Characteristic curve measurements are presented in Figure 10-33.

At first inspection, the curves in Figure 10-33(a) are a good match to the ideal of Figure 10-1. At higher values of V_{GS}, this transistor is capable of drawing amperes of

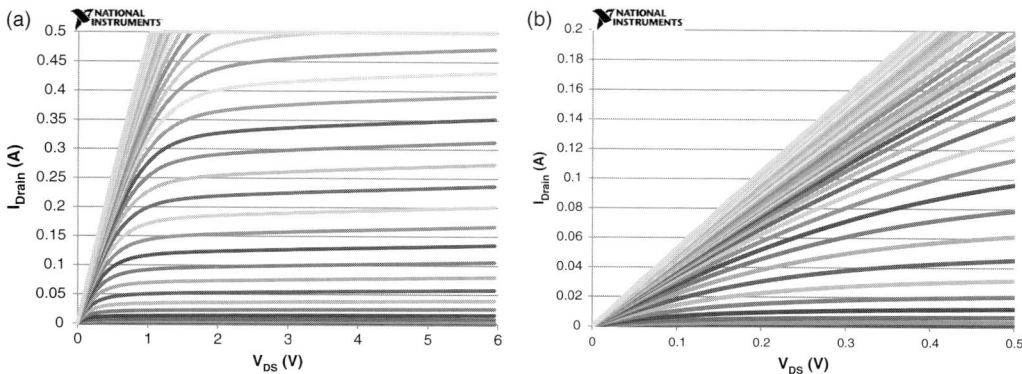

Figure 10-33 Characteristic curves for a silicon LDMOS RF power transistor: (a) measured curve family with stepped values of V_{GS}; (b) close-up view of these curves near the I–V plane origin.

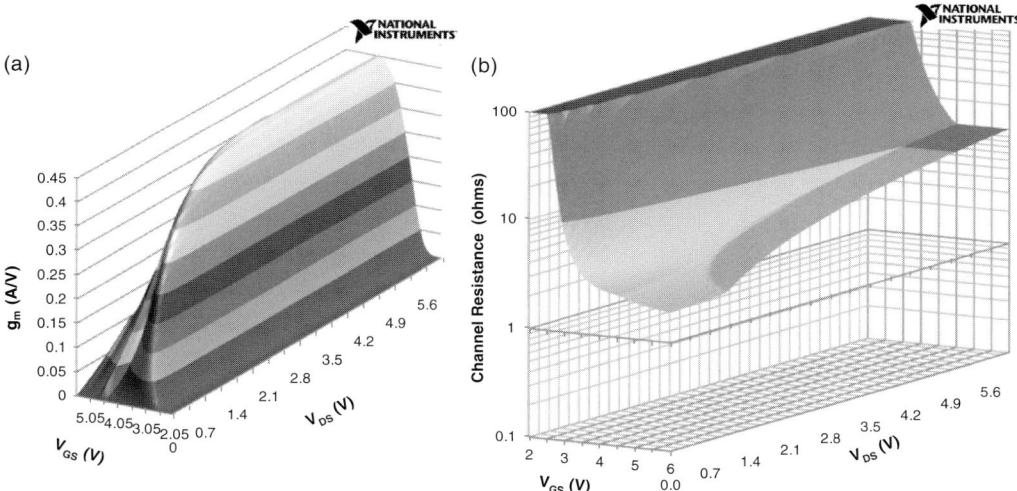

Figure 10-34 First-order analysis results from n-channel LDMOS characteristic curves: (a) transconductance (g_m) surface; (b) transistor absolute resistance surface.

current, beyond the 0.5A limit of this measurement equipment. There is plenty of power handling capability with this transistor. The close-up measurements of Figure 10-33(b) show horizontal slopes at low current limits, and show resistance (finite slopes) for most of the curves.

The transconductance and absolute ON resistance surfaces for this LDMOS transistor are shown in Figure 10-34. The transconductance surface shares many characteristics with the CMOS surface in Figure 10-27(a), but with lower values. Complete evaluation of transistor linearity is not finished due to the measurement system current limit at 0.5 A. ON resistance is very similar in shape to the CMOS data in Figure 10-27(b).

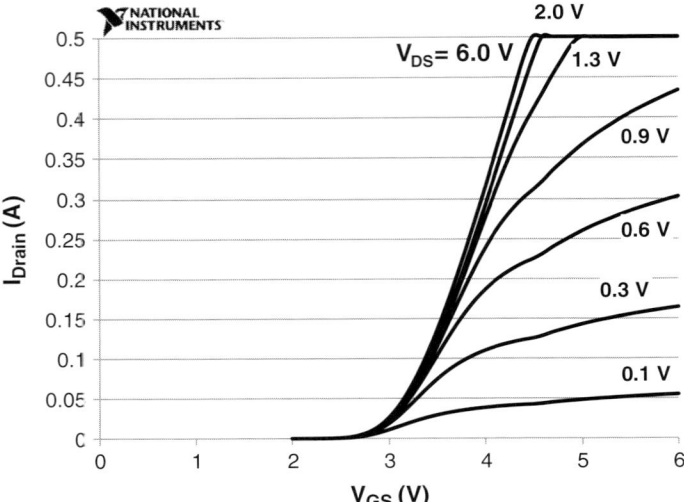

Figure 10-35   A subset of transfer functions corresponding to Figure 10-33(a). Gain is proportional to the slope of these curves, and linearity is related to their straightness. Threshold behavior is consistent across all applied V_{DS} supply values.

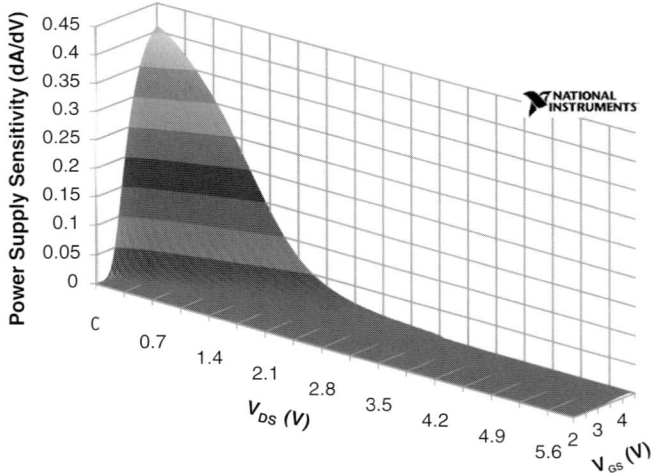

Figure 10-36   Power supply sensitivity surface for the example LDMOS RF power transistor

Transfer relationships in Figure 10-35 for this LDMOS transistor exhibit transistor linearity only with higher voltages across the device. Switching behavior is not so good, as seen from the ON parts of these curves not becoming horizontal, even with more than 3 V of channel enhancement. Threshold behavior is very consistent.

Figure 10-36 provides the PSS surface for this LDMOS transistor. The peak of the resistive operating region is at 0 V V_{DS} and high V_{GS}, with no offset. Transitioning to

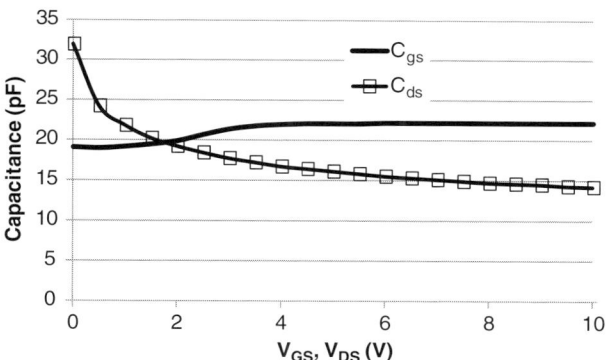

Figure 10-37 Measured parasitic capacitances for this LDMOS transistor at 100 MHz: input capacitance varies slightly as the channel forms above the threshold voltage, and output capacitance varies more than 2:1 following a varactor diode profile.

CCS operation is much more gradual as this transistor turns ON compared to the PSS surface of Figure 10-13.

Parasitic capacitances for this device are of very similar values at the input and output, as seen in Figure 10-37. The input capacitance is nearly constant, only showing a variation as the channel forms. Output capacitance begins at a higher value than the input capacitance, and falls to eventually being a lower value for the same applied voltage.

When evaluated for ET performance, this transistor also shows no region that meets (10.1). With no solution having 40 dB of power supply noise suppression, the knee voltage evaluation is first done for 30 dB as the boundary between resistive and CCS operating regions. This result is shown as the solid line in Figure 10-38. Along with this is the optimum bias trajectory that maximizes the ET dynamic range. Maintaining linearity is not as clear, given the bunching of characteristic curves at low drain currents.

Consistent with the PSS surface of Figure 10-36, the gradual transition from the resistive to CCS regions corresponds to a higher value for the knee voltage. This restricts the energy efficiency benefits available from the ET operation. Improved energy efficiency comes from polar operation, and the transition toward polar operation is shown in Figure 10-39 as power supply noise rejection is traded away. The knee voltage drops nearly by half.

When the transition to polar operation is completed, the modulation accuracy is seen in Figure 10-40. Accuracy is good for a nearly 30 dB dynamic range. From Figure 10-33(a), the knee voltage in polar operation stays below 0.5 V.

10.5.4 SiGe (silicon-germanium) HBT

Combining germanium with silicon in the base region forms a heterojunction bipolar transistor with excellent speed, at the cost of generally low operating voltage.

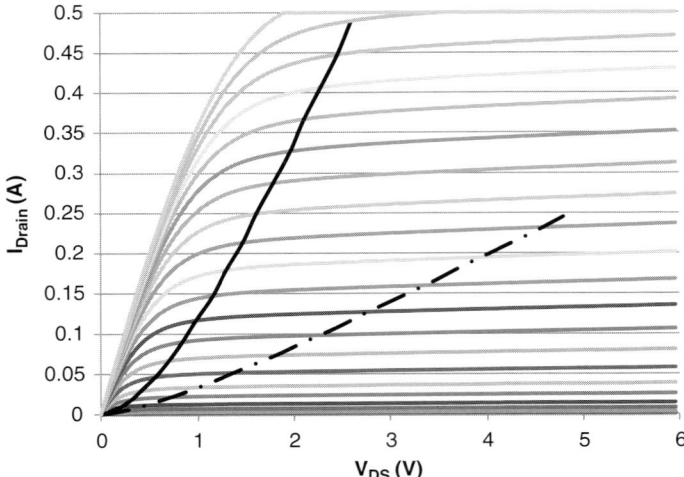

Figure 10-38 ET operation region for this LDMOS RF power transistor. This transistor approaches the performance of (10.1) but never quite gets there. The boundary for the CCS region meeting PSS = −30 dB is identified by the solid line. The dash-dot line shows the dynamic bias profile this transistor must follow to simultaneously maintain amplifier linearity and maximize the power control dynamic range.

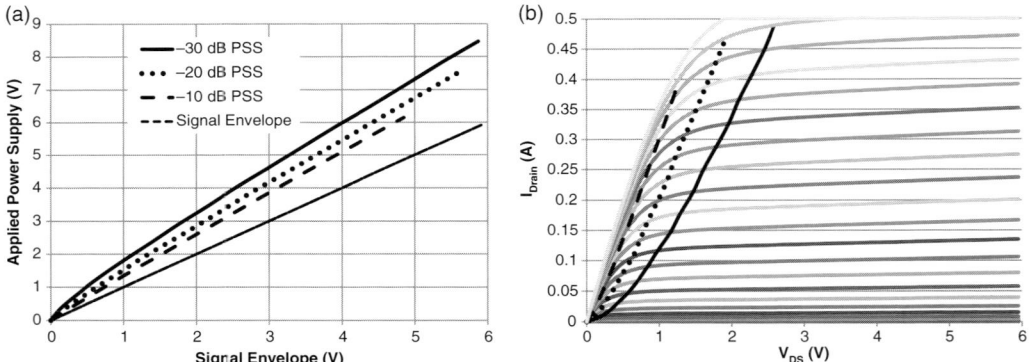

Figure 10-39 ET characteristics for this LDMOS RF power transistor: (a) DPS voltage profiles for power supply noise suppression of −30 dB, −20 dB, and −10 dB; (b) corresponding ET endpoint trajectories for these three values of power supply noise suppression.

Characteristic curve measurements for an example SiGe HBT are provided in Figure 10-41.

The most obvious difference between Figure 10-41(a) and Figure 10-1 is the upturn of each curve for the SiGe HBT. This acts as a dramatic increase in β at these higher voltages, and is not a good place to operate. Of particular importance is the close-up view provided in Figure 10-41(b). There is very little, if any, offset for

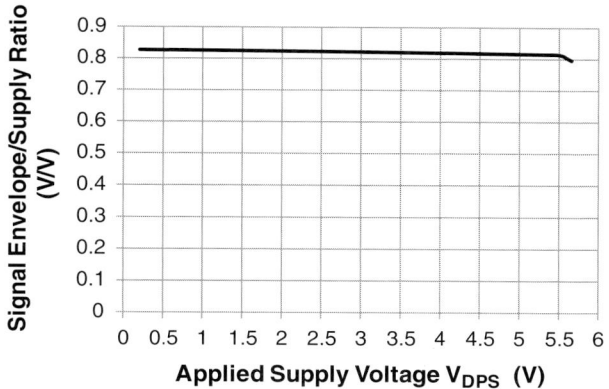

Figure 10-40   Envelope control precision and accuracy for the LDMOS RF power transistor when operating in polar DPST operation. Polar envelope control is accurate over a dynamic range of about 30 dB.

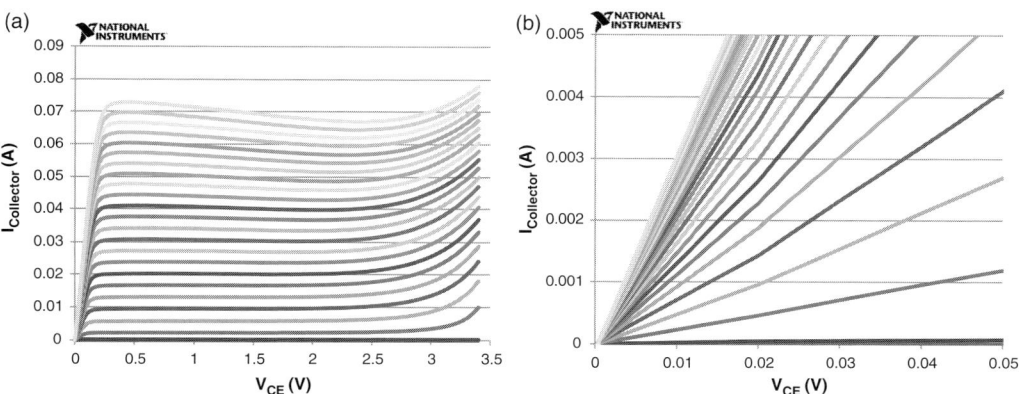

Figure 10-41   Characteristic curves for the silicon-germanium heterojunction bipolar RF power transistor: (a) measured curve family with stepped base current; (b) close-up view of these curves near the I–V plane origin.

this bipolar transistor. This is very different from the bipolar junction transistor in Figure 10-10(b).

Generating the β and ON resistance surfaces from these measurements provides the graphics in Figure 10-42. Two features are noteworthy in Figure 10-42(a). Most obvious is the high peak in the β surface at low input drive level and high V_{CE}. This is an operating region to avoid. Of greater practical interest is the wide flat region that is good for amplifier linearity, and that the β value of this surface is around 400, providing nearly 10 times more gain than the standard silicon bipolar junction transistor from Figure 10-11(a).

The absolute resistance surface in Figure 10-42(b) shows the channel resistance minimum very close to 0 V V_{CE}. Any actual offset is hard to discern on this surface.

(a)

(b)

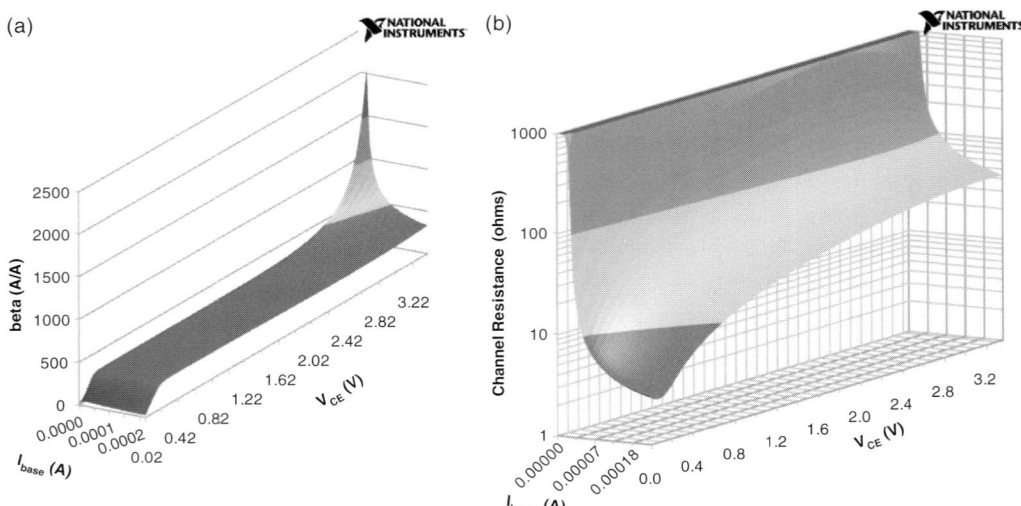

Figure 10-42 First-order analysis results from the SiGe HBT characteristic curves: (a) current gain β surface; (b) transistor absolute resistance surface.

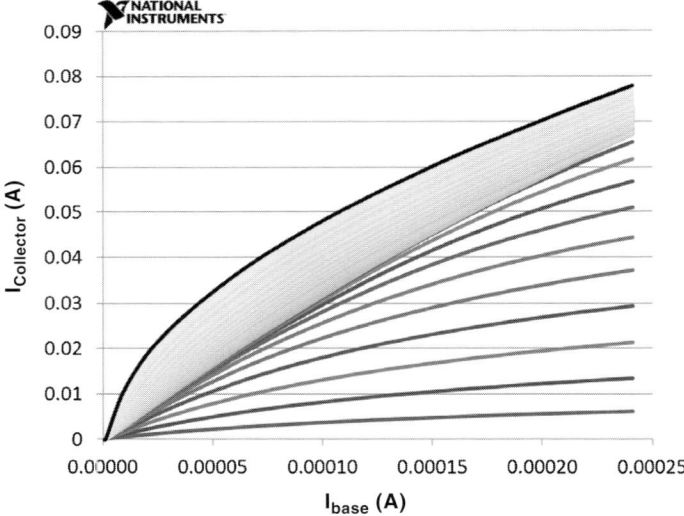

Figure 10-43 The set of current transfer functions corresponding to Figure 10-41(a). Gain is proportional to the slope of these curves, and linearity is related to their straightness. Gain is very high at low drive, and compresses quickly.

Fortunately, the more sensitive measure of offset is the PSS surface, presented in Figure 10-44.

The transfer relationships for this SiGe HBT across all the measured values of V_{CE} are calculated and presented in Figure 10-43. The interesting features are (1) this transistor under these conditions is not switching, determined by the nonhorizontal nature of all these curves; (2) greatest linearity is seen in the curves taken at V_{CE} near 2.5 V, and (3) at

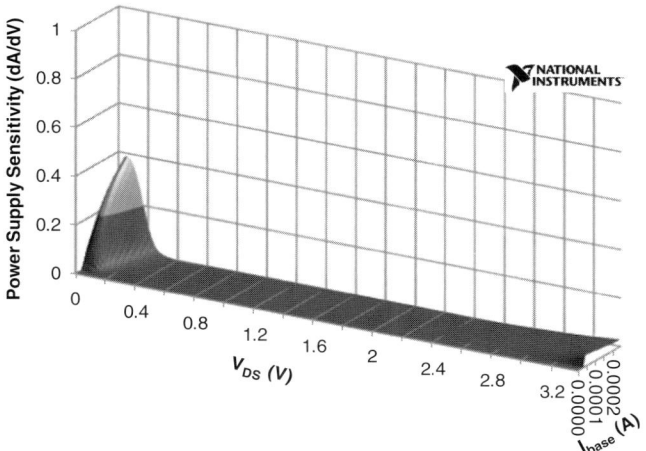

Figure 10-44 Power supply sensitivity surface for the SiGe HBT.

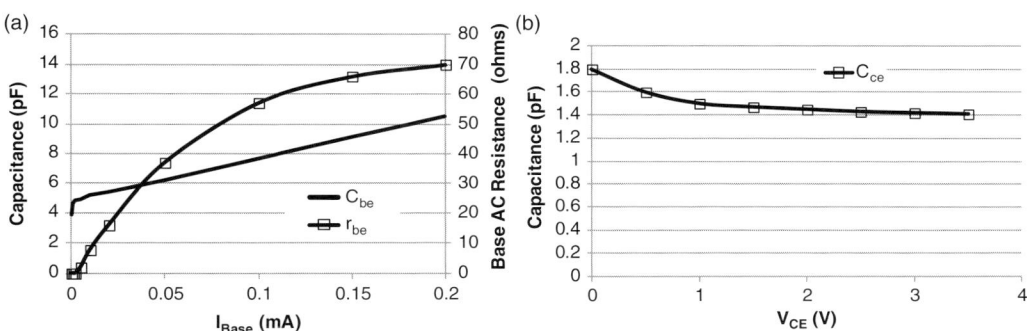

Figure 10-45 Measured parasitic capacitances for the SiGe HBT at 100 MHz: (a) input capacitance varies linearly as the transistor conduction increases; (b) output capacitance varies by less than 30% across its operating voltage range.

higher V_{CE} values there is significant nonlinearity from gain expansion at lower drive levels. This corresponds to the spike in the β surface of Figure 10-42(a). For best linearity directly from this transistor, the higher operating voltages are best avoided.

The PSS surface in Figure 10-44 clearly shows that there is indeed an offset in the peak of the transistor resistive region. There is also a second resistive region at $V_{CE} > 2.5$ V as the characteristic curves begin to turn upwards. This is the first transistor that actually bounds its CCS region at both lower and high supply voltages. Transition into the CCS region from the primary resistive region is rapid.

Input capacitance measurements for this SiGe HBT behave very differently from either of the silicon BJT devices examined earlier. No resonance effects are seen in Figure 10-45, and the base resistance increases with higher base current. Adding the germanium does dramatically alter this transistor's characteristics. Output capacitance changes very little, varying less than 30%.

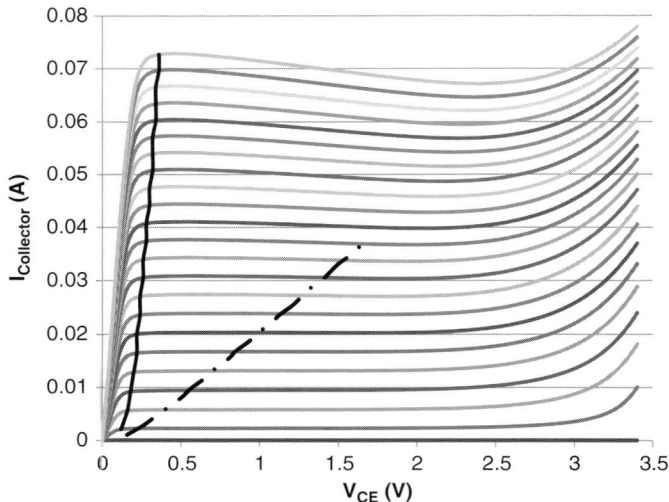

Figure 10-46 The ET operation region for the SiGe HBT RF power transistor. The boundary for the CCS region meeting (_0.1) is identified by the solid line. The dash-dot curve shows the dynamic bias profile this transistor must follow to simultaneously maintain amplifier linearity and maximize the power management dynamic range.

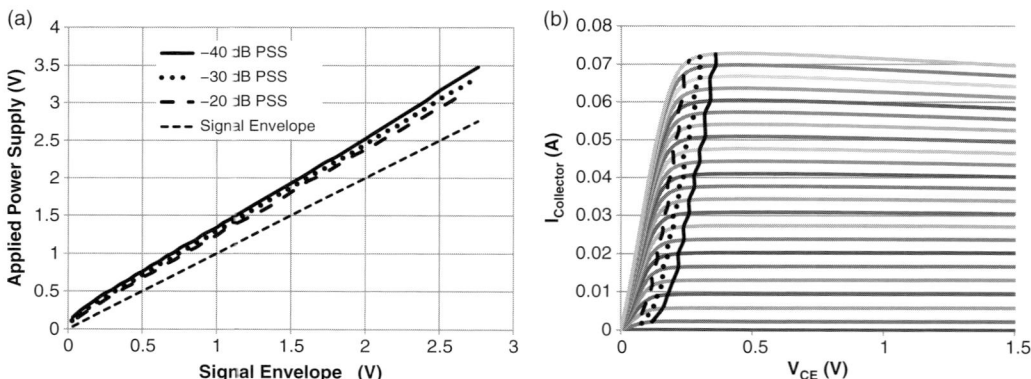

Figure 10-47 ET characteristics for the SiGe HBT RF power transistor: (a) DPS voltage profiles for power supply noise suppression of –40 dB, –30 dB, and –20 dB; (b) corresponding ET endpoint trajectories for these three values of power supply noise suppression.

This transistor easily meets the envelope tracking CCS boundary definition of (10.1). The resulting CCS boundary at 40 dB power supply noise rejection is shown as the solid line in Figure 10-46. The corresponding dynamic bias profile is also shown as the dash–dot curve. Note the offset at minimum voltage, where this profile avoids entering into P-mode operation.

Keeping the same analysis methods for this transistor, the profile of how this transistor transitions from ET toward polar operation is shown in Figure 10-47. As for the

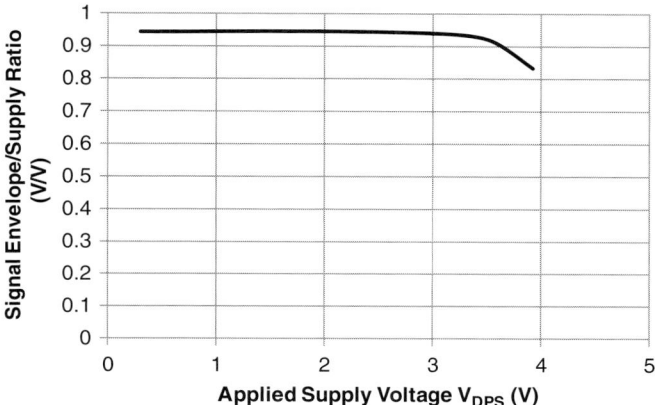

Figure 10-48 Envelope control precision and accuracy for the SiGe HBT in polar DPST operation. Polar envelope control is accurate over a dynamic range of about 20 dB. Distortion from curvature in the characteristic curves at low voltages (Figure 10-41(b)) is not evident when V_{AMO} is set to 0.003 V.

MS1649, this transition is very rapid. Any slight drop below the calculated knee voltage along this profile quickly loses power supply noise rejection.

At the limit where polar modulation is complete, the linearity of this modulation for the SiGe HBT is shown in Figure 10-48. With the best-fit modulation triangle at an offset of 3 mV, there is no apparent envelope distortion down to 300 mV. Above this lower limit, the polar envelope control is very accurate to 3 V, giving an accurate polar modulation dynamic range of 20 dB for this transistor.

It is interesting to note that the SiGe HBT is the only transistor (so far) that performs reasonably well in both envelope tracking and polar DPST operation. Having such a low available operating supply voltage is an unfortunate price for this generality.

10.5.5 Si MESFET

I have long maintained that the silicon technology base is in need of a fundamentally different transistor type in order to improve the practicality for RF power generation in more applications than the LDMOS supports at this moment. One interesting candidate for this new transistor is a MESFET, fabricated in standard silicon processes [10-1]. Characteristic curve measurements for an example silicon MESFET are presented in Figure 10-49. The characteristic curves show both the resistive and CCS regions, and the close-up view near the origin appears to have all of the resistive behavior pass through the origin with no offset.

The transconductance and channel resistance surfaces calculated from the characteristic curve data are shown in Figure 10-50. The slopes of the characteristic curves in the CCS region seen in Figure 10-49(a) are not perfectly parallel, which results in a transconductance that varies slightly with increasing power supply voltage. This explains the slope at the top of this transconductance surface. The value of this transconductance is very high, nearly triple that of the LDMOS example transistor.

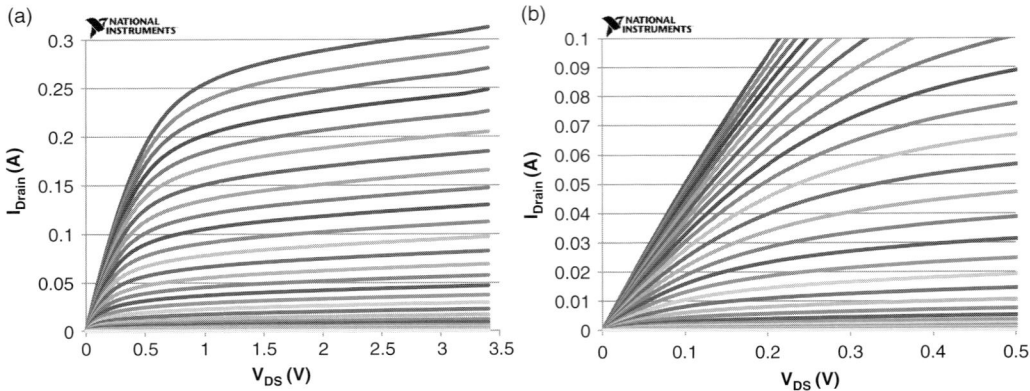

Figure 10-49 Characteristic curves for a silicon MESFET RF power transistor: (a) measured curve family with stepped V_{GS} values; (b) close-up view of these curves near the IV plane origin.

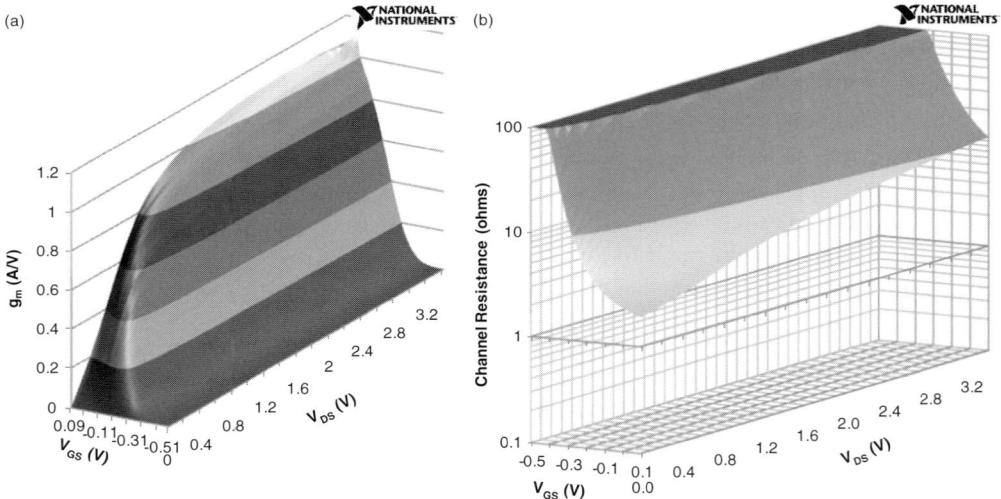

Figure 10-50 First-order analysis results from silicon n-MESFET characteristic curves: (a) transconductance (g_m) surface; (b) transistor absolute resistance surface.

This implies that the FET-activity range is not large for the silicon MESFET and that the pinch-off voltage is not very negative.

The channel resistance surface has its minimum at zero V_{DS} offset and highest V_{GS} value. This resistance is still dropping, indicating that this transistor is not yet switching. This conclusion is confirmed by looking at the transfer relation subset provided in Figure 10-51.

None of the transfer relation curves in Figure 10-51 is horizontal at high V_{GS}. This confirms that this transistor is not yet switching. At $V_{DS} > 0.8$ V, this transistor does exhibit linear behavior. As the power supply increases, the gain also increases, which is a distortion mechanism for the ET operation. Finally, when compared to the LDMOS

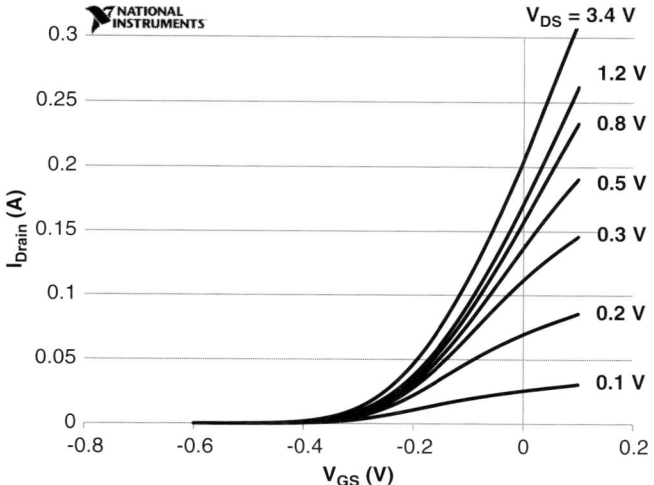

Figure 10-51   A subset of transfer functions corresponding to Figure 10-49(a). Gain is proportional to the slope of these curves, and linearity is related to their straightness. Threshold behavior is consistent across all applied V_{DS} supply values.

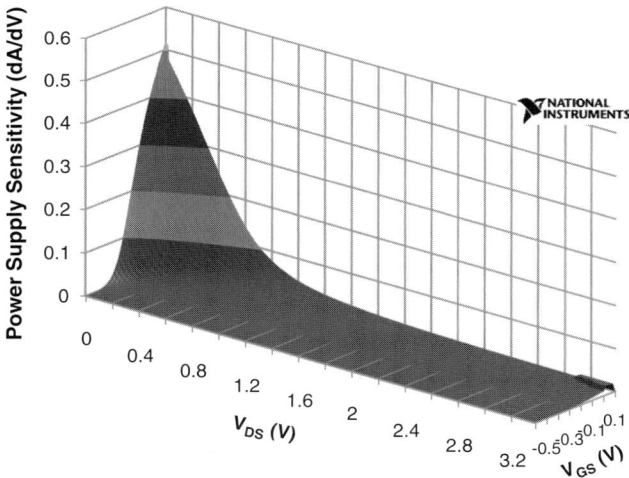

Figure 10-52   Power supply sensitivity surface for the silicon n-MESFET.

transfer characteristics in Figure 10-35, there appears to be a slight variation here of the pinch-off voltage as the power supply voltage is increased.

The PSS surface presented in Figure 10-52 looks very much like that for the LDMOS measurements in Figure 10-36. The peak of the resistive region is at 0 V V_{DS} and maximum V_{GS}. Transition from this resistive region peak down to the CCS region is gradual.

Parasitic capacitances for this depletion mode device are measured and reported in Figure 10-53. Variation of the input capacitance is close to 2:1. The value of the output

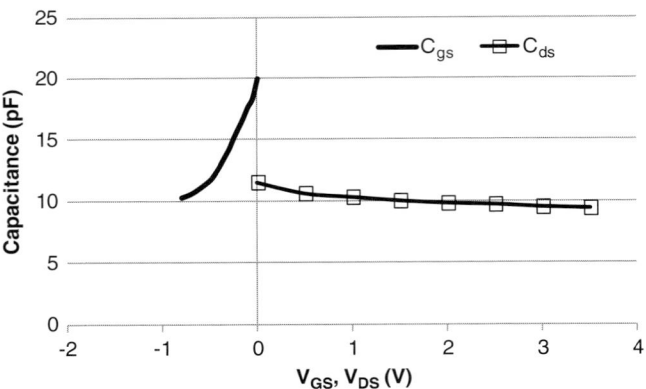

Figure 10-53   Measured parasitic capacitances for the silicon MESFET at 100 MHz: input capacitance varies by about 2:1 as the channel forms above the pinch-off voltage; output capacitance varies by about 20% across its operating voltage.

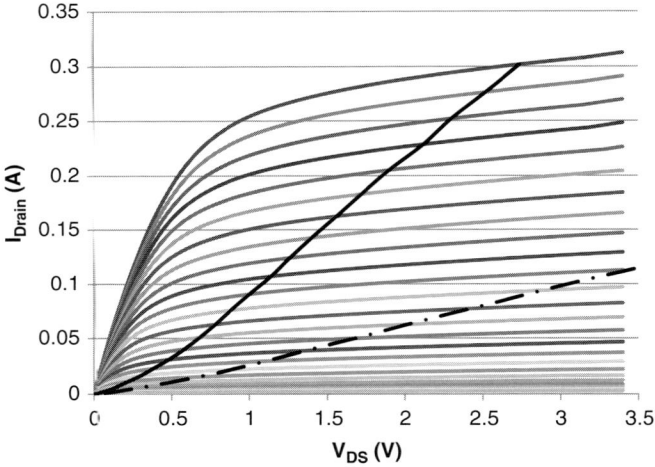

Figure 10-54   ET operation region for the silicon n-MESFET RF power transistor. This transistor approaches the performance of (10.1) but never quite gets there. The boundary for the CCS region meeting PSS = −30 dB is identified by the solid line. The dash–dot curve shows the dynamic bias profile this transistor must follow to simultaneously maintain amplifier linearity and maximize the power management dynamic range.

capacitance is much more stable, varying by less than 25% across this operating voltage range.

Evaluation of the DPST application performance capability of this silicon MESFET using the process from Section 7.6 begins with the result of Figure 10-54. This transistor closely approaches the CCS boundary definition (10.1), but does not quite get there. The back-off boundary that is evaluated is for 30 dB of power supply noise suppression. This boundary is shown by the solid line in Figure 10-54. This boundary gets very close to 0 V

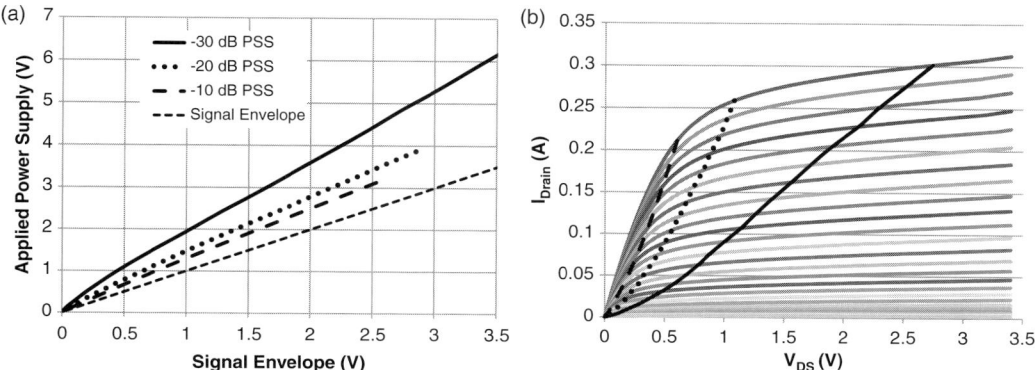

Figure 10-55 ET characteristics for the silicon n-MESFET RF power transistor: (a) DPS voltage profiles for power supply noise suppression of −30 dB, −20 dB, and −10 dB; (b) corresponding ET endpoint trajectories for these three values of power supply noise suppression, with their wide variation in effective knee voltages.

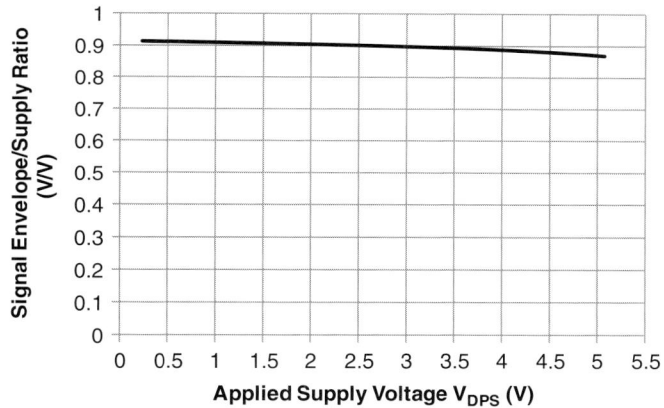

Figure 10-56 Envelope control precision and accuracy for the silicon n-MESFET in polar DPST operation. Polar envelope control is accurate over a dynamic range of about 30 dB.

at its minimum value. The corresponding optimum dynamic bias profile (dash-dot curve) is also shown in this figure.

The continued transition from envelope tracking to polar operation using this transistor is shown in Figure 10-55. This transition is very gradual, with wide power supply voltage spacing between having 30 dB of power supply noise suppression and losing noise suppression to 20 dB. Dropping the supply voltage for 10 dB of supply noise suppression is not as large a step.

At the completion of the transition to polar operation, the resulting modulation accuracy is shown in Figure 10-56. No offset is necessary in the optimum modulation triangle, confirming the observation from Figure 10-49(b). The available polar modulation dynamic range is close to 30 dB.

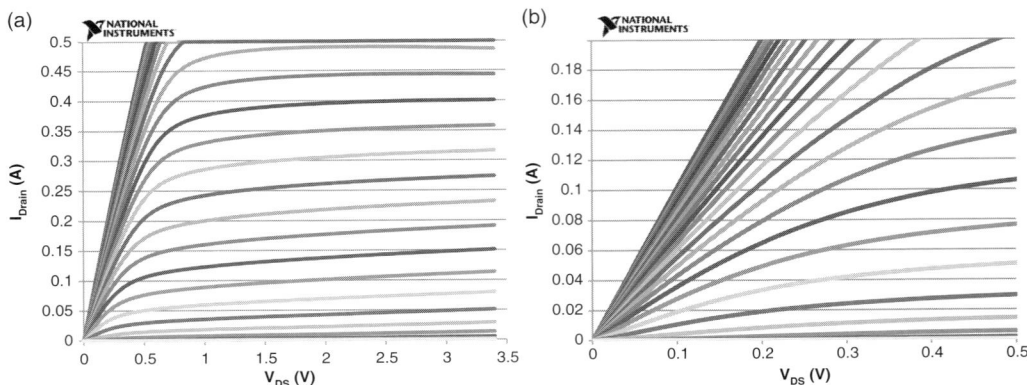

Figure 10-57   Characteristic curves for a GaAs MESFET RF power transistor: (a) measured curve family with stepped V_{GS} values; (b) close-up view of these curves near the IV plane origin.

10.6   (III/V) gallium arsenide technologies

Compound semiconductors such as gallium arsenide (GaAs) exist because they demonstrate much higher charge mobility and relatively smaller parasitic capacitances. The mobility increase is important because it lowers the available device resistance, allowing improved energy efficiency and higher frequency operation. Smaller parasitic capacitances also allow frequencies of operation to be higher than silicon devices can support. One generally finds GaAs technology-based devices in RF power applications where efficiency performance is very important at microwave frequencies. Most of the GaAs device varieties now available are FETs, both depletion mode and enhancement mode. Bipolar devices are now also available in GaAs and are widely used. With regard to total manufactured transistors using GaAs material, the bipolar HBT is likely the most numerous.

One unavoidable by-product of these transistors having high gain and wide bandwidth is that they are difficult to stabilize. Some examples of device oscillation are seen in the following discussions.

10.6.1   GaAs MESFET

The first GaAs device historically available for circuit use is the MESFET. Measured characteristic curves for an example device are presented in Figure 10-57.

The characteristic curves in Figure 10-57(a) are similar to the idealized transistor of Figure 10-1. Many of the lower resistance curves go straight up to the measurement current limit of 0.5 A, demonstrating the power capability of this sample device. The close-up view in Figure 10-57(b) shows a variable resistance behavior at low device voltages. This voltage controlled resistance is a property of P-mode operation.

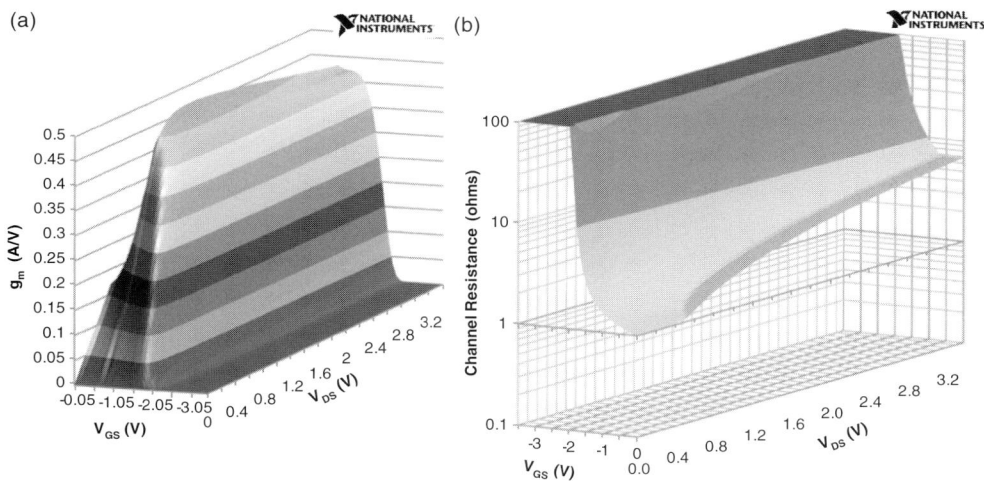

Figure 10-58 First-order analysis results from the GaAs MESFET characteristic curves: (a) transconductance (g_m) surface; (b) transistor absolute resistance surface.

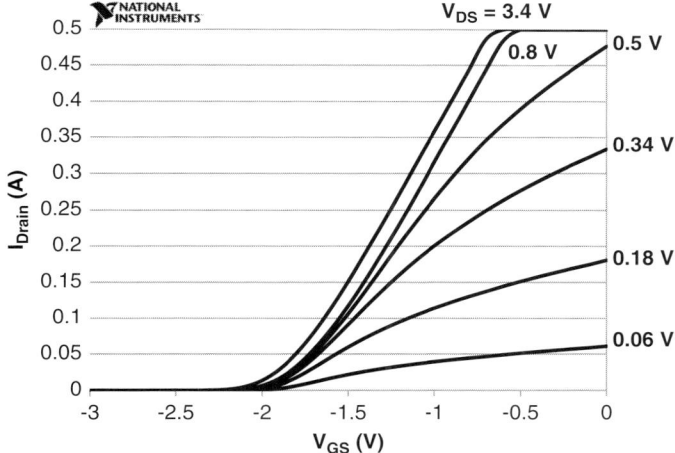

Figure 10-59 A subset of transfer functions corresponding to Figure 10-57(a). Gain is proportional to the slope of these curves, and linearity is related to their straightness. Threshold behavior shows a slight shift across all applied V_{DS} supply values.

Calculating the corresponding transconductance and channel resistance surfaces for these data obtains the results shown in Figure 10-58. The transconductance surface has a constant-value region at the higher supply voltages. Channel resistance falls as both V_{DS} goes to zero and V_{GS} increases. This curve does not show a resistance floor yet, so this transistor is not yet acting like a switch with these operating conditions.

The subset of transfer curves in Figure 10-59 confirms that this MESFET is not yet acting as a switch. Transfer curves from low V_{DS} values do show compression, but not the horizontal slope of a true switch. Of particular interest are the transfer curves for

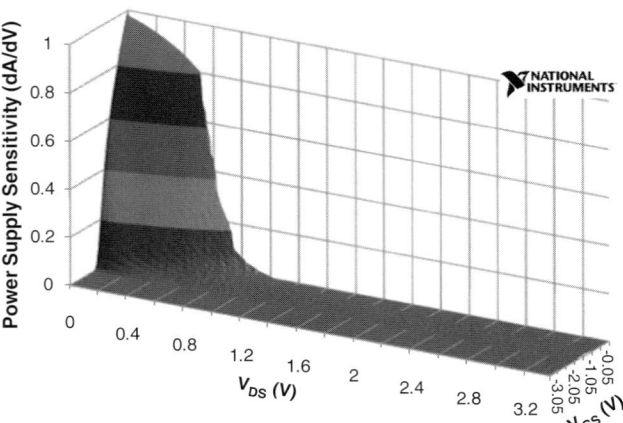

Figure 10-60   Power supply sensitivity surface for this GaAs MESFET. Current limits in the measurement restrict the completeness of this surface.

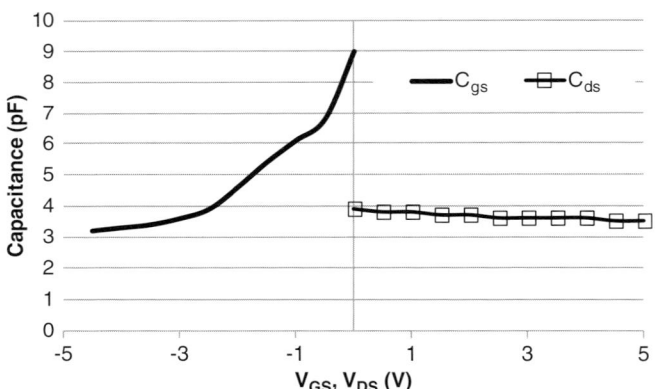

Figure 10-61   Measured parasitic capacitances for the GaAs MESFET at 100 MHz: input capacitance varies by nearly 3:1 as the channel forms above the pinch-off voltage, and output capacitance varies less than 12% across this operating voltage range.

$V_{DS} = 0.8$ V and $V_{DS} = 3.4$ V. Over this drain current range these two curves are parallel, which can only happen from a shift in the pinch-off voltage. This means that while the AC gain (transconductance) of this transistor does not vary with this supply value change, the drain current does change. This is a problem for linear amplifier use.

Calculating the PSS surface gives the result in Figure 10-60. The resistive region peak is at $V_{DS} = 0$, showing that there is no offset to the resistive characteristic of this transistor. Transition from the resistive region to the CCS region is reasonably fast, though the surface rendering is incomplete from the current limit of the measurement equipment used here.

Parasitic capacitance measurements for this depletion mode FET are presented in Figure 10-61. Input capacitance varies by nearly 3:1, but the output capacitance only varies by 11% across this voltage range. It is interesting to note that the value of the output

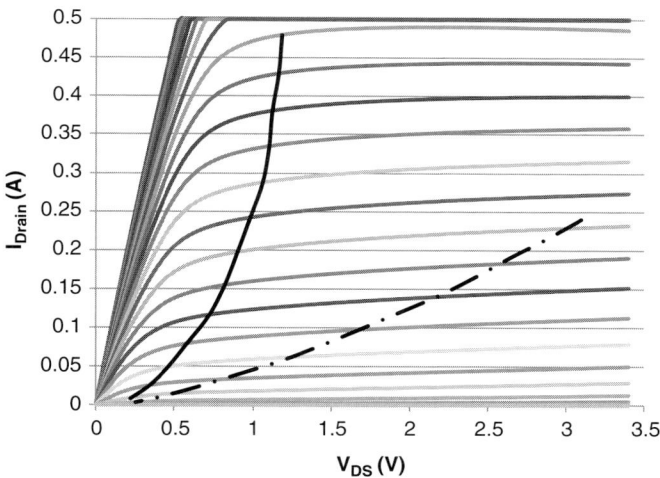

Figure 10-62 ET operation region for the GaAs MESFET RF power transistor. This transistor approaches the performance of (10.1) but never consistently gets there. The boundary for the CCS region meeting PSS = –30 dB is identified by the solid line. The dash–dot curve shows the dynamic bias profile this transistor must follow to simultaneously maintain amplifier linearity and maximize the power management dynamic range.

capacitance is near the minimum value of the input capacitance, enhancing the difference between the switching speed metric (8.13) and the common transistor speed metric f_T (8.15).

This characteristic curve set shows some locations where the CCS region boundary (10.1) is satisfied, but not a complete boundary. Therefore, the boundary for 30 dB of power supply noise rejection is evaluated and presented by the solid line in Figure 10-62. Also included is the associated dynamic bias profile that maximizes the ET dynamic range and corresponding amplifier linearity.

How this transistor transitions from envelope tracking to polar operation is investigated with the results presented in Figure 10-63. This transition is gradual, which does correspond to the slope of the PSS surface in Figure 10-60. Unlike most of the FET devices evaluated earlier, the knee voltage profile here has smaller variation.

When the transition to polar operation is complete, the modulation accuracy that results is shown in Figure 10-64. Accuracy is very consistent across the entire tested dynamic range of just over 27 dB.

10.6.2   GaAs HBT

Development of bipolar transistors in GaAs followed many years after availability of MESFET devices. Called heterostructure bipolar transistors (HBT), the measured characteristic curves from an example cell of a larger transistor are presented in Figure 10-65. The full characteristic curves are in Figure 10-65(a), and do resemble the idealized goal from Figure 10-1. Looking close-up at the IV origin, we see in

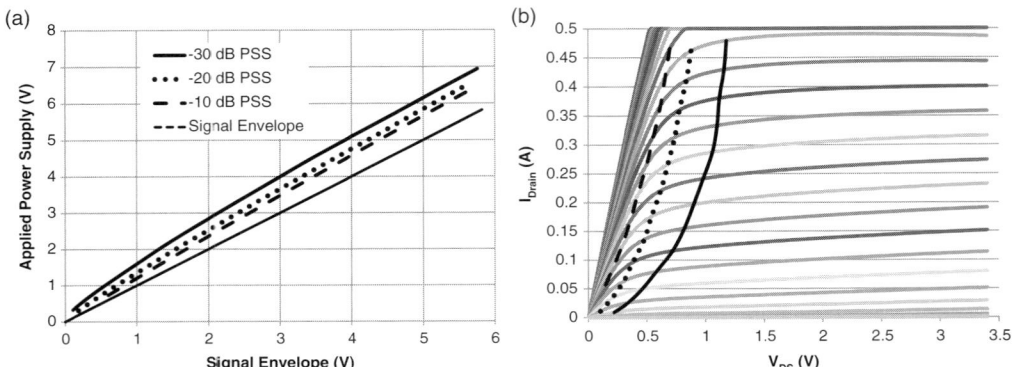

Figure 10-63 ET characteristics for the GaAs MESFET RF power transistor: (a) DPS voltage profiles for power supply noise suppression of –30 dB, –20 dB, and –10 dB; (b) corresponding ET endpoint trajectories for these three values of power supply noise suppression.

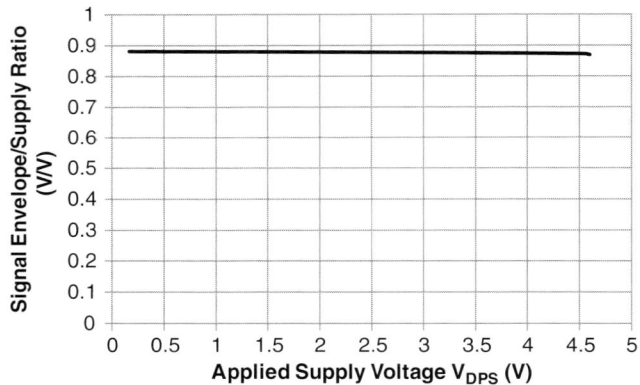

Figure 10-64 Envelope control precision and accuracy for the GaAs MESFET in polar DPST operation. Polar envelope control is accurate over the entire tested dynamic range of about 27 dB.

Figure 10-65(b) that there are significant differences from the idealized goal in Figure 10-2. These general differences have been seen before, in the silicon BJT data from Figure 10-10(b).

Calculating the β and channel resistance surfaces from this characteristic curve data provides the corresponding surfaces presented in Figure 10-66. The β surface is very flat, signifying that this transistor is carefully optimized for inherent linearity in amplifier applications. These β values are between those of the two silicon bipolar transistors evaluated in Sections 10.5.1 and 10.5.4. It is very important to also note that these β values are all measured at DC. At RF frequencies, the shunting effect of the HBT input capacitance described in Section 7.3.1 and shown in Figure 7-15 is active which reduces the β value realized at RF.

The channel resistance surface in Figure 10-66(b) shows a definite minimum at a voltage offset above $V_{CE} = 0$. In the minimum, it is evident that the resistance value is

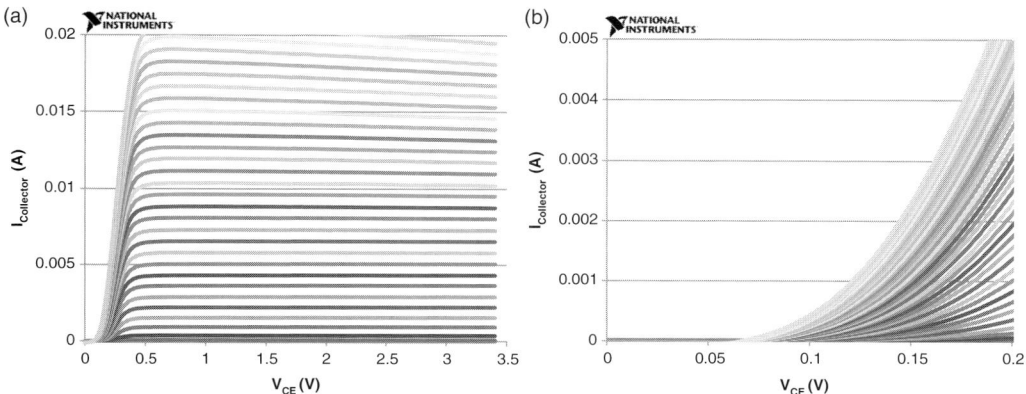

Figure 10-65   Characteristic curves for a GaAs HBT RF power transistor cell: (a) measured curve family with stepped base current values; (b) close-up view of these curves near the IV plane origin.

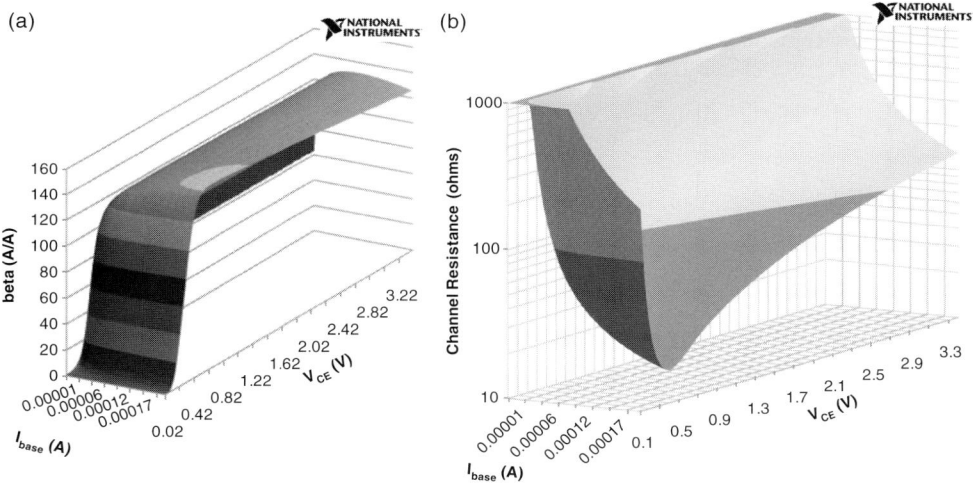

Figure 10-66   First-order analysis results from the GaAs HBT characteristic curves: (a) current gain β surface; (b) transistor absolute resistance surface.

still decreasing, showing that under these test conditions this transistor is not yet acting like a switch. This observation is confirmed with the transfer relation measurements in Figure 10-67 where none of the curves reach the horizontal slope at the highest test base current. Linearity and gain both appear to maximize near $V_{CE} = 2.5$ V. A turn-on offset is also present in all of these measurements.

The PSS surface calculated from these measurements is presented in Figure 10-68. This surface has characteristics that strongly resemble the result in Figure 10-13. Here though the resistive region peak is offset higher in voltage, and lower in peak magnitude. Transitioning from the resistive to the CCS region is similarly rapid.

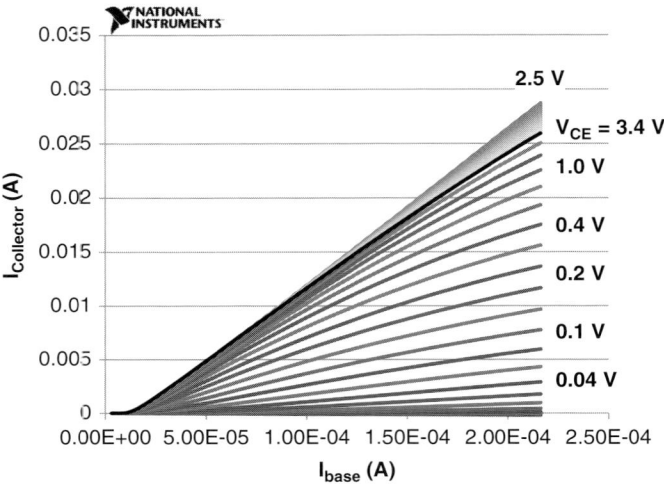

Figure 10-67   The set of current transfer functions corresponding to Figure 10-65(a). Gain is proportional to the slope of these curves, and linearity is related to their straightness. A small offset is evident as this transistor turns on.

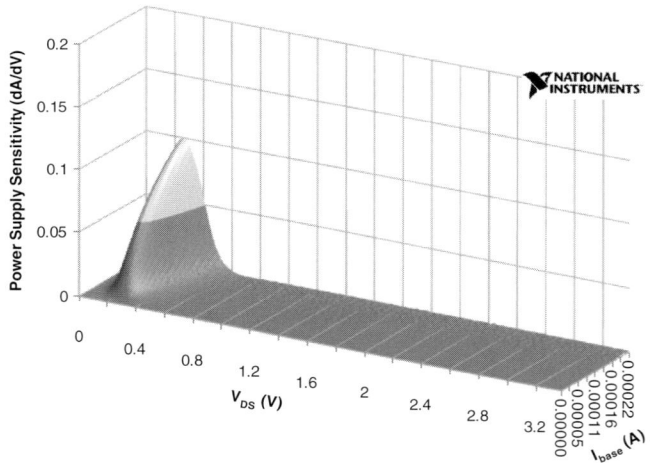

Figure 10-68   Power supply sensitivity surface for this GaAs HBT cell.

Evaluation of this transistor data for ET behavior readily finds the intended CCS region boundary of (10.1). This boundary is presented in Figure 10-69 by the solid line. The associated dynamic bias profile that maximizes linearity and control dynamic range is also shown here by the dash–dot curve. The offset needed to avoid P-mode operation is readily noted at the bottom end of both curves.

The corresponding DPS profile for the ET operation is shown in Figure 10-70(a) by the solid line. This predicted curve is very similar to the expected principle established in

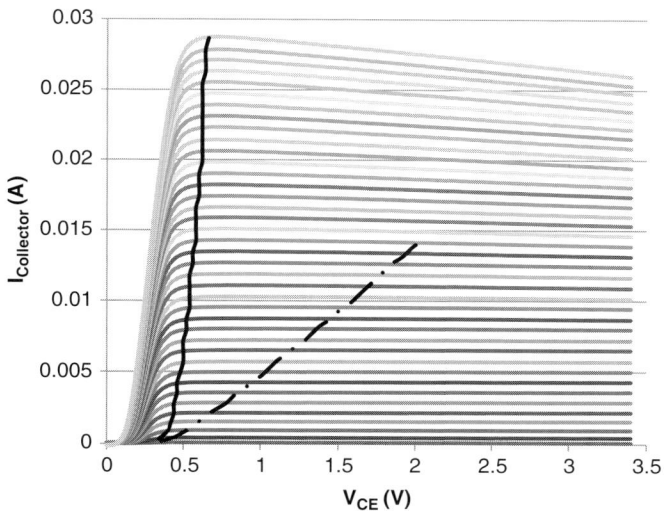

Figure 10-69 ET operation region for the GaAs HBT RF power transistor cell. The boundary for the CCS region meeting (10.1) is identified by the solid line. The dash-dot curve shows the dynamic bias profile this transistor must follow to simultaneously maintain amplifier linearity and to maximize the power management dynamic range.

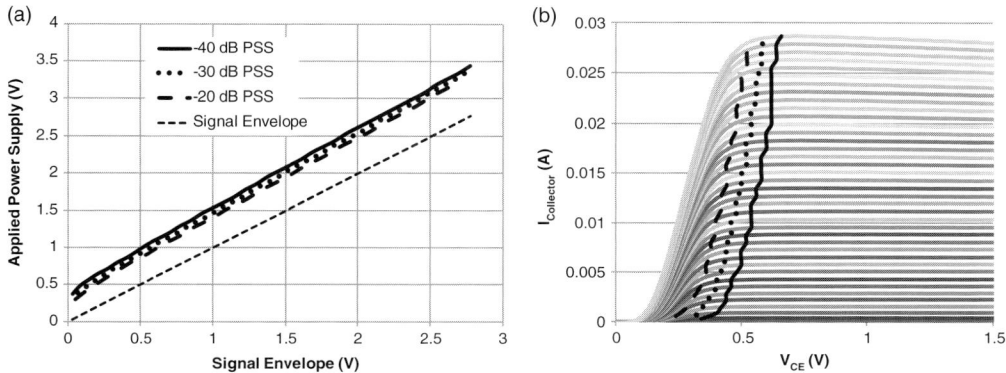

Figure 10-70 ET characteristics for the GaAs HBT RF power transistor cell: (a) DPS voltage profiles for power supply noise suppression of –40 dB, –30 dB, and –20 dB; (b) corresponding ET endpoint trajectories for these three values of power supply noise suppression.

Section 5.4.2 and shown in Figure 5-21. Evaluating the transition from envelope tracking to polar operation finds that the power supply noise suppression is very rapidly lost as the power supply value falls below this boundary.

When the conversion to polar operation is complete, according to Figure 10-70(a) the knee voltage falls by almost another 0.5 V and the PA energy efficiency increases a corresponding amount from the lower power dissipation. The best-fit modulation triangle requires an offset of 120 mV, the largest offset needed so far in this transistor technology survey. Modulation accuracy that follows from this best-fit triangle is shown in

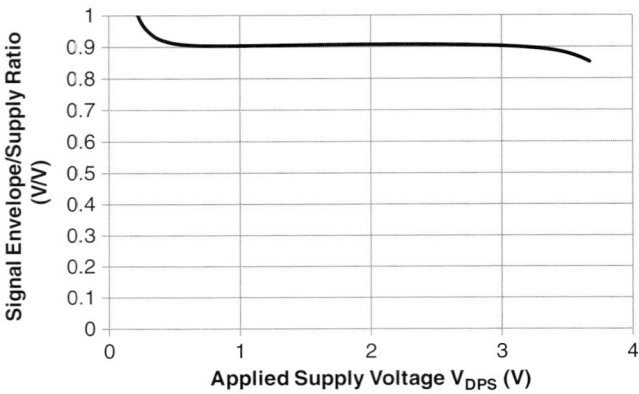

Figure 10-71   Envelope control precision and accuracy for the GaAs HBT in polar DPST operation. Polar envelope control is accurate over a dynamic range of about 15 dB. Distortion from curvature in the characteristic curves at low voltages (Figure 10-65(b)) is evident.

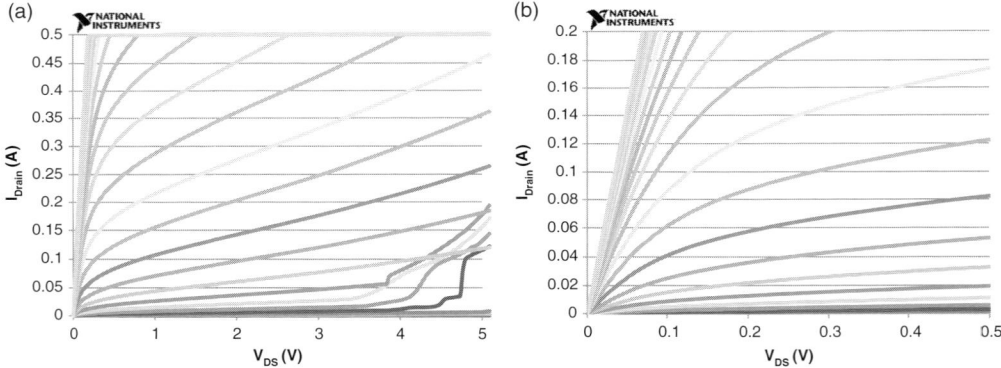

Figure 10-72   Characteristic curves for a GaAs pHEMT RF power transistor: (a) measured curve family with stepped V_{GS} values, including unstable behavior in the lower current/higher voltage region; (b) close-up view of these curves near the IV plane origin.

Figure 10-71. The accurate modulation dynamic range is 15 dB, limited on the low end by the characteristic curve curvature seen in Figure 10-65(b) and on the high end by the transition characteristics from resistive region operation toward CCS region operation.

10.6.3   GaAs pHEMT

The next GaAs FET developed after the MESFET is the pseudomorphic high electron mobility transistor (pHEMT). Characteristic curves from an example of a GaAs pHEMT are provided in Figure 10-72. The CCS region seen in Figure 10-72(a) does not exhibit the horizontal slope goal identified in Figure 10-1 for an idealized transistor for DPST application. Also exhibited in Figure 10-72(a) is the presence of device oscillation when

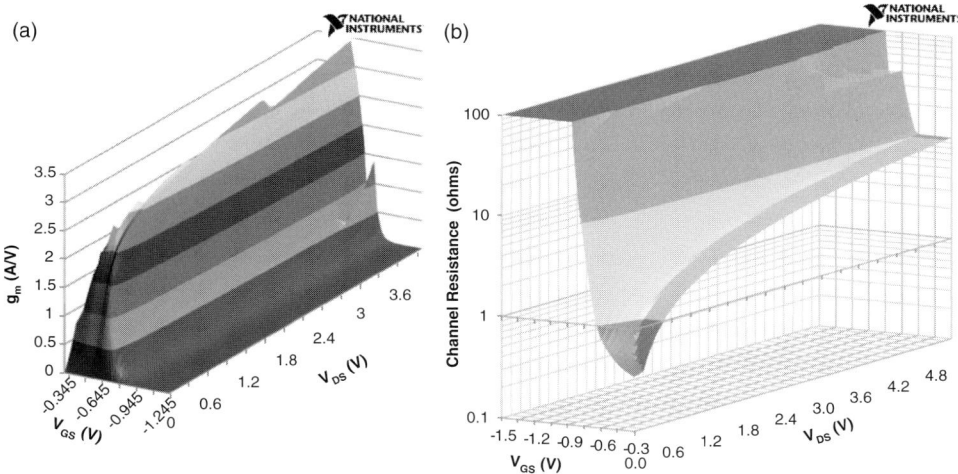

Figure 10-73 First-order analysis results from the GaAs pHEMT characteristic curves: (a) transconductance (g_m) surface, with restricted rendering from the current limiting in the measurement system; (b) transistor absolute resistance surface.

V_{DS} is high and V_{GS} is low. Removing these oscillations can be done, but it is informative here to present how the presence of such circuit instabilities are manifest in conventional characteristic curve measurements.

For the close-up look at the characteristic curves near the IV origin presented in Figure 10-72(b), the behavior is not affected by circuit instabilities. At higher V_{GS} values, this device acts as a voltage controlled resistor. At lower values of V_{GS}, the device appears to act more like a switch. This will be checked in later analyses.

Calculating the transconductance and absolute channel resistance surfaces from these characteristic curve measurements provides the graphics shown in Figure 10-73. The transconductance surface shows very high values but not much linearity. Distortions in what would otherwise be a smooth surface result from both the presence of circuit oscillation and encountering the current limiting in the measurement system at 0.5 A into the transistor.

The channel resistance of this transistor, seen in Figure 10-73(b), shows that the GaAs pHEMT gets to a relatively low ON resistance at the usual conditions of 0 V V_{DS} and higher values of V_{GS}. This channel resistance is not at a fixed value, so this transistor is not yet behaving as a switch.

Several of the calculated transfer relations for this transistor are shown in Figure 10-74. None of these curves reaches the horizontal slope at higher V_{GS} values, confirming that this transistor is not yet achieving switching behavior. It is interesting to note that as the applied power supply increases, the transconductance increases and also there is a small shift to the device pinch-off voltage. Both of these effects increase the drain current, and are amplifier distortion mechanisms.

The corresponding PSS surface is presented in Figure 10-75. The peak of the resistive region is at $V_{DS} = 0$ V which confirms that there is no voltage offset in this

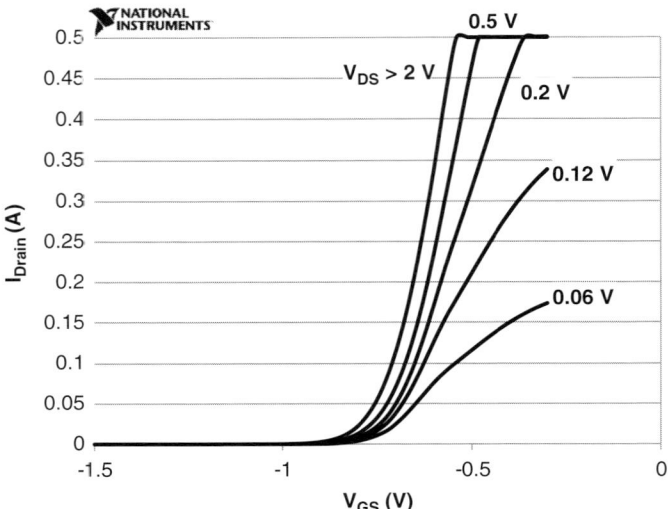

Figure 10-74   A subset of transfer functions corresponding to Figure 10-72(a). Gain is proportional to the slope of these curves, and linearity is related to their straightness. A small threshold shift is seen across these applied V_{DS} supply values.

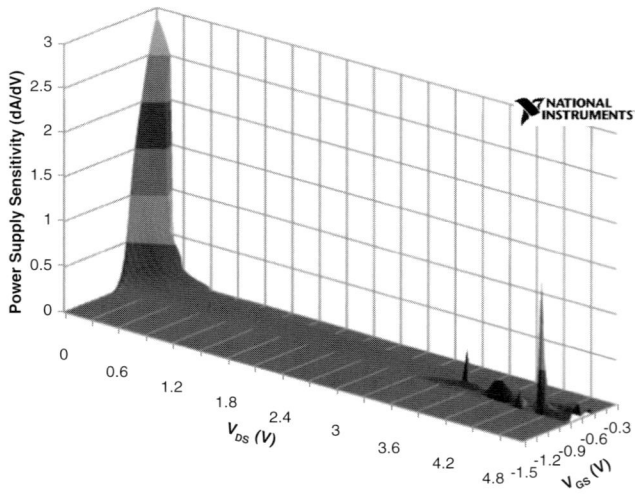

Figure 10-75   Power supply sensitivity surface for this GaAs pHEMT. Current limiting in the measurement system restricts the completeness of this surface. Note the instability at higher V_{DS} and lower V_{GS} which is also seen in Figure 10-72(a).

transistor to its resistive operation region. Spurious responses are present at the high device voltage and low control voltage condition that is already known to have circuit instability. These spurious responses make this instability very evident.

Parasitic capacitance measurements for this GaAs pHEMT are provided in Figure 10-76. As for the MESFET, the input capacitance varies widely, here more than 3:1 as the

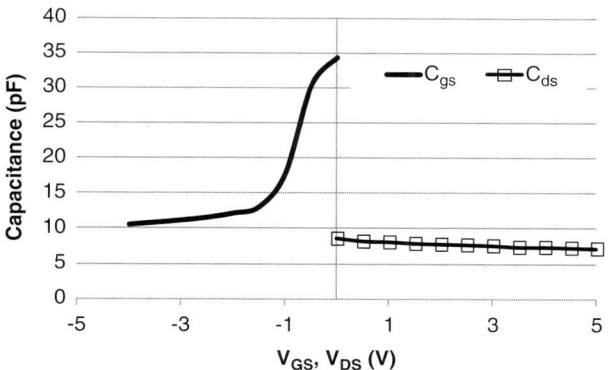

Figure 10-76   Measured parasitic capacitances for the GaAs pHEMT at 100 MHz: input capacitance varies by more than 3:1 as the channel forms above the pinch-off voltage, and output capacitance both varies by less than 20% across this operating voltage range and has a value always less than the minimum input capacitance.

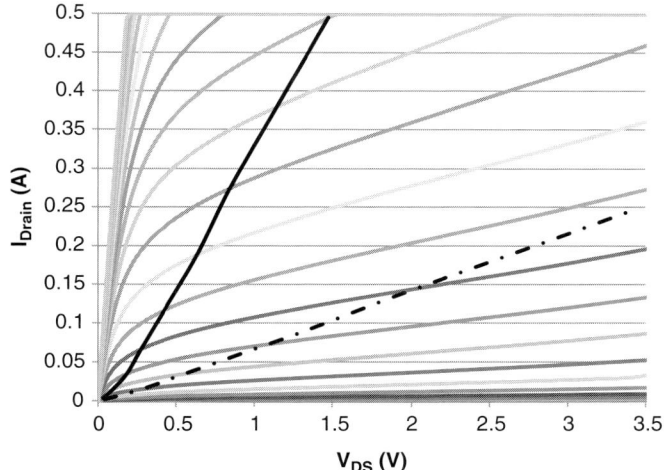

Figure 10-77   ET operation region for the GaAs pHEMT RF power transistor. This transistor approaches the performance of (10.1) but never quite gets there. The boundary for the CCS region meeting PSS = –30 dB is identified by the solid line. The dash-dot curve shows the dynamic bias profile this transistor must follow to simultaneously maintain amplifier linearity and maximize the power management dynamic range.

transistor turns on. Also like the MESFET, the output capacitance is almost constant, varying by less than 20% across the measured voltage range.

Without horizontal CCS region curves, there is no possibility that this transistor can meet the optimum ET power supply value independence defined in (10.1). This transistor actually barely meets a power supply noise suppression of 30 dB. The evaluated CCS region boundary identified by the solid line in Figure 10-77 is evaluated at this 30 dB power supply noise suppression performance value.

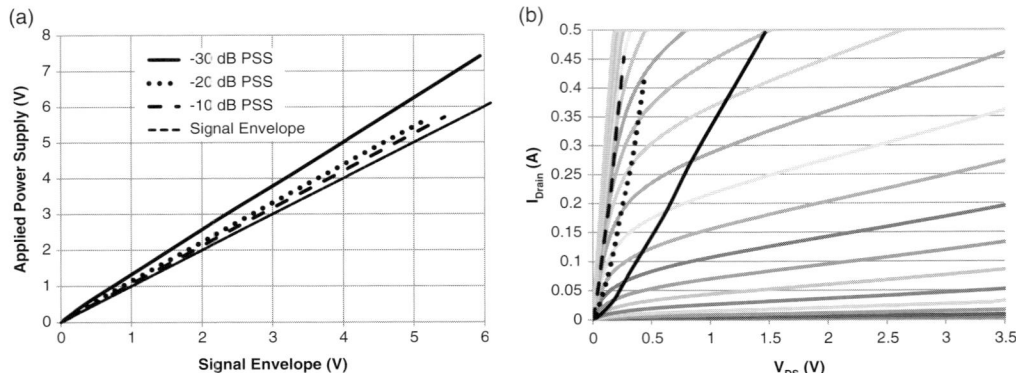

Figure 10-78 ET characteristics for the GaAs pHEMT RF power transistor: (a) DPS voltage profiles for power supply noise suppression of –30 dB, –20 dB, and –10 dB; (b) corresponding ET endpoint trajectories for these three values of power supply noise suppression.

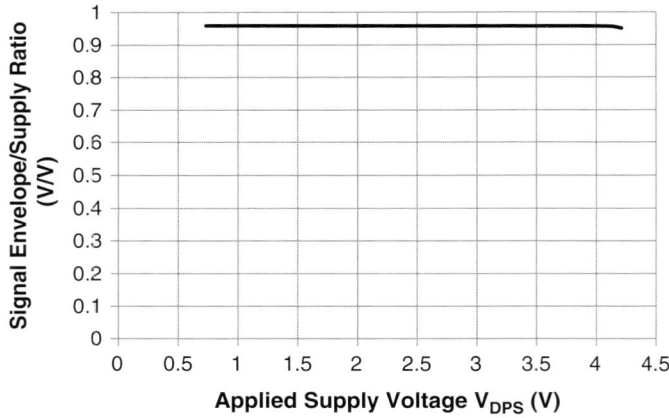

Figure 10-79 Envelope control precision and accuracy for the GaAs pHEMT in polar DPST operation. Polar envelope control is accurate over the entire tested dynamic range of about 26 dB.

Examining more of the transition from (barely) envelope tracking toward polar modulation is seen in the power supply voltage profiles in Figure 10-78(a). Unlike the rapid transition seen for the GaAs HBT, this transition is initially slow and then speeds up as polar modulation is reached. The corresponding boundaries at these reduced power supply noise rejection values are shown in Figure 10-78(b).

Once the transition to polar operation is completed, the resulting modulation accuracy is shown in Figure 10-79. Polar modulation accuracy is excellent across the entire tested dynamic range of 26 dB. With this excellent polar modulation performance, just keeping to polar modulation and avoiding the circuit instability problems in Figure 10-72 is an attractive possibility.

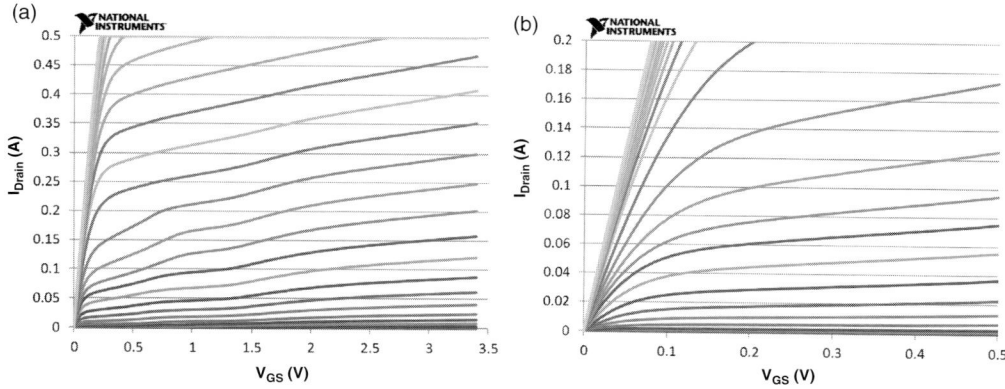

Figure 10-80 Characteristic curves for a GaAs EpHEMT RF power transistor: (a) measured curve family with stepped values of V_{GS}, hinting at some instabilities around 1 V for V_{DS}; (b) close-up view of these curves near the IV plane origin.

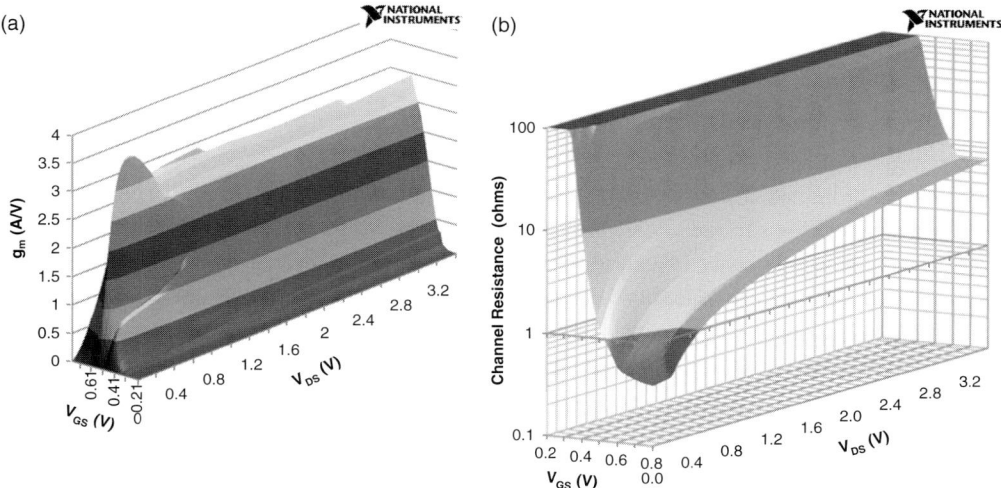

Figure 10-81 First-order analysis results from the GaAs EpHEMT characteristic curves: (a) transconductance (g_m) surface; (b) transistor absolute resistance surface.

10.6.4   GaAs EpHEMT

A more recent development is the enhancement mode GaAs pHEMT. Characteristic curve measurements are presented in Figure 10-80(a), with the corresponding close-up view of performance near the IV plane origin seen in Figure 10-80(b). Here the possible presence of circuit instabilities is not as obvious as in Figure 10-72(a), but undulations in curves around 1 V and below 0.25 A can be seen.

Calculating the surfaces for device transconductance and channel resistance from these measurements provides the results in Figure 10-81. The transconductance

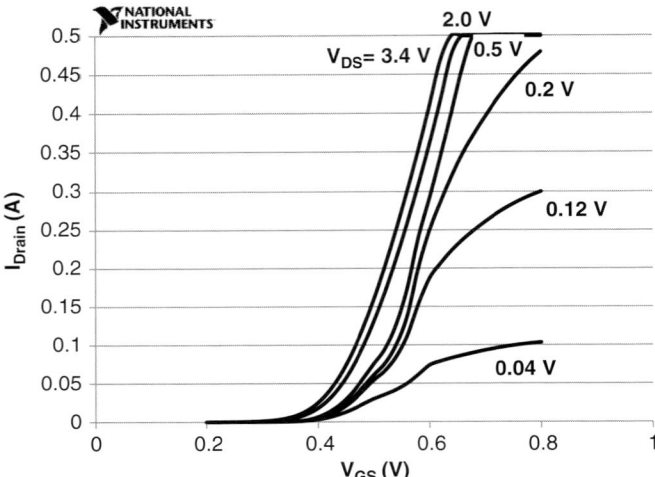

Figure 10-82   A subset of transfer functions corresponding to Figure 10-80(a). Gain is proportional to the slope of these curves, and inconsistent curvature as the transistor turns on at low applied V_{DS} hints at stability problems. Threshold behavior is consistent across all applied V_{DS} supply values.

shares similar high values as are seen for the GaAs pHEMT. Waves in this surface show the transconductance effect of the behavior noted in the characteristic curve measurements. Current limiting in the measurement system restricts complete rendering of this surface.

Channel resistance behavior of this transistor from Figure 10-81(b) shows minimum values at low V_{DS} and higher V_{GS} conditions. The surface gradient is not so high near this minimum, so this transistor is approaching switching behavior at these operating conditions. This observation is confirmed by the transfer relation measurements presented in Figure 10-82. At lower V_{DS} values, the transfer curves do begin to approach the horizontal slope indicative of switching behavior. Threshold shifting with varying V_{DS} values is also seen, which is not good for the linear amplifier performance needed by envelope tracking DPST designs.

Calculating the PSS surface provides the result presented in Figure 10-83. Variations in the characteristic curves are manifest as waves on this surface. Transitions from the resistive region to the CCS region are affected by these waves.

Parasitic capacitance measurements for this EpHEMT are provided in Figure 10-84. The input capacitance varies by 4:1 across the FET-active region, the largest variation seen so far. The output capacitance is comparatively benign, varying by only 33% across this operating voltage. It is interesting to note that the value of the output capacitance is always below the input capacitance value for this transistor. This helps its switching performance.

The presence of characteristic curve undulations is very obvious when this transistor is analyzed to find the optimum CCS region boundary for ET design. There is no possibility of meeting the goal of (10.1), so the first possible boundary is for 30 dB of

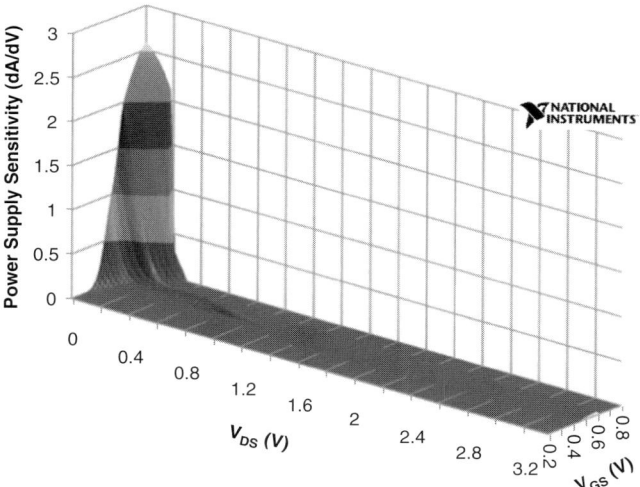

Figure 10-83 Power supply sensitivity surface for this GaAs EpHEMT. Curvature changes in the characteristic curves of Figure 10-80 manifest here as "waves" on the surface. Current limiting in the measurement restricts completeness of this surface at higher bias levels.

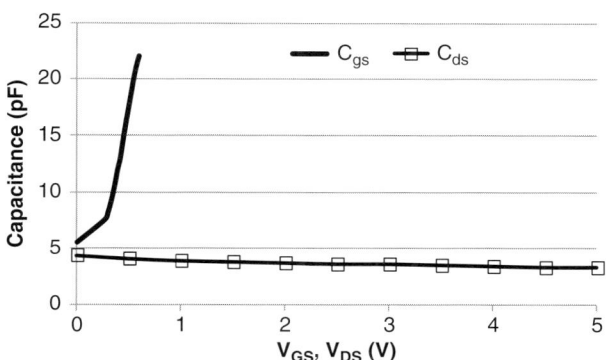

Figure 10-84 Measured parasitic capacitances for the GaAs EpHEMT at 100 MHz: input capacitance varies by nearly 4:1, and output capacitance both varies by about 30% across this operating voltage range and has a value always less than the minimum input capacitance.

power supply noise rejection. The evaluation algorithm faithfully follows the PSS surface waves, and the resulting boundary is shown in Figure 10-85. This result does look unusual, yet it is algorithmically correct. From these results, it is advisable to pay closer attention to whether these undulations are an actual artifact of the transistor or are a result of nonoscillation circuit instabilities.

Following the boundary search algorithm anyway, the transition from not quite envelope tracking toward polar modulation is seen in Figure 10-86. For the moment, ignoring the results at and below $V_{DS} = 3$ V, this transistor does exhibit a rather gradual transition between these DPST operating modes.

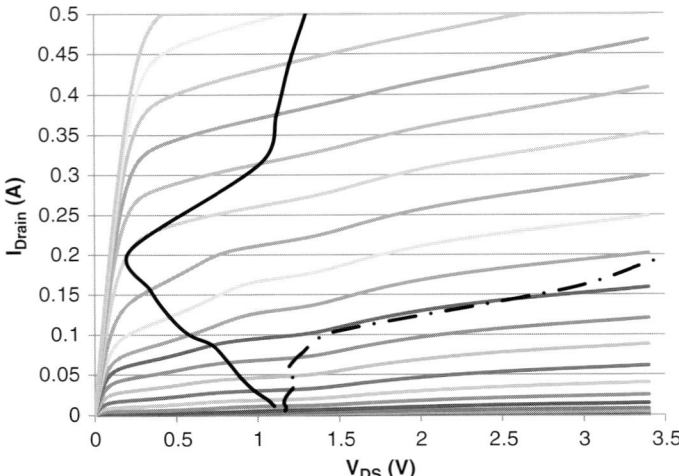

Figure 10-85   ET operation region for the GaAs EpHEMT RF power transistor. The boundary for the CCS region meeting PSS = −30 dB is identified by the solid line. The optimization algorithm faithfully follows the "waves" in the surface of Figure 10-83 The dash-dot line shows the corresponding dynamic bias profile this transistor must follow to simultaneously maintain amplifier linearity and maximize the power management dynamic range.

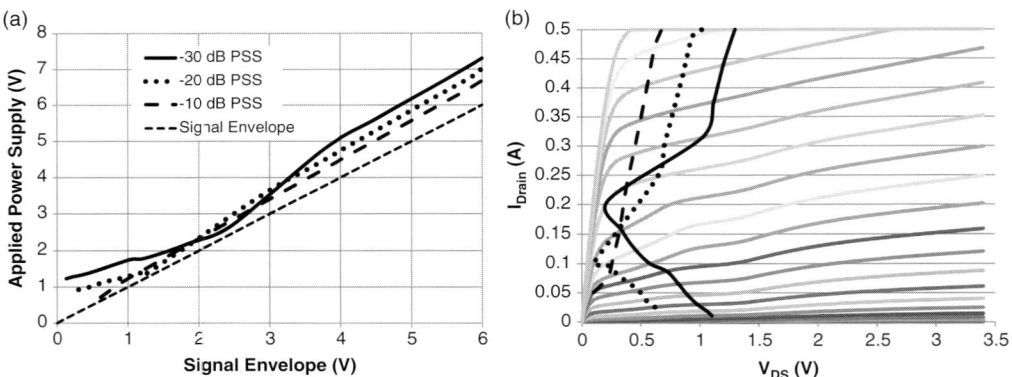

Figure 10-86   ET characteristics for this GaAs EpHEMT RF power transistor: (a) DPS voltage profiles for power supply noise suppression of −30 dB, −20 dB, and −10 dB; (b) corresponding ET endpoint trajectories for these three values of power supply noise suppression. Interaction with the device curvature changes is very different at these three boundaries.

When the transition to polar operation is complete, the resulting modulation accuracy is shown in Figure 10-87. This is excellent performance, and according to the polar principles of Chapter 6 is completely independent of any CCS region behavior. Thus, when used as a polar modulator this transistor eliminated any need for the engineering team to spend time investigating the unusual undulations in the characteristic curves. Even if wide power control dynamic range is required, when the technique of Section 6.12.3 is followed this transistor works very well as it is.

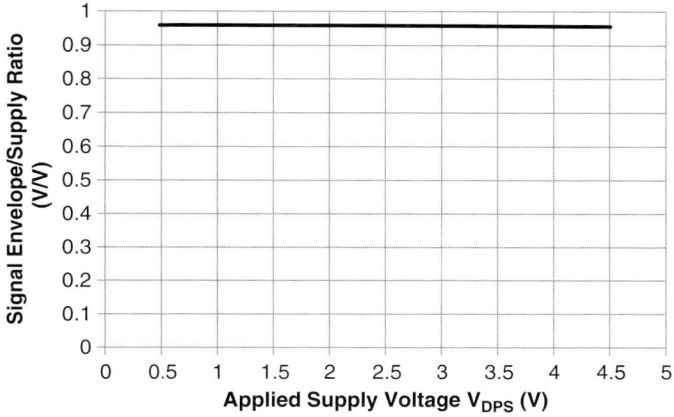

Figure 10-87 Envelope control precision and accuracy for the GaAs EpHEMT in polar DPST operation. Polar envelope control is accurate over the entire tested dynamic range of about 18 dB.

10.7   Gallium nitride (GaN) technologies

Gallium nitride is a relatively new material available to transmitter designers. It has two important properties that are important to this application: (1) the ability to operate at much higher voltages than any other technology except LDMOS, and (2) parasitic capacitance values that are a small fraction of the comparable values from LDMOS transistors. Unlike for GaAs technologies, the GaN community is beginning with HEMT device technology.

10.7.1   GaN HEMT

GaN HEMT devices are inherently depletion mode, similar in circuit design to the GaAs MESFETs. Characteristic curve measurements at lower operating voltages are presented in Figure 10-88. These both show promise of useful behavior for DPST use.

The calculated transconductance and channel resistance surfaces corresponding to the measured characteristic curves are presented in Figure 10-89. The transconductance rises rapidly, and then this transistor quickly draws more current than this measurement system can provide. The corresponding channel resistance data seen in Figure 10-89(b) shows both a minimum value at $V_{DS} = 0$ and high V_{GS}, and a flat floor in the resistance performance. This signifies that this transistor is operating as a switch under these conditions.

Looking at the transfer relationships in Figure 10-90 shows several important results. First, the transfer curves are nearly horizontal at higher V_{GS} values, confirming the switching behavior observation from Figure 10-89(b). Second, there is no discernible threshold shift with varying V_{DS}, which is good for the possible use of this device in ET designs.

Moving now to the PSS surface in Figure 10-91, the peak of the resistive region is at $V_{DS} = 0$, which confirms that there is no offset to the resistive region of this transistor. The transition to CCS operation is very gradual, and is not completely rendered here

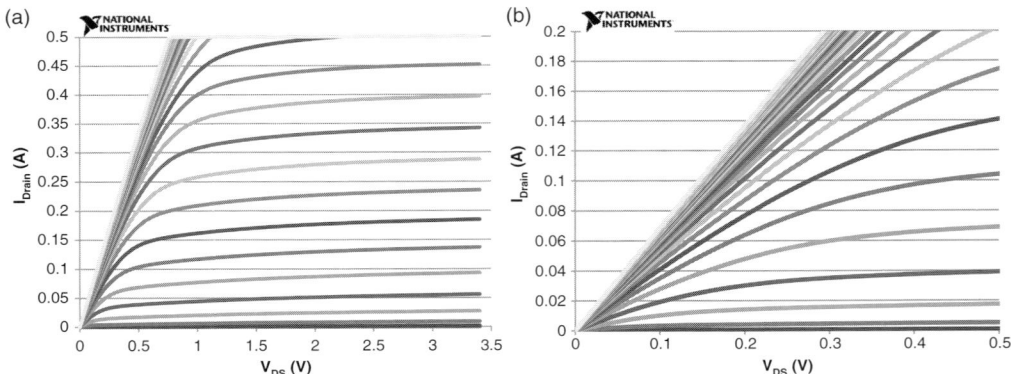

Figure 10-88   Characteristic curves for a GaN HEMT RF power transistor: (a) measured curve family with stepped V_{GS} values; (b) close-up view of these curves near the I–V plane origin.

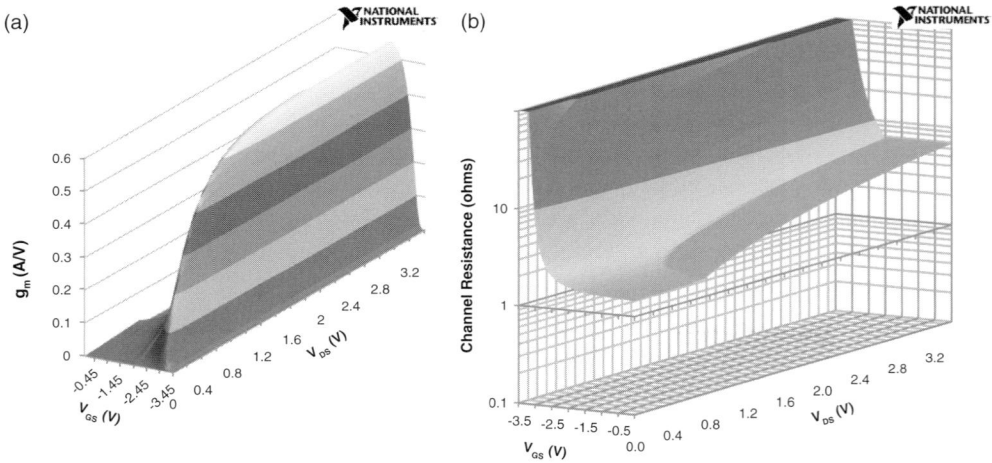

Figure 10-89   First-order analysis results from the GaN HEMT characteristic curves: (a) transconductance (g_m) surface; (b) transistor absolute resistance surface.

due to the current limiting of the test system used for the characteristic curve measurements.

Parasitic capacitance measurement results are provided in Figure 10-92. The input capacitance varies by just over 2:1, while the output capacitance stays within 12% of its nominal value across the tested voltage range. At all times, the output capacitance value is below the smallest input capacitance value. This says that the switching behavior of this transistor will be faster than would be expected from predictions based on the common (but not applicable here) transistor speed metric f_T, as discussed in Section 8.4.

This GaN HEMT does exhibit a boundary meeting (10.1), and this result is shown with the solid line in Figure 10-93. Along with this CCS region boundary is the optimum dynamic bias profile for the ET operation. There is a long shift in this boundary to a knee voltage exceeding 2 V, then the knee voltage increases much

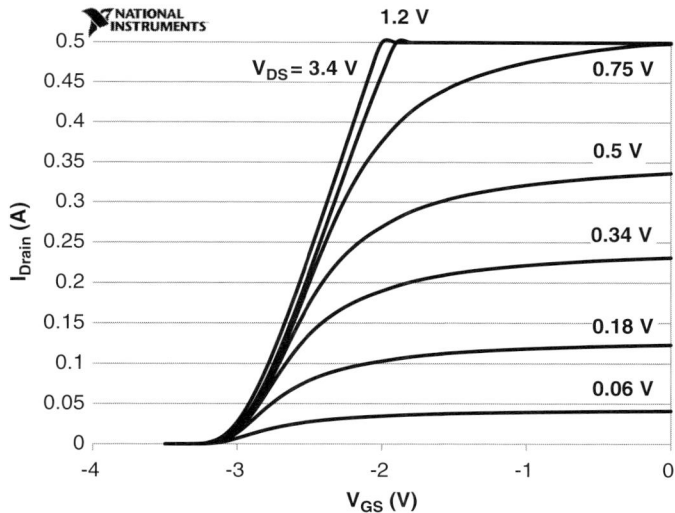

Figure 10-90 A subset of transfer functions corresponding to Figure 10-88(a). Gain is proportional to the slope of these curves, and linearity is related to their straightness. Threshold behavior is consistent across all applied V_{DS} supply values.

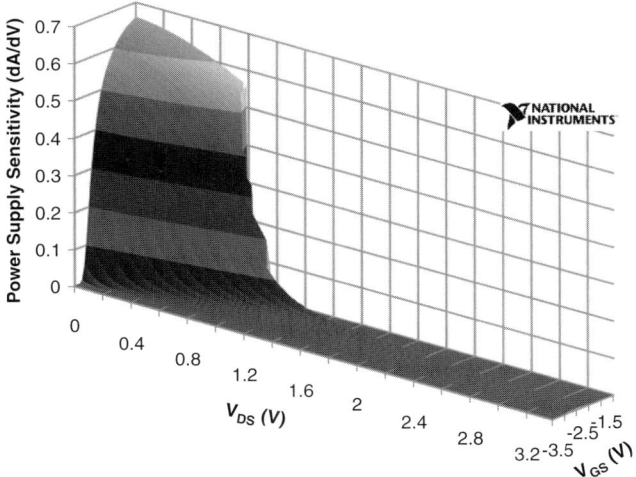

Figure 10-91 Power supply sensitivity surface for this GaN HEMT. Current limiting in the measurement system restricts completeness of this surface at higher bias levels.

more slowly. Inspecting the associated characteristic curves in Figure 10-93, we note that there is a slope change in the characteristics at voltages below this boundary. This slope change reduces the power supply noise suppression enough to force the ET optimum boundary out this far.

Evaluating the transition from this envelope tracking operation toward polar operation provides the results in Figure 10-94. Reducing the power supply noise suppression to 30

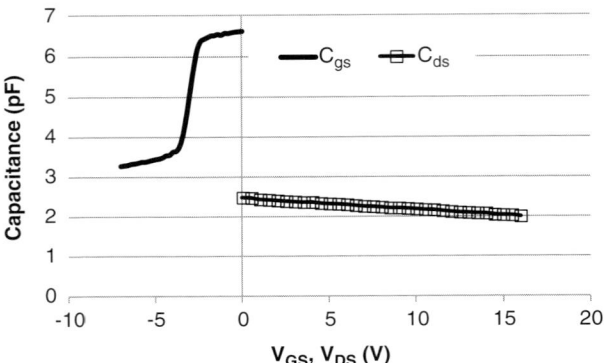

Figure 10-92   Measured parasitic capacitances for the GaN HEMT at 100 MHz: input capacitance varies by more than 2:1 as the channel forms above the pinch-off voltage, and output capacitance both varies by about 20% across this operating voltage range and has a value always less than the minimum input capacitance.

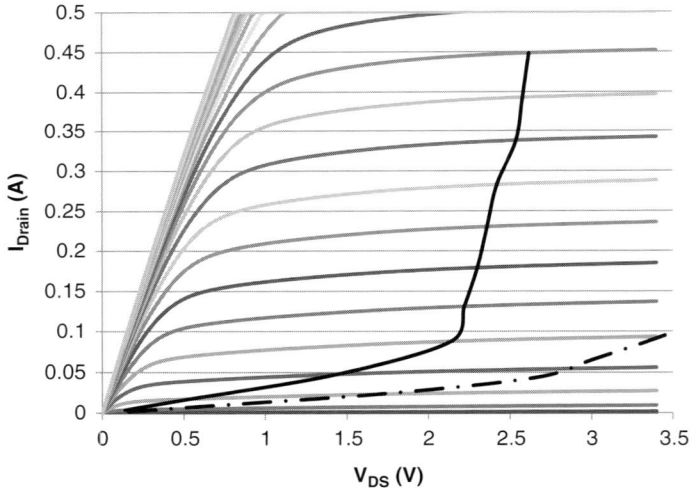

Figure 10-93   ET operation region for the GaN HEMT RF power transistor. The boundary for the CCS region meeting (10.1) is identified by the solid line. The dash-dot line shows the dynamic bias profile this transistor must follow to simultaneously maintain amplifier linearity and maximize the power management dynamic range.

dB eliminates the long shift seen in the "official" CCS region boundary. This provides a dramatic drop to the knee profile, at the price of greater power supply sensitivity. Continuing on to even lower supply suppression performance shows a gradual transition toward to polar modulation.

Once the transition to polar modulation is complete, accuracy of the polar modulation is very good as seen in Figure 10-95. This polar modulation remains at a very high quality across the entire tested dynamic range, which here is nearly 30 dB.

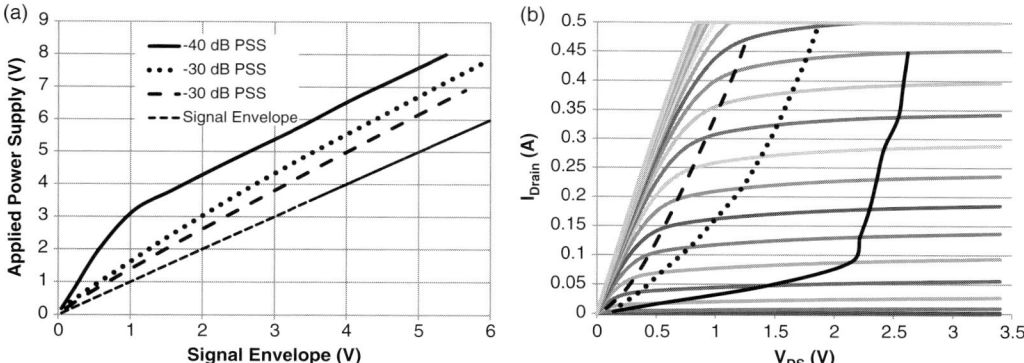

Figure 10-94 ET characteristics for the GaN HEMT RF power transistor: (a) DPS voltage profiles for power supply noise suppression of –40 dB, –30 dB, and –20 dB; (b) corresponding ET endpoint trajectories for these three values of power supply noise suppression.

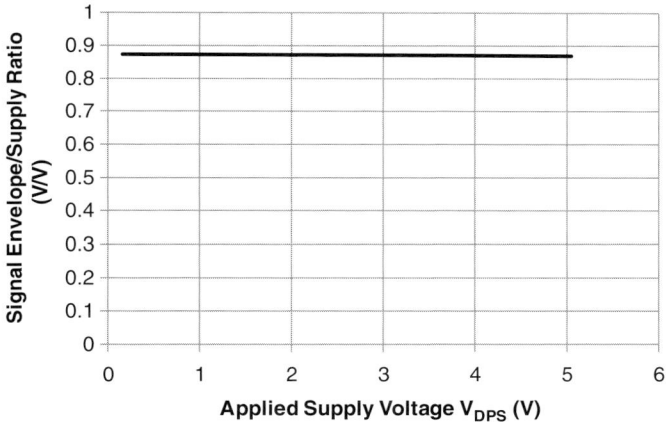

Figure 10-95 Envelope control precision and accuracy for this GaN HEMT in polar DPST operation. Polar envelope control is accurate over the entire tested dynamic range of about 30 dB.

The GaN devices are designed to operate at fixed power supply voltages between 20 and 50 V. Characterizations of two other commercial GaN HEMT devices out to 28 V are presented in Figure 10-96 and Figure 10-97. These extended characteristic curves are very different from each other. The transistor used for the measured data in Figure 10-96(a) exhibits dramatic modulation in the channel characteristics across both V_{DS} and V_{GS}. Gain is much higher in the higher voltage region, and with nearly equal spacing between these individual curves one expects that the amplifier would be very linear. The effective knee voltage at highest output is close to 10 V, which may be acceptable for fixed supply linear amplifier operation. If DPST operation is desired, the knee voltage profile will actually increase toward 15 V as the supply is lowered. This will work in principle, but the knee voltage is so large that baseline power dissipation will

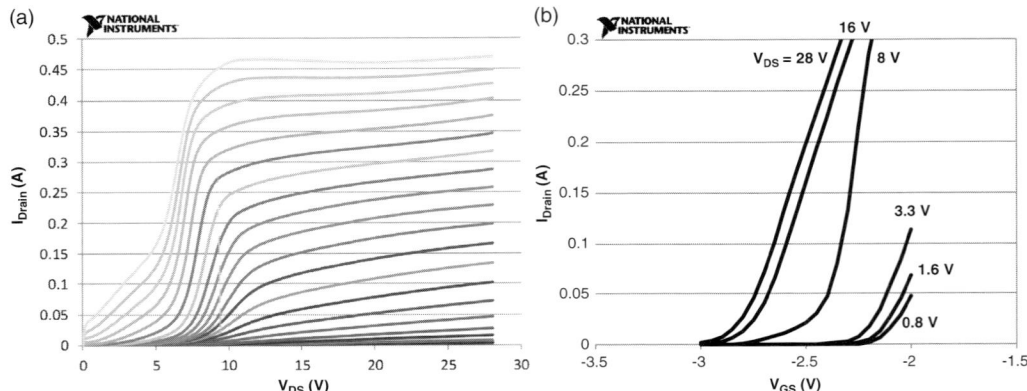

Figure 10-96 Higher voltage characterization of a GaN HEMT: (a) IV characteristic curves showing very significant channel modulation of some sort; (b) shift in device threshold as the power supply changes.

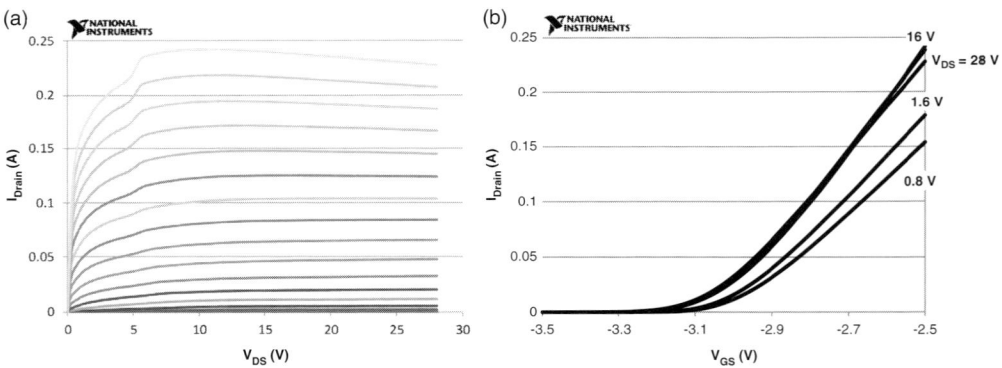

Figure 10-97 Higher voltage characterization of a different GaN HEMT from that in Figure 10-96: (a) IV characteristic curves showing much less channel modulation; (b) shift in device threshold, as the power supply changes, is present but is much less than that in Figure 10-96.

remain very high. This is not a good device to select for ET or polar DPST operation.

The threshold shift exhibited by this GaN transistor is shown in Figure 10-96(b). With this type of behavior, it is extremely difficult to predict how an ET amplifier will behave with this transistor.

All GaN transistors are not created equal, as seen by measuring a different transistor (from a different manufacturer) with results presented in Figure 10-97. The characteristic curves in Figure 10-97(a) look almost reasonable. They still have some form of channel variations, but these are much smaller than the measurements in 10-96(a). The transfer relation data in Figure 10-97(b) also show threshold shifting, though at a much smaller amount. For ET operation, this transistor might be usable, but there will be significant distortion that may require elaborate linearization techniques to correct.

Operating this transistor in a polar DPST does avoid most of these anomalies. With polar modulation, the threshold shift in Figure 10-97(b) has no impact. Nonideal behaviors in the CCS region also do not matter.

10.7.2   GaN E-HEMT

To date, the maturing GaN technology is all depletion mode. Enhancement mode GaN devices, call them eGaN or GaN eHEMT, are new developments that show a lot of promise.

At the time of this writing, the development of eGaN for RF use is too new to be included in this technology survey.

10.8   Backgating/output lag

Output lag is a memory effect in some transistors where the power available from specific bias conditions depends on what the output power has been in the past. Output lag is a serious problem in any DPST design. Experience so far shows that output lag is a particularly serious problem with transistor types manufactured with ion implant techniques, and is not as much of a problem with heterostructure devices.

One measurement of output lag behavior is shown in Figure 10-98, in this case measured on a GaN HEMT. Following eight high power pulses, the lower power pulses immediately following the high power pulses are slightly suppressed from their intended value. The lower value pulse magnitude does eventually recover to the intended value. In this measurement, the recovery time is about 45 milliseconds.

Not all transistor types exhibit output lag effects. This is not a conventional transistor characterization parameter, so it must be measured on each candidate device before that transistor can be considered for DPST design.

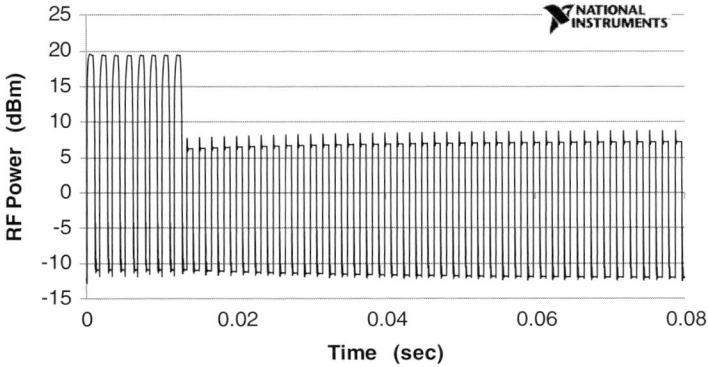

Figure 10-98   Transistor output lag temporarily suppresses lower power signals following high power outputs. These data are measured from a GaN HEMT.

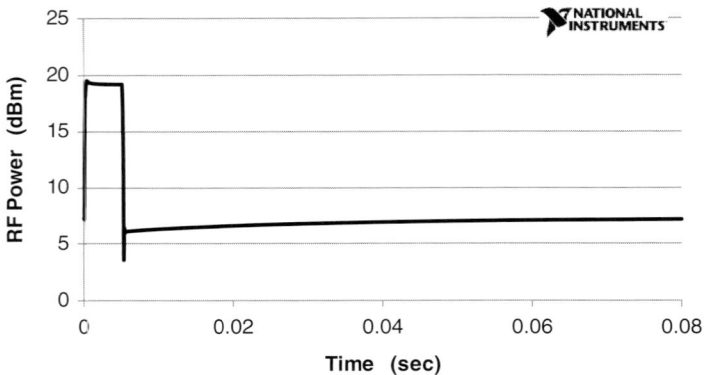

Figure 10-99 Using a single long pulse for the measurement of output lag instead of multiple shorter pulses as in Figure 10-98 incurs heating droop in the high power interval. Measurement results for the lag characteristic remain similar to the pulsed measurement.

One natural question is whether the repetitive pulse technique used in Figure 10-98 is necessary to obtain the correct output lag characteristic. It is easier to use only one pulse instead of many, all else being equal. This experiment is done and the result is provided in Figure 10-99. The output lag data do appear as desired with apparently the same recovery time. A different behavior is seen in the high power pulse from the pulse sequence in Figure 10-98. In Figure 10-99, a power droop is apparent as the transistor power dissipation continues. This power droop is not present in the pulsed measurement. The pulsed measurement seems to be safer for the transistor. There is plenty of opportunity to optimize this test procedure.

10.9   Comparison discussion

Which technology is best? Usually this is a loaded question, because many organizations have preferred technologies for economic reasons. In essence, "if you only have a hammer, everything looks like a nail." In this regard, all of these transistor technologies will function in both envelope tracking and polar modulation DPST designs. Performing well in a particular DPST design is another matter entirely. Fortunately, now there is enough data available to make meaningful progress on this comparison question.

Operating frequency is always a challenge for truly switching amplifiers. Using the technique developed in Section 8.4, the τ_{ON} values for each of these transistors is evaluated and plotted in Figure 10-100. Contours of identical time constant values are also included. Using the criteria from Section 8.4, that for acceptable switching performance the operating frequency must remain below $1/(10*\tau_{ON})$, the corresponding frequencies to the time constant values are added across the top of the figure.

Transistors that emphasize low capacitance (e.g. SiGe HBT) or low resistance (e.g. GaAs pHEMT) and yet have very similar τ_{ON} values are readily identified on this chart. Achieving high efficiency switching action above a few GHz is very

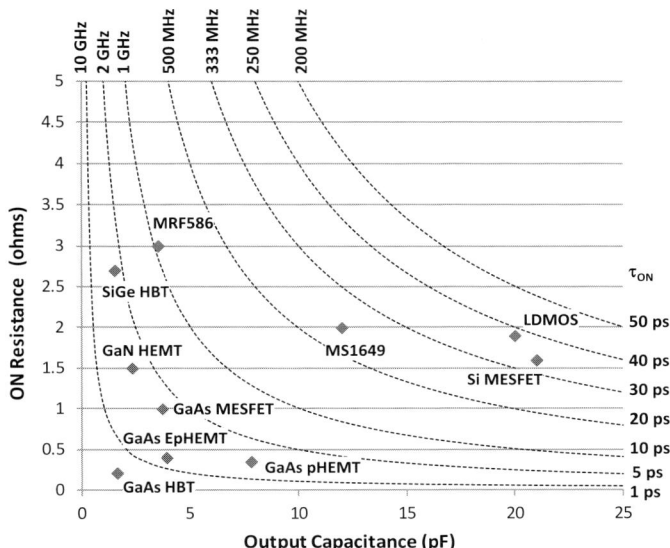

Figure 10-100 Compilation of τ_{ON} (8.13) values for the evaluated transistors in this chapter.

difficult, and effectively impossible, with this set of transistors. And in particular the frequency limitation of LDMOS is readily understandable from its relatively large τ_{ON} value.

Characteristics that are particularly important for either envelope tracking or direct polar DPST applications are explored in the following sections.

10.9.1 Preferred characteristics for envelope tracking

The most important characteristic for a successful envelope tracking application is to meet the requirement of (7.1), which is consistent with the measurable characteristic of (10.1). Reviewing the CCS boundary results for each of the measured transistor types provides the results shown in Figure 10-101. A dashed line is included at the CCS region boundary defined by (10.1), and there are four transistor types that meet it: all three bipolar transistor types, and GaN HEMT. It may be possible that other transistor examples may provide slightly different results, but the results in Figure 10-101 are consistent with this author's experience.

Having high PSR is necessary for successful ET design, but it is not sufficient. In addition, the transistor knee voltage must be small so that the energy efficiency goals can be achieved. Reviewing the ET power supply profiles evaluated in this chapter, they are further analyzed to show the effective knee voltage profiles for each transistor type. All of these knee voltage profiles are overlaid in Figure 10-102. This looks like a mess, but important results are seen here.

The most important result is that the smallest knee voltages are exhibited by the bipolar transistors. In fact, separating these transistors in Figure 10-103 for closer

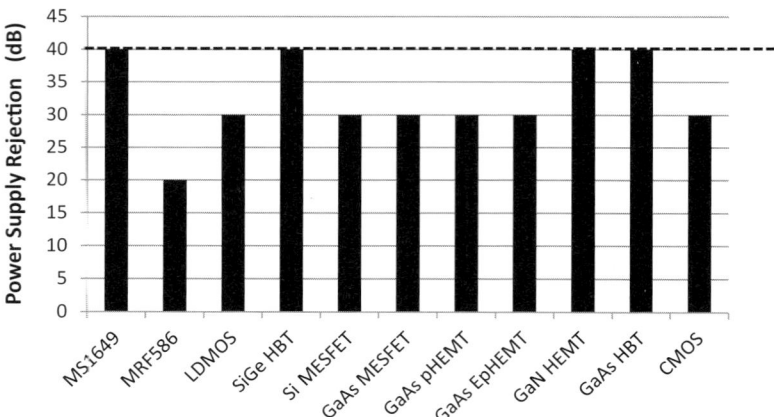

Figure 10-101 Transistors that meet (10.1) (dashed line) for successful ET use are the bipolar types, and GaN HEMT.

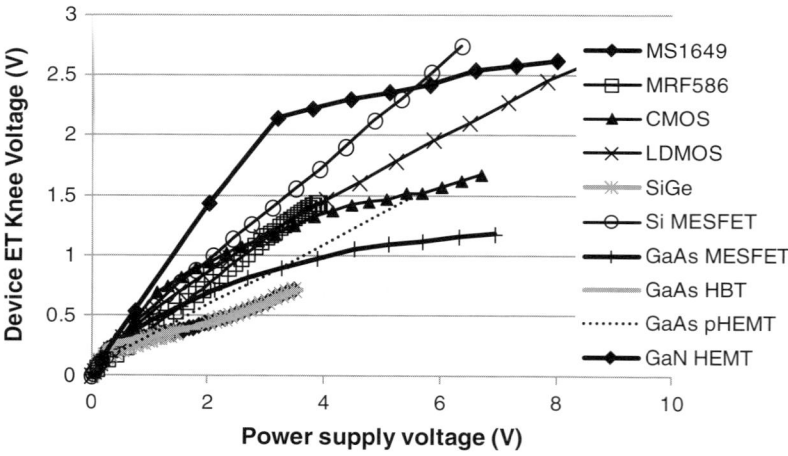

Figure 10-102 Overlay of the knee voltage profiles for all the evaluated transistor technologies. Lower knee voltages are preferable for ET power amplifiers.

inspection shows that these knee voltage profiles lie nearly on top of each other, particularly at envelope peak voltages. This is a fortuitous, but not at all intuitive, result for such a wide variation of bipolar transistor technologies. Knowing that the bipolar transistors have both excellent PSR and low knee voltages makes them the most appropriate for envelope tracking DPST designs.

10.9.2   Preferred characteristics for DP modulation

Objectives for successful polar modulation are very different. Power supply rejection has no meaning in polar DPST operation (6.13), so the comparison of Figure 10-101 is

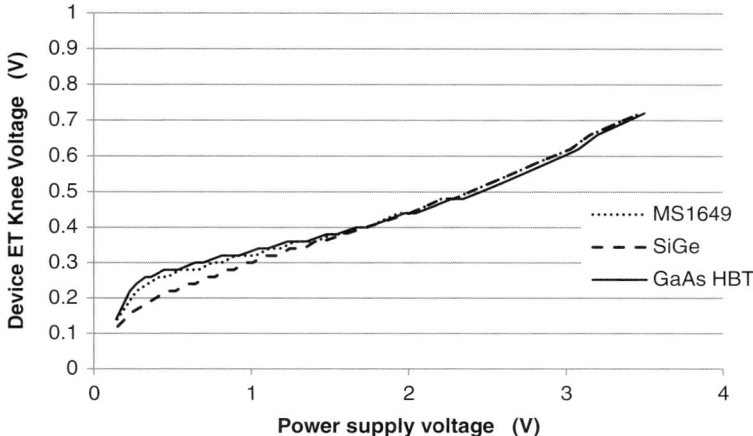

Figure 10-103 Knee voltage requirements for the bipolar transistors that meet the envelope tracking CCS region requirement (10.1) from Figure 10-101.

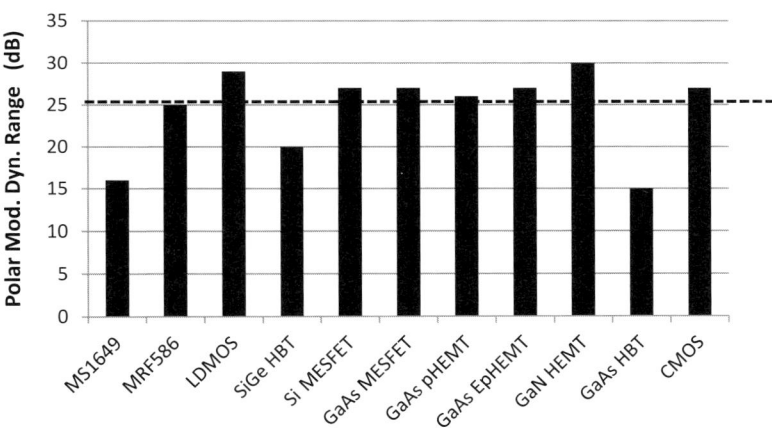

Figure 10-104 All of the FET technologies meet polar DPST modulation requirements. Distortion from bipolar technologies at low output magnitudes makes these unsuitable.

not important here. What is important is the polar modulation dynamic range presented in each transistor type discussion. These dynamic ranges are collected together into Figure 10-104 for convenient comparison.

A modulating dynamic range of 25 dB or higher is a useful beginning for polar DPST application. Figure 10-104 has a dashed line just above a 25 dB modulating dynamic range, and we note that all of the FET technologies surpass this limit. Equally noteworthy is that all of the bipolar technologies from Figure 10-103 lie well below this dynamic range, even though they are already corrected for the voltage offset inherent in bipolar characteristics. An initial conclusion is that FET technologies are more amenable to polar modulation than bipolar technologies are.

Evaluating this initial conclusion highlights why the bipolar transistors perform so poorly when polar modulated. The main culprit is the curvature of the transistor characteristics at low currents, which is a significant distortion mechanism that FET devices do not share. FET devices can be polar modulated much further down than high supply rejection bipolar transistors will support.

This places a spotlight on the present state of DPST design – the quandary. If envelope tracking is all that is desired, then it is best to select a high supply rejection bipolar technology that is capable of supporting the necessary RF output power. But do not try to push such a design into polar operation to achieve improved energy efficiency because the polar performance is horrible. On the other hand if good polar modulation is the goal, then an FET technology consistent with the RF power requirements is the best choice. But if some operating conditions require that the DPST leave the switching operation and revert to linear envelope tracking, then the ET operation will not have the power supply independence and low signal distortion expected of that technique.

Presently there is no transistor type that works universally well in all DPST applications. Hopefully the transistor development community can work on these differences and provide more applicable options in the future.

10.10   Reference

[10-1] W. Lepkowski, S. J. Wilk, J. Kam, and T. J. Thornton, "40 V MESFETs Fabricated on 32nm SOI CMOS," *Proceedings of the 2013 IEEE Custom Integrated Circuits Conference (CICC)*, San Jose.

11   Hybrid system combinations

The properties of envelope tracking DPS transmitters from Chapters 5 and 7 and those of direct polar DPS transmitters from Chapters 6 and 8 are often in opposition to each other. A concise summary of the defining properties for the envelope tracking and DP subsets of DPS transmitters is presented in Figure 11-1. The transition between these two DPST operating modes is initially explored in Chapter 10, which shows that the transition between ET and DP is strongly dependent on the transistor technology being used. Indeed, some transistor technologies never actually reach the ET operation but rather stay in this transition region.

When one or more operating modes of any circuit exhibit widely separated properties, a natural thought is to find methods to combine the operations so that the most desirable properties are achieved, or that undesirable properties are avoided. This directly leads to hybrid designs, and is the topic of this chapter.

11.1   Hybrid operation overview

Chapter 6 showed that optimizing an ET design for higher energy efficiency necessarily results in polar operation. To get the RF PA at its highest energy efficiency, it *must* shift from linear to compressed operation. Through this transition, the power supply noise suppression is eliminated from the increasingly nonlinear power amplifier and the precise value of the power supply becomes critical as it comes to directly control the envelope modulation. Signal drive must also increase as the gain drops due to circuit compression to maintain the output power. Such is the price of high efficiency operation.

Transitions from linear through ET to DP and back are summarized at a high level in Table 11-1. Activity requirements on the DPS and the associated PA linearity are well discussed in the prior chapters. The important additions in Table 11-1 are the associated modulation transfer functions for each of the DPST operating modes. The differences among these transfer functions points out the major challenge in getting any hybrid DPST implementation to work. As the modes transition from one to another, the values for the associated transfer functions need to match in order to avoid distortion on the output signal.

Published literature on DPS transmitters refers to the "power supply profile," or sometimes the term "shaping table" is used. Both of these terms refer to the relation between the actual output signal envelope and the voltage applied to the RF PA. This

Table 11-1 Transitions among the DPST operating modes

	DPS	PA linearity	Transfer function
DP	Varies matching the signal envelope	None	$g_{DPS\_AM}(V_{DPS})$
ET	Varies near the signal envelope	Yes	$g_{RP}(P_{IN}, V_{DPS})$
Linear	Constant value	Yes	$g_{RP}(P_{IN})$

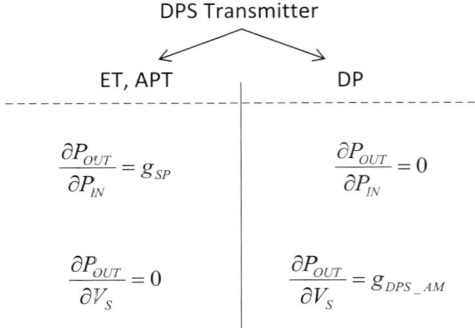

Figure 11-1   Concise mathematical summary of the definitions for so-called average power tracking (APT), envelope tracking (ET), and direct polar (DP) DPST operation.

type of profile is exactly what was shown in Figure 5-19 and Figure 5-21. Real-world profiles derived from specific transistor characteristics are provided throughout Sections 10.5 through 10.7. These power supply profiles are each developed by following the procedure from Figure 5-18, based on specific values of power supply noise tolerance required by the application.

In the literature, a typical calculated power supply profile curve is defined by [11-1]

$$V_{DPS}(V_{env}) = V_{MAX} \cdot \left(\frac{V_{env}}{V_{MAX}} + b \cdot e^{-\left(\frac{V_{env}}{V_{MAX}}\right)/b} \right). \tag{11.1}$$

For convenience, (11.1) is called the exponential DPS profile. Adding this profile to the minimum ET profile from Figure 5-19 allows the comparison in Figure 11-2. The operation of a DPST that follows this profile is a hybrid of all three modes listed in Table 11-1. The regions of linear, envelope tracking, and polar modulation operation are shown in the diagram. This transmitter also has an extended transition region between its envelope tracking and polar modulation operations.

It is important to remember that gain, no matter the functional definition from Section 4.1.5, is a scalar quantity relating one input to the signal output. If more than one input influences the desired output then there are multiple transfer functions active in the transmitter operation. We may consider the concept of *partial gains*, meaning the scalar partial derivatives of the output with respect to each individual influencing input. I personally prefer to keep things simple and describe what actually is happening: there are multiple transfer functions which are not independent. The discussion then proceeds

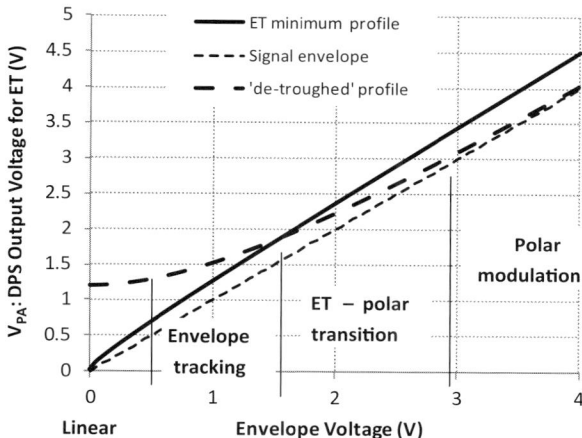

Figure 11-2   Comparison of a published power supply profile (long-dashed curve) with the minimum ET profile (solid curve) and polar modulation profile (short-dashed line) for a simple FET model. Regions of actual circuit performance are identified.

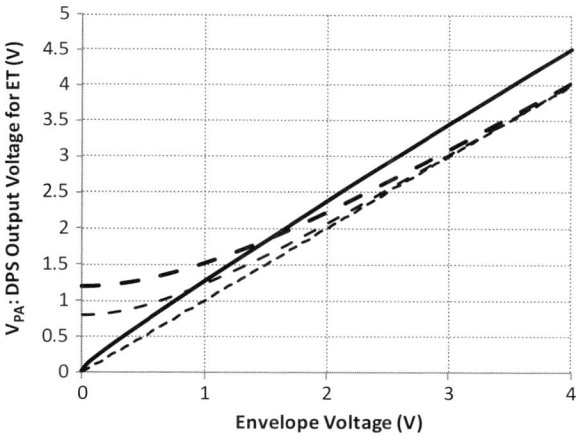

Figure 11-3   Examples of exponential DPS voltage profiles for hybrid envelope tracking and polar DPST operation.

to describe these transfer functions, but the word "gain" no longer has any meaning and should not be used. The actual action of interest is accuracy of the output signal envelope (and phase).

Three commonly used hybrid DPST operations are presented and discussed in the following sections. First is the strategy to operate using polar modulation at the signal peaks where efficiency is most important. Second the technique called "constant gain" (more accurately referred to as a constant joint transfer function) is investigated, where we learn that gain is not at all constant even though the output envelope is reasonably well controlled. Third is the maximal energy efficiency strategy, which we find is really an extension of the first strategy.

11.2   Polar at signal peaks, ET, or linear in signal valleys

This first hybrid operation takes advantage of high operating efficiency when the transmitter output power is greatest. This is naturally interesting because energy efficiency is most valuable when the output power is high. This requires polar modulation, which has great difficulty producing low output envelope values, as is shown in Chapter 6. But linear operation has no problem with low output signal envelope values and great difficulty with high energy efficiency. The strategy then is to operate in polar modulation at times with high output envelope values and as a linear amplifier at minimum output envelope values.

As explained in Section 4.1.6 on P-mode, when amplifier operation is sensitive to both the input signal power and the applied supply voltage, the output signal envelope is severely distorted when both inputs are varied. Hence it is important to not use ET into the P-mode amplifier region. Proper ET design therefore must know where the P-mode boundary is for any particular amplifier, and any design must respect that boundary by staying above it.

One easy way to identify the P-mode boundary is to evaluate the amplifier gain flatness characteristic. The rapid roll-off of gain, the so-called "gain collapse" characteristic of amplifiers at low values of applied supply voltage, is indicative of P-mode operation at the RF power transistor. We just need to quantify when this transition begins. Using a small GaN HEMT amplifier as an example, from its measured Booth chart the relationship between amplifier ratiometric power gain and the applied supply voltage is shown in Figure 11-4 by the solid line. The corresponding first derivative of this gain relationship is shown by the dashed curve. At high supply values, this slope value never goes to zero, indicating that this amplifier does not

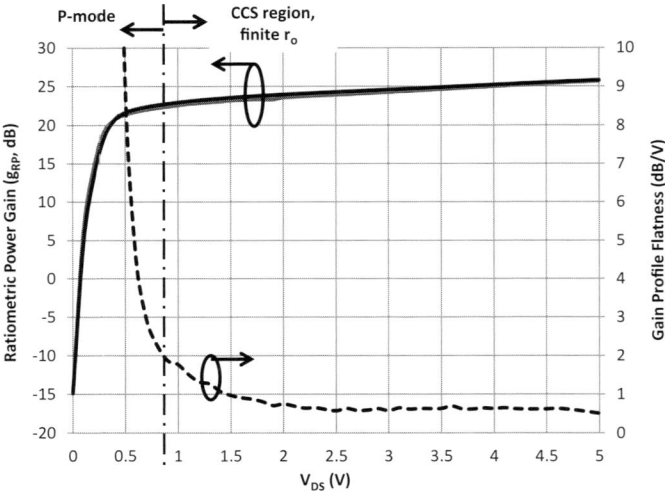

Figure 11-4   Determining a practical P-mode boundary for an example GaN HEMT amplifier by evaluating the slope (first derivative) of the g_{RP} vs. supply voltage curve.

support an ideal ET operation (7.1). Fortunately, with this hybrid design we want to use polar modulation at these supply values, so this ET limitation is not of immediate concern.

The transition to P-mode is gradual, so the establishment of a firm boundary needs additional considerations. A purist would select where the upward trend begins, here near 2.5 V. But we are engineering a transmitter that must operate with high efficiency, so that the P-mode boundary needs to have the smallest practical value. The decision here is to set the boundary when the gain sensitivity crosses 2 dB/V. The result sets the minimum applied supply voltage at 0.8 V.

11.2.1 Power supply profile

To identify the various regions of hybrid operation for this amplifier, the information in Figure 11-2 specific to this GaN amplifier are shown in Figure 11-5(a). The exponential power supply profile (11.1) is scaled for its minimum value at the P-mode boundary identified above at 0.8 V, and for a maximum supply voltage of 5 V. Using this exponential power supply profile along with the ET boundary for this transistor and amplifier load resistance, this design does operate with polar modulation at any applied power supply voltage above 2.1 V. Transition between envelope tracking and polar modulation occurs at power supply voltages between just below 1 V and 2.1 V. The actual ET operation occurs when the power supply is between 0.8 V and 1 V. It may be legitimately debated whether this is ET or simply linear PA operation at these low supply values, since the power supply variation is very small.

For the two-tone DSB-SC-AM signal from Figure 6-45, the corresponding hybrid DPS power supply is shown in Figure 11-5(b) by the dashed curve. The power supply overlays the envelope completely when its value exceeds 2.1 V, directly showing the

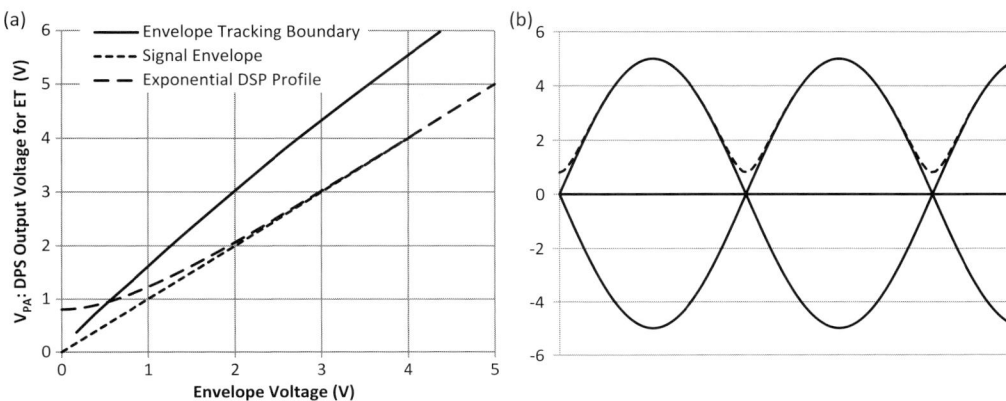

Figure 11-5 DPS profile using (11.1) for the example GaN amplifier, matched to the P-mode boundary from Figure 11-4: (a) Exponential DPS profile compared to polar (signal envelope) and ET boundaries; (b) time waveforms of the signal envelope and the corresponding exponential DPS profile.

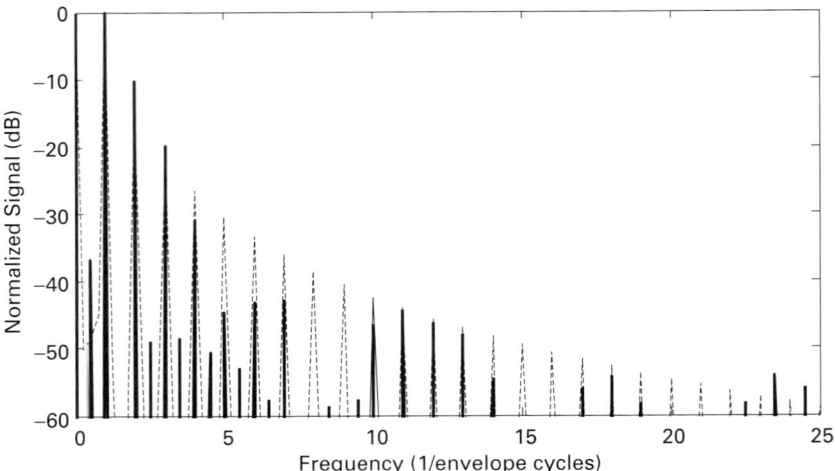

Figure 11-6   Envelope signal bandwidth reduction by applying the exponential hybrid profile.

action of polar modulation (6.15). The power supply never goes below 0.8 V, so the envelope valley is "filled in." This is particularly evident whenever the signal envelope is below 1 V. Keeping the power supply up during an envelope zero event maintains linear PA operation, making the zero valued envelope straightforward to achieve.

It is interesting to observe that the onset of polar modulation seen in Figure 11-5(a) coincides with the transition of the gain profile in Figure 11-4 to a fixed slope. This is purely coincidental since Figure 11-4 shows measured data and Figure 11-5 is two calculated curves which are only related by the scaling of the exponential hybrid profile to the P-mode boundary chosen from Figure 11-4.

Of greater immediate interest is the comparison of the envelope bandwidth to the bandwidth of the DPS waveform following this exponential hybrid profile. The result is seen in Figure 11-6 where the dashes represent the frequency characteristics of the signal envelope, and the solid line shows the frequency characteristics of this DPS waveform. Both of these envelope signals have the same power at the fundamental frequency, and both spectra are normalized to this power. To a −60 dBr bandwidth, the expansion of the unmodified envelope is 25x, while the bandwidth of this DPS waveform is below −60 dBr by 9x. By not having to accurately trace the envelope minimum and its derivative discontinuity at zero magnitude, the required bandwidth of the DPS is reduced by nearly a factor of 3.

11.2.2   Gain variation

Now we look at how the gain reacts to this hybrid DPST operation. Since the ratiometric gain of any amplifier in compression is less than the same gain when operating linearly, it is expected that the gain must increase as the operation of this

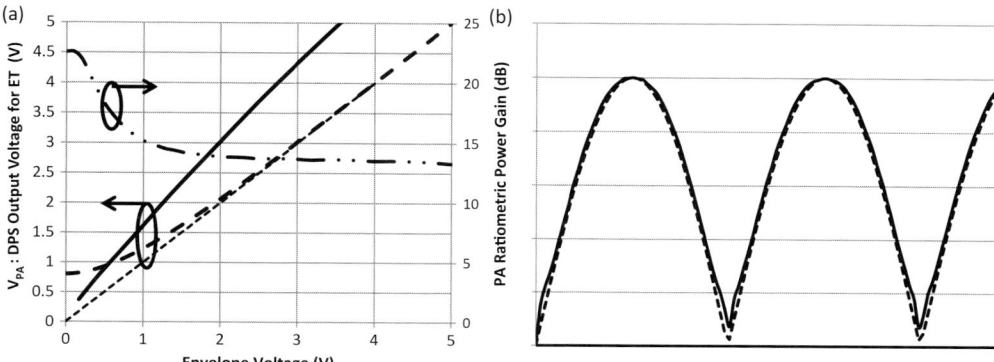

Figure 11-7   Gain increases as the PA transitions from C-mode to L-mode: (a) for this amplifier, the gain change is nearly 10 dB; (b) envelope distortion occurs at all values, but is most pronounced at lower envelope values.

hybrid DPST shifts from polar to envelope tracking. Evaluation of the Booth chart data for this amplifier show that this is indeed what happens. The corresponding ratiometric power gain profile, reported in dB, is added to Figure 11-7(a) and associated to the secondary axis scaling on the right.

This gain shift is not small. From a value of 13 dB for these compression levels, the gain increases as the amplifier comes out of compression until it gets to 22.5 dB at low supply values. Without constant gain across all signal levels, there must be distortion in the output waveform. This distortion is mainly at the lower envelope values as seen in Figure 11-7(b). The gain difference is a measure of the compression that the amplifier output is driven to, which here is nearly 10 dB. This is deeply compressed, from which it is expected that the output efficiency will be high. More about this follows in Section 11.2.5.

The physical processes involved are a change in transfer functions as presented in Table 11-1. When polar modulation is active, the governing transfer function setting the output envelope level is from the applied power supply to the output (6.15), with negligible influence from input signal envelope variations. During the transition from polar to envelope tracking (linear) PA operation, the output envelope is influenced by both the input signal envelope and the PA supply voltage. Once the ET operation is reached, the influence from the PA supply voltage becomes negligible and the output envelope is controlled by the envelope on the RF input signal. Finally, when the variation on the power supply stops at a low voltage, the actual operating transfer function is not necessarily linear and must be tested to resolve the ambiguity: whether it is L-mode at this low supply (PA has supply-noise suppression) or P-mode (PA has no supply-noise suppression).

11.2.3   Stability issues

When operating in C-mode, the RF amplifier is unconditionally stable because the slope gains g_S and g_{SP} are very small. Thus, the amplifier does not satisfy

Barkhausen's gain criterion for oscillation. Once the slope gains (4.8) and (4.11b) increase to unity as the PA comes out of compression, the consideration of circuit stability becomes vital. Any linear operation forces this design task into the project.

In this hybrid operation, evaluation of circuit stability is actually a very difficult task. While the usual stability evaluation of a linear amplifier follows the process from Section 4.3.1, here this must be evaluated separately for all of the conditions the PA encounters in which it has slope gain at or above unity. All of the s-parameters in (4.22) are functions of the applied PA supply voltage. They also vary depending on how compressed the amplifier operation is.

Therefore, the evaluation conditions for (4.22) change significantly at the various points in Figure 11-7(a) when the PA supply voltage is below 2 V. Each of them must be first evaluated individually. Then these individual stability evaluations must be combined to establish what circuit designs are available to the PA designer to achieve the power and linear efficiency goals. There is no guarantee that an acceptable solution to all of these design conditions exists for any particular transistor type.

11.2.4   Load variation to the DPS

When the PA operates in C-mode, the load it presents to the DPS is given in (6.17), which shows that the load behaves as a simple resistance. On the other hand, with the PA operating in L-mode the load it presents to the DPS follows (5.20), which approaches 0 ohms at 0 V. Plotting SSR for all available conditions of drive power and PA voltage provides a surface describing the possible load variations that the particular PA being evaluated will present to the DPS. This SSR surface for the GaN PA in these examples is provided in Figure 11-8(a).

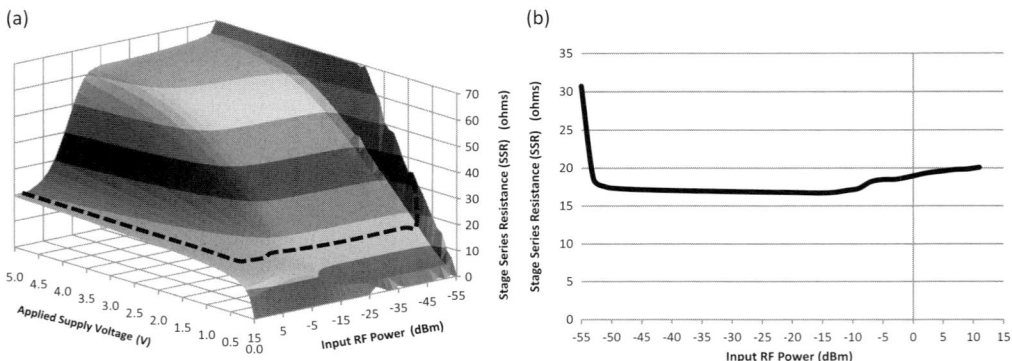

Figure 11-8   Absolute resistance of the example PA which is the load presented to the DPS: (a) complete PA SSR surface with the trajectory of Figure 11-4 shown (dashed curve); (b) R_{PA} experienced by the DPS with this profile.

C-mode operation, closely following (6.17), is seen at the edge of this surface where the input RF power is maximum. As the drive is reduced, the behavior quickly transitions to follow (5.20). Knowing the relationship between input RF power and the associated applied voltage to the PA allows determination of the load this PA will provide to the DPS by following the corresponding path along this surface. This track is added to Figure 11-8(a) as the dashed curve. The resulting characteristic of SSR vs. input RF power is extracted and presented in Figure 11-8(b).

Having selected 0.8 V for the minimum PA supply voltage, the SSR track shows that the very low DPS load impedance at low supply voltage is avoided. In this particular case, the load impedance experienced by the DPS varies by about 25%. This is a good situation for the DPS design.

11.2.5   Energy efficiency profile

The entire purpose for using DPS transmitters is to capture most or all of the available PA energy efficiency. With the top 60% of the available envelope range operating in C-mode as seen in Figure 11-7, it is expected that the energy efficiency of this hybrid operation will be close to the available maximum for this PA. This is true for the example PA, which is seen in Figure 11-9(a).

Some description of Figure 11-9 is in order. It is common practice to plot efficiency data against output power, not the input power curves seen here. While the standard practice is certainly legitimate, I personally do not find it valuable in providing design insight toward the success of a DPST design. This is simply because PA output power is not a design input parameter. PA input power and PA supply voltage are the two

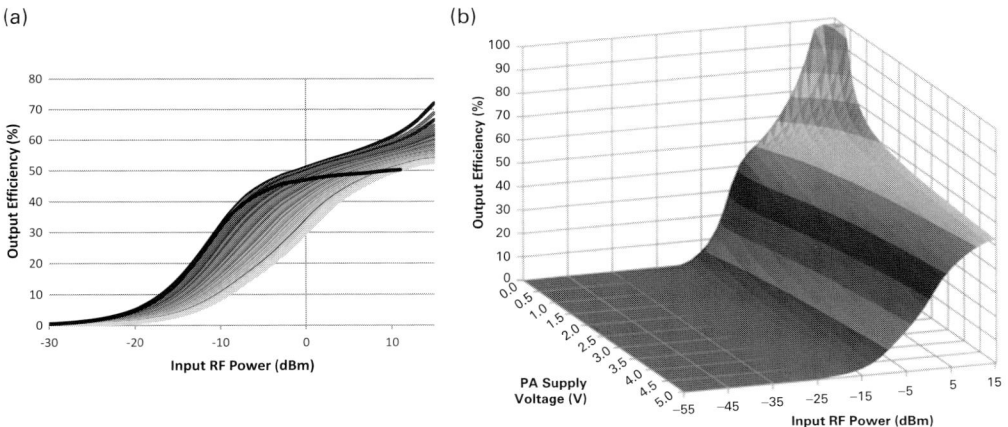

Figure 11-9   Energy efficiency profile of this hybrid polar/envelope tracking/linear PA operation: (a) profile (solid line) overlaid on the amplifier output efficiency curve family, showing that much of the available energy efficiency is being accessed; (b) surface of the curve family to clearly show how the efficiency profile actually behaves.

available controllable inputs to the design. Output power is a result. Energy efficiency is also a result. We earnestly strive to improve both energy efficiency and output power. But we cannot improve energy efficiency by manipulating output power directly. Therefore, the standard plots are a good means to check on progress toward meeting the design goals, but they are of little help in providing design insight toward what changes can be made for improving performance. For those, we need to have plots against the controllable input parameters.

With the curves in Figure 11-9(a) extending above the evaluated overall PA efficiency profile, the question comes up as to whether the available energy efficiency is being captured. Additional insight toward the answer is gained from Figure 11-9(b) where the overall output efficiency measurements are plotted as a surface. Here we observe that the higher efficiency measurements occur when the input power is large and the PA supply voltage is small. This is a distortion of the efficiency calculations caused by leakage of the input drive signal significantly contributing to the measured output power. This efficiency peak is not real. The shown efficiency capture curve overlaid on Figure 11-9(a) is correct.

Applying this exponential hybrid operating profile on to the Booth chart for the example PA obtains the solid line in Figure 11-10. This shows that P-mode is avoided, but just barely as the operating ratiometric power gain is reduced from using low supply voltages in the presence of finite transistor output resistance (5.23).

At higher input RF power, the trajectory in Figure 11-10 shifts to the right, deeper into C-mode. The output power drops, as is required as the PA goes further into compression. It also intersects all of the higher voltage curves in the Booth chart which describes the influence of the power supply variation on the output power.

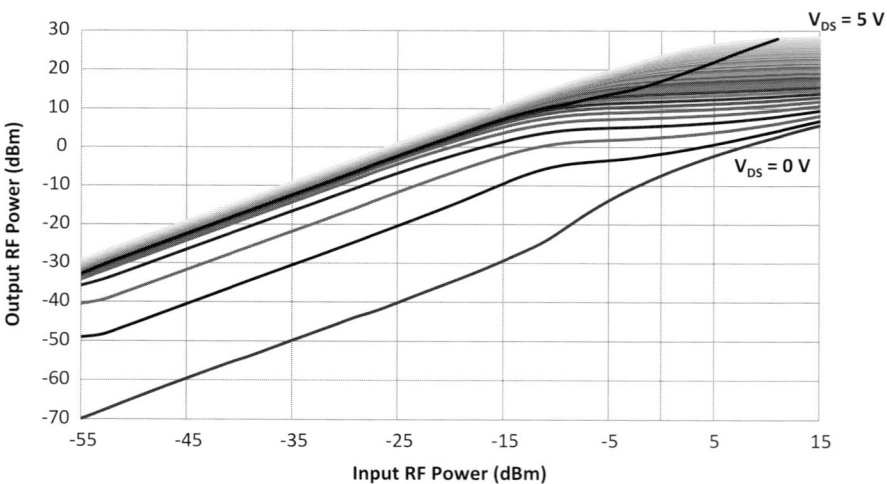

Figure 11-10   Path of the exponential DPS profile along the corresponding Booth chart, showing the curve bending into the C-mode region at higher input powers.

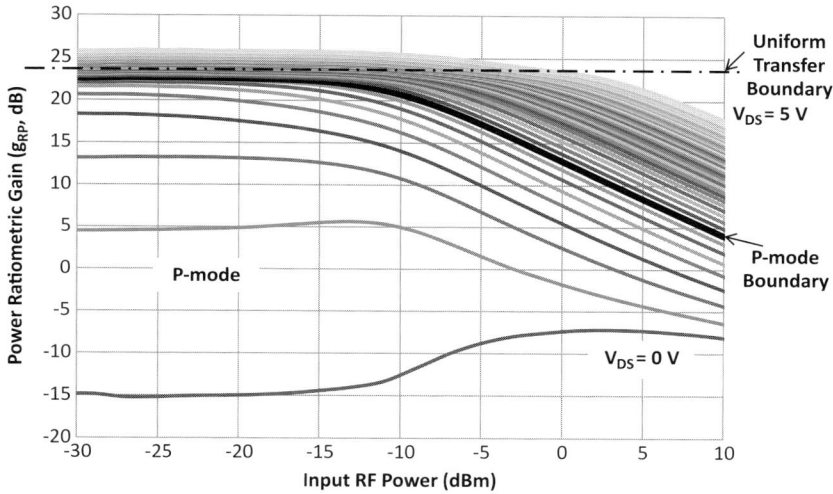

Figure 11-11   Amplifier gain profile vs. applied supply voltage with a boundary drawn at a fixed value of effective gain, called the uniform transfer boundary.

11.3   Constant joint transfer function

Having a widely varying gain is a significant problem for the exponential supply profile DPS hybrid technique, as described in Section 11.2.2. It is possible to design a DPS hybrid technique that optimizes for no "gain" change at all across the control range, following the idea of Figure 11-11. Here a line is drawn across the family of ratiometric power gain curves at a fixed value of g_{RP}. Simple, yes?

This line is labeled the uniform transfer boundary (UTB) in Figure 11-11 for a specific reason. According to the chart, this line is at a fixed g_{RP} value, so does it also represent a constant *gain*? According to all the history of what the word gain means, the answer actually is no. "Gain" always is used to refer to a *dimensionless scalar* metric relating a single output to a single input. Thus, each of the individual curves in Figure 11-11 is a gain. A profile that involves a collection of such curves adds a second control parameter in order to realize the desired result. This result is now neither dimensionless nor scalar. It is now something else than a simple gain. And we in the engineering community must not be lazy and describe it as it truly is so that we can access all the information actually present, with the very important side benefit of avoiding ambiguities with not only the present design activities but also with the historical record in the archive literature.

What actually is happening is a recognition that the truly important metric of any transmitter is the accurate construction of the output signal, here focused on accuracy in the envelope variation of the output signal. We can achieve this output accuracy through a gain process. We may also use a combination of physical processes to achieve this accuracy. Hence the name UTB: along this boundary, the transfer of input conditions to the output is uniform. For the present interest, the input conditions are RF input power and PA supply voltage, and the output of interest is consistent output

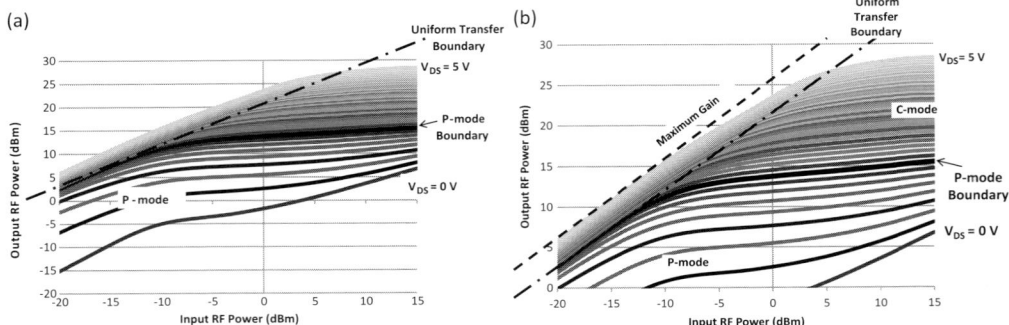

Figure 11-12   Equivalent path of the UTB to implement the constant joint transfer function hybrid profile: (a) on the corresponding Booth chart; (b) close-up view identifying C-mode, P-mode, and how far this UTB is below the maximum g_{RP} value for this amplifier.

magnitude. This leads to a more accurate descriptive name for this technique as the constant joint transfer function profile

Again the Booth chart becomes useful to aid understanding of what actually is going on. The gain chart of Figure 11-11 is actually derived from Booth chart data, and the corresponding foundation of Booth chart measurements is shown in Figure 11-12. The UTB is now a line with unit slope (1 dB/dB), with a vertical offset defined by the desired output power for each input power value. These data are for the same example amplifier as that used for Section 11.2, so the P-mode boundary from Figure 11-4 is also identified.

11.3.1   Intrinsic output magnitude accuracy

This technique is introduced in Section 7.4.3, where the analysis shows that this technique does not qualify as ET because it violates the fundamental definitions of ET, particularly (7.1). It indeed is a hybrid DPST technique.

By design, the output magnitude accuracy is good. The major problem of the exponential DPS hybrid profile of a widely varying gain is eliminated. There are consequences to this selection, of course, because nothing comes for free. One of the more significant consequences is seen in Figure 11-13(a), which shows the UTB overlaid on the Booth chart data scaled here for signal voltages instead of signal power. The UTB line only overlays amplifier data for an envelope range of 0 V to 4.2 V. Even though this amplifier can provide peak envelope voltages nearing 6 V, these higher output voltages are only available from compression modes and higher input signal magnitudes that are not accessed by the UTB.

The UTB can be selected to have a slope to access these higher envelope values, but this necessitates a lower ratiometric gain and moves this PA operation even further away from desired linear operation. It also forces operation within the P-mode region, where any DPST operation must be done very carefully. In any case, the following discussion of this technique assumes that the dynamic bias required to achieve these control dynamic ranges is successfully implemented.

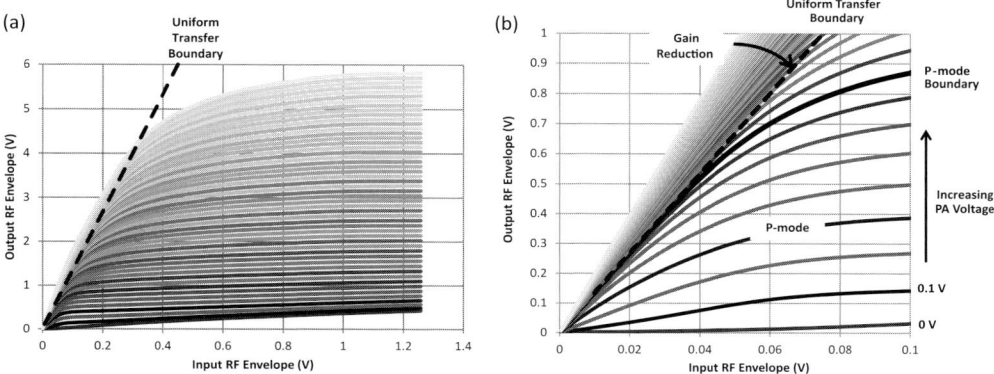

Figure 11-13 Showing the uniform transfer boundary on a linear plot of the Booth chart: (a) most of the amplifier operation is compressed, but not accessed by the constant transfer function profile; (b) close-up view near the origin showing the selected P-mode boundary and the gain reduction associated with the uniform transfer boundary.

11.3.2 Power supply profile

The Booth charts of Figure 11-12 and the ratiometric power gain chart of Figure 11-11 all show that some combination of coordinated input power and PA supply voltage variation is required to realize the constant joint transfer function operation needed to get the desired accurate output magnitude. The most direct way to quantify this joint transfer function is to use the ratiometric power gain data from Figure 11-11 and plot it as a surface. This corresponding surface is presented in Figure 11-14(a). Then this surface is truncated at the desired value of the joint transfer function. The value selected in Figure 11-11 at the UTB is 22.5 dB. This truncated surface is shown in Figure 11-14(b), rotated to make the result easier to read. The edge of this truncated surface is the required joint transfer function of input RF power and applied PA supply voltage needed to meet this design objective.

The chart in Figure 11-15(a) is the extracted joint transfer function of PA supply voltage and input RF power that provides the chosen UTB value of 22.5 dB. Below an input RF power of –17 dBm, the PA is held constant at the P-mode boundary of 0.8 V. In this region, it is legitimate to call the circuit operation a gain. At higher values of input power, there are unique pairings of input RF power and applied PA supply voltage that must be followed to obtain the desired output. In this operating region, this is not a gain, and is actually a closely choreographed dance between the two controllable input parameters.

In Figure 11-15(b), this same information is plotted on an horizontal axis, rescaled for output signal envelope voltage instead of input RF power in dBm. This matches the plot conditions of Figure 11-7(a). But here the constant joint transfer function hybrid strategy is seen to remain within the transition region between the L-mode of ET and the C-mode of polar modulation. When the PA supply voltage is fixed at its minimum value, the PA operates as a "conventional" linear amplifier though with very small supply voltage.

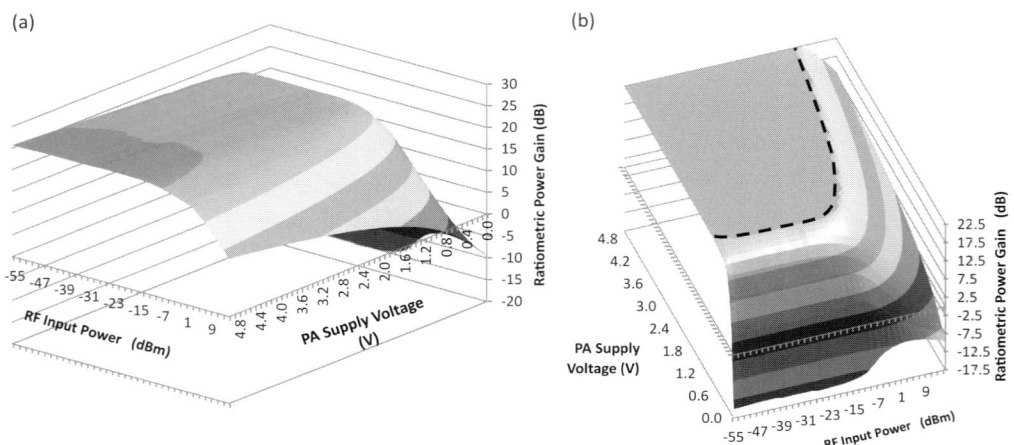

Figure 11-14   Ratiometric power gain surfaces for the example GaN HEMT amplifier: (a) full surface of gain vs. RF input power and PA supply voltage variations; (b) the same surface limited to the selected uniform gain of 22.5 dB, rotated to show the relationship between RF input power and PA supply voltage.

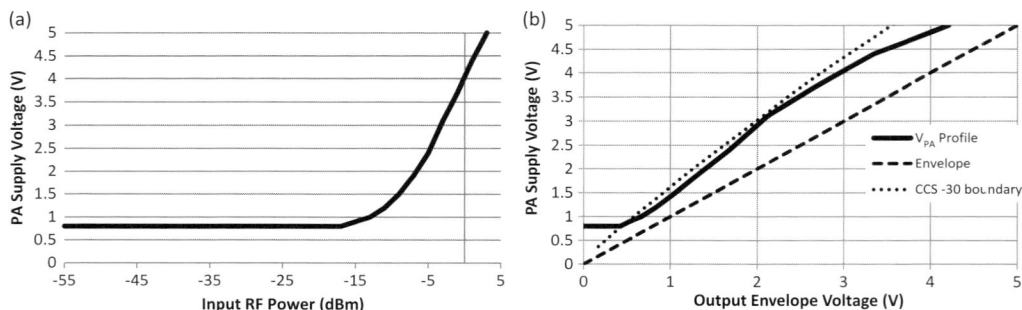

Figure 11-15   Power supply voltage profiles for constant transfer function hybrid operation on the example GaN HEMT amplifier: (a) joint transfer function of PA supply voltage and input RF power that provides the chosen UTB; (b) converted to the conventional view of PA supply voltage with respect to the desired output envelope value.

All of these results inherently assume that the dynamic bias in the PA necessary to achieve this envelope control dynamic range is properly implemented.

How the DPS output voltage directly compares to a varying signal envelope is shown in Figure 11-16. At no point is the DPS output voltage equal to the output envelope, so (6.15) is never satisfied and there is no polar modulation. This performance is very different from that seen in Figure 11-5(b). Yet this is a direct consequence of the choice to establish a UTB, which cannot tolerate the gain compression of C-mode operation. To keep the joint transfer function at a constant value, the PA supply voltage is inherently restricted to remain outside of C-mode operation.

The bandwidth characteristics of the DPS output waveform are shown in Figure 11-17. This signal has frequency components at both harmonic and

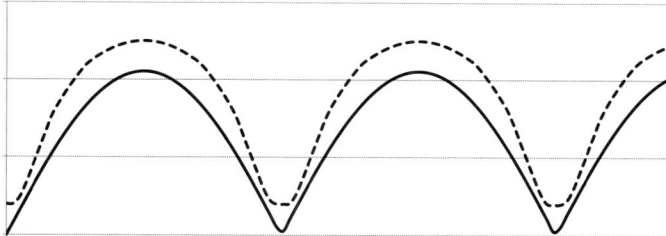

Figure 11-16 Constant joint transfer function power supply profile is close to ET in that there always is headroom above the actual output signal envelope. At no time can CJTF operation tolerate the gain drop of output compression necessary for C-mode operation.

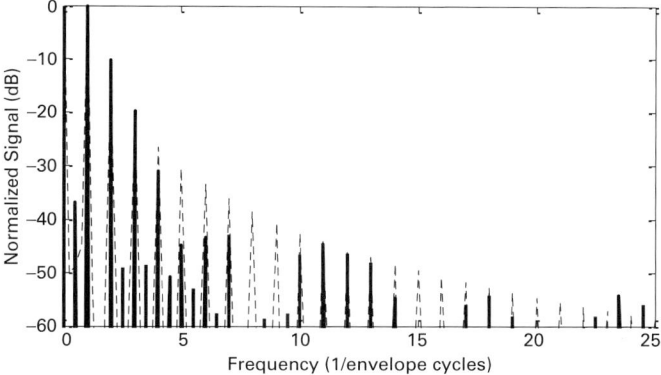

Figure 11-17 Envelope signal bandwidth for the constant joint transfer function hybrid profile of Figure 11-16 (solid line) compared to the envelope bandwidth characteristics (dashed line).

half-harmonic frequencies. And while some envelope frequency harmonics are suppressed, other higher order ones are not. The DPS circuit bandwidth necessary to implement this constant joint transfer function hybrid DPST operation is at least double that seen for the exponential profile hybrid DPST operation.

11.3.3   Energy efficiency profile

Because the inherent operation of the constant joint transfer function hybrid profile prohibits using the gain drop inherent in compressed C-mode operation, the overall efficiency is expected to be less than that achieved by the exponential hybrid DPST profile. Figure 11-18 presents the energy efficiency characteristic of the constant joint transfer function hybrid DPST profile. This curve is indeed lower than that in Figure 11-9, by more than ten efficiency points.

   This points out the major trade-off for this particular hybrid DPST operation: assured output magnitude accuracy comes at the cost of reduced energy efficiency.

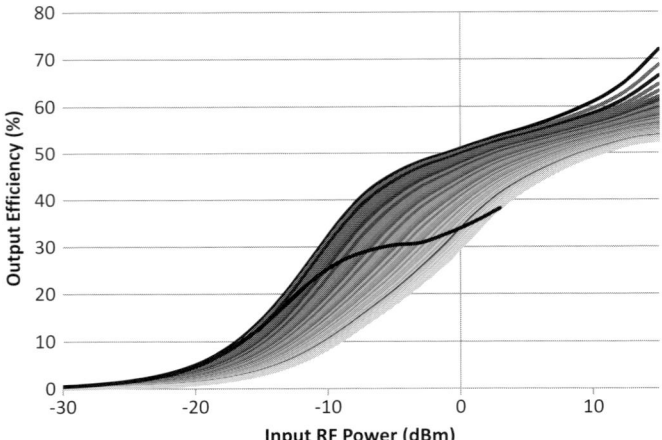

Figure 11-18   Energy efficiency of the constant joint transfer function hybrid polar/ET/linear PA operation overlaid on the same amplifier output efficiency curve family from Figure 11-9, showing that without the switching operation of the RF transistor much of the available PA energy efficiency is not being accessed.

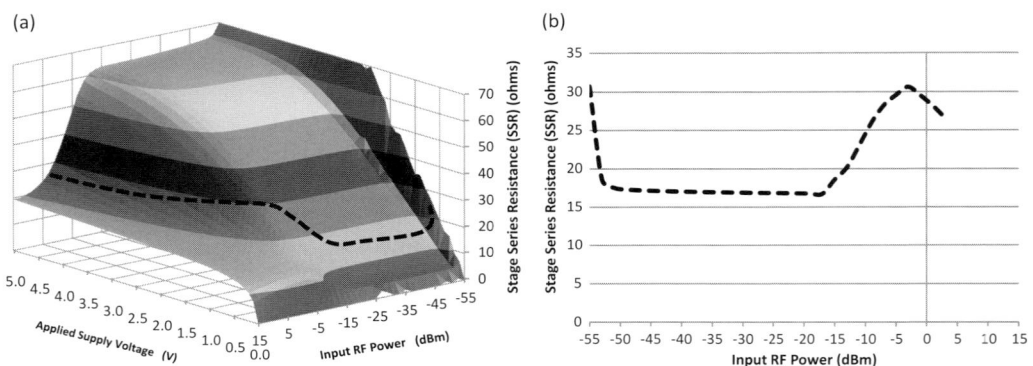

Figure 11-19   Load characteristics of this GaN HEMT PA to the DPS when operated along the selected constant transfer function profile: (a) CJTF path along the SSR surface; (b) extracted R_{PA} experienced by the DPS with this control profile.

11.3.4   Load to the DPS

The very tight correlation between input RF power and applied PA supply voltage necessary to maintain output accuracy also sets the load profile that a PA operated this way presents to the DPS. For the GaN example amplifier used throughout this chapter, the corresponding load profile to the DPS is presented in Figure 11-19. Using the same SSR surface of Figure 11-8, the different profile along this surface for the constant joint transfer function is shown by the dashed curve in Figure 11-19(a). Extracting the associated SSR values across the controlling parameter of input RF power provides the curve in Figure 11-19(b), which is the load variation that this control technique causes the PA to present to the DPS.

This load profile stays nearly constant at low input power values as long as this PA remains at the edge of L-mode operation along the P-mode boundary. When the gain of the PA begins to drop with this low supply voltage value, the supply voltage is increased sufficiently to maintain the desired output magnitude corresponding to the input power. From the measured data for this example amplifier, the absolute resistance also increases while the PA transistor operates in the transition region between envelope tracking and polar modulation. Figure 11-15(b) shows that the DPS begins to move closer to polar operation at the peak available envelope (limited as seen in Figure 11-13(a)), which corresponds here to the peak and drop of the PA SSR presented to the DPS near signal peaks. The net variation in load resistance seen by the DPS is nearly 2 to 1; from 16 to 31 ohms.

11.4   Maximum efficiency strategy

We see in Section 11.3 that in order to maintain a constant joint transfer function between the input RF power and the output power, the realizable energy efficiency of the PA is reduced. This goes against the primary reason to adopt any DPST architecture. The next question then is: what are the properties of a hybrid DPST operating strategy that is selected for maximum energy efficiency of the RF PA?

The first step to answer this question is to generate a plot for the target PA of energy efficiency in CW operation against the PA output power. While the discussion in Section 11.2.5 about the limited use of this plot remains valid, here is a legitimate design use for this plot presented in Figure 11-20. The parameter along each of these curves is a sweep of input RF power for a particular value of the PA supply voltage. The immediate

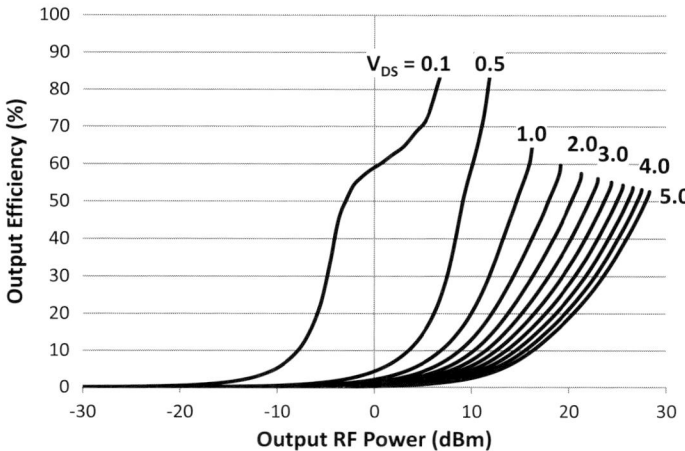

Figure 11-20   Measured output efficiency of this GaN HEMT PA, one curve for each sweep across input power for supply voltages stepping down every 0.5 V from 5 V.

conclusion from Figure 11-20 is that the best energy efficiency from this particular amplifier always occurs at the maximum input RF power. This conclusion is also supported from a review of the energy efficiency surface for this PA in Figure 11-9(b).

11.4.1   Intrinsic magnitude accuracy

The conclusion from Figure 11-20 is that for maximum energy efficiency the input power must remain at a fixed and high level. This removes one of the controllable input parameters for the PA, leaving only the PA supply voltage. Looking again at the linear plot of the Booth chart from Figure 11-13(a), a plot of this control profile is added to achieve the chart in Figure 11-21(a). The profile is the dotted vertical line along the maximum input-level value, with corresponding values of PA supply voltage.

Extracting the data showing the output signal magnitude across this variation of PA supply voltage obtains the data shown by the dotted line in Figure 11-21(b). This is predominantly linear, but when compared with the dashed line of (6.15) there are two significant differences. One is the offset of this control characteristic, and the other is a higher slope value. The offset is due to leakage of the large input signal to the output of this PA as seen in the bottom curve of the family in Figure 11-21(a). The different slope is actually a measure of how the load resistance at the drain of this GaN HEMT in this amplifier is different from 50 ohms. Using (5.7), we can actually determine that the design load resistance at the PA transistor is 43 ohms.

Using the energy efficiency curve family for this PA from Figure 11-9(a), with the constant input at maximum power strategy from Figure 11-20, we see the profile in Figure 11-22. This is not useful for several reasons, but the most important reason is the output signal dynamic range limitation to 21 dB caused by the leakage of the large input signal. The DPS transmitter is only useful when it can provide all the required output envelope values across the required output power control dynamic range. Solving this problem requires the input RF power to vary along with the PA supply voltage.

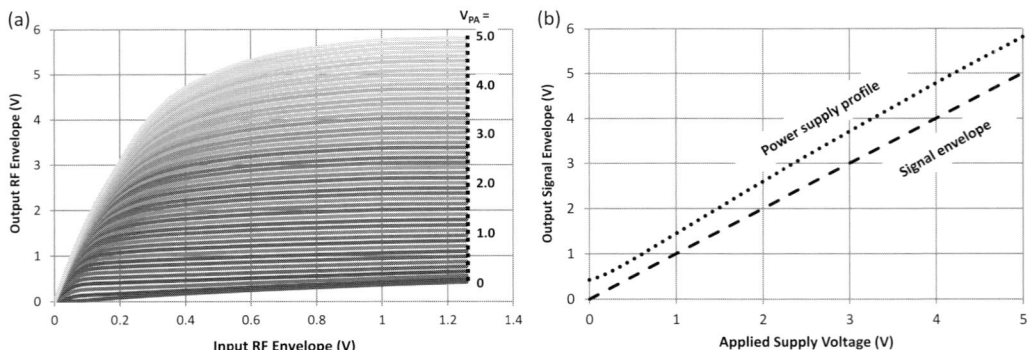

Figure 11-21   Power supply profile (dotted line) that follows the maximum efficiency points from Figure 11-15: (a) along the linear scaled Booth chart; (b) envelope (dashed line) vs. supply voltage profile from the dotted line in (a).

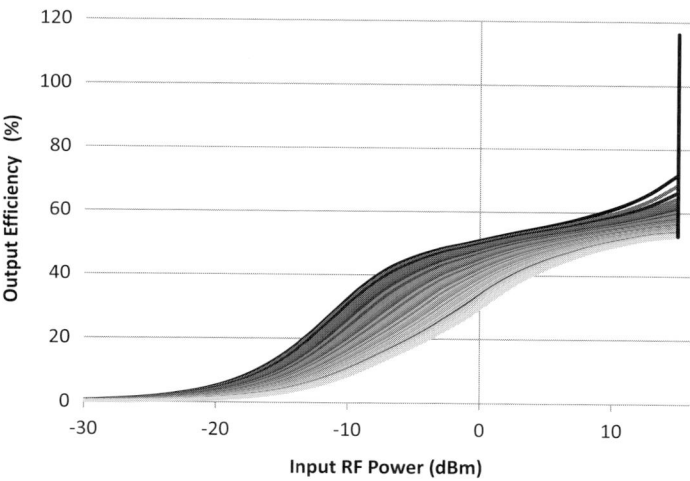

Figure 11-22 Maximum energy efficiency requires a constant high input power, but the efficiency measures above 100% at lower output power due to drive signal leakage to the output.

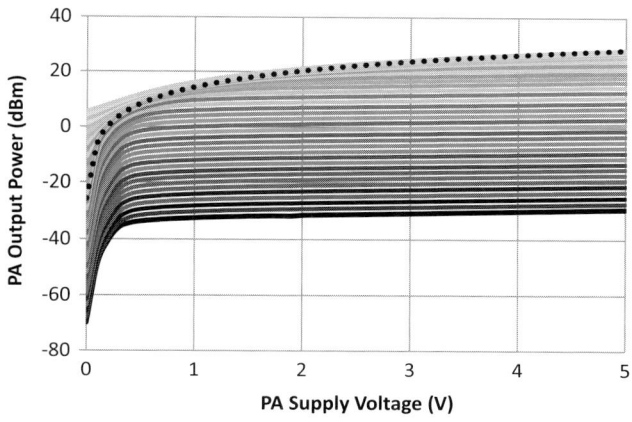

Figure 11-23 Ideal square-law power control (6.25) profile (dotted curve) overlaid on the PA measurements of RF output power vs. PA supply voltage family for various values of input RF power.

11.4.2   Power supply profile

Varying the input RF power with the PA supply voltage variation also manages the leakage problem because leakage power distortion reduces directly with input RF power reduction as discussed in Section 6.11.3. The ideal objective here is to match the ideal square-law power control profile (6.25). This is readily accomplished with the chart shown in Figure 11-23. The ideal power control profile is shown by the dotted curve. Measured data above this curve are from undesired leakage. The desired control profile is the intersection of each of these family curves with the ideal power control profile.

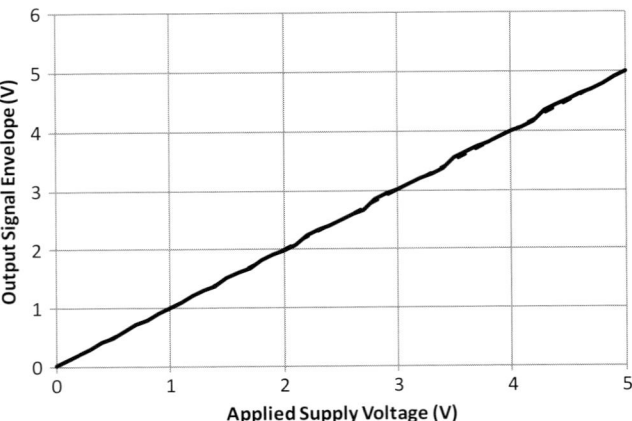

Figure 11-24 Output signal envelope vs. applied supply voltage shows a linear relationship (a constant value for g_{DPS_AM}), proving polar modulation is active.

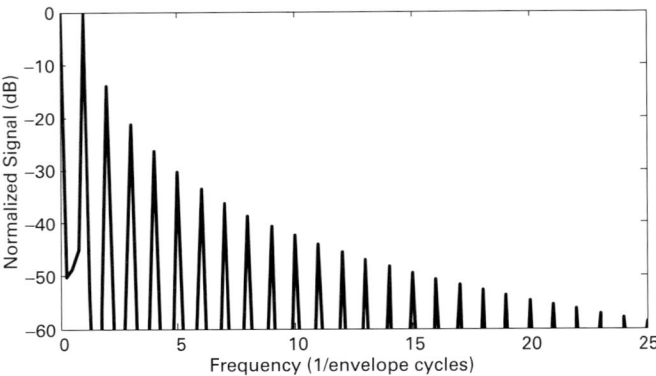

Figure 11-25 Envelope signal bandwidth for the maximum energy efficiency hybrid DPST strategy shows that no envelope signal bandwidth reduction is available.

Plotting the values of the closest curve intersections of measured output power with the calculated ideal power control provides the power supply profile shown by the solid line in Figure 11-24. This line is not smooth because of the granularity in the measured power. It is seen that this power supply profile now stays very close to the $V_{PA} = V_{env}$ (6.15) relationship shown by the dashed line in Figure 11-24. Improving the measurement granularity does bring these two curves into alignment. By meeting the condition of (6.15), this proves that this power supply profile is polar modulation, because the PA supply voltage is directly controlling the signal output envelope (2.9).

Because the power supply profile in this maximum efficiency strategy matches the signal envelope, the frequency-domain characteristics of the DPS output waveform also match those of the ideal envelope. This is shown in Figure 11-25, where here the DPS waveform spectral characteristics directly overlay those of the ideal

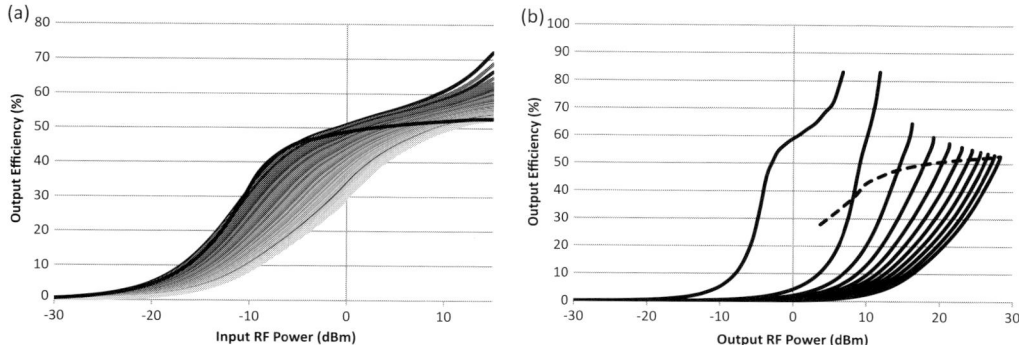

Figure 11-26 Energy efficiency profile when input drive level is gradually reduced to keep the leakage down: (a) all the available output efficiency is realized (heavy line) overlaid on the efficiency curve family from Figure 11-9; (b) corresponding output efficiency profile with input signal reduction to manage leakage and control output envelope accuracy.

envelope. By adopting a maximum energy efficiency strategy, the ability to reduce the bandwidth of the envelope signal at the DPS is eliminated.

11.4.3   Energy efficiency profile

Now that an input RF power profile ($\gamma(t)$ in (2.8)) is established, the comparable energy efficiency profile for this hybrid strategy can be evaluated. The result is presented in Figure 11-26(a) by the solid curve overlaying the measured family of output efficiency curves. This result is slightly better than that for the exponential supply profile in Figure 11-9(a), which is expected since here the primary objective is to maximize the PA energy efficiency.

Because the RF input power has to be gradually reduced in order to manage leakage to the output and to expand the available dynamic range from the transmitter, this strategy leads to a *maximum available* energy efficiency result. The actual energy efficiency that can be realized is shown in Figure 11-26(b). Any higher output efficiency incurs output magnitude error from signal leakage in this example amplifier. With the proper transistor technology, this leakage can hopefully be reduced and allow more of the available energy efficiency to be actually realized in the final transmitter.

11.4.4   Load to the DPS

Just as in Sections 11.2.4 and 11.3.4, here the load presented to the DPS by the PA being operated in this maximum available energy efficiency strategy is evaluated using the SSR surface of the PA. This is shown in Figure 11-27(a) where the trajectory shown by the dashed curve follows the profile of input RF power and PA supply voltage determined from Figure 11-23. Extracting the load provided to the DPS from this operation style across the input RF power dynamic range gives the results in Figure 11-27(b).

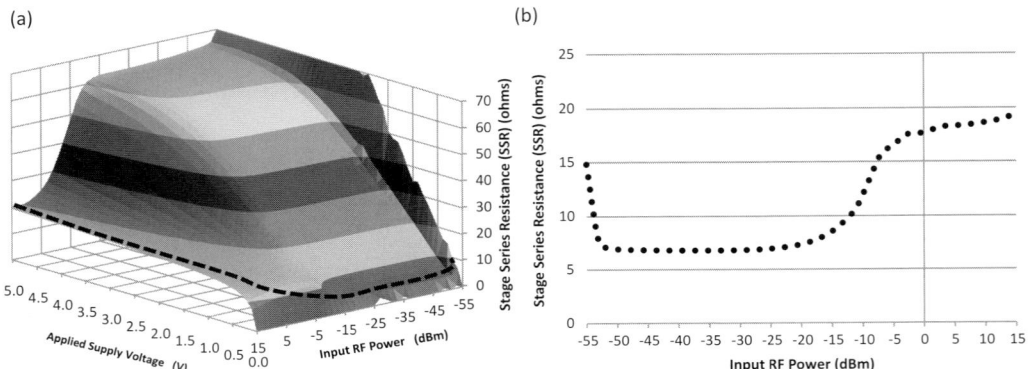

Figure 11-27   Load to the DPS for this GaN HEMT PA operated for maximum available efficiency: (a) profile along the SSR surface; (b) actual SSR value as the input RF power varies, showing both C-mode and P-mode amplifier operating regions.

The trajectory in Figure 11-27(a) is clearly in C-mode at the higher input powers and across most of the range of the PA supply voltage. Then a quick transition occurs to P-mode operation for all of lower input (and output) powers. This transition happens when the power supply value is between 0.4 V and 0.1 V, all of which is well below the P-mode boundary set from Figure 11-4. The low value of PA load resistance in this region is consistent with (5.20).

11.5   Comparisons

All of the performance charts in this chapter are from using the exact same PA. The performance differences are due only to the different control strategies applied to this PA. This means that the design choices made during DPST development have considerable influence on how the amplifier behaves, and also on the requirements for DPS design.

The largest differences are on the load the PA presents to the DPS. By changing the PA control strategy on its two inputs, RF power and supply voltage, the load experienced by the DPS can vary widely. All three of the DPS load profiles from the three hybrid operating strategies are collected and overlaid in Figure 11-28. In addition, the PA operating modes identified by the individual analyses for each hybrid operation are also listed in this figure.

Looking at the SSR value variations, the most consistent load is presented by the hybrid operation with the most inconsistent gain. The other two hybrid operation types have very consistent transfer functions controlling the output power, one by design (uniform transfer) and the other by result (maximum efficiency). Both of these have large variations in resistance presented to the DPS, 2:1 for the uniform transfer and 3:1 for the maximum efficiency designs. The C-mode PA SSR is slightly lower for the

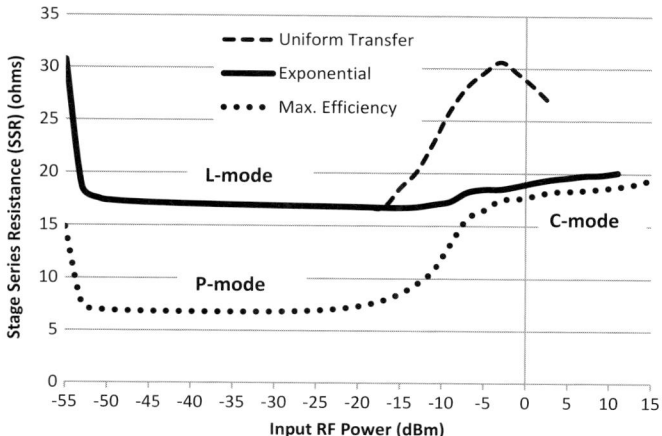

Figure 11-28 Load presented to the DPS by this GaN PA for each of the hybrid operating profiles of this chapter. The most consistent SSR behavior corresponds to the profile with the greatest gain variation.

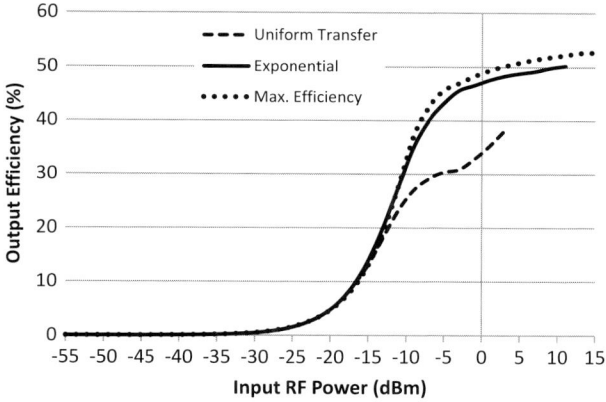

Figure 11-29 The maximum efficiency profile does live up to its name, but the exponential profile is close until it reaches its maximum available power limit. The constant transfer function profile provides the lowest realized energy efficiency and the lowest available power limit of this set of hybrid DPST profiles.

maximum efficiency design because it has larger input RF power for each value of output power than the exponential profile provides for.

Comparing the energy efficiency performance of these three hybrid operating strategies obtains the result in Figure 11-29. Here the maximum efficiency does come out on top, as we expect. It is interesting to see how close the exponential DPS profile design comes to the maximum efficiency performance. If all else were equal, then this difference would probably not be material to the final product. The efficiency difference to the uniform transfer design is greater than 10 percentage points, which can be a very important difference. But, of course, any economic value decision is strictly governed by the application the design is targeted to serve.

Table 11-2 Characteristic Comparison of these three Hybrid DPST Profiles

	Exponential	Constant Joint Transfer Function	Maximum Energy Efficiency
Overall efficiency	High	Moderate	Highest
Gain variation	Large	None	None
Envelope signal bandwidth	1/3	Some reduction	No reduction
PA load resistance variation	25%	2:1	2.8:1
PA operating modes	C-mode ↔ L-mode	Transition region	C-mode ↔ P-mode

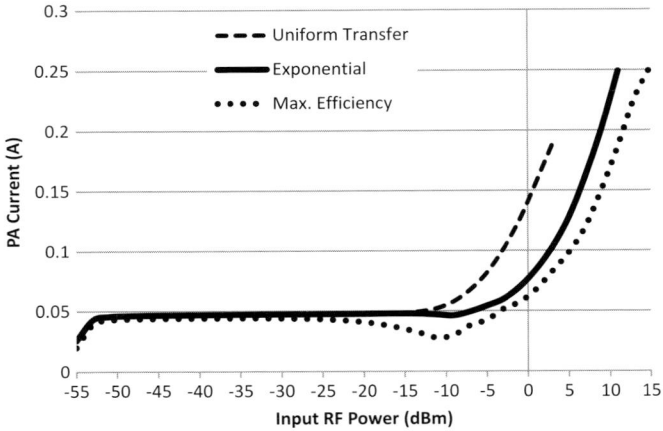

Figure 11-30 Comparison of the PA supply current characteristics shows two of these that exhibit negative dynamic resistance within the control dynamic range.

It is important to also stay with the basics, and to specifically check what the corresponding PA currents are for any hybrid profile being designed. Surprises may lurk in the data that would be most unpleasant if first encountered in the test lab – or not. For the three profiles in the present comparison, this PA current data are presented in Figure 11-30. And here we do observe two instances of negative dynamic resistance (voltage goes up and the current goes down). This effect is most noticeable in the maximum efficiency profile, though with careful observation this is also seen in the exponential profile. The uniform transfer profile does not show any negative dynamic resistance behavior.

The major performance similarities and differences among these three hybrid strategies are collected together in Table 11-2. In addition to the points discussed above, the table also includes the comparison of envelope signal bandwidth reduction. Only the exponential profile approach provides significant bandwidth reduction in this particular comparison. These results are only representative because different amplifier designs and different modulated signals used in an evaluation like this will likely provide results that differ in specific details. Still the general trend remains: if you want an accurate transfer function to inherently have good output accuracy along with the highest available energy efficiency, you will not get much envelope signal bandwidth reduction – if any.

It is physically inconsistent to claim both tolerance of power supply variations and ability to perform gain adjustment with the supply. Only one of these capabilities can be possible at a time.

Choose the problem you want to deal with. No single approach is uniformly ideal.

Excellent performance results are reported in the journals for transmitters using hybrid dynamic power supply transmitter techniques, whether or not the authors recognize that their designs are actually hybrid combinations. Particularly noteworthy examples include [11-2] [11-3] [11-4] and [11-5]. All of these papers use data based (empirical) design methods but do not connect their DPST performance measurements directly to transistor characteristic curves. Of these, the most complete recognition of DPS transmitter unique effects is reported in [11-5].

11.6   References

[11-1] Z. Wang, *Envelope Tracking Power Amplifiers for Wireless Communications*, Artech House, Boston, 2014.

[11-2] I. Kim, J. Kim, J. Moon, and B. Kim, "Optimized Envelope Shaping for Hybrid EER Transmitter of Mobile WiMAX – Optimized ET Operation," *IEEE Microwave Wireless Component Letters*, vol. 19, no. 5, pp. 335–337, 2009.

[11-3] H. Nemati, C. Fager, U. Gustavsson, R. Jos, and H. Zirath, "Characterization of Switched Mode LDMOS and GaN Power Amplifiers for Optimal Use in Polar Transmitter Architectures," *2008 IEEE MTT-S International Microwave Symposium Digest* (IMS2008), pp. 1505–1508.

[11-4] J. Jeong, D. Kimball, M. Kurak, P. Draxler, C. Hsia, C. Steinbeiser, T. Landon, O. Krutko, L. E. Larson, and P. M. Asbeck, "High-Efficiency WCDMA Envelope Tracking Base-Station Amplifier Implemented with GaAs HVHBTs," *IEEE Journal of Solid-State Circuits*, vol. 44, no. 10, Oct. 2009, pp. 2629–2639.

[11-5] J. Hoversten, S. Schafer, M. Roberg, M. Norris, D. Maksimovic, and Z. Popovic, "Co-design of PA, Supply, and Signal Processing for Linear Supply-Modulated RF Transmitters," *IEEE Trans. Microware Theory and Techniques*, vol. 60, no. 6, pp. 2010–2020, June 2012.

12   Multistage modulation

DPS transmitters are not restricted to only having DPS operation at the final transmitter stage. Under the right conditions there can be significant advantages to operating multiple stages in the transmitter with dynamic power supplies. And under the wrong conditions, an attempt to operate multiple stages as a DPS transmitter can be disastrous. In this chapter, both situations are encountered.

The basic architecture is straightforward, and a three-stage version is shown in Figure 12-1. The most flexibility is provided by simply cascading DPS stages. There are two primary motivations to consider the complexity of extending DPS operation to driver stages. Further improvement of the overall transmitter energy efficiency is the primary motivation. Particularly when polar modulation is used for maximum energy efficiency, DPS operation of earlier stages extends the dynamic range for output envelope accuracy.

12.1   Interstage signal magnitude management

Multistage design exists because overall gain requirements in product design exceed the gain capabilities of a single stage. The power consumption of earlier stages must be progressively lower for this strategy to make sense. Ohm's Law shows that there are not many degrees of freedom to achieve this goal: if the supply voltage is unchanged, then the load resistances must progressively increase. If the stage load resistances stay the same, then the supply voltages must progressively decrease.

Examination of the interstage design issues points out significant differences in design objectives. When a bipolar transistor is being driven, the driving stage must provide sufficient base current to (1) drive the required collector current through the transistor β action, and (2) provide the C_{BE} current that shunts the transistor action as discussed in Section 7.3.1 and shown in Figure 7-15. Along with V_{BE} and r_π, this current establishes the input impedance of the transistor at the operating frequency. Design of the interstage matching network (InMN) in Figure 12-2(a) must convert the signal at the output of the driver to this resistive and reactive current and voltage.

When an FET device is being driven, as long as the gate waveform stays within the FET active area the input impedance is reactive and not resistive, as shown in Figure 8-16. Without a resistive component to the input current, in principle there is no input power. The driver stage sees a capacitive load, and all of the voltage developed

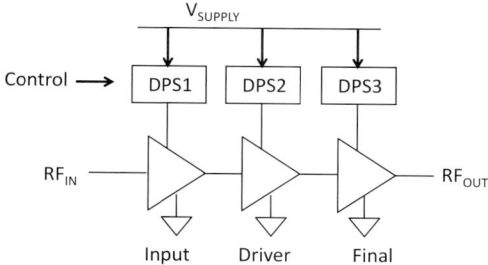

Figure 12-1 Multistage DPS operation using three stages.

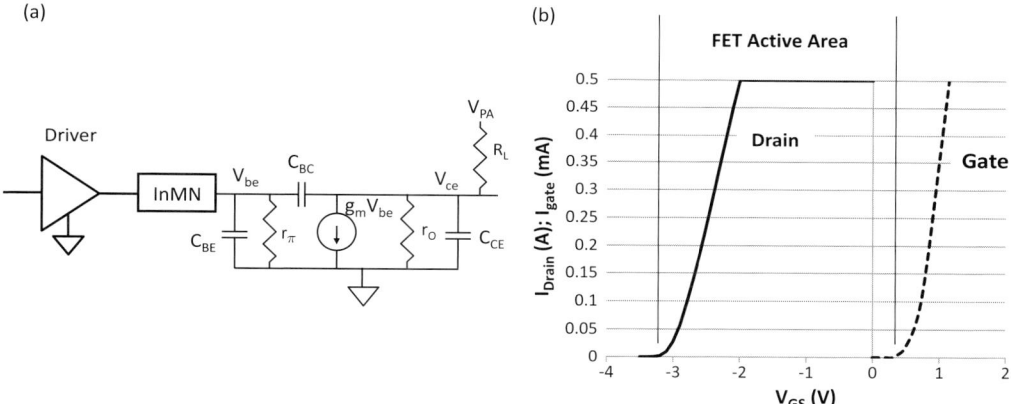

Figure 12-2 Interstage design goals are very different depending on the following stage technology: (a) driving a bipolar transistor; (b) driving an FET by staying within the FET active area (here for a GaN HEMT).

across the input capacitance does appear as V_{GS} and directly influences the FET drain current without the shunting effect seen in the bipolar transistor.

Treating the input capacitance of the FET as this single lumped element is valid only when the amplifier construction is no larger than 5% of the wavelength of the signal frequency in the material used to construct the amplifier. This lumped element approximation is generally valid on a monolithic RF integrated circuit up to the middle microwave frequencies. At higher frequencies or for other construction techniques, it is necessary to use transmission lines for interconnection of PA circuit blocks. In these cases, the capacitance of an FET input is a severe mismatch to the transmission line impedance and standing waves must result on the interconnecting transmission line (not good). Adding resistance is usually needed to "tame" these circuits, which works fine but reduces energy efficiency (and also usually has lower gain) from the resistive power dissipation.

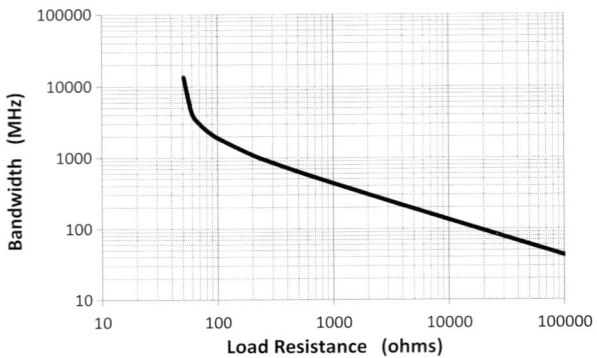

Figure 12-3   Circuit bandwidth is restricted when the load resistance is greater than 50 ohms. Compare this result with Figure 8-27.

The bandwidth impact of having amplifier load resistances lower than 50 ohms when the matching networks are built using passive reciprocal circuit elements (standard inductors, capacitors, and transmission lines) is shown in (8.34) and Figure 8-27. When the amplifier load resistance is greater than 50 ohms, there is a symmetrical effect, as the matching network transforms the higher resistance to the smaller 50 ohm output resistance, based on the quality factor Q of the passive matching network, to result in

$$BW_{HL} = \frac{f_0}{\sqrt{\dfrac{R_L}{50} - 1}}, \tag{12.1}$$

where the variable name here has the subscript HL to distinguish it from (8.34). Following this pattern, the dependent variable in (8.34) can best be named BW_{LH} to emphasize that this formula works only for low to high resistance steps. This bandwidth effect is graphically shown in Figure 12-3. Going to high resistance also makes the presence of small stray capacitances more important because the RC time constants get large from the large resistance value.

12.1.1   Dynamic range extension

In Section 6.11.3, the output power control dynamic range of the C-mode stage is seen to extend when the leakage through the C-mode stage is managed. This means, of course, that the input signal magnitude must be reduced just enough to suppress the undesired leakage while still maintaining the C-mode core principle of (8.1). The input RF power profile that manages the leakage signal from Figure 11-23 while still maintaining C-mode operation by always satisfying (8.1) is shown in Figure 12-4. This curve is very device-type specific, and depends strongly on the specific construction of the FET. Experience shows that once the FET is designed, this leakage characteristic is very repeatable and the leakage management curve does not need to be calibrated on the production line.

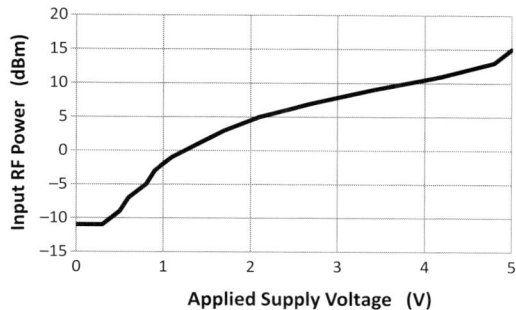

Figure 12-4   The leakage signal management curve for the transistor used in Figure 11-23 to maintain C-mode operation across a 45 dB polar modulation dynamic range.

When FET devices are used, the range of the input voltage swing is the same for any particular technology, no matter whether the transistor size is tiny or huge. This means that the voltage range of the input signal at the gate of the FET never really shrinks when the switching operation is required. Smaller transistors have commensurately small reverse transfer capacitances C_{DG}, so the inherent isolation does improve with smaller transistors. At some point, the input signal is just too large and there is no more benefit to extending the DPS technique to more stages earlier in the amplifier chain. Again this is not something that (so far) has any theory to establish this DPS stage-count practicality limit, but experience hints that this limit is three stages.

12.1.2   Distortion reduction

Distortion reduction can only occur when the ET region is no longer being used, meaning that the DPS transmitter operation does not satisfy (7.1). Once the transition region is entered and there is a trade-off established between the input RF power and the power supply on setting the RF output envelope, then we can use DPS operation to improve output signal accuracy. Indeed, this is exactly the idea used in Section 11.3. One trade-off for input RF power and supply voltage to maintain the required output magnitude profile is shown in Figure 11-15(a). Another view of managing input RF power and supply voltage to minimize distortion is seen in both parts of Figure 6-38.

12.1.3   Joint operation

With independent control of the DPS for each stage, the DPST designer can choose if each stage operates in envelope tracking, polar modulation, in the transition region between these two, or simply in conventional L-mode. An example strategy that operates one or more stages in all of the available PA operating modes is shown in Figure 12-5. Another possible strategy is to operate all stages in ET. Yet another is to use

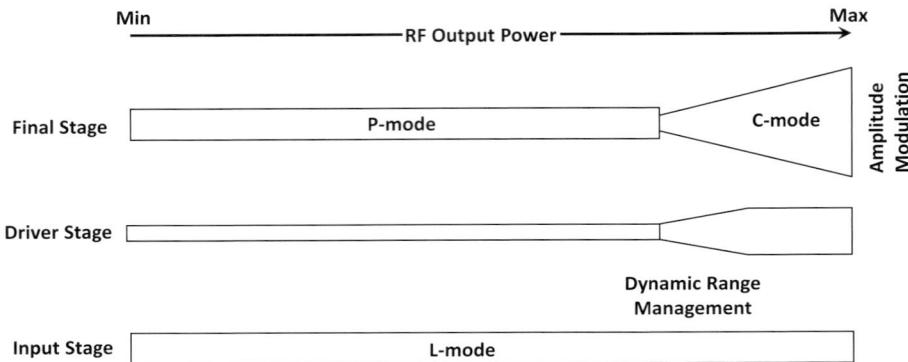

Figure 12-5   A different graphical interpretation for the wide dynamic range control method of Section 6.12.3 and Figure 6-40. For a 3-stage PA, it is most common to operate the input stage in L-mode at all times.

polar modulation in the final stage and ET in the driver. Or polar operation can be used in all stages. The available varieties are numerous.

12.2   DPS applied at driver stages

The core principle of ET (7.1) requires that when a DPST is operating as ET there is no effect from the DPS in any way on the output signal quality. A significant improvement in overall transmitter energy efficiency is realized when the dynamic bias is done properly at all stages. This is particularly true when stage gain is not very high and the power required in and from the driver is relatively large.

12.2.1   Independent supplies

Using a separate DPS for each stage is inherently stable as long as the associated RF PA stage does not individually exhibit negative dynamic resistance (see Figure 9-3). Figure 12-1 shows this architecture, which provides for highest performance because each stage can be individually tuned for top performance. The need to use multiple DPS circuits does increase cost, returning yet again to the common trade-off for performance vs. cost. Highest performance never comes from the minimum cost implementation. As in business, you do not get what you do not pay for.

12.2.2   Common supply connection

Operating all PA stages from a single DPS is very attractive from a low-cost perspective. Since most of the current supplied to a PA flows through the final stage, a reasonably margined DPS designed for that final stage is very likely to also be able to supply current

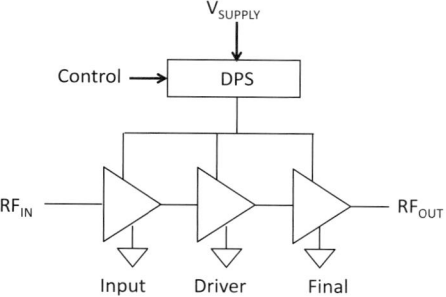

Figure 12-6   Reducing cost by using a single DPS to operate a PA.

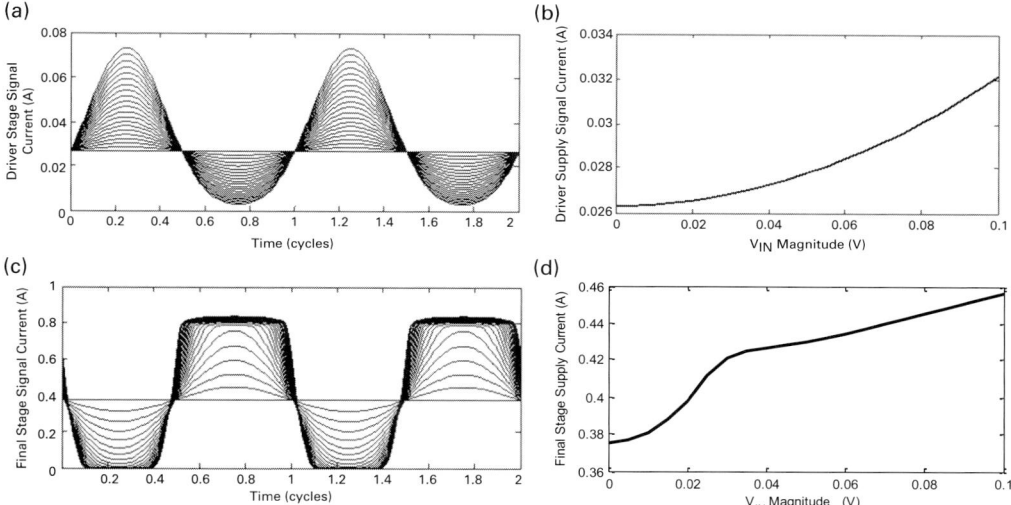

Figure 12-7   Two stage model for a PA with the output stage intended to operate in C-mode, as the input signal magnitude varies: (a) overlay of driver output signal current waveforms; (b) current drawn by the driver stage; (c) overlay of output (final) stage signal current waveforms; (d) current drawn by the final stage.

to the entire PA. Then the architecture changes from that of Figure 12-1 to the simplification shown in Figure 12-6.

Performance simulations of this FET-based amplifier cascade are provided in Figure 12-7. The first two graphics show how the driver stage behaves in this design as the magnitude of the input RF signal is swept from 0 to 100 mV. An overlay of all the drain current waveforms for the driver is seen in Figure 12-7(a). At low input magnitudes, this stage behaves as a nice linear amplifier. As the input signal increases further, the positive current swings are greater than the negative swings about the bias (quiescent) current. This signifies class AB bias with respect to the largest input signals. The current drawn by this stage as the input magnitude varies is shown in Figure 12-7(b). At zero input, only the quiescent current flows. As the input signal magnitude increases, the overall current also increases as more output signal is produced.

Output stage performance is seen in Figure 12-7(c). The drain current waveform also starts out linear, and transitions into the compressed OFF and ON waveform required for C-mode operation as the input signal increases. The quiescent current is near the midpoint of the current extremes, which is the desired condition from the discussion in Section 6.8. The output stage is therefore biased for class A operation, though here there is no intention to ever operate this stage with no RF signal present. Indeed, the definition of a polar signal process is that the control input (power supply) must be zero for a zero-magnitude output signal, a condition that guarantees no quiescent current will ever flow.

The current drawn by this output stage across the varying input signal magnitudes is shown in Figure 12-7(d). As for the driver, at zero input signal the only current that flows is the quiescent current. As the input signal increases, the current drawn increases rapidly, until the onset of clipping on the output waveform. As clipping of the output waveform increases, the current drawn slows down. The current drawn stops increasing when the output signal becomes a square wave, which for this set of conditions does not happen because of the finite rise and fall times seen in Figure 12-7(c).

12.2.3   Negative dynamic resistance DPS load effect

Any polar transmitter must operate in C-mode, and as we see in Figure 12-7 the output stage is well within C-mode at the maximum input signal evaluated in this simulation. Choosing this input drive level, the next step is to replace the fixed power supply with the DPS. First, the DPS is only connected to the output (final) stage, leaving the driver stage powered by the original nominal supply voltage. Sweeping the value of the DPS output from zero to 3.6 V, the currents drawn by the driver and the total PA (driver + final) are

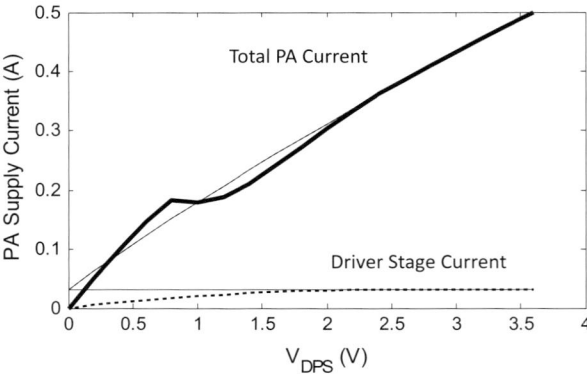

Figure 12-8   Load to the DPS varies significantly if both the driver and final stages are varied together: (dashed lines) voltage on the driver is held at the nominal PA value, and only the final stage power supply is varied; (solid lines) both the final stage and driver stage power supplied are swept together.

reported by the dashed lines in Figure 12-8. Driver current is constant at just over 30 mA, consistent with the result for maximum input signal from Figure 12-7(b).

Drawing current when the DPS output is zero is something we really do not want. The natural thought is to simply connect the DPS output to the driver stage too. In this connection, the current draw must go to zero when the DPS output is zero because no drain current can flow in either stage. This does indeed happen, as seen by the solid lines in Figure 12-8. The driver current does fall to zero as the DPS output voltage goes below 2 V. But, as the driver current falls, the current in the output stage exhibits some very strange behavior. It is now far from a linear relationship with V_{DPS} the DPS output voltage. What happened?

The answer comes from Figure 4-23. When the driver voltage goes to zero, its RF output signal also goes to zero. Figure 4-23 shows that there is no way to operate in C-mode when the input falls much below the value needed to get the output transistor to switch, even when V_{DPS} is very small. Here, when the voltage on the driver stage is low, the output stage is actually operating in P-mode. It is not drawing much current because its power supply is also low. But it definitely is not operating in the desired C-mode.

As V_{DPS} continues to increase, the output from the driver stage starts to increase. This eventually begins to drive the output stage toward compression, where the current draw begins to slow down its rate of increase. If the driver output signal increases fast enough with changes in V_{DPS}, the total current drawn by the PA can actually go down with an increase in V_{DPS}. This does happen here. Given the apparently reasonable design from the results in Figure 12-7, when the driver and final stages are both connected to the DPS for power then the current drawn from the DPS goes *down* when the DPS increases from 0.8 V to 1.1 V. This PA provides a *negative dynamic resistance* load to the DPS over this region. This is very bad for the DPS.

From the point of view of the DPS, the PA in this connection is acting somewhat like a tunnel diode. Applying a voltage across a tunnel diode that places its characteristic curve in its similar negative slope region is well known to support oscillations. The same behavior can, and does, happen with this polar circuit. Overall oscillation of the DPS and PA combination when the DPS output voltage is within this negative slope region does indeed happen. It is essential to design the transmitter so that this condition never occurs under any circumstances.

The only solution is to return to having the driver stage and the output stage powered by independent power supplies. Both of them can be dynamic. In fact, this is the best situation as derived in Section 6.8.3. The input to the final stage, which must operate in C-mode for polar operation, is optimum when it varies nonlinearly with the changes in V_{DPS} applied at the output stage. If the driver stage is also operated in C-mode for best energy efficiency, then its independent DPS can be used to set the proper RF magnitude into the final stage for best performance and overall efficiency. It is also possible, and a good design practice, to vary the input to the overall PA stage to the minimum necessary to guarantee C-mode operation of the necessary PA circuit stages. This is consistent with (2.8) and perfectly aligned with polar operation as long as (2.9) also applies to the output signal.

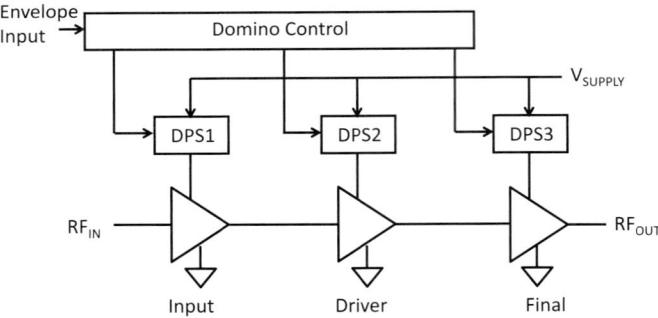

Figure 12-9   Domino PA operation requires the PA stages to be turned ON in sequence as the output power requirement grows. OFF stages draw no supply current which improves energy efficiency.

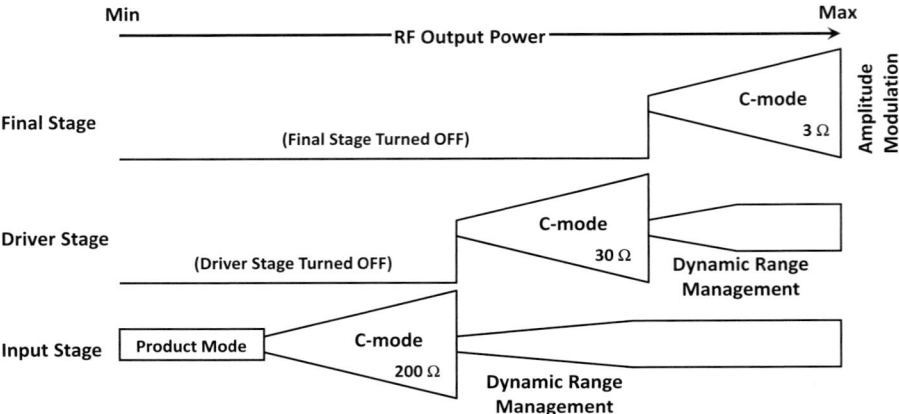

Figure 12-10   Domino operation calls for each stage to operate in C-mode, bringing the input stage up first, then the driver stage, and last the final stage. Earlier stage output signals reach the output either through the leakage of the following stages, or by adding shunt switches to bypass the turned-OFF subsequent stages.

12.3   Domino operation

Using DPST techniques allows a sequential technique called Domino operation. The Domino idea is to only power up the amount of the PA that is needed to provide the necessary output envelope power. This approaches the objective of only having the PA provide signal power by bringing bias power down to a minimum, even while always operating class A. This technique has particular advantage when all stages operate in C-mode.

The top-level architecture for Domino operation is shown in Figure 12-9, with the operating idea illustrated in Figure 12-10. Starting from maximum output power, as the required output envelope power falls both the supply and the bias are reduced on the final stage using DPS3. This can be done with either an envelope tracking or a polar modulation profile. The principles of the Domino operation are independent of the PA operating mode. The control dynamic range of the final stage is limited, and when it gets to its minimum it is powered OFF and the output signal magnitude control is transferred to the driver stage. The driver stage signal can reach the output of the PA through the leakage path of the final stage, making this leakage path actually useful instead of a problem. Alternatively, the PA can be designed with an internal switch that provides a low impedance path to the output from the driver, though the design of such a switch is quite a challenge because it must remain OFF with negligible parasitic reactances when the final stage is operating normally.

This principle continues while the driver stage provides the PA output power. DPS2 manages the supply to the driver stage until it also reaches its minimum output capability. Then the driver stage is powered OFF by DPS2 and the input stage continues providing the PA output power. DPS1 manages the supply to the input stage. If the input stage is operated in C-mode and it reaches its minimum output power, then it continues to lower power outputs by converting to P-mode operation. As output power increases the stages are powered up in consecutive order, as when resetting up a string of dominos.

One of the advantages of the Domino operation is the ability to take advantage of C-mode modulation accuracy across a much wider dynamic range than a single stage of polar modulation normally provides. For this to be the intended operation of the PA when the transistors are scaled down for each earlier stage, it is important to keep the C-mode design rules from Section 6.7.5 satisfied at all times, particularly (6.11).

For system applications where maintaining the phase of the output signal across the power control dynamic range is important (e.g. 3GPP WCDMA), the Domino approach may be a problem. This problem arises because each stage is a nominal phase inverter. As the Domino stages step from one to another there is a phase shift on the nominal output signal that must be accounted for.

12.4   Power quadrature modulation (power-QM)

Traditional quadrature modulation designs combine the modulation components at a very low output power and then use a conventional amplifier chain to bring the modulated signal up to full power. It is possible to amplify the individual in-phase and quadrature-phase modulation components and then perform the signal sum later at full output power. There are two power amplifiers needed, one for the in-phase modulation component and another for the quadrature-phase component. Neither PA has phase-modulated inputs since they only see the in-phase or quadrature-phase local oscillator (LO) carriers. All the output signal phase modulation appears as a result of the output quadrature summation.

Multistage modulation can also happen in parallel. In this case, the power amplifier is divided in half and moved within the quadrature modulator function as shown in Figure 12-11. Signal summation occurs at full power in a quadrature hybrid.

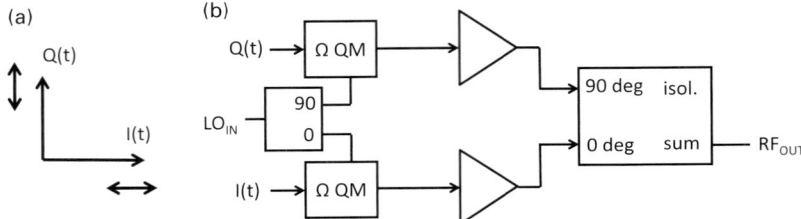

Figure 12-11 Power quadrature modulation applies a PA to the $I(t)$ and $Q(t)$ signals individually, and then combines the two signals at full power in a quadrature hybrid: (a) vector diagram; (b) architecture block diagram.

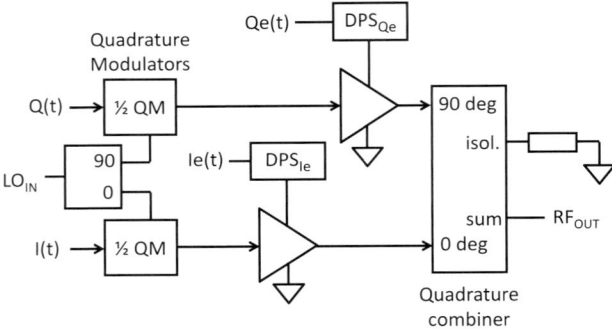

Figure 12-12 Using ET for the power-QM technique.

12.4.1   Envelope tracking in each arm

Another option to do power-QM is to apply ET to each of the PAs in Figure 12-11, resulting in the block diagram shown in Figure 12-12. The basic operation of Figure 12-11 does not change, but with ET the energy efficiency of Figure 12-12 is greatly improved.

12.4.2   Direct polar in each arm

It is also possible to operate C-mode polar modulators in each arm of the power-QM, as shown in Figure 12-13. Here the multiplication happens in the C-mode PA in accordance with the modulation triangle, so the input modulator block from Figure 12-11 and Figure 12-12 is eliminated here. There is no phase modulation of the input of these C-mode power stages, which greatly simplifies their circuit implementation. To handle phasors in all four quadrants, a phase-inverting technique is required, such as those presented in Sections 6.13.2 and 6.13.3.

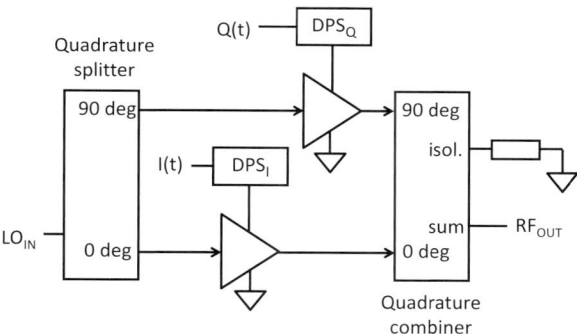

Figure 12-13   Using polar modulation in the power-QM technique.

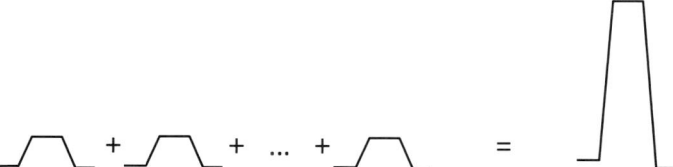

Figure 12-14   Parallel operation of amplifiers producing equal power allows coherent power combining to generate higher output power than any individual PA can provide.

12.5   Higher output power

Higher output powers (10 W and more) are often attempted by building the high power amplifier with larger and more powerful transistors. While large transistors are capable of providing more power, they also have larger parasitic capacitances and inductances which both restrict their operating bandwidth and draw currents that lower available energy efficiency.

Generating high output power by combining the outputs of multiple amplifiers is also a well-used technique. If the application requirement calls for wide tuning bandwidth, this combining technique is usually preferred.

DPS techniques fit well with combining techniques for high output power. One technique is parallel combining, where all of the amplifiers generate exactly the same power. These powers are summed into the actual output as shown in Figure 12-14.

It is also possible to add signals in series. In effect, the first PA provides as much power as it can, and then the second PA is activated. When that PA reaches its power limit, then the third PA is activated, while both the first and second PAs continue providing their saturated power. Peak envelope power capability is defined by the total power from all the available power amplifiers. This sequential action is illustrated in Figure 12-15.

Both of these techniques work fine with linear amplifiers. If ET is used, the output power characteristics are the same as when linear amplifiers are used, but the ET action

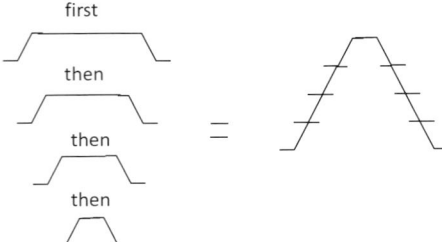

Figure 12-15 Series combining of power amplifiers to get higher output power than any individual PA can provide involves turning PAs ON and OFF in sequence as the output signal envelope requires.

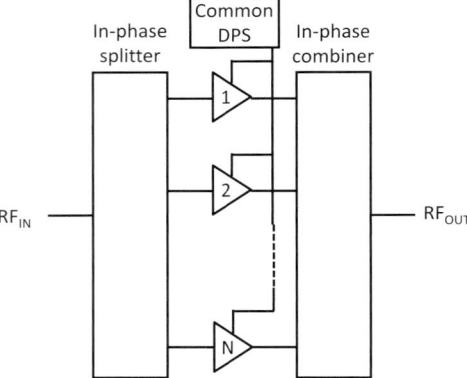

Figure 12-16 Parallel power combining with multiple power stages all supplied in common from one large DPS.

significantly reduces power dissipation. These techniques become particularly interesting when the individual power stages are operated with polar modulation.

12.5.1   Parallel polar modulation

Power combining is at its most efficient when all the inputs are fully coherent, meaning that each input signal has identical phase and magnitude. Matching the individual signal magnitudes is usually more difficult to ensure across amplifiers that must have identical gains. This difficulty is nearly eliminated when the individual amplifiers are operated in C-mode, where the output magnitudes are not dependent on the gain of each individual amplifier but are set by the tightly controllable DPS output voltage according to (6.15). Using the architecture in Figure 12-16, all of the individual C-mode stages are supplied by one common DPS which sets all output magnitudes to the same value. As long as the in-phase power splitter maintains accurate phases into each of the C-mode stages, the output in-phase combiner becomes very efficient with nearly no loss.

For N identical C-mode stages, the output power is N times the power from each stage. The DPS driving all of the stages together also therefore controls the total output envelope [12-1]. This structure implements the situation described in Figure 12-14.

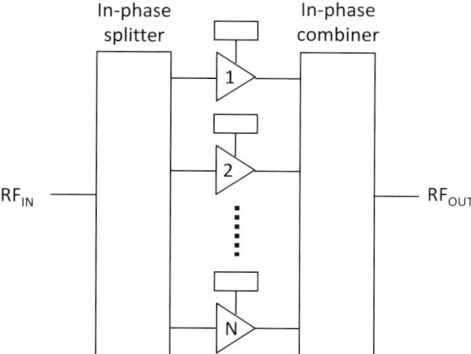

Figure 12-17   Parallel power combining with multiple instances of power stages each with its own DPS.

Having one large DPS has the same problems as having one large power amplifier: large transistors with significant parasitic reactances limiting the DPS bandwidth and efficiency. Taking note of the bandwidth benefit from dividing the PA into N smaller PA cores, similar advantages in bandwidth capability are available by dividing this large DPS into a set of N smaller DPS cores. This in effect makes the core unit of this polar modulator a pairing of the core PA with one of the core DPS units. This modified architecture is presented in Figure 12-17. Matching among the DPS cores is very important to get the most efficient performance from the N-way power combiner, to be sure that the output from each of the C-mode PA cores actually does have the same magnitude as all the others.

A common characteristic of both of these architectures is that the power control dynamic range is no greater than the power control dynamic range of any of the PA cores. If the overall output power is required to be small, then each of the PA cores must provide $1/N$ of this small output power. The design of each PA core must emphasize a power control dynamic range that exceeds the PCDR required of the entire combined PA. Leakage of the input signal across the C-mode transistor must be carefully minimized.

This discussion so far assumes that the output combiner is operated for absolute minimum loss, which fundamentally restricts all PA core outputs to be exactly identical. The consequence of this is an N times expansion of the DPS step size between each PA core and the combined output signal. This resolution can be completely recovered, at the price of increased loss in the output combiner. By removing the restriction that each PA core has exactly the same output magnitude, we can allow very small differences between the PA core power values. Such an approach requires the architecture of Figure 12-17. The strategy of how to step each DPS to get the desired output resolution with the objective of minimizing the combiner loss actually has a large amount of flexibility. Investigating these various strategies is an interesting research topic.

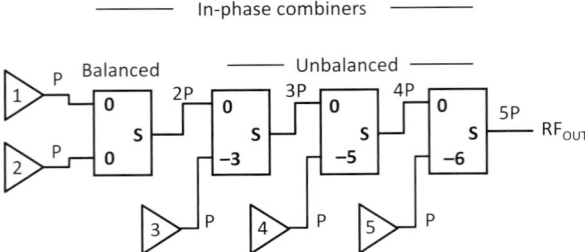

Figure 12-18 Series combining using a sequence of unbalanced power combiners to provide the best available efficiency. Inputs on the combiners are listed by the nominal dB difference between the input power levels.

12.5.2   Series stacking for power

While the parallel power combiners of Section 12.5.1 use the C-mode PA cores together to get a multiplicative power generation effect, it is also possible to use these core amplifiers one at a time. Using a combination of balanced and unbalanced in-phase combiners as shown in Figure 12-18, the power levels from each amplifier combine linearly [12-1]. This is demonstrated in this figure by defining the unit power available from each core PA as P. With 5 PA cores in this example, each with power output capability P, the power output from each successive combiner is a linear summation of the input powers. And the power at each successive combiner step is one unit power higher than the previous output. It is vitally important to maintain phase alignment among all of these constituent amplifiers in order to have the individual output powers add as expected.

By sequentially activating these power amplifiers as all of the earlier PAs reach their saturation levels, this structure can be used to get higher output powers while maintaining nearly all of the efficiency provided by each PA individually [12-2].

In some sense, this is an extension of the Doherty idea, but here the implementation does not use load modulation between the individual PAs. But the idea to let one PA operate at the lowest output envelope levels until it compresses, and then to have another PA add in to provide output power above what the original PA is capable of, is followed here. This approach can be described as N-way PA summing, similar in power behavior to an N-way Doherty. This is definitely not a true Doherty approach from the circuit operation perspective, but the top-level operating principle is very similar: add power from another amplifier if the signal peak power gets higher than the present power amplifier(s) can provide.

When each of the amplifiers in Figure 12-18 is a C-mode DPS stage, and therefore a polar modulator, the resolution of the final output power is equal to the resolution of each individual DPS. This is different from the parallel architecture of Figure 12-16, where the output power minimum step is N-times the step of the DPS. In a sense, this strategy can be viewed as an RF digital to analog converter.

In the 1980s, this idea was taken to an extreme in the construction of AM broadcast transmitters [12-3]. The architecture used is to have 256 very efficient amplifiers (90% or better) that are operated either ON or OFF. No power control of any constituent PA is allowed, to maintain some measure of simplicity. The challenge is in the 256:1 combiner, a challenge that is successfully solved. With each amplifier capable of providing 500 W, the total output power ranged, with 8-bit resolution, from 0 to 128 kW at well over 80% overall energy efficiency. And this approach also generates these very large powers with unconditional stability because all constituent PAs operate in C-mode and are well outside the Barkhausen boundary.

12.6   References

Note: Patent references are provided for bibliographic use only. Citation of specific patents here is not indicating any view on priority issues.

[12-1] K. Russell, "Microwave Power Combining Techniques," *IEEE Transactions on Microwave Theory and Techniques*, vol. MTT-27, no. 5, May 1979, pp. 472–478.

[12-2] E. McCune, "High Efficiency Modulation Amplifier," US Patent 6377784, issued April 23, 2002.

[12-3] H. Swanson, "Digital AM transmitters," *IEEE Transactions on Broadcasting*, vol. 35, no. 2, Feb. 1989, pp. 131–133.

Part III

Testing and manufacturability

13 Testing and calibration techniques

It is unfortunately very easy to set up a DPS transmitter design project that is immediately doomed to failure. The reason is that if the implementation technology selection is not carefully matched to the project performance objectives, then, when it comes time to connect the new DPS with the RF PA it will be used with and to do full testing, the chances of it all working together are actually quite small.

The way to succeed is no different from any other complicated project: plan ahead and do the required component and subsystem testing so that a substantial well-understood foundation is established for the system operation. For DPST operation, the testing is much more complicated than for conventional designs because the variable power supply adds a new degree of freedom. Familiar analysis methods based on curves must change to operate on surfaces, for example. Add to this a completely new interface that is fundamentally not well-behaved and the project gets even more interesting. Careful characterization of this new interface is vitally important.

This chapter provides significant detail on the testing required in DPS transmitter development in order to have near certainty of project success. This test profile is the result of 20 years' experience this author has with making DPST designs work. Each of these tests has its important place on the path to project success. Many of the unusual tests in these lists are the result of unpleasant surprises in the test lab.

13.1 Characterization planning

The three major blocks of characterization testing required for any successful DPS transmitter design are: (1) the DPS itself, (2) the RF power amplifier on its own, and (3) the transistor used within the RF PA. These blocks are illustrated in Figure 13-1. At this writing, the manufacturers of dynamic power supplies and RF power amplifiers are different companies. This necessitates their interconnection to be considered as a new interface. Much more effort is required to establish standards through which the interconnection of any DPS with any RF PA is likely to work. In any case, the ultimate responsibility for operation of this interface falls on the transmitter designer.

The DPS is initially characterized on its own, often by the DPS product manufacturer but always verified by the DPST designer. When the RF PA is purchased from an outside vendor, then there is restricted ability to evaluate the transistor used inside. The designer

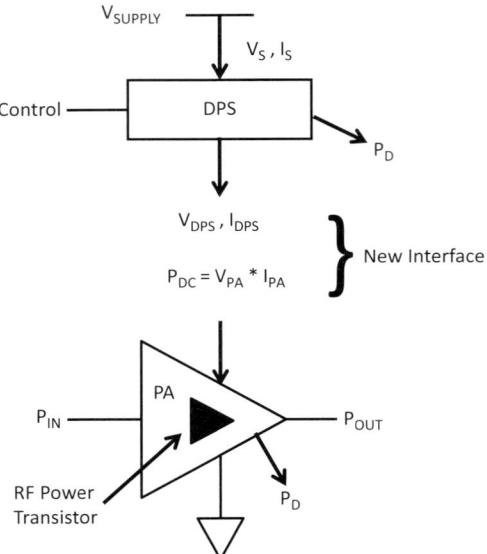

Figure 13-1   Testing block diagram for DPST design.

of the RF PA must very carefully evaluate the transistor technology selection using the tests in Section 13.4 as a minimum.

13.2   DPS characterization

Initial DPS characterization must be done stand-alone, independent of any PA attached as a load. This means that a test "dummy" load, or set of loads, must be attached to the DPS output during these characterizations.

13.2.1   List of DPS tests

Independent characterization tests for any DPS must include the following as a minimum. Some of these are conventional and hopefully familiar characterizations of voltage regulators. It is important that all of the following information be provided by the DPS manufacturer. If so, then all the DPST designer needs to do is verify the DPS performance. If any of this information is absent, then the DPST designer must thoroughly evaluate the missing data.

- **Input voltage range** ($V_{S,MIN}$; $V_{S,MAX}$) – The voltage range acceptable on the input of the DPS must include the nominal value of the top supply, as well as more than the entire possible excursion of this top supply.
- **Output voltage range** ($V_{DPS,MIN}$; $V_{DPS,MAX}$) – The maximum available output voltage from the DPS must exceed the needs of the target PA to achieve its required

PEP. For DPST use, the lower end of the available DPS output is usually more interesting and difficult. But this is where the greatest overall energy efficiency gain is likely to occur, so it is extremely important to characterize and understand this lower limit.

- **Output current requirements and limits** ($I_{DPS,MIN}$; $I_{DPS,MAX}$) – Both limits of output current are important in DPST operation. At the high end, $I_{DPS,MAX}$ must exceed the PA current requirements at PEP power and with any output mismatch at the phase of minimum load resistance value (the maximum current value). Some DPS architectures require a minimum amount of load current to flow for the regulators to be stable. In these cases, $I_{DPS,MIN}$ will not be zero.

- **Power dissipation limits** – Operating temperature of the DPS follows (3.11), driven by the internal power dissipation, the thermal resistance of the materials used in constructing the DPS, and the ambient temperature of the active application

$$P_D = V_S I_S - V_{DPS} I_{DPS}. \tag{13.1}$$

The resulting temperature from (3.11) must be below manufacturer specification limits to maintain reliability.

- **DC transfer function** – The fundamental operation of any linear DPS is a linear transfer function with an offset

$$V_{DPS}\left(V_{control}\right) = K_{DPS} V_{control} + V_{offset}. \tag{13.2}$$

Validating the voltage transfer gain K_{DPS} and the operation of the fixed output offset process are what is important here.

- **AC transfer function** – The ability of the DPS output to accurately follow the waveform applied to its input is vital for the highest efficiency polar modulation (8.2), and is less so for ET (7.1). Precision output waveforms require that both the magnitude and phase relationships of the input waveform are maintained at the DPS output (Sections 3.6 and 9.6), meaning that the DPS circuitry *phase* response actually sets the DPS bandwidth. Measuring a complete Bode plot, or any equivalent, is necessary here.

- **Output impedance vs. V_{DPS}** – Low frequency stability of the PA depends on the output impedance of the DPS being low enough that the DPS serves the function of the conventional large bypass capacitor (Section 7.2.2). The DPS output impedance does in general vary with the value of V_{DPS}, so the output impedance must be evaluated at each possible output voltage value.

- **Output impedance vs. frequency** – Besides needing a small value for the DPS output impedance to establish low frequency stability in the PA, this low impedance must be maintained across the bandwidth of the PA bias network. For any PA intended for wide bandwidth signal use in ET, this will be 10s of MHz. Evaluation from "DC" out to 100 MHz is often sufficient, but this must always be validated for each specific PA to be used.

- **Slew rate vs. V_{DPS}: up and down** – Slew rate capability of the DPS, in both rising and falling directions, must be characterized. This slew rate performance is measured two ways: (1) from $V_{DPS,MIN}$ up to $V_{DPS,MAX}$, and back (full-range slew) and (2) partial range slews, first from $V_{DPS,MIN}$ up to $0.5*V_{DPS,MAX}$, and back; then from $0.5*V_{DPS,MAX}$ up to $V_{DPS,MAX}$, and back. The measured slew rates must exceed the peak values from the necessary signal slew rate histograms.

- **Signal slew rates supported** (profiles) – Before slew rates required from the DPS can be established, it is necessary to get the slew rate characterizations from all of the signals that the DPS transmitter is required to support. Ideally, these signal slew rate characterizations are of the form in Figure 9-5 because this includes slew rate and the DPS output voltage(s) that the slew rates occur at. At a minimum, the generic slew rate histograms of the style shown in Figures 9-34 through 9-36 are needed.

- **Requirements on load resistance/impedance** – Many DPS designs have constraints on the load applied to the DPS. Particularly for DPS designs using the LDO architecture of Figure 9-13, the load resistance must not get too high, or the resulting output stage gain makes the feedback loop unstable. Such load requirements must be identified and the limits carefully characterized.

- **Tolerable output load impedances vs. V_{DPS}** – Any envelope-tracked PA presents a widely varying load to the DPS described by (5.20) and Figure 5-33(b). Can the DPS properly operate into this load at all possible V_{DPS} values? If not, the restrictions on what the acceptable V_{DPS} values are must be identified.

- **Conversion efficiency** – Keeping the power dissipation (13.1) low means keeping the conversion efficiency (3.2) high. Conversion efficiency is not constant with output power across a wide range, so the variation of conversion efficiency across the DPS output power range must be characterized.

- **Output noise density (nV/sqrt(Hz))** – When the PA operates in ET, it suppresses any DPS output noise by at least 40 dB (10.1). This noise suppression is reduced as the PA operation transitions from the ET operation toward polar operation, where the noise suppression is eliminated. At that point, the allowable DPS output noise density must be less than the limit calculated with the procedure in Section 6.2.4.

- **Load effect on bandwidth** – The value of the load impedance to the DPS often has an effect on the AC transfer function, and therefore a corresponding effect on the DPS operating bandwidth. For all possible values of the PA apparent load impedance seen by the DPS, the AC transfer function must be measured at each load value (e.g. Figure 11-28) to be sure that the required operating bandwidth is always available.

- **PSRR** – Power supply rejection ratio is the ability of the DPS to suppress any AC variations that may be present on its input supply V_S. Particularly during times of polar modulation, any distortion on the DPS output directly transfers to the PA output signal envelope.

- **Line regulation** – This is the ability of the DPS to hold the correct output (13.2) as the input voltage V_S may vary, such as during battery discharging and charging.

Again, the measurements listed above are only a minimum set.

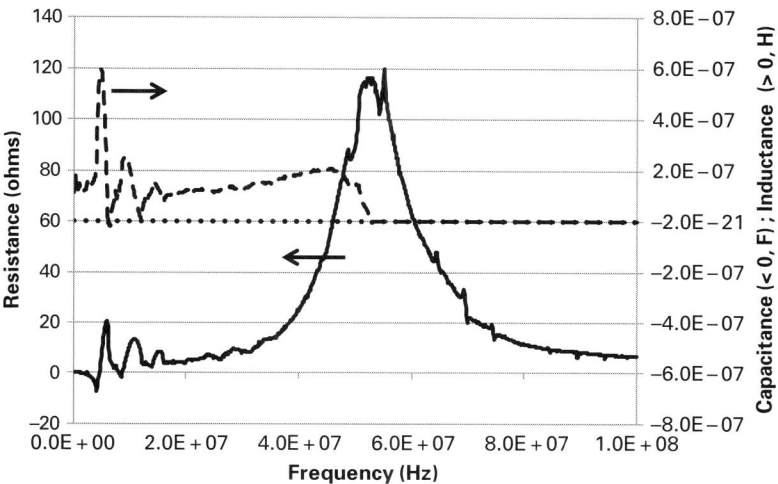

Figure 13-2 Example DPS output impedance measurement across 0.03–100 MHz. Near DC, the output resistance is near zero as desired (Section 7.2). This DPS has an inductive phase shift, and some frequency bands where the output resistance goes negative: a warning of possible oscillation in the presence of appropriate capacitance on the DPS load.

13.2.2 Conditions and sample results

One illustrative example of a DPS output impedance measurement is presented in Figures 9-16 through 9-18. Another DPS output impedance measurement comparable to Figure 9-17 is presented in Figure 13-2. This measurement is performed with the same technique as a traditional s_{22} measurement using a vector network analyzer (VNA). Data from the VNA are provided in the form $u + jv$, which represents the plotting location of the point (u, v) in conventional coordinates on the instrument screen. When these data represent an impedance measurement $R + jX$ on a Smith chart normalized to resistance $R_0,$ the conversion equations are [2-1]

$$R = R_0 \frac{1 - \left(u^2 + v^2\right)}{\left(1 - u\right)^2 + v^2},$$ (13.3)

$$X = R_0 \frac{2u}{\left(1 - u\right)^2 + v^2}.$$ (13.4)

Knowing the measurement frequency and complex impedance values using (13.3) and (13.4) at that frequency, it is straightforward to calculate the corresponding values of resistance, inductance, and capacitance for the DPS output impedance for each measurement point across the frequency span.

This impedance profile is the source impedance experienced by the PA as it draws its varying current governed by the signal envelope.

13.3   RF amplifier characterization

All RF power amplifiers that are intended to be used with a DPS, whether the ultimate intention is to operate in envelope tracking, polar modulation, or a hybrid mode, must be fully characterized as the 3-port signal processing blocks they are. Characterization test results are best presented as either a family of curves or as a surface. Some test results present themselves better in one form or the other, and occasionally both formats prove useful. These are highlighted in this section.

It is important to perform all PA 3-port characterizations independently of their target DPS. The data provided by test equipment are necessary to establish the properties and behaviors of the PA itself, not masked by possible (and likely) interactions between the PA and the DPS. These latter issues need separate understanding, and are covered in Section 13.5 on the DPS to PA interface.

13.3.1   List of 3-port RF PA tests

Independent characterization for any DPS-compatible RF PA must include the following tests as a minimum. Some of these are conventional and hopefully familiar characterizations of RF power amplifiers. It is preferable that all of the following information be provided by the PA manufacturer. If it is, then the DPST designer simply needs to verify the PA performance. If any of this information is absent, then the DPST designer must thoroughly evaluate the missing data.

All of the familiar two-port PA tests are still required:

- output power vs. input RF power (g_{RP}); $s_{21}{}^2$,
- peak output power,
- linearity,
- power dissipated in the PA,
- PA distortion components AM-AM and AM-PM,
- PA RF port impedances at the input (s_{11}) and output (s_{22}),
- reverse isolation (s_{12}),
- PA currents drawn from the power supply,
- output stage efficiency,
- power added efficiency.

New and possibly less obvious tests that are now required include:

- V_{PA} impedance restrictions,
- dynamic bias and its available bandwidth,
- new transfer functions DPS-AM and DPS-PM,
- input signal leakage and its relative phase,
- voltage slope gain g_S, and the Barkhausen boundary,
- amplifier operating mode identification (C-, L-, and P-mode),
- ET operation boundary profile for V_{PA},
- ET-DP transition region characteristics.

Fortunately, many important characterization results can be derived from a focused set of measurements, which makes this detailed characterization project more manageable and faster to complete. For PA characterization, the key measurements needed are those for the Booth chart. Additional measurements are needed to complete the needed information. All of these characterization steps are detailed in the following sections.

Characterization tests for the transistor(s) used in the PA design itself are largely based on conventional device characteristic curves in the I–V plane. This set of characterizations follows the discussion of PA measurements.

13.3.2   Booth chart set

One major set of PA characterization results are based on the measurements comprising the Booth chart (Section 4.1.6, Figure 4-22). The following set of tests comprises the Booth chart measurements and nearly 20 additional insightful results achieved by further analysis of the Booth chart data.

- **Booth chart** – The Booth chart is designed to make identification of the three PA operating modes (Section 4.1.6) easy to identify, along with the 3-port PA input conditions that lead to these three operating modes. By definition, the Booth chart is a family of g_{RP} curves (4.12), each curve taken at a different value of applied supply voltage. The Booth chart measurements used for the examples in this section are presented in Figure 13-3. Identification of the operating regions for this PA is done following the procedure from Figure 4-23.

It is certainly possible to present the Booth chart as a surface, giving the result in Figure 13-4.

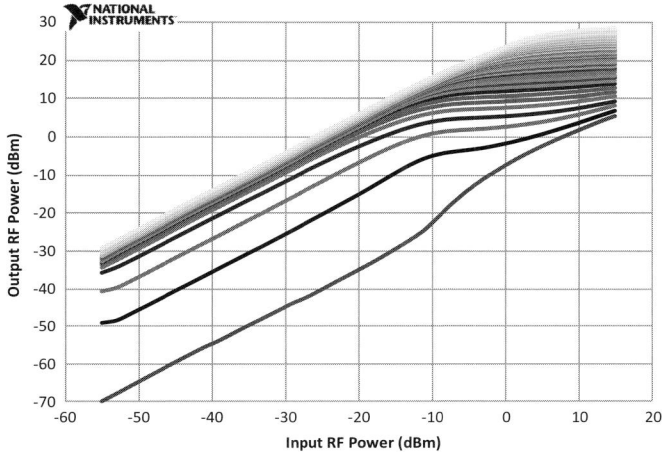

Figure 13-3   Booth chart measurements used for the examples in this section.

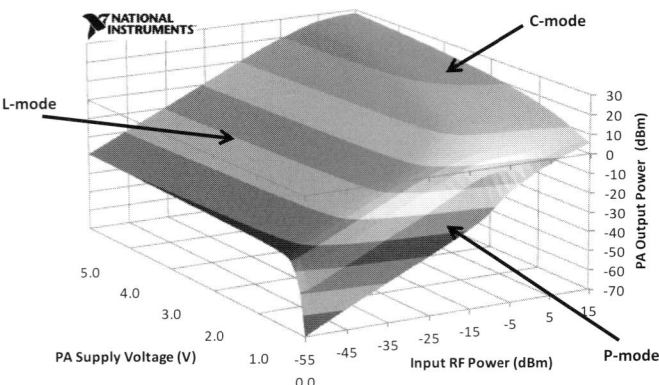

Figure 13-4   Booth chart measurements from Figure 13-3 presented as a surface plot (the Booth surface), showing the relative extents of the L-mode planar region, the C-mode region, and the relatively small extent of the P-mode region.

- **Leakage** – Leakage of the input signal to the output of a linear PA is usually never a concern. At least, it generally is never paid attention to in linear amplifier design, largely because the leakage is always much smaller than the output signal since the output signal and the input signal have a fixed relative ratio based on the PA gain. While this remains true for the ET operation, once compression is entered into, the leakage signal becomes very important to understand. This is because it drives the DPS-PM distortion process described in Section 6.11.2.

  Leakage of the input signal is directly read from the Booth chart as the curve representing the power at the output when the applied power supply is 0 V. This is the bottom-most curve in Figure 13-3.

- **Linear Booth chart** – Changing the axes from decibel power to signal peak volts provides the linear-scaled Booth chart shown in Figure 13-5. This points out that with respect to the actual voltage range of this amplifier, most of the available operation is actually in compression. Even at the highest available supply voltage (top-most curve), compression begins before half of the available output signal swing is used. In terms of output back-off for linear operation, the restriction in the output signal voltage swing is very significant – about 40%.

  As before, leakage at the output port is shown by the bottom curve. In Figure 13-4, this apparently increases linearly, implying a fixed coupling for the leakage path to the output. We also get a view into the linearity of the DPS-AM transfer function by examining the spacing between each family curve at its right-hand edge.

- g_{RP} **gain chart** – Decibel differences of output and input power provide measures of ratiometric power gain g_{RP}. Evaluating these differences for each curve in the Booth chart of Figure 13-3 provides the ratiometric power gain chart shown in Figure 13-6. This chart is widely seen in the recent literature on ET design. It is important to remember that this measure of gain is not directly representative of waveform accuracy because it is based on the power transfer function, not the voltage transfer

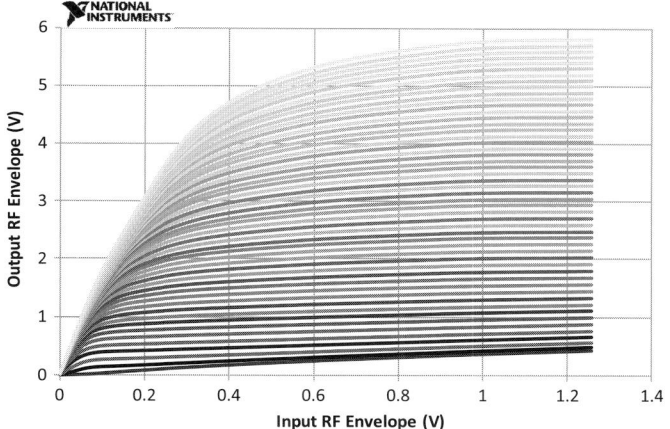

Figure 13-5   Linear-scaled Booth chart measurements from Figure 13-3, showing that most of this amplifier's available operation is within or transitioning to the C-mode region.

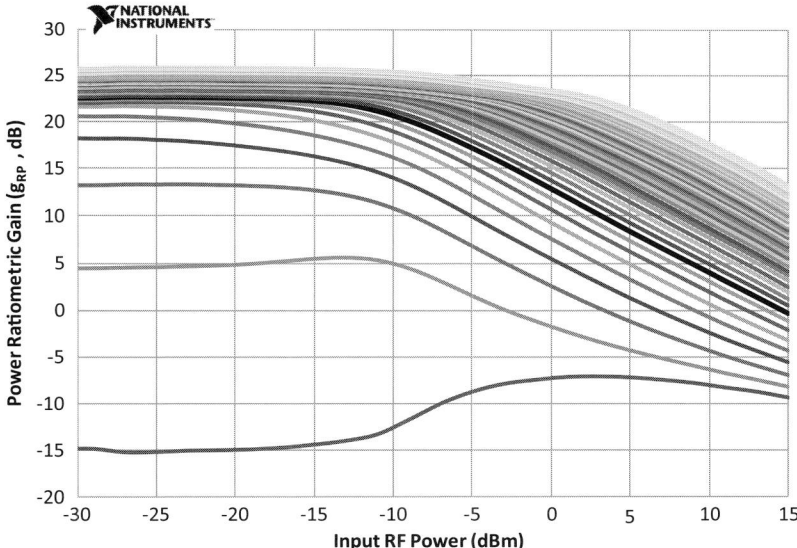

Figure 13-6   The traditional amplifier gain chart derived from the Booth chart measurements from Figure 13-3 and scaling as shown in Figure 13-4. The appropriate gain measure is ratiometric power gain. The boundary of the P-mode operating region is shown with the bold black curve.

function (Section 4.1.5). This is not inherently a good chart to base the design of linearized amplifiers on.

- g_{RP} **gain surface (dB)** – As for the Booth chart, the data in this gain chart can also be plotted as a surface across the two input variables of input RF power and applied supply voltage. The result is presented in Figure 13-7.

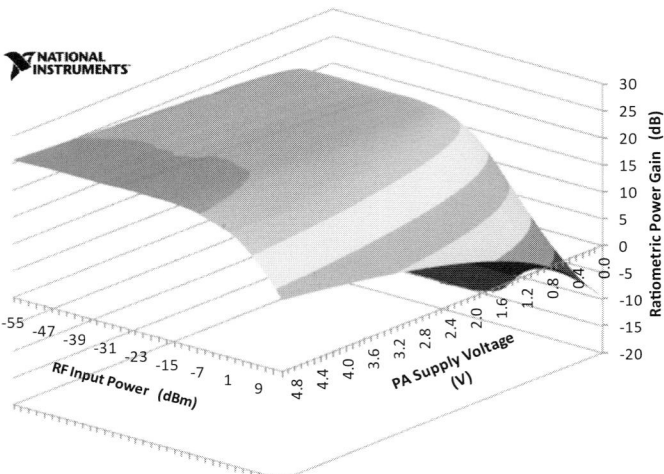

Figure 13-7   The corresponding surface plot for the data from the curve family in Figure 13-6. Depending on the particular interest of the designer, different rotations of this surface may be more practical.

- **Truncated for a fixed CJTF** – One particularly useful result from the g_{RP} surface plot of Figure 13-8 is to define the particular input signal combination needed to produce a constant joint transfer function (CJTF) used to produce a well-controlled output signal magnitude (Section 11.3). Such a CJTF value must be below the peak of the gain surface. For this example, the CJTF value selected is 22.5 dB.
- g_R **gain surface** – Just as the Booth chart can be rescaled for peak signal voltages, so can the ratiometric power gain surface. This new surface is the ratiometric voltage gain (g_R) profile for this amplifier and more easily shows two important aspects of amplifier performance: (1) how constant the amplifier gain actually is (or is not) across the variation of the amplifier inputs, and (2) any regular measurement errors from the test equipment. The rescaled surface of Figure 13-9 illustrates both of these effects. This amplifier exhibits a gain that varies linearly with changing supply voltage (not good for ET), and there are variations in the surface that correlate with particular input envelope values. These ripples imply that there are variations in the calibration of the signal generator providing the amplifier input RF power.
- g_S **surface** – The slope gain of an amplifier is directly representative of waveform accuracy, because it clearly identifies when the amplifier output stage enters voltage clipping. This makes the g_S surface the most appropriate starting place for amplifier linearizer design. The corresponding example g_S surface for this example amplifier is shown in Figure 13-10. We note that in linear amplifier operation the values for the g_R and g_S surfaces are the same. Important differences are present when the amplifier leaves linear operation either toward compression (C-mode) or gain collapse (P-mode).

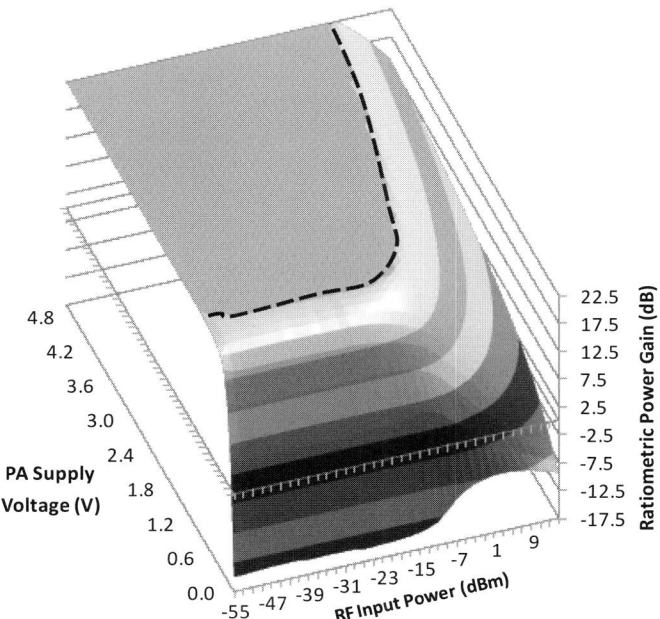

Figure 13-8   Truncating the ratiometric power gain surface at a selected CJTF value immediately provides the required combination of amplifier inputs needed to achieve this constant joint transfer function value. Here the CJTF value is selected to be 22.5 dB. This resulting combination of input RF power and applied supply voltage was shown in Figure 11-15.

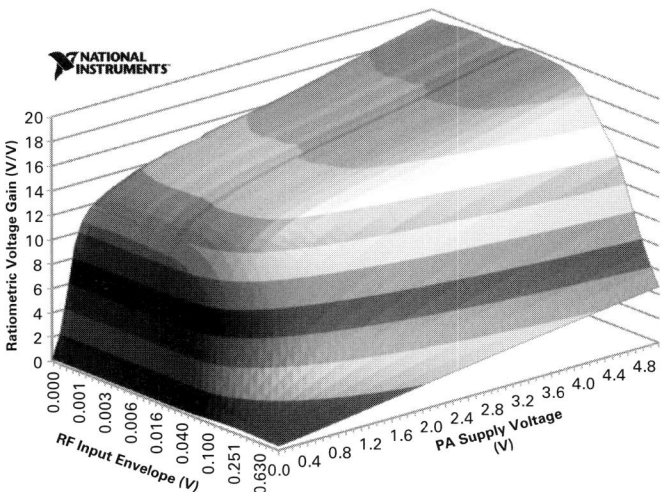

Figure 13-9   Rescaling the ratiometric power gain surface for amplifier voltage response provides insight on amplifier gain consistency across variations of the amplifier inputs. Here we also see that the RF signal generator has some small calibration variations across this range.

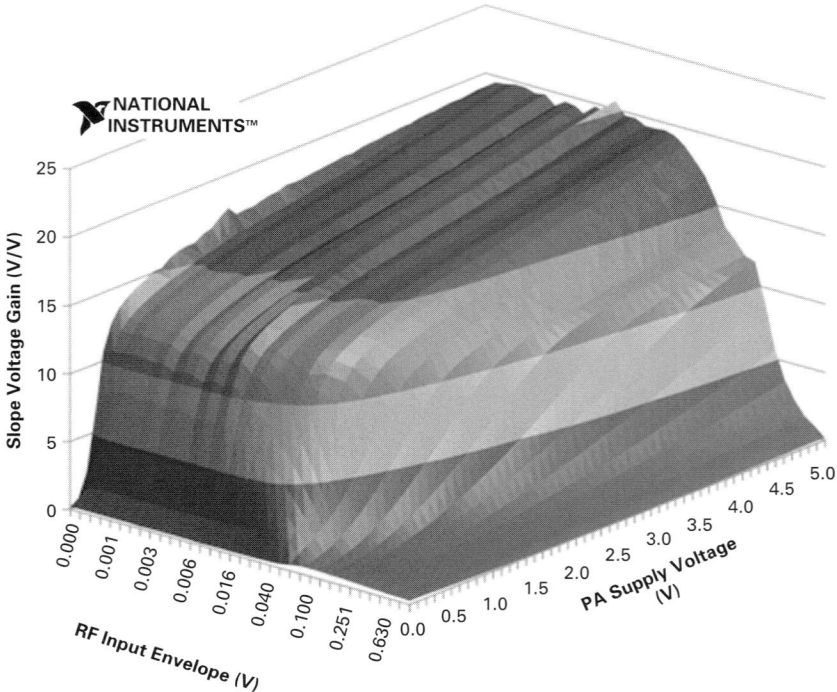

Figure 13-10 The slope voltage gain (g_S) surface for amplifier voltage response provides insight on actual waveform distortion for this amplifier. Calibration errors of the signal generator are more pronounced on this surface than in Figure 13-8.

- **Barkhausen boundary** – The Barkhausen gain condition for oscillation requires that the circuit gain is > 1. The circuit gain of interest here is g_S, the slope voltage gain. Finding the boundary between when oscillation is possible and when stability is unconditional is found at the profile where the g_S surface is truncated to 1 V/V. A rotated view of truncating the surface in Figure 13-10 to 1 V/V is presented in Figure 13-11.

Two views of this Barkhausen boundary that are useful for design are shown in Figure 13-12. Both charts show exactly the same information, only with different scaling of the axes. The version in Figure 13-12(b) has the highest dynamic range, and also presents the input RF power in its usual design form.

- g_{RP} **flatness** – Constant gain eliminates output signal distortion by guaranteeing that the output waveform is exactly a scaled replica of the input waveform (4.1). It is therefore a fundamental goal of circuit design to achieve constant gain, particularly in precision signaling applications. Achieving inherently constant gain in an amplifier is very difficult, particularly when the supply voltage is also an input variable.

     Which gain shall be evaluated? The most common gain measure is ratiometric power gain, shown in Figure 13-6. One way to evaluate the consistency of amplifier

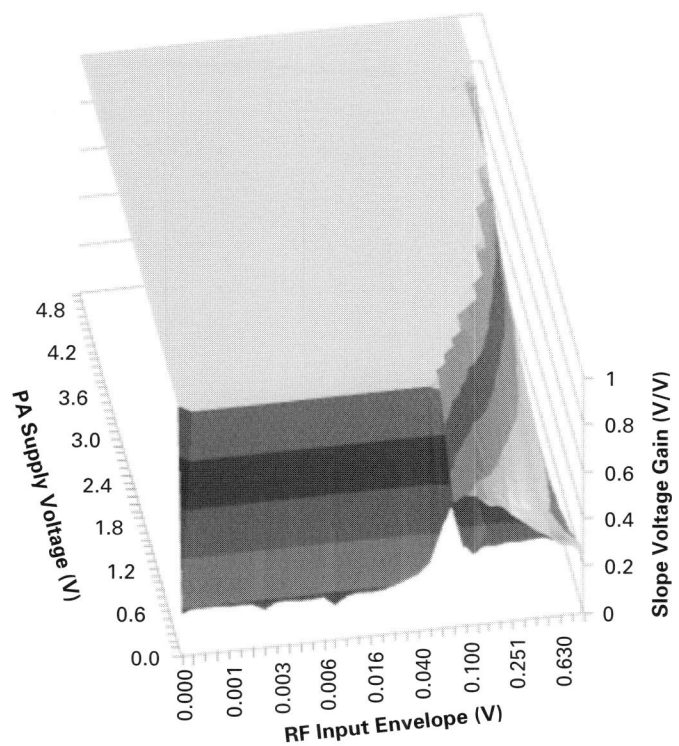

Figure 13-11 Truncating the slope voltage gain surface at 1 V/V identifies the Barkhausen boundary. At all values of g_S less than 1, oscillation is not possible. For all operating conditions where $g_S > 1$, then circuit stability must be carefully evaluated.

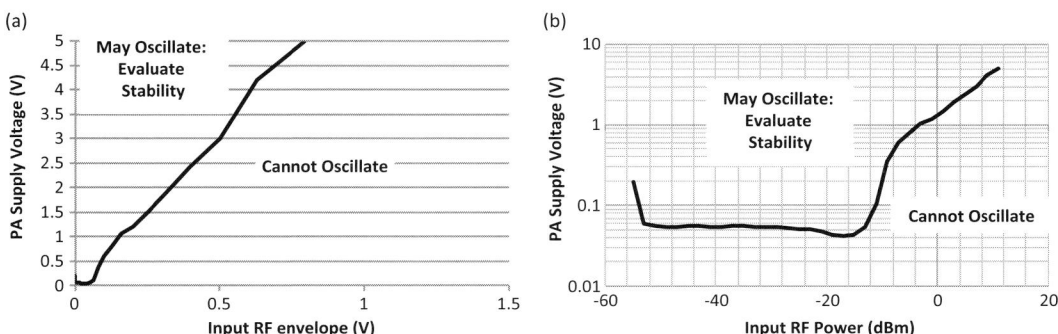

Figure 13-12 The Barkhausen boundary identified in Figure 13-10 presented in two formats: (a) linear scaling for both amplifier inputs; (b) logarithmic scaling for both amplifier inputs.

gain is to take a vertical "slice" of Figure 13-6 to show how g_{RP} varies with supply voltage. Such a plot is in Figure 13-13, where the solid curve is taken below the transition into compression. Then taking the first derivative of this curve provides the desired measure of gain flatness. Ideally, the value of this derivative

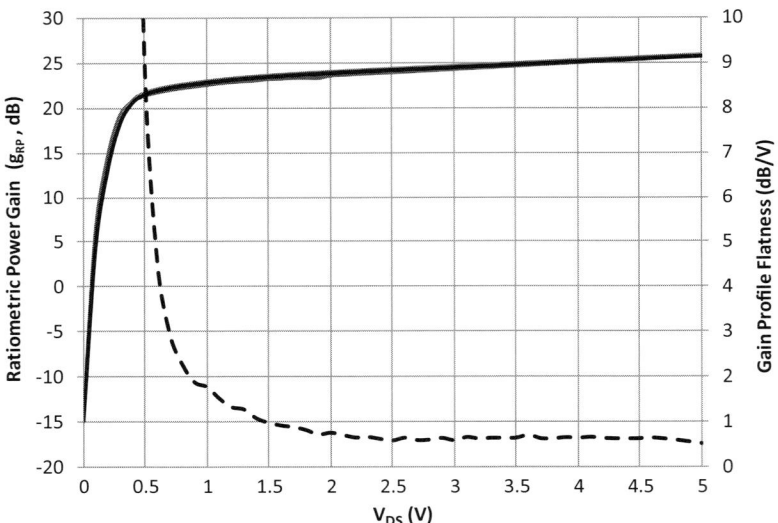

Figure 13-13   Small signal g_{RP} gain flatness evaluation for this amplifier. Measured g_{RP} (solid line) shows both supply dependence when in CCS operation and the "gain collapse" of P-mode operation. The first derivative of this measurement curve (dashed line) helps identify when the gain flatness deviates sufficiently to declare the P-mode boundary.

(dashed curve in Figure 13-13) is zero. Positive values represent gain expansion, and negative values show gain compression. Nonconstant values of this derivative represent gain curvature, a real problem for linear applications.

- V_{AMO} **chart** – Plots of PA output power vs. the applied supply voltage for each value of input power are used to determine the value of V_{AMO} (6.16). This calculation is discussed in Section 8.6.6. This chart is also very useful to identify when leakage from the input signal becomes a problem and needs to be managed by careful reductions. This is shown in Figure 13-14 by the added dotted line that follows the ideal C-mode power control square-law characteristic (6.1). If the measured output power is above this reference curve, then input signal leakage is dominating and must be reduced.

Intersections of the ideal profile with the family of curves in Figure 13-14 identify the maximum allowed input power for that C-mode output power. The profile identified from Figure 13-13 is shown in Figure 13-15. This profile is not linear. Neither is it easy to predict, because this strongly depends on detailed transistor capacitance behaviors.

- **PA current surface** – By also measuring currents into the PA along with the RF power measurements needed for the Booth chart, it becomes possible to also evaluate bias behavior, amplifier efficiencies, and PA port impedances. The PA current measurements provide the surface shown in Figure 13-16.

At minimum input RF power, the bias current dominates the PA current drawn. Current source operation of this transistor is seen to hold down to very small supply voltage levels, where this current surface exhibits a sharp corner just before turning off

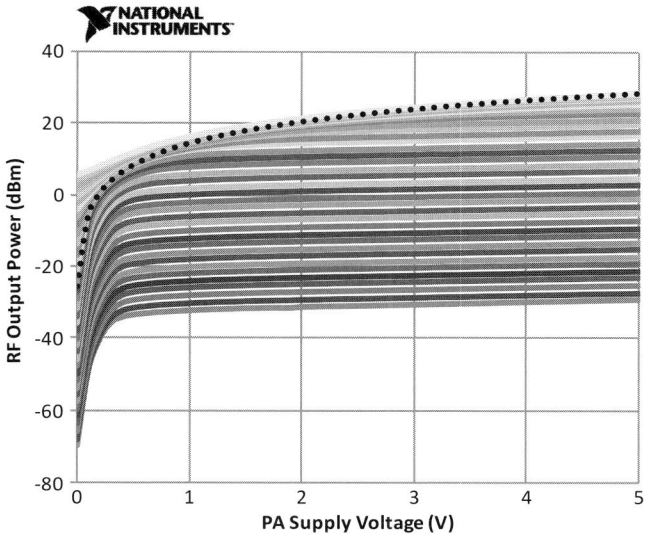

Figure 13-14 This family of output power vs. applied supply voltage for each value of input RF power is called the V_{AMO} chart. The dotted curve is an ideal square-law power control profile. Each curve corresponds to a different value of RF input power.

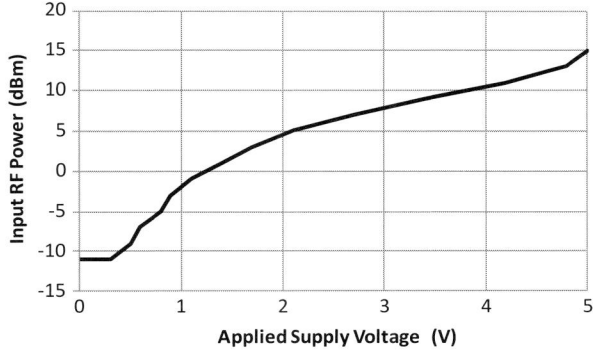

Figure 13-15 Evaluating the intersections of the ideal square-law power control profile with the curve family in Figure 13-14 provides the required coordination between input RF power and the supply voltage to manage leakage of the drive signal to the RF output.

(Figure 5-33(a)). Near –20 dBm, the input signal changes how the bias current operates, eliminating the sharp corner as the low-supply characteristics compress and transition into C-mode operation. C-mode operation is demonstrated by the resistive slope at that and higher input signal power in accordance with (6.17).

• **Output efficiency** – Once the PA current drawn is also known, the supplied power can be calculated and efficiencies can be determined. For the example PA, the output stage efficiency data are provided in Figure 13-17. Two different formats are useful during DPST design. The family of curves in Figure 13-17(a) is useful for checking efficiency

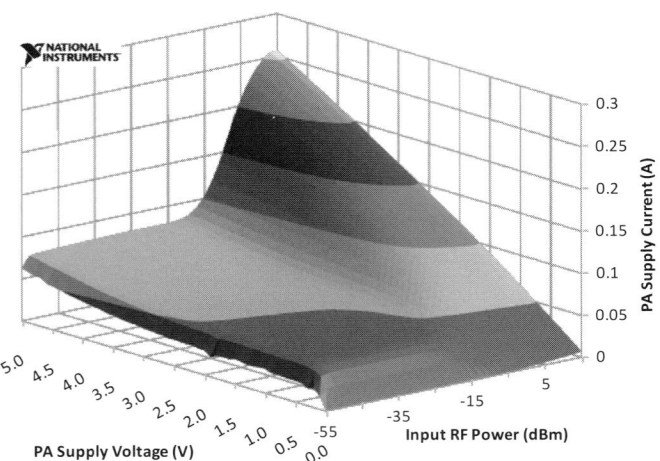

Figure 13-16 Supply current surface for this example PA. Note that the bias is affected by the input signal above −20 dBm. Bias current is also influenced by the PA supply voltage. A rise in the PA current for large output signals is clearly evident.

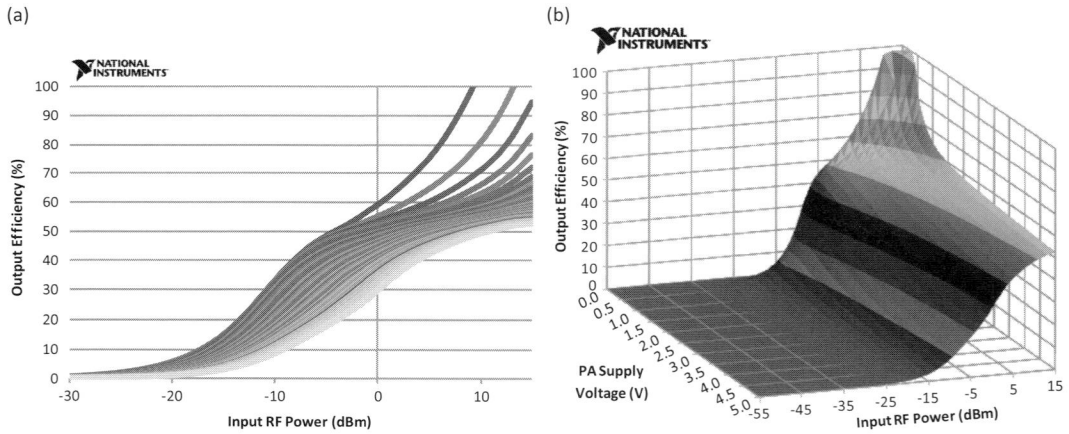

Figure 13-17 Measured amplifier output efficiency: (a) family of curves; (b) surface plot showing that the efficiency rise is due to input signal leakage to the output.

performance based on differences in power supply profiles applied to the PA. But there are some curves in the family that head towards and exceed 100% efficiency. How is this possible?

The answer is provided by the surface plot version of these same data. This region of very high efficiency occurs at the corner of very low power supply and very high input power. These are the conditions for maximum leakage of the large input signal to the output, where it contributes to measured output power without drawing on any corresponding supply current. This is why the efficiency appears to be so high.

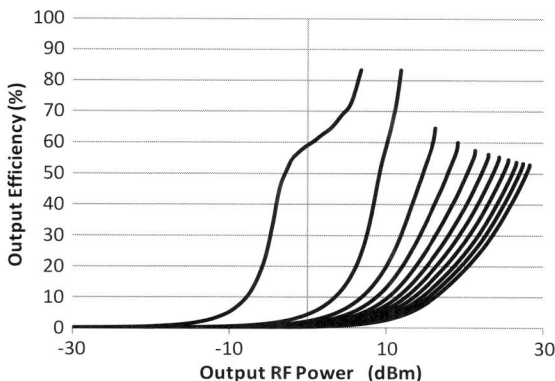

Figure 13-18   Measured amplifier output efficiency plot against output power (from Figure 11-20). This chart is generally difficult to design with because output power is not a controlled input to the transmitter.

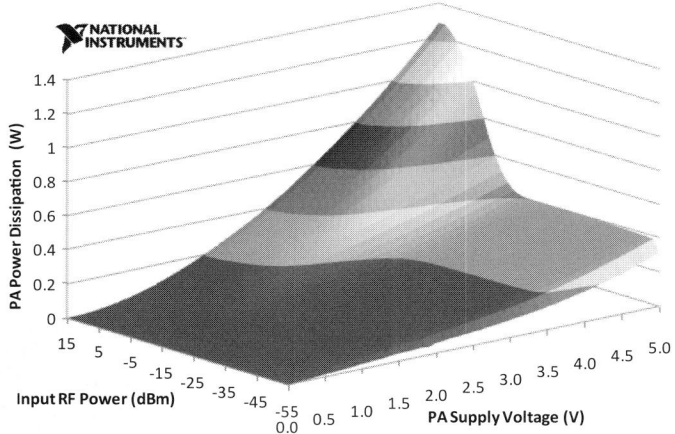

Figure 13-19   Measured total amplifier power dissipation due to both DC and generation of the RF output power.

- **Output efficiency vs. output power** – This plot is given separate consideration because it is very prevalent in the present literature. However it is not based on controllable amplifier inputs, which makes this chart relatively useless for design activities. It is useful as a final report on what the ending circuit performance is. Though if there is something that needs to be changed in order to improve circuit performance, this chart in Figure 13-18 does not provide it. The charts in Figure 13-17 are much more useful in design.
- **Power dissipation surface** – Power dissipation is the difference between the power supplied to the PA and the power supplied from the PA. This power difference stays in the PA and is a heat flow driving a temperature rise (Figure 3-4). For this example amplifier, the corresponding power dissipation surface is presented in Figure 13-19.

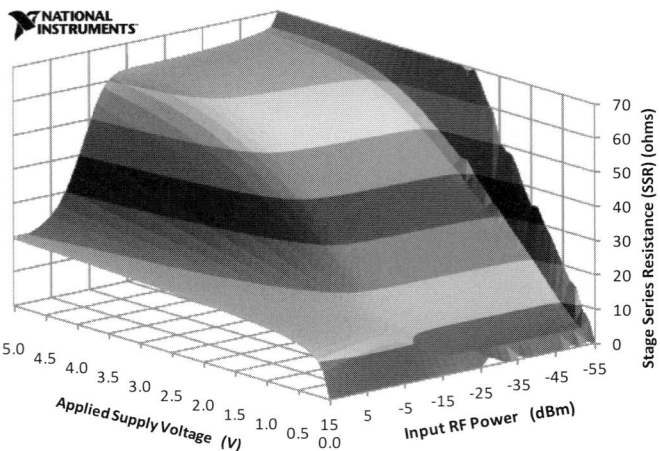

Figure 13-20   Measured amplifier output stage series resistance, which is the absolute resistance presented to the DPS as its load.

Several important design characteristics are evident in Figure 13-19. First, the peak power dissipation corresponds to the peak output power and the highest true energy efficiency (not distorted by input signal leakage). Second, the benefit of ET is apparent by the slope in this power dissipation surface. Since energy efficiency is improved by reducing power dissipation, how this mechanism operates for this amplifier is open and evident. Third, the quiescent power dissipation that linear amplifiers are famous for shows up clearly when the supply voltage is high and stays fixed.

- **SSR surface** – The absolute resistance (9.1) that the PA presents as the load to the DPS is provided by the stage series resistance surface of the PA. Depending on the amplifier operating mode, this surface is modeled by either (5.20) or its C-mode simplification (6.17). The SSR surface for this example PA is presented in Figure 13-20.

- **Dynamic resistance surface** – The load seen by small AC components on the PA supply voltage from the DPS is the PA supply dynamic resistance (9.2). Being a first derivative of the SSR surface, any small variations in the SSR surface are much more evident. Such variations are very evident in Figure 13-21.

Of particular interest in the dynamic resistance surface in Figure 13-21 is its dynamic range. This is actually expected, because the dynamic resistance of a current source is ideally infinite, while its absolute resistance can actually be fairly low. Thus, when the amplifier transistor is operating as a CCS (in L-mode) the dynamic resistance is very high. When the amplifier operates in C-mode, the dynamic and absolute resistance are the same. Other changes in dynamic resistance do provide information valuable to the DPST designer.

- **PSS surface** – From the PA point of view, sensitivity to variations in the power supply appear as changes in output power. The PA power supply sensitivity (PSS) surface therefore represents changes in output power (dB units are most useful) for changes in

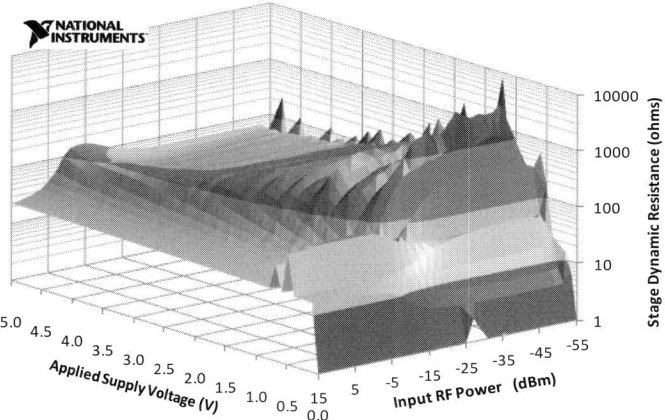

Figure 13-21 Measured amplifier output stage dynamic resistance, which is highest when the amplifier operates linearly (L-mode). This is the DPS load to AC components.

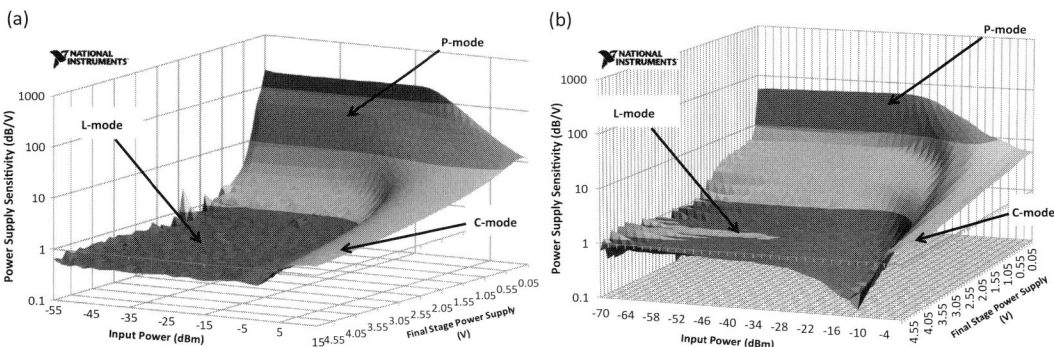

Figure 13-22 Measured amplifier PSS surfaces: (a) for the example amplifier; (b) for an amplifier designed for optimized linearity at large output signals.

supply voltage. Any amplifier that is not sensitive to changes in supply voltage value will have low values of the PSS surface. Large values of this surface show operating conditions where there is high sensitivity to supply voltage variations. Two examples of measured amplifier PSS surfaces are presented in Figure 13-22.

All amplifiers exhibit all three operating modes. But how those modes behave, interact, and transition from one to another is very visible from the amplifier PSS surface. The higher the quality (meaning supply voltage independence (7.1)) of L-mode operation, the lower the value of the L-mode region will be. This reduction in power supply sensitivity does not have to be at all operating conditions, as the amplifier evaluated in Figure 13-22(b) shows. This amplifier is designed for best linear operation near its maximum available output power, just before compressing and heading into C-mode operation. Clearly, this design objective is achieved.

13.3.3   Additional set of PA tests

Any effect that has an impact on the output waveform must be characterized. From the PA point of view, there are many more important tests that must be performed to help ensure success for the DPST development project. These include the following 13 additional characterizations:

- output lag,
- leakage signal relative phase,
- supply complex impedance $Z_{PA}(f)$,
- dynamic bias performance,
- wideband noise,
- conventional distortion transfer functions AM-AM and AM-PM,
- new DPST transfer functions (DPS-AM; DPS-PM),
- supply-noise rejection profile: ET operation boundary,
- reverse isolation,
- reverse intermodulation,
- RF power transient (0 V → ON, ON → 0V)
- transistor output characterizations via bias inputs,
- operating bandwidths: power, small signal.

Each of these additional tests is briefly discussed below.

Output lag – This is a memory effect where the PA output power depends also on what the output power was in the past, along with what the amplifier controlling inputs require right now. This distortion mechanism is discussed in Section 10.8 and applies to all DPST modes. Output lag testing is most reliably done with a bursting test (see Figure 10-98).

Leakage signal relative phase – Leakage of the driver input signal through a C-mode stage is the dominant cause of phase distortion, as well as contributing to magnitude distortion, as presented in Section 6.11. While the magnitude distortion is a weaker function of the leakage phase, the output phase distortion is strongly dependent on this parameter. Measurement of this phase is relative to the phase of the peak PA output power, as discussed in Section 6.11.2. With this phase measurement, the DPS-PM transfer function is readily modeled.

Supply complex impedance $Z_{PA}(f)$ – It is nice to think that the input impedance at the PA supply is resistive, but it is not. The reactive and frequency dependent aspects of the PA supply impedance are important to the DPS, which also has reactive aspects to its output impedance. It is well known that these reactances can interact at this interface and result in oscillation at the PA supply. This must be avoided by design, founded on careful characterization.

Conventional distortion transfer functions – The distortion mechanisms familiar to linear amplifier operation are the AM-AM and AM-PM transfer functions. These both relate solely to the RF path through the PA, and therefore can be mitigated by pre-distorting the input signal appropriately. These mechanisms continue to operate with ET as long as (7.1), or more particularly (10.1), are satisfied. When the amplifier is

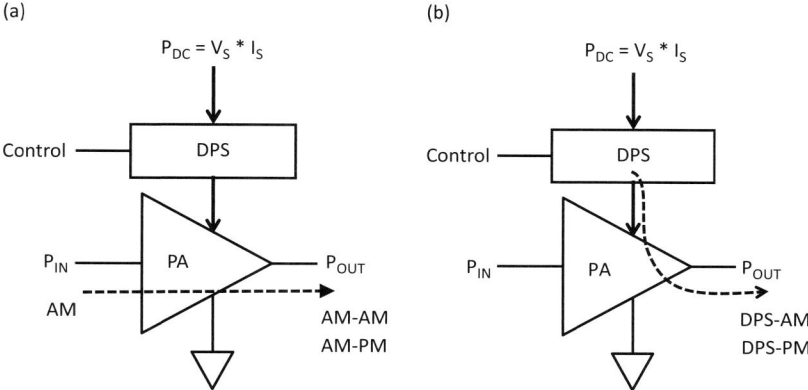

Figure 13-23 The different distortion transfer functions in DPST designs: (a) conventional linear AM-AM and AM-PM which operate with L-mode; (b) the new DPS-AM and DPS-PM transfer functions that operate when the PA is in C-mode.

operated into compression such that the linear operation is suppressed, then these distortion models fail, along with their mitigation techniques. The expansion of conventional AM-AM and AM-PM transfer function characterizations for DPST use is necessary to identify the boundaries of this applicability.

New DPST transfer functions – When the linear amplifier distortions of AM-AM and AM-PM no longer operate, new distortion mechanisms rise to take their place. These new transfer functions are presented in Figure 13-23, operate between different ports, and therefore need different strategies to mitigate the resulting distortions. These new transfer functions are nonidealities in the DPS-AM (5.5) process used in C-mode to set the signal envelope, and any impact attributable to DPS variations on the output signal phase, called DPS-PM.

Dynamic bias performance – All DPS transmitters, both envelope tracking and polar, require dynamic bias on the power transistor to achieve the dynamic ranges required of most communication systems. Bias networks are usually designed to have very firm and nonvarying behavior, so this requirement is a big change from conventional practice. The bandwidth of the bias network must at least equal that of the envelope modulation. Measuring this is a significant challenge.

Wideband noise – All amplifiers provide noise at their output that is spread across a very wide bandwidth. This noise can interfere with the local receiver, and/or with other nearby receivers through a process called near-far interference (NFI). The concept of noise figure is usually used to analyze this effect. Noise figure is frequency dependent, as was shown in Section 7.8. With increasing requirements on both signal bandwidth and the need for operation on multiple band allocations, the importance of this test is growing.

Supply-noise rejection profile and the ET operation boundary – One of the key operating parameters separating the ET operation from polar operation is the PA supply-noise rejection performance. Envelope tracking has plenty (7.1) and polar has none at all (8.2). Any attempt at understanding how a DPS transmitter is actually operating must measure the profile of how the PA suppresses noise on the power supply.

One good way to do this is with the method outlined in Section 6.2 and shown in Figure 6.8. This is a direct measurement of the desired effect. One result is presented in Figure 6-10, though the complete characterization will provide a complete surface related to those in Figure 13-22.

From the supply-noise rejection surface for the PA, a trajectory is then drawn at the −40 dBr level in accordance with (10.1). This trajectory is the ET operation boundary for that particular PA. Operating in conditions with greater supply-noise suppression satisfies (7.1) and is envelope tracking. Any operation with less supply-noise suppression enters the transition region toward polar modulation. The shape and characteristics of this transition region are evident from this surface.

Reverse isolation – Amplifiers are assumed to be RF "diodes," in that the RF signal only flows in one direction: from input to output. Sadly, this is not universally true. The test called reverse isolation characterizes this backwards behavior. This is important because the input impedance of an amplifier does depend on the signals and impedances present on the amplifier output. Circuits such as oscillators are very sensitive to the load impedance they are provided, so management of amplifier input impedance is important.

This test is even more important in the coming era of multiple transmitter operation, whether it is under the name of carrier aggregation, phased array, spatial signal processing, or even MIMO. All of these systems operate more than one transmitter at a time, and their separate outputs do come back and interact with each other.

Reverse intermodulation – In any environment where there are multiple transmitters, and also multiple antennas, the different transmit antennas will also receive all signals from the other transmitters. This presents a situation where the output of any PA has the intended signal being generated for transmission, and also has the already transmitted signals from other transmitters coming backwards and interacting with the desired signal. Discussed in Section 6.11.4, this test has importance in any situation where the spatial density of transmitters is high, such as all of the scenarios listed for the reverse isolation test.

RF power transient – Whenever the signal being transmitted has a high peak-to-average power ratio (PAPR), the transmitter on occasion must quickly ramp up from low to high power, and then return quickly back to the lower average power. This rapid change in output power puts a strain on bias networks. In conventional linear amplifier design, the bias network to support such signals is usually designed to have a very low impedance so that the rapid rise in current requirements at these signal peaks can be met.

This bias design strategy is no longer possible when the dynamic bias required in any high dynamic range DPST must operate. Therefore, a new test is required to validate that the transmitter can support the rapid rise and fall in a signal peak with precision. This is the purpose of the RF power transient test.

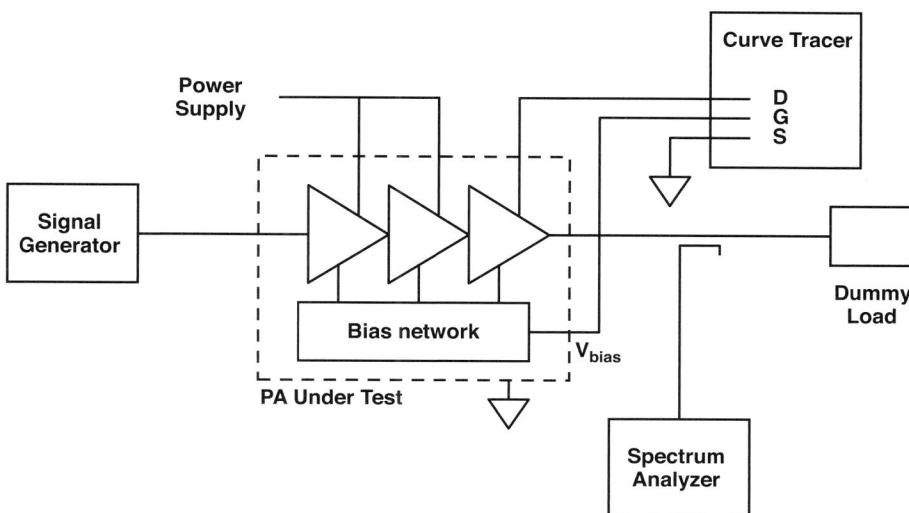

Figure 13-24 Partial characterization of an integrated PA transistor through coordinated control of the PA bias input. This provides a view into the output characteristics of the transistor such as they are made available by the bias network.

Transistor output characterizations via bias inputs – When access to a transistor inside an existing PA is not available, it is still important to characterize that transistor to the greatest extent that control access is available. For many power amplifiers, this is possible with the technique shown in Figure 13-24. While such a characterization is not complete, it is very useful regarding characterization of transistor ON resistance, linearity of that ON resistance, and measurement of any voltage offset on that transistor that the DPST control algorithms must take into account.

Operating bandwidths – As operating, signal, and tuning bandwidth specifications increase the characterization of amplifier bandwidth is also of increasing importance. There are two measures of particular interest which are measured in different ways.

Power bandwidth is a measure of the frequency range over which an amplifier provides a specified amount of power into the defined load. This is shown as a plot of measured RF power across the tested frequency range. Often this measurement is made with the amplifier operating in compression so that it is near, or at, output power saturation (see Figure 4-18). Therefore, this test applies to polar modulation since at P_{SAT} the amplifier certainly is operating in its C-mode region.

The second bandwidth measure is sometimes called small signal, and is reported as a gain across the tested frequency range. For an amplifier operating in L-mode, the signal does not actually have to be small for this test. But the amplifier does have to be operating in its CCS region, and therefore this test does apply to ET operation.

13.4   RF power transistor characterization

The foundation of all circuitry is the underlying technology. What the system specifications are has little meaning if the transistor technology selected to implement the circuitry is fundamentally incapable of meeting those specifications. There are a multitude of transistor technologies in existence for a reason, which is that not all applications can be implemented from just one transistor technology. Choice of transistor technology is critical to project success. Choose correctly and when implemented well the project will succeed. Choose wrong and there is no way that the project can ever succeed, no matter the brilliance of the implementation. This choice is ours.

This is exactly the reason why this book is tied strongly to what the RF power transistors are actually doing in DPST designs. It is from the transistor technology that the circuit design can succeed. To prove, or disprove, that any particular transistor technology can meet DPST requirements, it must be appropriately characterized, at a minimum by following this process. Many results of such characterizations are presented in Chapter 10.

13.4.1   List of transistor tests

As in Section 13.3 for PA characterization tests, the testing needed for the transistors is a mixture of conventional and new procedures. The familiar transistor characterization tests are all required:

> IV characteristic curves,
> parasitic capacitances,
> transfer function,
> threshold,
> transconductance or β, as appropriate,
> transition frequency f_T,
> breakdown voltage.

New transistor tests for DPST applications include:

> dynamic bias profile,
> polar modulation accuracy range,
> modulation triangle offset,
> PSS-IV surface,
> PSR boundaries,
> output offset characterization,
> turn-on time constant,
> FET active region.

If the transistor is ever to be used in L-mode operation, then s-parameter data are extremely useful. When the transistor is to be operated in C-mode, then any s-parameter data are completely useless and this aspect of the design must be performed in the time

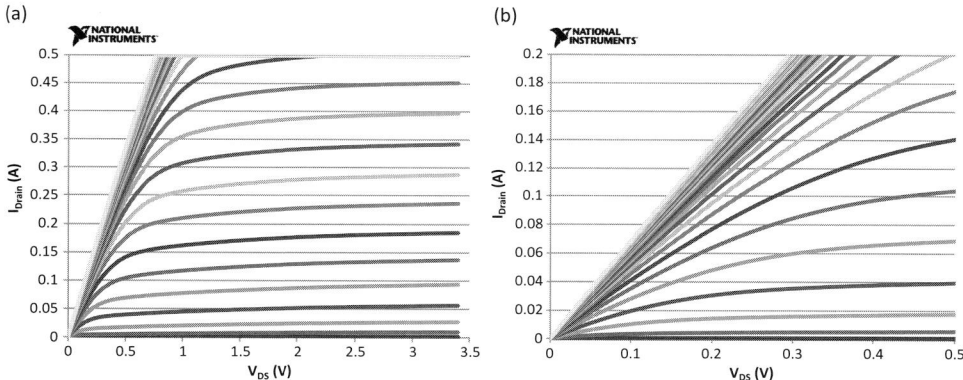

Figure 13-25 FET characteristic curves: (a) full span of the IV plane application space; (b) close-up to the IV plane origin for output offset evaluation (from Figure 10-57).

domain. Here the particularly important part of design modeling is complete incorporation of the voltage variability of all transistor capacitances. It is vitally important that the transistor model ensures that charge is always conserved (which unfortunately is not necessarily the case with some capacitance-based transistor models).

13.4.2   IV curve set

- **DC IV curves** – The characteristic curves of a device are independent of any circuit that it is used in. Even just sitting on a table, any transistor still maintains its characteristic curves. In any effort to understand how a transistor operates in a circuit, knowing its DC characteristic curves is fundamentally important.

  A summary set of example characteristic curves is provided here for completeness. Figure 13-25 is a repeat of Figure 10-57 to provide an example of FET IV characteristic behavior. For an example of bipolar transistor IV characteristics, Figure 13-26 repeats the data provided in Figure 10-65.

- **Transfer and compression** – In essence, any transistor is designed to have a current flow between two of its three ports (through the "channel") controlled by a parameter applied to the third port. For an FET, the controlling parameter is a voltage, and for a bipolar transistor, the controlling parameter is a current. This sets up a transfer function between the controlling parameter and the controlled current. This transfer function depends on both the controlling parameter and the applied supply voltage, seen by the examples provided in Figure 13-27.

  This set of transfer functions readily shows if this particular transistor is switching or not. To be considered an ON switch, the channel current must be constant with continued changes in the controlling parameter (8.1). Without this property, the designer of a polar stage must pay particular attention to (6.11) to decide if the transistor is "ON enough."

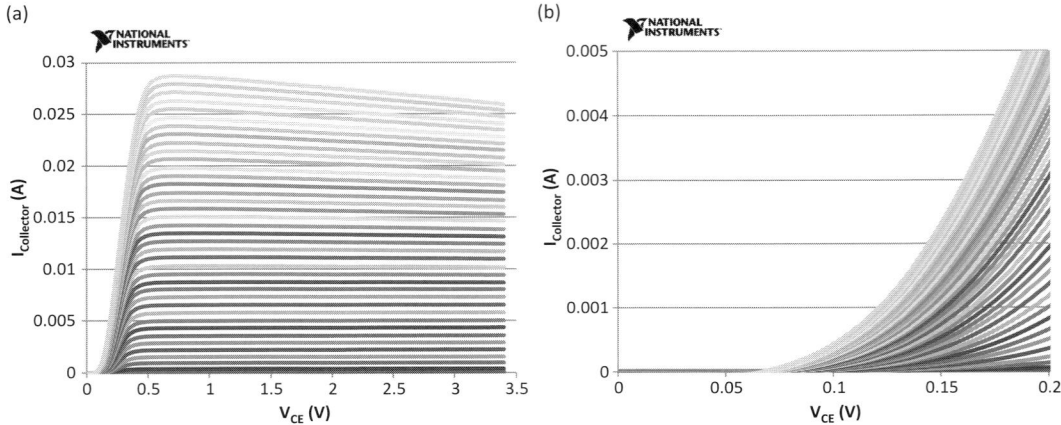

Figure 13-26   HBT characteristic curves: (a) full span of the IV plane application space; (b) close-up to the IV plane origin for output offset evaluation (from Figure 10-65).

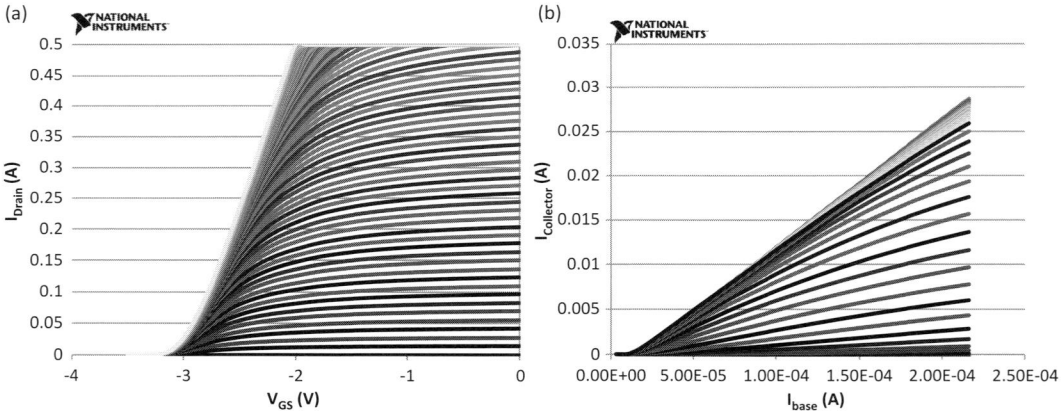

Figure 13-27   Transistor transfer functions evaluated at separate supply voltages: (a) depletion mode FET, here a GaN HEMT with 0.05 V steps on V_{DS}; (b) GaAs HBT with stepped base current.

The transistor of Figure 13-27(a) does exhibit this inherent ON behavior only at higher V_{GS} and low V_{DS} values. The transistor in Figure 13-27(b) never exhibits switching behavior. This appears to be very common: bipolar transistors are very difficult to drive into the switching operation. This is likely to be the case because they are not designed to be switches, but are intended to be used for linear amplifiers – which is something bipolar transistors do very well.

Published transfer functions in transistor data sheets traditionally assume that the supply voltage is high enough and fixed so that only one curve is reported. This is not sufficient for DPS transmitter design because the intentionally varying supply has a large effect on device transfer behavior, as seen in Figure 13-27. This is important 3-port information and must be made available to the designer of any PA to be used in ET.

Much discussion in the literature focuses on class B operation of a transistor. By definition, this means that the transistor is biased right at its threshold. Both charts in Figure 13-27 show that there is significant curvature to the transfer function just as the channel current goes to zero. This effect is more pronounced for the FET than for the bipolar transistor, but it is there. This curvature leads to waveform distortion that is very difficult to remove. Even for class AB bias, when the transistor enters cutoff it passes through this transfer function curvature and again generates distortion on the output signal.

Another important transistor characteristic is a requirement that the threshold value of its transfer function not change with any value of the supply voltage. For any transistor to be successfully used in an ET design, this threshold must be constant – if it is not constant, the load current has yet another mechanism to vary with changes in the supply voltage (beyond output resistance (5.23)) and to therefore not meet the fundamental ET property (7.1). Figure 13-28 presents two examples of measured threshold behavior in GaN HEMTs. The threshold in Figure 13-28(a) is stable with varying supply, while the threshold in Figure 13-28(b) definitely is not stable with varying supply. Any attempt to use the transistor in Figure 13-28(b) in an ET design will be disastrous.

- g_m/β **surfaces** – Continuing with the 3-port device characterization, the evaluation of transconductance (for an FET) and β (for a bipolar transistor) also needs to be shown as a surface. For the design of linear amplifiers, this surface should show a flat top. In general, there is a small region that meets this criterion for FET devices as seen in Figure 13-29.

One example β surface for in this case a GaAs HBT is shown in Figure 13-30. Here a wide region of nearly constant β shows that an amplifier with inherent linearity (before

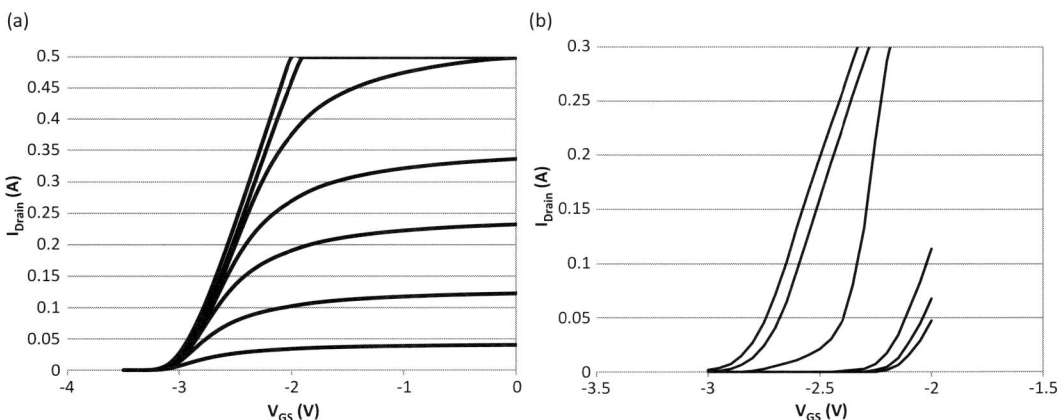

Figure 13-28 Transistor transfer function subsets: (a) FET (GaN HEMT) from Figure 10-90(a); (b) a different and particularly troublesome GaN HEMT (from Figure 10-96(b)).

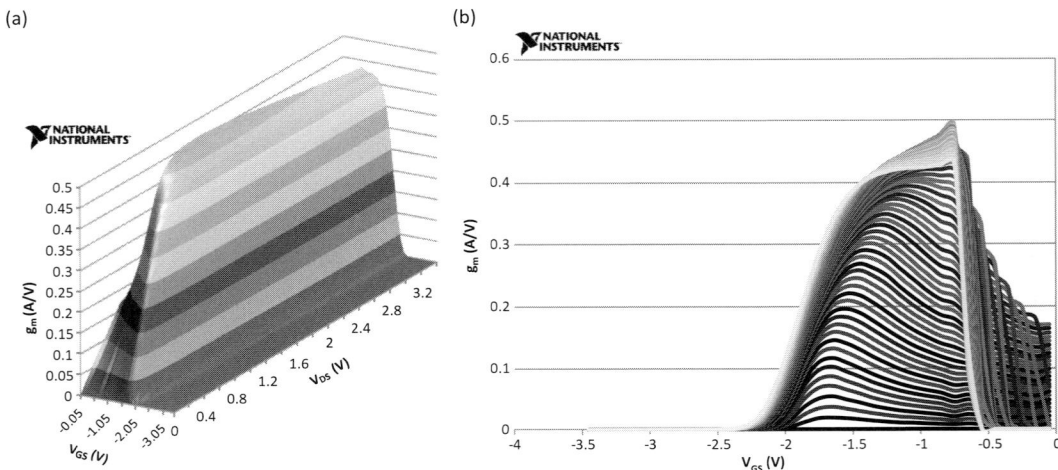

Figure 13-29   FET transconductance measures: (a) transconductance surface; (b) curve family for part (a), the sharp drop at high V_{GS} is due to current limiting in the measurement system.

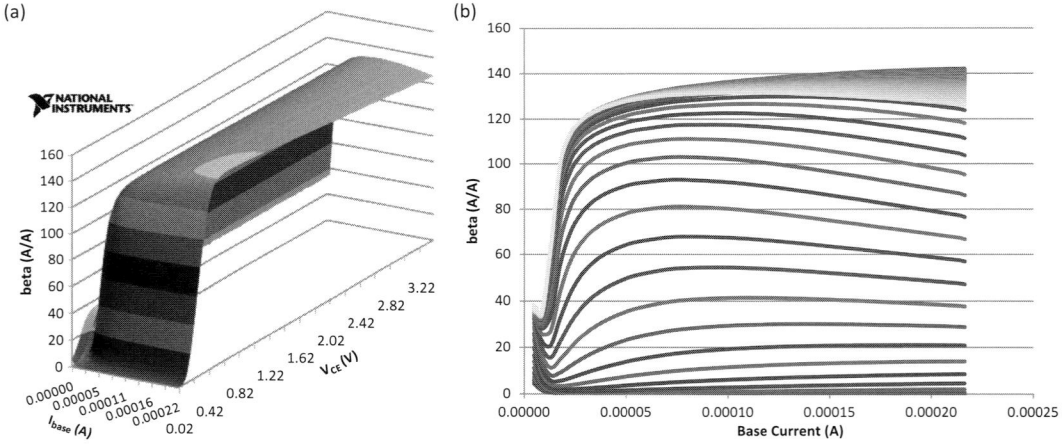

Figure 13-30   HBT current gain measures: (a) β surface; (b) curve family for part (a).

the addition of circuit techniques to enhance this) can be made using this transistor, and that that linearity is maintained across a fairly wide range of supply voltage. This transistor will make a very good envelope tracking PA. However according to Figure 13-27(b), it will not perform so well if pressed further toward C-mode operation.

- R_{ON} **surfaces** – Channel ON resistance is the fundamental parameter of C-mode design as noted in (6.11). The 3-port evaluation of channel resistance generates another surface, two examples of which are presented in Figure 13-31.

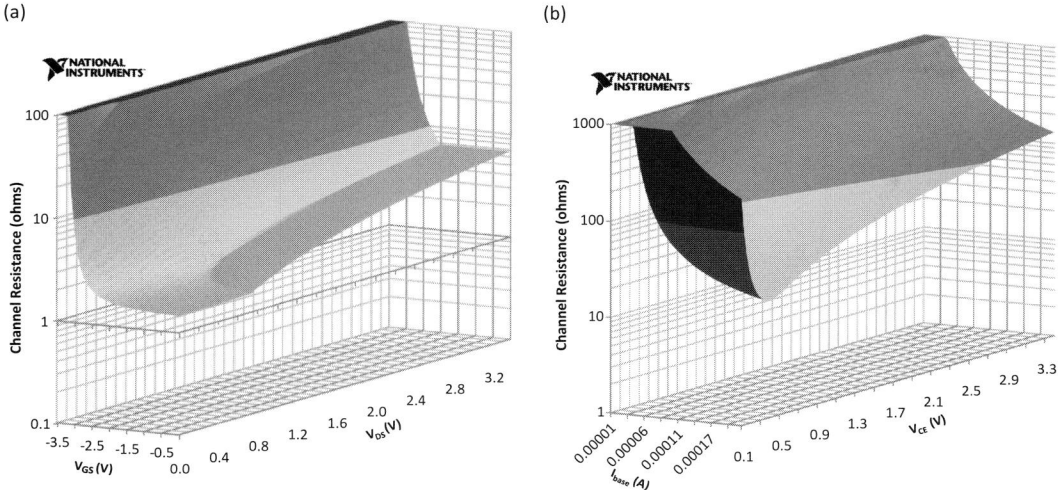

Figure 13-31 Channel resistance surfaces: (a) FET; (b) HBT.

To be considered a switch for the purposes of Section 6.7.5, the best dynamic range is achieved when the lowest ON resistance appears where the supply voltage is zero (6.12) and has a value much less than the load resistance at the transistor (6.11). Whether the ON resistance of the transistor in Figure 13-31(a) is acceptable for the amplifier it is to be used in is a decision for the designer to make. But it does exhibit a constant value of ON resistance around and at $V_{DS} = 0$, and also shows what the transistor controlling parameter must stay within to get this behavior. For the transistor of Figure 13-31(b), an offset from $V_{CE} = 0$ is clearly evident, placing restrictions on V_{CE} to operate to satisfy (6.12) when C-mode operation is attempted.

- **Polar modulation accuracy** – The mechanism to evaluate the accuracy of a C-mode stage performing polar modulation is most readily viewed as the modulation triangle presented in Section 8.5. The polar modulation accuracy measures how closely the resistive region characteristic curves align with the corresponding side of this triangle. Figure 13-32 provides two typical examples from FET and bipolar transistor types. In particular, bipolar transistors exhibit modulation inaccuracies at low magnitudes due to the collector current curvature seen in Figure 13-26(b). No known FET has a similar curvature. FET-based polar transmitters do successfully generate accurate envelopes with 20 mV magnitudes and smaller.
- **PSR surfaces** – A different power supply sensitivity measure is applicable to the transistor directly, separate from the power supply sensitivity surfaces defined for amplifiers shown in Figure 13-22. The surfaces in Figure 13-33 show the change in device current for a given change in applied supply voltage, measured across all values of supply voltage and controlling parameter. Representative surfaces are presented for FET and bipolar transistors.

When this measured surface is normalized to its peak value, it represents the available suppression of noise on the power supply supported by the transistor. Near

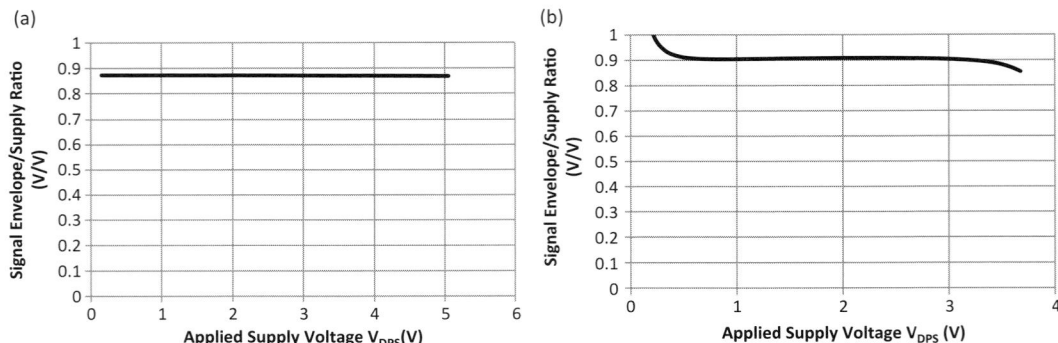

Figure 13-32 Polar modulation accuracy curves: (a) FET; (b) HBT.

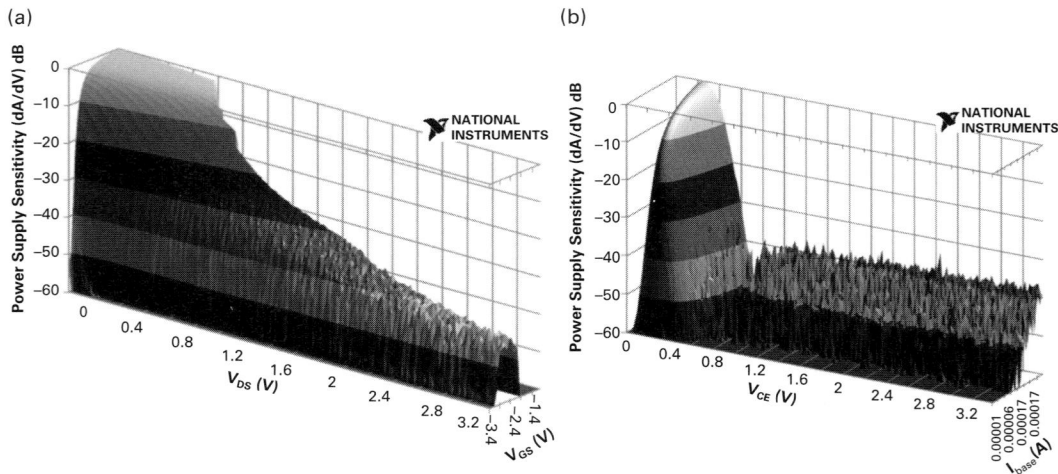

Figure 13-33 PSR surface representative samples: (a) FET (here LDMOS); (b) bipolar (here GaAs HBT).

the peak of this PSR surface, the transistor operates in C-mode and the DPST behavior is polar. When these surfaces are at or below –40 dB of supply noise suppression, the DPST behavior is envelope tracking.

One particular use of the PSR surface is a measure for the size of the transition region between polar operation and envelope tracking. This transition is very gradual for the FET device in Figure 13-33(a). In contrast, the transition for the bipolar transistor in Figure 13-33(b) is rather sharp. Because the knee voltage is the contour of the PSR surface at –40 dB, this graphic shows easily how this knee voltage varies across the transistor operating condition options.

- **PSR boundaries** – The ET boundary (10.1) is defined at the PSS-IV surface contour at –40 dB. Following this contour, it is possible to define the knee voltage behavior of the transistor, which can be overlaid on top of the characteristic

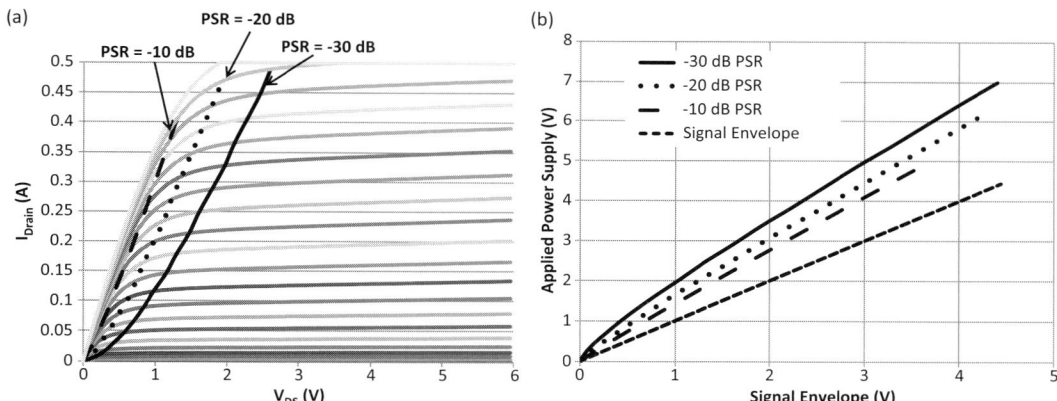

Figure 13-34 PSR boundary examples based on data from Figure 10-39: (a) Boundaries overlaid on the transistor characteristic curves; (b) corresponding power supply voltage profiles for a load line of 9 ohms. Line styles have the same meanings in both graphics.

curves to illustrate the PSR boundaries such as those shown in Figure 10-39(b) and similar figures. By following other contours of the PSR surface, the knee voltage evaluation changes. This procedure can provide PSR boundaries that represent different values of supply noise rejection, typically evaluated at the ET boundary of −40 dB and within the transition region toward polar modulation at −30 dB, −20 dB, and −10 dB. One example is shown in Figure 13-34(a).

Later in the PA design process when the load line design is complete, this load line can be combined with the appropriate PSR boundary to determine the power supply profile that corresponds to that value of power supply noise rejection. A typical result of this analysis is seen in Figure 13-34(b), which is closely related to Figure 10-39(a) and similar figures.

13.4.3   Additional set of transistor tests

Beyond the transistor characterization test curves and surfaces described in Section 13.4.2, there are four additional tests that are known to be particularly useful to the design of PAs for DPST use.

- **FET activity region** It is particularly advantageous to the bias circuitry for an FET device to not have any conduction current in the gate. We might think that this is intuitively obvious, but for many FET transistor types it is very likely that gate conduction will occur under operating modes used in DPST designs. This is particularly true when amplifier compression is used to get high energy efficiency.

  The concept of the FET activity region is introduced in Figure 10-7, which is copied here as Figure 13-35 for local reference. It is determined by overlaying two

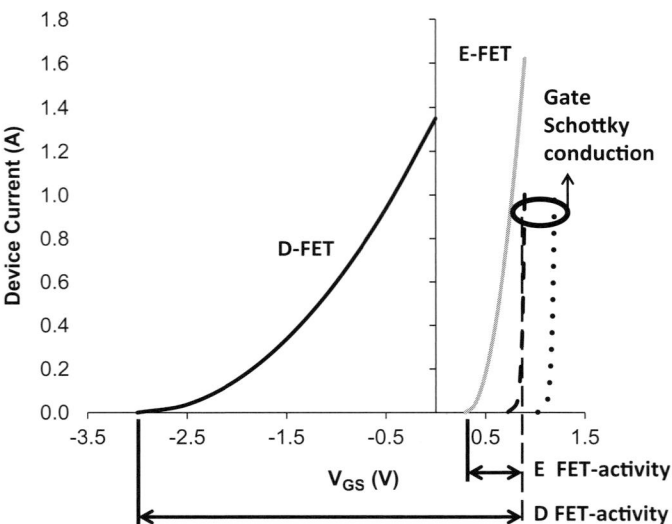

Figure 13-35   FET activity region is the range of V_{GS} values between threshold (or pinch-off) and the onset of gate current conduction.

measurements on the same graph: (1) measure the $I_D(V_{GS})$ transfer function at sufficiently high V_{DS} to not see compression over the drain current range of interest, then (2) measure the gate current ($I_G(V_{GS})$ transfer function) at a much more sensitive current scale so that the transistor is not damaged. Plotting both of these measurements along the same V_{GS} horizontal axis exposes the FET activity region, which is the V_{GS} range over which the drain current is controlled but there is no gate current, or otherwise where the gate current is kept below a specified small value for reliability.

- **Pin capacitances and integrated charge**   Even if an FET operates with no conduction current, the driver must repeatedly supply charge on to and remove charge from the capacitances present in the transistor and its mounting structure. At the output, the transistor must do the same for the load circuitry and output structure.

  Example capacitance and charge measurements for a set of GaN HEMTs are provided in Figure 13-36. The same markers identify the corresponding input charge transfer and ON resistance data for a transistor.

- **Switching characteristics (τ_{ON})**   If the transistor is to be operated in C-mode, then the discussion in Section 8.4 about switching speed is particularly relevant. It is shown there that while the usual transistor speed metrics of f_{MAX} and transition frequency f_T do not directly apply to transistor switching speed, a metric defined as τ_{ON} is a useful transistor oriented characterization.

  According to (8.13), the measurements for ON resistance and output capacitance are a possible metric for comparing transistors on expected output switching time performance. Since both of these performance parameters are voltage-varying, one proposal is to use the minimum value of R_{ON} and the maximum value of C_{DS} for this

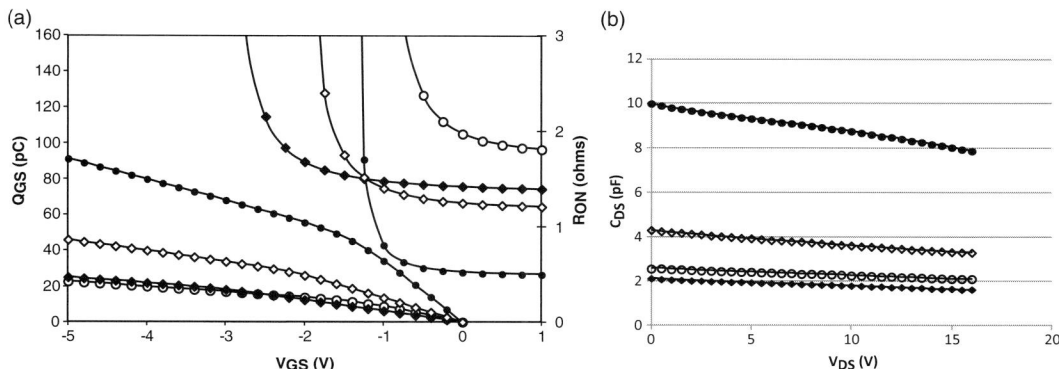

Figure 13-36 Measured capacitances and selected integrated charge profiles for a set of GaN HEMT transistors: (a) input integrated charge and associated channel resistance; (b) measured output capacitances for these same parts, where markers on the curves correspond to the same transistor.

time constant metric. Further experience with this metric will likely provide many ideas for improvement and refinement to make it more useful.

- **Dynamic bias profile**   For both envelope tracking and polar operation, the analyses show that dynamic bias in the PA is needed to achieve the needed control dynamic range. This dynamic bias characteristic is tied to the applied supply voltage as an input, so the step in Figure 13-33(b) must be completed first. Such profiles are usually not linear, such as the dynamic bias profile presented in Figure 10-38 and similar figures. Compromise is likely between accuracy leading to maximum control dynamic range and complexity in implementing an optimum but nonlinear bias control profile.

No mention is made here on the circuit design challenges in stabilizing the amplifier with a dynamic bias in place that must have sufficient bandwidth to keep up with the envelope variation, or at least with the DPS variation. A minimum of a full chapter in another book is needed to do justice to this complicated topic.

13.5   Power supply interface characterization

The new interface in DPS transmitters (see Figure 13-1) can be characterized with a selected subset of the measurements defined above for the DPS and the PA. This list of specifications only concerns guaranteeing the stability of this interface. It does not include all the specifications necessary to make the entire transmitter operate as required. Specifically, this subset includes:

from the DPS side,
output voltage range,
output current range,
output current limits (maximum and minimum (if any)),

output impedance: vs. V_{DPS}; and across frequency at each V_{DPS} value,
slew rate,
load impedance range: resistive (minimum and maximum); reactive (minimum and maximum).

From the PA side:

supply voltage bounds (minimum and maximum),
input current range,
maximum sinewave ripple,
voltage supply port impedance: vs. V_{PA}; and across frequency at each V_{PA} value.

It is strongly hoped that interested parties in both the DPS community and the associated PA community will get together to develop a specification that can work for all parties, providing high confidence that this interface will be stable no matter which DPS is connected to which PA. Until this specification exists, it is the responsibility of the DPST designer to collect all of the data from measurements described in this chapter to determine for themselves whether the anticipated pairing(s) will be stable and operate as intended.

13.6   DPST full-up characterization

Once the RF PA is generating power, is unconditionally stable in all its operating conditions and modes, and modulating with sufficient accuracy, the DPS is well understood, not oscillating with the PA connected as its load, and meeting all of its requirements, and the interface between the DPS and the PA is stable under all conditions, *then* the actual envelope tracking and/or polar transmitter testing can begin.

As a system the DPST must be characterized as providing sufficient signal quality under all specified signal parameters and all required environmental conditions. There is still a significant amount of testing to be done, in addition to all the testing and characterization already undertaken to reach this point in the project. To date, nearly all of this testing has been ad-hoc, meaning that operating conditions are selected largely arbitrarily, tried on the DPST, and evaluated on how well it performs.

Applying the information in earlier chapters, this ad-hoc approach can be rightfully abandoned for a truly science-based technique where the extensive 3-port characterizations show closely what the operating conditions must be to achieve the desired response. Then, the full-up hardware characterization points out the places needing refinement before the project can be completed.

Remember: *only hardware provides 100% of the physics 100% of the time.*

If measurements do not match the simulations, it is not the hardware that is at fault, but the simulations. The simulator models have not taken into account enough of the effects present in the physical implementation– a long as, of course, the hardware construction is done according to the input provided to the simulator. Simulator models are hence only useful in so far as they represent all of the physics which are important across the frequency range and dynamic range of interest. Knowing the limits of all models being used in the simulators is vital if they are to be used properly. Most simulation problems occur at the extreme limits of the models, whether frequency (e.g. millimeter wave) or dynamic range (e.g. phase noise).

13.6.1 Assurance of the operating mode(s) desired

Before much else happens, it is most important to verify that the PA operating modes along the design voltage profile of the DPS output corresponding to the PA output powers are as intended. This is most accurately done by stepping along the DPS and input RF power design profile while doing the AM sideband test described in Figure 6-8. Make sure that linear operation really is happening where it should be, that deep compression is happening by validating polar modulation, and then profile the transition between these endpoints.

A far less desirable alternative is to perform a SSR test of the PA along the DPS output voltage profile. This will clearly point out any C-mode operation that corresponds to polar modulation. However, the SSR test does not clearly discern between true L-mode operation and the more distorting transition region among the PA operating modes.

13.6.2 ACLR and EVM metrics

Output signal quality is validated with two measurements, one in the frequency domain and one in the time domain. They are *both* required. The frequency-domain test will be a check of the modulated output against a spectral mask, or with some standards this test is a check of spectral distortion called adjacent channel power ratio (ACPR) or adjacent channel leakage ratio (ACLR). The time-domain test is nearly always one of the various error vector magnitude (EVM) tests, though for some older modulations some other measure of rms modulation path error may be used.

Having one of these tests in compliance with the standard is not a guarantee that the other test will also be in compliance. The frequency-domain and time-domain tests measure different things. In particular, passing the frequency-domain test has absolutely no assurance that the time-domain test is close to compliance. The easiest way to visualize this is to apply a 64 QAM signal to a receiver expecting a 16 QAM signal. With identical symbol times and bandlimiting filtering, the signal spectra will be exactly the same. But a 16 QAM demodulator will have no idea how to handle the 48 additional signal states it sees, so it will report a huge EVM problem.

Having a good EVM is also not a predictor of having good frequency-domain performance. While this direction is more likely to be good, the EVM test limits are many percent which is far too coarse to also guarantee good spectral performance.

13.6.3 DPS accuracy

If testing shows that the value of the DPS is critical to the output signal accuracy, then the PA is not in L-mode operation, the operation is not ET, and the DPS profile must be showing that polar modulation is expected. This can be checked against the operating mode profile generated from the testing called for in Section 13.6.1. The fundamental principle of ET in (7.1) directly means that the value of the DPS output voltage is not critical. Therefore, if the voltage value is critical, then the operation is not ET. This behavior must match the design expectations. Otherwise something is very, very wrong.

13.6.4   DPS time alignment

A close corollary to the need (or not) of DPS voltage accuracy is the need (or not) for accurate time alignment of the envelope waveform. According to Figure 5-25, if the time alignment tolerance is zero, then the operation is not ET, but actually is polar. The tolerance of envelope signal timing variation steadily increases as the voltage offset increases. Eventually, if the DPS voltage offset exceeds the envelope peak, then the timing of the envelope signal is of no matter at all.

This allows the following two statements to be made for a troubleshooting tree:

- DPS signal time alignment is not critical *if and only if* the operating mode is ET,
- DPS signal time alignment is critical *if and only if* the operating mode is polar modulation.

Discussion of the circuit operating physics in Chapters 5 through 8 make clear that this relationship is symmetrical.

13.6.5   Use of adaptive digital pre-distortion

Particularly in present academic environments, it is considered commonplace to simply wrap any circuit that is not yet meeting performance specifications with an adaptive digital pre-distortion (A-DPD) system and then all is well. Maybe. In industry, A-DPD is expensive and is something to be avoided, particularly in products that must be manufactured in huge quantities. As a business manager, I don't want to pay for it. But, if I do pay for it, it had better be worth it.

I view this widespread use of A-DPD as largely an abdication of general engineering responsibility to understand what our circuits are doing and making sure that the underlying circuit is really doing what it needs to be doing. It is only after all corrections to bias and signal coordination are implemented properly and there are still any small inaccuracies remaining that there is any need for A-DPD to be discussed – but it also must be *economically* justified.

One argument heard is that implementing A-DPD is so straightforward today that just doing that is faster than actually fixing the problem. This may be true. But it also means that the product in production is much more expensive, and more power hungry in the field than a properly designed product. A-DPD is not a guarantee of better competitiveness. Count me among the strong skeptics, particularly for mobile device applications.

Even when the use of A-DPD is economically justified, the application of conventional algorithms to DPST designs is not at all straightforward. A-DPD algorithms are designed for use with linear (usually multicarrier) amplifiers that always operate in L-mode. This means that the distortion mechanisms that need to be corrected are those shown in Figure 13-22(a). These distortion mechanisms are driven by signal conditions at the RF input to the PA, so it is very appropriate to fix them by applying pre-distortion to the RF input signal.

But as DPST operation is pushed toward C-mode to achieve improved PA energy efficiency, this distortion mechanism actually stops. Particularly when the high efficiency C-mode is actually reached, property (8.1) says that any RF input signal envelope pre-distortion no longer has any effect. The algorithm must change to inject the pre-distortions in two places, and only in polar coordinates. Envelope modulation corrections must be entered through the DPS path as shown in Figure 13-22(b). Phase component pre-distortions must be applied to the RF path. For a DPST design that follows an envelope voltage profile such as those in Figure 11-3, the complexity of a successful A-DPD is much greater than for a conventional design.

13.6.6   Wideband noise

Wideband noise at the output of the DPST can come from many sources. The most obvious wideband noise source is the PA itself, as described in Section 7.8. If the PA is operating in L-mode and the PSR is –40 dB or better, it is very likely that this will be the only noise source. When the PA PSR is less than –40 dB, then any noise on the DPS output begins to contribute. It is straightforward to test for this by first measuring output noise when the PA is powered from a low noise source (e.g. a battery) and then repeating the noise measurement when the DPS powers the PA. It is vital to remember that this test is only valid when an input signal is present. One would hope that the PA never compresses on its own noise output!

Any quantization effects on the envelope signal, such as aliasing from discrete time signal processing, or the wideband noise property of sigma-delta digital to analog conversion, will also transfer to the PA output whenever the PA is operated at a PSR less than –40 dB. This is another check on the operating mode of the PA: if noise effects from the envelope path are present at the PA output, then the PA is not operating in ET. It is either transitioning to polar modulation or is already completely there.

13.7   Calibration principles

In a manufacturing environment, even when the manufacture quantity is only one, the input values of all components are not precisely known. Calibration is the traditional method to manage statistical problems. The objective of calibration is to achieve a bounded output from semi-random and possibly time-varying (nonstationary) inputs.

Second-order statistics (mean (μ) and variance (σ^2)) are all that are fundamentally needed for this discussion on calibration. Any set of components X used in product manufacture can be described as having a sample mean and a sample variance, which is written as

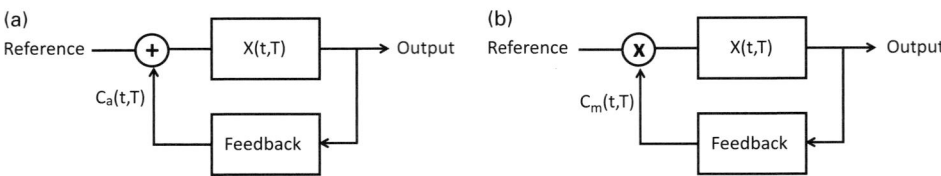

Figure 13-37   Feedback structures to apply control theory as an adaptive calibration: (a) additive feedback; (b) multiplicative feedback.

$$X\left(\mu, \sigma^2\right). \tag{13.5}$$

But each individual component has its particular value X_n, which is not random. Once the manufacture is complete, the components inside it are not random – and neither is the output. But this output may not be within specifications, requiring calibration to bring the output within the allowable error ε of the desired value.

Design often happens assuming that the component value will be at its sample statistical mean, such that $X_n = \mu$. Calibration therefore must handle the variance present in the set of components used. Calibration C can be additive (C_a), such that (Figure 13-37(a))

$$\mu - \varepsilon \leq X_n + C_a \leq \mu + \varepsilon, \tag{13.6}$$

or multiplicative (C_m), such that(Figure 13-37(b)

$$\mu - \varepsilon \leq C_m X_n \leq \mu + \varepsilon. \tag{13.7}$$

The range of C depends completely on the properties of the component sample variance. The design of any calibration system is fundamentally based upon knowing the range and characteristics of the input sample variances.

The calibration problem gets more complicated when the component values are not constant over time or over environmental conditions such as temperature

$$X\left(\mu(t, T), \sigma^2(t, T)\right). \tag{13.8}$$

A fixed calibration is not useful in this situation. It is far more common to apply control theory to these nonstationary situations, which means to apply feedback around the varying circuit block.

All feedback systems need to have a reference applied which the feedback network uses to measure changes in the output signal from which to determine what the feedback control value needs to be. For an amplifier, this reference would likely be the input signal. For a modulator or complete transmitter, the reference is usually the modulation baseband components.

In all cases, the use of feedback control requires a circuit to measure the controlled output and convert it into the form that can be directly compared with the reference. This conversion of the output signal into a reference-comparable form must happen with greater precision than the overall control system must perform to. Otherwise the control loop also adds the inaccuracies of the feedback on to the output signal, which is something that is definitely not desired.

13.8 Calibration at design

Calibration at design is only possible when the variances on the components used in the design are small enough that they can be neglected. If at design there is some interaction among the components that provides an undesired result, but that interaction is stable such that the signals around it can be manipulated so as to cancel the undesired result, then a characterization at design time [13-1] allows a calibration to be done that will hold in the presence of manufacturing variations. One major example of this is the DPS-PM transfer function from Section 6.11.2, which can be characterized and calibrated using the technique in Section 6.11.3 at design time. In a well-designed C-mode transmitter, this design-time calibration is known to hold across both manufacturing variations and temperature swings [6-8].

13.8.1 L-mode PA operation

Suppression of noise on the DPS output is definitely a design-time calibration that works for L-mode amplifiers, as long as the ET core principle (7.1) is implemented within the bounds of (10.1). When the bias and applied supply voltage operate the transistor outside its –40 dB PSR boundary, this noise suppression is considered a design-time calibration.

Another design-time calibration for L-mode amplifiers is the assurance of circuit stability. All L-mode amplifiers operate within the Barkhausen boundary of Figures 13-10 and 13-11 or it is not possible for them to ever have useful gain. Accordingly, the stability evaluations of (4.22) and particularly of (7.7) are considered a design-time calibration.

For any L-mode operation, the transistor regulates the current flow through the load. All the parameters in the transistor current equations for FET devices (4.23) or bipolar transistors (4.24) apply and invoke their effects. This means that the statistical description of (13.8) applies to these designs. A complete calibration at design time that holds across all manufacturing is difficult. With so many variables, it is far more likely to use feedback control to adapt to all of the variables and therefore keep the output signal within the allowable error.

This, of course, is exactly what A-DPD does. Within its control range, it operates to hold a constant gain across its controlled circuit in spite of time, temperature, and manufacturing component variations. The challenge is to provide the A-DPD with enough gain and dynamic range so that it maintains a stable control loop around its controlled circuit, particularly across the required output power control dynamic range (PCDR). The algorithm must correct for compression through the amplifier innate AM-AM transfer function. It is more difficult to develop an algorithm that also corrects for the amplifier's innate AM-PM distortion, because this is often considered a memory effect. This is so far commonly done. In an ET transmitter, these algorithms also must correct for the variations of all of these distortion processes as the applied supply voltage varies on the amplifier.

DPS output stability when connected to the PA also must be assured at design time. This is largely done by meeting the DPS output specification of maximum output capacitance for no oscillation. These values are typically at or below 100 picofarads, making the task of ensuring low frequency stability of an envelope tracking PA more challenging to achieve.

13.8.2   C-mode PA operation

C-mode PA operation, when properly implemented according to the design rules in Section 6.7.5, is known to have innate stabilities to temperature swings as described in Section 6.15.3 and manufacturing variations as described in Section 6.15.4. Circuit stability is also inherent at all frequencies because C-mode circuits operate outside the Barkhausen boundary. With this collection of properties, it is usual practice to take advantage of them and do most of the necessary circuit calibrations at design time.

Specific characterizations of C-mode design include managing drive signal leakage as shown in Figures 13-13 and 13-14. This step also inherently extends the polar modulation dynamic range as presented in Section 6.11.3, and reduces the polar phase distortion transfer function DPS-PM. If the required PCDR is large enough to require the use of P-mode as presented in Section 6.12.3, then there are two additional calibrations necessary. First the transition point must be aligned as shown in Figure 6-40 so that the highest P-mode power and the smallest C-mode power are the same. Then the attenuator controlling the RF input during P-mode must be calibrated across the remainder of the needed PCDR.

Stability of the DPS to PA interface is also a task of design-time calibration. As for the L-mode PA, the C-mode PA needs to present a small capacitance to the DPS to ensure that its output will not oscillate. Fortunately, with C-mode operation there is no issue about ensuring low frequency stability.

13.9   **Production floor calibration**

In any high volume production process, the only way to reach high unit volumes is to use less and less time to make and test each unit. Test time is extremely precious, and each second of calibration time multiplies by the number of manu-factured units to the total time spent doing calibration instead of making more units. The product designer must strive to not need any calibration time on the production floor.

This is usually impossible, since even for C-mode operation there are parameters that are subject to wide enough component variance that production calibration is necessary. In particular, the actual value of the PA load resistance R_{PA} in (6.17) is subject to manufacturing variations. Though for each unit, once its components are in place, the randomness is gone, and a one-time calibration for output power finishes that task. Also the product must stay calibrated forever once all manufacturing steps are completed.

Production calibration complexity increases if the manufacturing variations of transistors must be calibrated, simply because each transistor has many variables in its operation. Building look-up tables is a common approach used here, but this is time consuming if the table size is large. Procedure complexity is therefore generally bad, even if the resulting calibration is very good, simply because such procedures take more time to do.

This leads to a fairly obvious question: is it possible to have the units calibrate themselves?

13.10   Self-calibration

Self-calibration of high volume products is very enticing. Once the product is manufactured, you turn it on and let it take care of itself. Such capability has existed in laboratory test equipment for many years. Can this also be done with consumer wireless products? Eventually, I certainly hope so.

Two things are essential for self-calibration to become practical. Most important is the availability of an internal measurement reference with sufficient accuracy to meet regulatory requirements at the final calibration. To date, these measurement references are large and expensive. The second essential element needed for self-calibration is time: sufficient time that the calibration can complete to the needed accuracy. If the calibration must be done off-line, meaning that the unit cannot perform its designed function while calibrating, then this self-calibration time is very critical to become negligibly small. In instances where the calibration can happen "in the background," while normal use continues to happen, then calibration time is not as critical.

With the progression of Moore's Law for silicon technology, the ability to add thousands of logic functions within an RF PA is getting easier. When sufficient on-chip sensors are available, this leads to the growing practicality of what is called digital-assisted RF. For example, in a PA a temperature sensor may observe that the PA transistor is getting hot and its threshold may be changing. This can alert the logic to send a number to a digital to analog converter to adjust an voltage offset (for example in accordance with (13.6)) to make the port behavior of the PA act as if nothing is changing.

This type of operation is also useful in a polar DPS transmitter when it is necessary to begin using P-mode to extend the power control dynamic range to lower output power. P-mode is inherently less accurate (4.18) than C-mode (6.10). Using the accuracy of C-mode as the internal reference, a polar transmitter can self-calibrate to match the power from P-mode operation to that from C-mode at the transition boundary. This operation is called a power alignment loop (PAL) [13-2].

13.11   References

Note: Patent references are provided for bibliographic use only. Citation of specific patents here is not indicating any view on priority issues.

[13-1] S. Schell, "Method and Apparatus for Accurate Measurement of Communication Signals," US Patent 6724177, issued April 20, 2004.

[13-2] D. Flowers, "Mode Shift Calibration in Power Amplifiers," US Patent 8000663, issued August 16, 2011.

Appendix Switching transistor evaluation metrics across technologies

It is common for specifications for FET devices intended for use in switching applications to include the gate charge Q_G necessary to get the FET to its specified ON condition, the drain-source resistance achieved when this charge is in place. There is an associated voltage present between the gate and source terminals V_{GS}. Basic physics establishes the relation between voltage and charge as capacitance, which here is the equivalent gate to source capacitance, C_{EGS}, given by the large-scale relation

$$C_{EGS} = \frac{Q_G}{V_{GS}}. \tag{A.1}$$

All large transistors are constructed as a collection of many smaller transistor devices called unit cells. Starting from the FET model in Figure 8-16, the model for a unit cell within the entire FET that incorporates (A.1) is given in Figure A-1. The channel resistance R_{DS} is shown as a function of only the input voltage V_{GS}.

When N instances of this unit cell are connected in parallel to form the large transistor, the resulting effective input capacitance, channel ON resistance, and gate charge are found (to first order) to be

$$C_{EGS} = \sum_{k=1}^{N} C_{EGS,u} = N C_{EGS,u}, \tag{A.2}$$

$$R_{DS}\bigg|V_{GS,ON} = \left(\sum_{k=1}^{N} \frac{1}{R_{DS,u}\big|V_{GS,ON}} \right)^{-1} = R_{DS,u}\frac{V_{GS,ON}}{N}, \tag{A.3}$$

and

$$Q_G = \frac{C_{EGS}}{V_{GS}} = \sum_{k=1}^{N} \frac{C_{EGS,u}}{V_{GS}} = N Q_{G,u}. \tag{A.4}$$

The input capacitance (A.1) and total gate charge are both N-times their corresponding unit cell values, and the total channel ON resistance is $1/N$-times its unit cell value. This fits with the known characteristic that larger FET devices have lower ON resistance and higher input capacitance than a smaller transistor will have.

One attractive FET figure of merit is the (artificial) time constant TSW, defined by

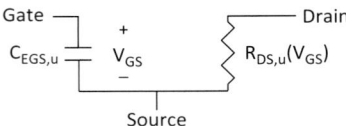

Figure A-1   FET unit cell model for switching characterization.

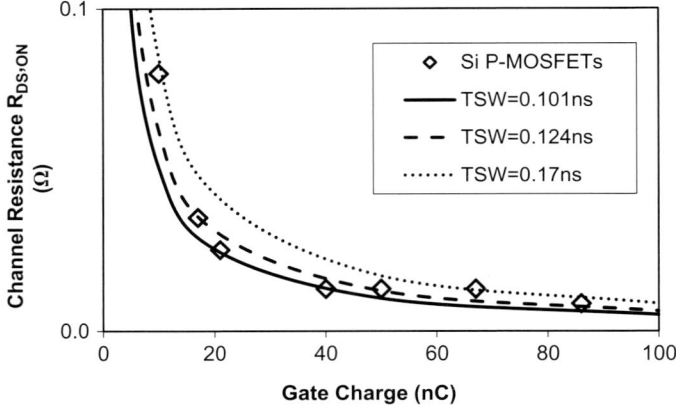

Figure A-2   Fitting (A.5) to a set of transistors from a single manufacturer shows that this set of transistors consists of members from three different families, each based on three different unit cells.

$$TSW \equiv C_{EGS}\left(R_{DS}\middle|V_{GS,ON}\right) \text{ sec.} \tag{A.5}$$

It is desired for a high-speed switching operation that both C_{EGS} and $R_{DS,ON}$ be minimized. This means that as a figure of merit, the time constant TSW indicates that smaller values are better. One particularly interesting feature of TSW is that when substituting (A.2) and (A.3) to see what the equivalent value for the unit cell is, the result is

$$TSW = C_{EGS,u}\left(R_{DS,u}\middle|V_{GS,ON}\right) \tag{A.6}$$

which is exactly the same as (A.5). This makes sense because the scaling applied to $C_{EGS,u}$ and that applied to $R_{DS,u}$ are mutually reciprocal and therefore cancel out. This also means that macro measurements of any transistor that is constructed from multiple unit cells will all evaluate to the same value of TSW, no matter how many unit cells are used in the particular transistor. In this way, it is possible to identify what the unit cell is for any particular transistor family. It is also straightforward to identify which transistors of a set are members of which family. One example that shows three families of transistor types is presented in Figure A-2.

Evaluating publically available data sheets from several manufacturers for transistors that may or may not be related by design provides one sample of parts evaluated in Figure A-3. The silicon MOSFETS in Figure A-3(a) include both n-channel and p-channel parts. Each set of transistors evaluates closely to one value of TSW, strongly implying that they are likely constructed from the same unit cell design.

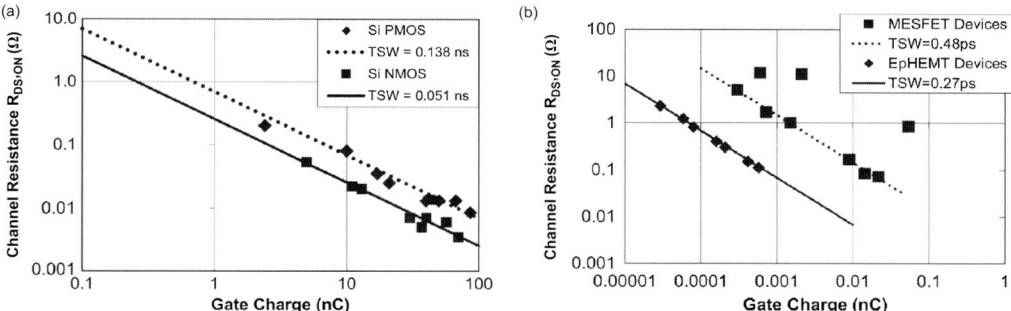

Figure A-3   Applying TSW evaluation to collections of transistors using very different technologies and materials: (a) silicon MOSFETs; (b) GaAs MESFET and EpHEMT.

In addition, the two TSW values shown in Figure A-3(a) are related by a ratio of 2.7. This is close to the mobility ratio of 3 between electrons and holes in silicon, implying only that the design of these transistor device families likely happened at the same time and used very similar technology generations.

In Figure A-3(b), two sets of GaAs FET devices are evaluated, here from different manufacturers. The TSW model still applies to identify which transistor sets are made from different scaling from the same unit cells. These TSW values are more than 1000x smaller than those seen in the silicon MOSFET data. But there is something else curious in the data of Figure A-3(b). The transistor data points plot in locations separated by more than 10x, but the TSW values differ by less than 2:1. This points out that the TSW model (A.6) is not a good descriptor of the first-order physics of these parts.

Reviewing the data sheets between these GaAs parts, the largest difference is in their relative transconductances. The EpHEMT transistors require much smaller V_{GS} values to reach their specified $R_{DS,ON}$ values. Looking back at (A.1), we observe that if both gate charge and gate-source voltage change in the same direction, the value of C_{EGS} may not change significantly. We are very interested in having smaller gate charge requirements, so TSW is not a good figure of merit. To capture the gate charge and small V_{GS} values, the figure of merit is changed to

$$FSW = R_{ON}Q_G(V_{GS} - V_{Th}). \tag{A.7}$$

Each term in (A.7) is desired to be small, so evaluations of FSW are useful as a new figure of merit in that smaller values are better.

The units of FSW are V^2 sec, which is not as intuitive as the simple seconds for TSW. Still, FSW captures more important features of the FET devices and is more likely to be useful across multiple and very different technologies. One test of this is to re-evaluate the transistors in Figure A-3, now using FSW as the metric instead of TSW. The result of this comparison is shown is shown in Figure A-4.

This FSW evaluation set includes three major classes of materials: silicon, GaAs, and GaN. All of these are clearly separated which highlights their individual characteristics. Several conclusions can be drawn:

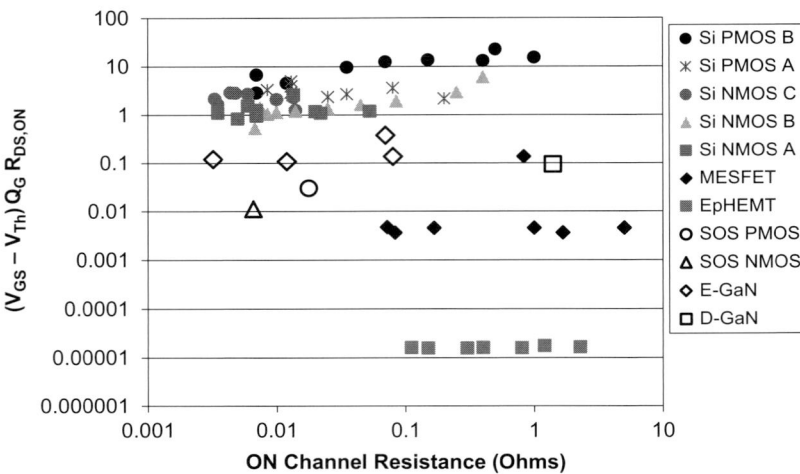

Figure A-4   Evaluation of multiple transistor technologies using FSW as the figure of merit.

- Silicon devices reach very low ON resistance values, but the drive requirements (input charge) are not changing commensurate with this reduced ON resistance. These transistors have the highest FSW values.
- The lowest FSW values are from GaAs EpHEMT devices. These are presently designed for RF amplifier use, which does not require low ON resistances.
- GaN HEMT devices have FSW values below the best silicon devices and above the GaAs MESFET devices. ON resistance performance matches that of silicon devices.
- Silicon on Sapphire (SOS) MOSFETS have FSW values lower than the best present GaN technologies.

Comparing these two figures of merit, we note that TSW is simple and has intuitive units. This figure of merit is useful when the gate characteristics V_{GS} and V_{Th} are constant among the devices being compared. Across different semiconductor technologies, the gate control characteristics vary widely, leading to the development of the FSW figure of merit. FSW is more general, and demonstrates its usefulness for cross-technology comparisons. Its units are not intuitive, but this is not a flaw.

Additional discussion on the use of these figures of merit in the design of drive circuits for switching transistors is provided in [A-1], which is also listed as [9-5].

Reference

[A-1] E. McCune, "Process- and Technology-Independent Power Switching Transistor Figures of Merit," *Proceedings of the 2008 IEEE Radio and Wireless Symposium (RWS)*, Jan. 2008, p. 195.

Index

Dynamic Power Supply Transmitters

Learn how envelope tracking, polar modulation, and hybrid designs using these techniques really work. The first physically based and coherent book to bring together a complete overview of such circuit techniques, this is an invaluable resource for practicing engineers, researchers, and graduate students working on RF power amplifiers and transmitters.

Create more succesful designs:

- Step-by-step design guidelines and real-world case studies show you how to put these techniques into practice
- A survey of various transistor technologies will help you to choose which type of transistor to use for best results
- Details on testing and measurement of all aspects of these designs explain how to measure what the circuit is actually doing and how to interpret measurement results

Earl McCune is a practicing engineer and Silicon Valley entrepreneur. A graduate of UC Berkeley, Stanford University, and UC Davis, he has over 35 years of post-graduate industry experience in wireless communications circuits and systems and more than 70 issued US patents. Now semi-retired, he has founded two successful start-up companies in addition to working in medium and very large corporations. He is also the author of *Practical Digital Wireless Signals* (Cambridge University Press).

THE CAMBRIDGE RF AND MICROWAVE ENGINEERING SERIES

Series Editor

Steve C. Cripps, Professor, University of Cardiff and Hywave Associates

Peter Aaen, Jaime Plá and John Wood, *Modeling and Characterization of RF and Microwave Power FETs*

Dominique Schreurs, Máirtín O'Droma, Anthony A. Goacher and Michael Gadringer, *RF Amplifier Behavioral Modeling*

Fan Yang and Yahya Rahmat-Samii, *Electromagnetic Band Gap Structures in Antenna Engineering*

Enrico Rubiola, *Phase Noise and Frequency Stability in Oscillators*

Earl McCune, *Practical Digital Wireless Signals*

Stepan Lucyszyn, *Advanced RF MEMS*

Patrick Roblin, *Nonlinear RF Circuits and the Large-Signal Network Analyzer*

Matthias Rudolph, Christian Fager and David E. Root, *Nonlinear Transistor Model Parameter Extraction Techniques*

John L. B. Walker, *Handbook of RF and Microwave Solid-State Power Amplifiers*

Anh-Vu H. Pham, Morgan J. Chen and Kunia Aihara, *LCP for Microwave Packages and Modules*

Sorin Voinigescu, *High-Frequency Integrated Circuits*

Richard Collier, *Transmission Lines*

Valeria Teppati, Andrea Ferrero and Mohamed Sayed, *Modern RF and Microwave Measurement Techniques*

Nuno Borges Carvalho and Dominique Schreurs, *Microwave and Wireless Measurement Techniques*

David E. Root, Jason Horn, Jan Verspecht and Mihai Marcu, *X-Parameters*

Earl McCune, *Dynamic Power Supply Transmitters*

Forthcoming

Richard Carter, *Theory and Design of Microwave Tubes*

Hossein Hashemi and Sanjay Raman, *Silicon mm-Wave Power Amplifiers and Transmitters*

Isar Mostafanezad, Olga Boric-Lubecke and Jenshan Lin, *Medical and Biological Microwave Sensors*